THE DESCENT OF MAN

AND SELECTION IN RELATION TO SEX

BY CHARLES DARWIN

A Digireads.com Book
Digireads.com Publishing

The Descent of Man; and Selection in Relation to Sex
By Charles Darwin
ISBN: 1-4209-3399-X

Please visit *www.digireads.com*

CONTENTS

PART I. THE DESCENT OR ORIGIN OF MAN.

The Evidence of the Descent of Man from some Lower Form.

Nature of the evidence bearing on the origin of man—Homologous structures in man and the lower animals—Miscellaneous points of correspondence—Development—Rudimentary structures, muscles, sense-organs, hair, bones, reproductive organs, etc.—The bearing of these three great classes of facts on the origin of man.

On the Manner of Development of Man from some Lower Form.

Variability of body and mind in man—Inheritance—Causes of variability—Laws of variation the same in man as in the lower animals—Direct action of the conditions of life—Effects of the increased use and disuse of parts—Arrested development—Reversion—Correlated variation—Rate of increase—Checks to increase—Natural selection—Man the most dominant animal in the world—Importance of his corporeal structure—The causes which have led to his becoming erect—Consequent changes of structure—Decrease in size of the canine teeth—Increased size and altered shape of the skull—Nakedness —Absence of a tail—Defenceless condition of man.

Comparison of the Mental Powers of Man and the Lower Animals.

The difference in mental power between the highest ape and the lowest savage, immense—Certain instincts in common—The emotions—Curiosity—Imitation—Attention—Memory—Imagination—Reason—Progressive improvement —Tools and weapons used by animals—Abstraction, Self-consciousness—Language—Sense of beauty—Belief in God, spiritual agencies, superstitions.

Comparison of the Mental Powers of Man and the Lower Animals—continued.

The moral sense—Fundamental proposition—The qualities of social animals—Origin of sociability—Struggle between opposed instincts—Man a social animal—The more enduring social instincts conquer other less persistent instincts—The social virtues alone regarded by savages—The self-regarding virtues acquired at a later stage of development—The importance of the judgment of the members of the same community on conduct—Transmission of moral tendencies—Summary.

On the Development of the Intellectual and Moral Faculties during Primeval and Civilized times.

Advancement of the intellectual powers through natural selection—Importance of imitation—Social and moral faculties—Their development within the limits of the same tribe—Natural selection as affecting civilized nations—Evidence that civilized nations were once barbarous.

On the Affinities and Genealogy of Man.

Position of man in the animal series—The natural system genealogical—Adaptive characters of slight value—Various small points of resemblance between man and the Quadrumana—Rank of man in the natural system—Birthplace and antiquity of man—Absence of fossil connecting-links—Lower stages in the genealogy of man, as inferred firstly from his affinities and secondly from his structure—Early androgynous condition of the Vertebrata —Conclusion.

On the Races of Man.

The nature and value of specific characters—Application to the races of man—Arguments in favour of, and opposed to, ranking the so-called races of man as distinct species—Sub-species—Monogenists and polygenists—Convergence of character—Numerous points of resemblance in body and mind between the most distinct races of man—The state of man when he first spread over the earth—Each race not descended from a single pair—The extinction of races—The formation of races—The effects of crossing—Slight influence of the direct action of the conditions of life—Slight or no influence of natural selection—Sexual selection.

PART II. SEXUAL SELECTION.

Principles of Sexual Selection.

Secondary sexual characters—Sexual selection—Manner of action—Excess of males—Polygamy—The male alone generally modified through sexual selection—Eagerness of the male—Variability of the male—Choice exerted by the female—Sexual compared with natural selection—Inheritance at corresponding periods of life, at corresponding seasons of the year, and as limited by sex—Relations between the several forms of inheritance—Causes why one sex and the young are not modified through sexual selection—Supplement on the proportional numbers of the two sexes throughout the animal kingdom—The proportion of the sexes in relation to natural selection.

cannot be accounted for on the principle of protection—Male fishes building nests, and taking charge of the ova and young. AMPHIBIANS: Differences in structure and color between the sexes—Vocal organs. REPTILES: Chelonians—Crocodiles—Snakes, colors in some cases protective—Lizards, battles of—Ornamental appendages—Strange differences in structure between the sexes—Colors—Sexual differences almost as great as with birds.

CHAPTER XXI.—P. 483

General Summary and Conclusion.

Main conclusion that man is descended from some lower form—Manner of development—Genealogy of man—Intellectual and moral faculties—Sexual selection—Concluding remarks.

Supplemental Note.

PREFACE TO THE SECOND EDITION.

During the successive reprints of the first edition of this work, published in 1871, I was able to introduce several important corrections; and now that more time has elapsed, I have endeavoured to profit by the fiery ordeal through which the book has passed, and have taken advantage of all the criticisms which seem to me sound. I am also greatly indebted to a large number of correspondents for the communication of a surprising number of new facts and remarks. These have been so numerous, that I have been able to use only the more important ones; and of these, as well as of the more important corrections, I will append a list. Some new illustrations have been introduced, and four of the old drawings have been replaced by better ones, done from life by Mr. T.W. Wood. I must especially call attention to some observations which I owe to the kindness of Prof. Huxley (given as a supplement at the end of Part I.), on the nature of the differences between the brains of man and the higher apes. I have been particularly glad to give these observations, because during the last few years several memoirs on the subject have appeared on the Continent, and their importance has been, in some cases, greatly exaggerated by popular writers.

I may take this opportunity of remarking that my critics frequently assume that I attribute all changes of corporeal structure and mental power exclusively to the natural selection of such variations as are often called spontaneous; whereas, even in the first edition of the 'Origin of Species,' I distinctly stated that great weight must be attributed to the inherited effects of use and disuse, with respect both to the body and mind. I also attributed some amount of modification to the direct and prolonged action of changed conditions of life. Some allowance, too, must be made for occasional reversions of structure; nor must we forget what I have called "correlated" growth, meaning, thereby, that various parts of the organization are in some unknown manner so connected, that when one part varies, so do others; and if variations in the one are accumulated by selection, other parts will be modified. Again, it has been said by several critics, that when I found that many details of structure in man could not be explained through natural selection, I invented sexual selection; I gave, however, a tolerably clear sketch of this principle in the first edition of the 'Origin of Species,' and I there stated that it was applicable to man. This subject of sexual selection has been treated at full length in the present work, simply because an opportunity was here first afforded me. I have been struck with the likeness of many of the half-favourable criticisms on sexual selection, with those which appeared at first on natural selection; such as, that it would explain some few details, but certainly was not applicable to the extent to which I have employed it. My conviction of the power of sexual selection remains unshaken; but it is probable, or almost certain, that several of my conclusions will hereafter be found erroneous; this can hardly fail to be the case in the first treatment of a subject. When naturalists have become familiar with the idea of sexual selection, it will, as I believe, be much more largely accepted; and it has already been fully and favourably received by several capable judges.

DOWN, BECKENHAM, KENT, *September*, 1874.

First Edition February 24, 1871.
Second Edition September, 1874.

THE DESCENT OF MAN;

AND

SELECTION IN RELATION TO SEX.

INTRODUCTION.

The nature of the following work will be best understood by a brief account of how it came to be written. During many years I collected notes on the origin or descent of man, without any intention of publishing on the subject, but rather with the determination not to publish, as I thought that I should thus only add to the prejudices against my views. It seemed to me sufficient to indicate, in the first edition of my 'Origin of Species,' that by this work "light would be thrown on the origin of man and his history;" and this implies that man must be included with other organic beings in any general conclusion respecting his manner of appearance on this earth. Now the case wears a wholly different aspect. When a naturalist like Carl Vogt ventures to say in his address as President of the National Institution of Geneva (1869), "personne, en Europe au moins, n'ose plus soutenir la création indépendante et de toutes pièces, des espèces," it is manifest that at least a large number of naturalists must admit that species are the modified descendants of other species; and this especially holds good with the younger and rising naturalists. The greater number accept the agency of natural selection; though some urge, whether with justice the future must decide, that I have greatly overrated its importance. Of the older and honoured chiefs in natural science, many unfortunately are still opposed to evolution in every form.

In consequence of the views now adopted by most naturalists, and which will ultimately, as in every other case, be followed by others who are not scientific, I have been led to put together my notes, so as to see how far the general conclusions arrived at in my former works were applicable to man. This seemed all the more desirable, as I had never deliberately applied these views to a species taken singly. When we confine our attention to any one form, we are deprived of the weighty arguments derived from the nature of the affinities which connect together whole groups of organisms—their geographical distribution in past and present times, and their geological succession. The homological structure, embryological development, and rudimentary organs of a species remain to be considered, whether it be man or any other animal, to which our attention may be directed; but these great classes of facts afford, as it appears to me, ample and conclusive evidence in favour of the principle of gradual evolution. The strong support derived from the other arguments should, however, always be kept before the mind.

The sole object of this work is to consider, firstly, whether man, like every other species, is descended from some pre-existing form; secondly, the manner of his development; and thirdly, the value of the differences between the so-called races of man. As I shall confine myself to these points, it will not be necessary to describe in detail the differences between the several races—an enormous subject which has been fully described in many valuable works. The high antiquity of man has recently been demonstrated by the labors of a host of eminent men, beginning with M. Boucher de Perthes; and this is the indispensable basis for understanding his origin. I shall, therefore, take this conclusion for granted, and may refer my readers to the admirable treatises of

Sir Charles Lyell, Sir John Lubbock, and others. Nor shall I have occasion to do more than to allude to the amount of difference between man and the anthropomorphous apes; for Prof. Huxley, in the opinion of most competent judges, has conclusively shown that in every visible character man differs less from the higher apes, than these do from the lower members of the same order of Primates.

This work contains hardly any original facts in regard to man; but as the conclusions at which I arrived, after drawing up a rough draft, appeared to me interesting, I thought that they might interest others. It has often and confidently been asserted, that man's origin can never be known: but ignorance more frequently begets confidence than does knowledge: it is those who know little, and not those who know much, who so positively assert that this or that problem will never be solved by science. The conclusion that man is the co-descendant with other species of some ancient, lower, and extinct form, is not in any degree new. Lamarck long ago came to this conclusion, which has lately been maintained by several eminent naturalists and philosophers; for instance, by Wallace, Huxley, Lyell, Vogt, Lubbock, Büchner, Rolle, etc. [1], and especially by Häckel. This last naturalist, besides his great work, 'Generelle Morphologie' (1866), has recently (1868, with a second edition in 1870), published his 'Natürliche Schöpfungsgeschichte,' in which he fully discusses the genealogy of man. If this work had appeared before my essay had been written, I should probably never have completed it. Almost all the conclusions at which I have arrived I find confirmed by this naturalist, whose knowledge on many points is much fuller than mine. Wherever I have added any fact or view from Prof. Häckel's writings, I give his authority in the text; other statements I leave as they originally stood in my manuscript, occasionally giving in the foot-notes references to his works, as a confirmation of the more doubtful or interesting points.

[1] As the works of the first-named authors are so well known, I need not give the titles; but as those of the latter are less well known in England, I will give them:—'Sechs Vorlesungen über die Darwin'sche Théorie:' zweite Auflage, 1868, von Dr L. Büchner; translated into French under the title 'Conférences sur la Théorie Darwinienne,' 1869. 'Der Mensch im Lichte der Darwin'sche Lehre,' 1865, von Dr. F. Rolle. I will not attempt to give references to all the authors who have taken the same side of the question. Thus G. Canestrini has published ('Annuario della Soc. d. Nat.,' Modena, 1867, page 81) a very curious paper on rudimentary characters, as bearing on the origin of man. Another work has (1869) been published by Dr. Francesco Barrago, bearing in Italian the title of "Man, made in the image of God, was also made in the image of the ape."

During many years it has seemed to me highly probable that sexual selection has played an important part in differentiating the races of man; but in my 'Origin of Species' (first edition, page 199) I contented myself by merely alluding to this belief. When I came to apply this view to man, I found it indispensable to treat the whole subject in full detail. [2] Consequently the second part of the present work, treating of sexual selection, has extended to an inordinate length, compared with the first part; but this could not be avoided.

[2] Prof. Häckel was the only author who, at the time when this work first appeared, had discussed the subject of sexual selection, and had seen its full importance, since the publication of the 'Origin'; and this he did in a very able manner in his various works.

I had intended adding to the present volumes an essay on the expression of the various emotions by man and the lower animals. My attention was called to this subject many years ago by Sir Charles Bell's admirable work. This illustrious anatomist maintains that man is endowed with certain muscles solely for the sake of expressing his emotions. As this view is obviously opposed to the belief that man is descended from some other and lower form, it was necessary for me to consider it. I likewise wished to ascertain how far the emotions are expressed in the same manner by the different races of man. But owing to the length of the present work, I have thought it better to reserve my essay for separate publication.

PART I. THE DESCENT OR ORIGIN OF MAN.

CHAPTER I.

THE EVIDENCE OF THE DESCENT OF MAN FROM SOME LOWER FORM.

Nature of the evidence bearing on the origin of man—Homologous structures in man and the lower animals—Miscellaneous points of correspondence—Development—Rudimentary structures, muscles, sense-organs, hair, bones, reproductive organs, etc.—The bearing of these three great classes of facts on the origin of man.

He who wishes to decide whether man is the modified descendant of some pre-existing form, would probably first enquire whether man varies, however slightly, in bodily structure and in mental faculties; and if so, whether the variations are transmitted to his offspring in accordance with the laws which prevail with the lower animals. Again, are the variations the result, as far as our ignorance permits us to judge, of the same general causes, and are they governed by the same general laws, as in the case of other organisms; for instance, by correlation, the inherited effects of use and disuse, etc.? Is man subject to similar malconformations, the result of arrested development, of reduplication of parts, etc., and does he display in any of his anomalies reversion to some former and ancient type of structure? It might also naturally be enquired whether man, like so many other animals, has given rise to varieties and sub-races, differing but slightly from each other, or to races differing so much that they must be classed as doubtful species? How are such races distributed over the world; and how, when crossed, do they react on each other in the first and succeeding generations? And so with many other points.

The enquirer would next come to the important point, whether man tends to increase at so rapid a rate, as to lead to occasional severe struggles for existence; and consequently to beneficial variations, whether in body or mind, being preserved, and injurious ones eliminated. Do the races or species of men, whichever term may be applied, encroach on and replace one another, so that some finally become extinct? We shall see that all these questions, as indeed is obvious in respect to most of them, must be answered in the affirmative, in the same manner as with the lower animals. But the several considerations just referred to may be conveniently deferred for a time: and we will first see how far the bodily structure of man shows traces, more or less plain, of his descent from some lower form. In succeeding chapters the mental powers of man, in comparison with those of the lower animals, will be considered.

THE BODILY STRUCTURE OF MAN.

It is notorious that man is constructed on the same general type or model as other mammals. All the bones in his skeleton can be compared with corresponding bones in a monkey, bat, or seal. So it is with his muscles, nerves, blood-vessels and internal viscera. The brain, the most important of all the organs, follows the same law, as shown by Huxley and other anatomists. Bischoff [1], who is a hostile witness, admits that every chief fissure and fold in the brain of man has its analogy in that of the orang; but he adds that at no period of development do their brains perfectly agree; nor could perfect agreement be expected, for otherwise their mental powers would have been the same. Vulpian [2], remarks: "Les différences réelles qui existent entre l'encéphale de l'homme et celui des singes supérieurs, sont bien minimes. Il ne faut pas se faire d'illusions à cet égard. L'homme est bien plus près des singes anthropomorphes par les caractères anatomiques de son cerveau que ceux-ci ne le sont non seulement des autres mammifères, mais même de certains quadrumanes, des guenons et des macaques." But it would be superfluous here to give further details on the correspondence between man and the higher mammals in the structure of the brain and all other parts of the body.

[1] 'Grosshirnwindungen des Menschen,' 1868, s. 96. The conclusions of this author, as well as those of Gratiolet and Aeby, concerning the brain, will be discussed by Prof. Huxley in the Appendix alluded to in the Preface to this edition.
[2] 'Leç. sur la Phys.' 1866, page 890, as quoted by M. Dally, 'L'Ordre des Primates et le Transformisme,' 1868, page 29.

It may, however, be worth while to specify a few points, not directly or obviously connected with structure, by which this correspondence or relationship is well shown.

Man is liable to receive from the lower animals, and to communicate to them, certain diseases, as hydrophobia, variola, the glanders, syphilis, cholera, herpes, etc. [3]; and this fact proves the close similarity [4] what I have here said with much severity and contempt; but as I do not use the term identity, I cannot see that I am greatly in error. There appears to me a strong analogy between the same infection or contagion producing the same result, or one closely similar, in two distinct animals, and the testing of two distinct fluids by the same chemical reagent.] of their tissues and blood, both in minute structure and composition, far more plainly than does their comparison under the best microscope, or by the aid of the best chemical analysis. Monkeys are liable to many of the same non-contagious diseases as we are; thus Rengger [5], who carefully observed for a long time the *Cebus Azaræ* in its native land, found it liable to catarrh, with the usual symptoms, and which, when often recurrent, led to consumption. These monkeys suffered also from apoplexy, inflammation of the bowels, and cataract in the eye. The younger ones when shedding their milk-teeth often died from fever. Medicines produced the same effect on them as on us. Many kinds of monkeys have a strong taste for tea, coffee, and spiritous liquors: they will also, as I have myself seen, smoke tobacco with pleasure. [6] Brehm asserts that the natives of north-eastern Africa catch the wild baboons by exposing vessels with strong beer, by which they are made drunk. He has seen some of these animals, which he kept in confinement, in this state; and he gives a laughable account of their behaviour and strange grimaces. On the following morning they were very cross and dismal; they held their aching heads with both hands, and wore a most pitiable expression: when beer or wine was offered them, they turned away with disgust, but relished the juice of lemons. [7] An American monkey, an Ateles, after getting drunk on brandy, would never touch it again, and thus was wiser than many men. These trifling facts prove how similar the nerves of taste must be in monkeys and man, and how similarly their whole nervous system is affected.

[3] Dr. W. Lauder Lindsay has treated this subject at some length in the 'Journal of Mental Science,' July 1871; and in the 'Edinburgh Veterinary Review,' July 1858.

[4] A Reviewer has criticised ('British Quarterly Review,' Oct. 1st, 1871, page 472

[5] 'Naturgeschichte der Säugethiere von Paraguay,' 1830, s. 50.

[6] The same tastes are common to some animals much lower in the scale. Mr. A. Nichols informs me that he kept in Queensland, in Australia, three individuals of the *Phaseolarctus cinereus*; and that, without having been taught in any way, they acquired a strong taste for rum, and for smoking tobacco.

[7] Brehm, 'Thierleben,' B. i. 1864, s. 75, 86. On the Ateles, s. 105. For other analogous statements, see s. 25, 107.

Man is infested with internal parasites, sometimes causing fatal effects; and is plagued by external parasites, all of which belong to the same genera or families as those infesting other mammals, and in the case of scabies to the same species. [8] Man is subject, like other mammals, birds, and even insects [9], to that mysterious law, which causes certain normal processes, such as gestation, as well as the maturation and duration of various diseases, to follow lunar periods. His wounds are repaired by the same process of healing; and the stumps left after the amputation of his limbs, especially during an early embryonic period, occasionally possess some power of regeneration, as in the lowest animals. [10]

[8] Dr. W. Lauder Lindsay, 'Edinburgh Vet. Review,' July 1858, page 13.

[9] With respect to insects see Dr. Laycock, "On a General Law of Vital Periodicity," 'British Association,' 1842. Dr. Macculloch, 'Silliman's North American Journal of Science,' vol. XVII. page 305, has seen a dog suffering from tertian ague. Hereafter I shall return to this subject.

[10] I have given the evidence on this head in my 'Variation of Animals and Plants under Domestication,' vol. ii. page 15, and more could be added.

The whole process of that most important function, the reproduction of the species, is strikingly the same in all mammals, from the first act of courtship by the male [11], to the birth and nurturing of the young. Monkeys are born in almost as helpless a condition as our own infants; and in certain genera the young differ fully as much in appearance from the adults, as do our children from their full-grown parents. [12] It has been urged by some writers, as an important distinction, that with man the young arrive at maturity at a much later age than with any other animal: but if we look to the races of mankind which inhabit tropical countries the difference is not great, for the orang is believed not to be adult till the age of from ten to fifteen years. [13] Man differs from woman in size, bodily strength, hairiness, etc., as well as in mind, in the same manner as do the two sexes of many mammals. So that the correspondence in general structure, in the minute structure of the tissues, in chemical composition and in constitution, between man and the higher animals, especially the anthropomorphous apes, is extremely close.

[11] Mares e diversis generibus Quadrumanorum sine dubio dignoscunt feminas humanas a maribus. Primum, credo, odoratu, postea aspectu. Mr. Youatt, qui diu in Hortis Zoologicis (Bestiariis) medicus animalium erat, vir in rebus observandis cautus et sagax, hoc mihi certissime probavit, et curatores ejusdem loci et alii e ministris confirmaverunt. Sir Andrew Smith et Brehm notabant idem in Cynocephalo. Illustrissimus Cuvier etiam narrat multa de hâc re, quâ ut opinor, nihil turpius potest indicari inter omnia hominibus et Quadrumanis communia. Narrat enim Cynocephalum quendam in furorem incidere aspectu feminarum aliquarem, sed nequaquam accendi tanto furore ab omnibus. Semper eligebat juniores, et dignoscebat in turbâ, et advocabat voce gestûque.

[12] This remark is made with respect to Cynocephalus and the anthropomorphous apes by Geoffroy Saint-Hilaire and F. Cuvier, 'Histoire Nat. des Mammifères,' tom. i. 1824.

[13] Huxley, 'Man's Place in Nature,' 1863, p. 34.

EMBRYONIC DEVELOPMENT.

Man is developed from an ovule, about the 125th of an inch in diameter, which differs in no respect from the ovules of other animals. The embryo itself at a very early period can hardly be distinguished from that of other members of the vertebrate kingdom. At this period the arteries run in arch-like branches, as if to carry the blood to branchiæ which are not present in the higher Vertebrata, though the slits on the sides of the neck still remain (see f, g, fig. 1), marking their former position. At a somewhat later period, when the extremities are developed, "the feet of lizards and mammals," as the illustrious Von Baer remarks, "the wings and feet of birds, no less than the hands and feet of man, all arise from the same fundamental form." It is, says Prof. Huxley [14], "quite in the later stages of development that the young human being presents marked differences from the young ape, while the latter departs as much from the dog in its developments, as the man does. Startling as this last assertion may appear to be, it is demonstrably true."

[14] 'Man's Place in Nature,' 1863, p. 67.

As some of my readers may never have seen a drawing of an embryo, I have given one of man and another of a dog, at about the same early stage of development, carefully copied from two works of undoubted accuracy. [15]

[15] The human embryo (upper fig.) is from Ecker, 'Icones Phys.,' 1851-1859, tab. xxx. fig. 2. This embryo was ten lines in length, so that the drawing is much magnified. The embryo of the dog is from Bischoff, 'Entwicklungsgeschichte des Hunde-Eies,' 1845, tab. xi. fig. 42B. This drawing is five times magnified, the embryo being twenty-five days old. The internal viscera have been omitted, and the uterine appendages in both drawings removed. I was directed to these figures by Prof. Huxley, from whose work, 'Man's Place in Nature,' the idea of giving them was taken. Häckel has also given analogous drawings in his 'Schöpfungsgeschichte.'

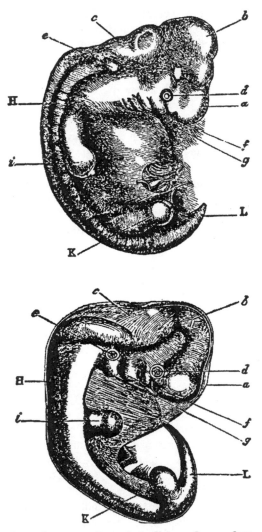

FIG. 1.—Upper figure human embryo, from Ecker. Lower figure that of a dog, from Bischoff.

a. Fore-brain, cerebral hemispheres, etc.
b. Mid-brain, corpora quadrigemina.
c. Hind-brain, cerebellum, medulla oblongata.
d. Eye.
e. Ear.

f First visceral arch.
g. Second visceral arch.
H. Vertebral columns and muscles in process of development.
i. Anterior } extremities.
K. Posterior }
L. Tail or os coccyx.

After the foregoing statements made by such high authorities, it would be superfluous on my part to give a number of borrowed details, showing that the embryo of man closely resembles that of other mammals. It may, however, be added, that the human embryo likewise resembles certain low forms when adult in various points of structure. For instance, the heart at first exists as a simple pulsating vessel; the excreta are voided through a cloacal passage; and the *os coccyx* projects like a true tail, "extending considerably beyond the rudimentary legs." [16] In the embryos of all air-breathing vertebrates, certain glands, called the corpora Wolffiana, correspond with, and act like the kidneys of mature fishes. [17] Even at a later embryonic period, some striking resemblances between man and the lower animals may be observed. Bischoff says that "the convolutions of the brain in a human fœtus at the end of the seventh month reach about the same stage of development as in a baboon when adult." [18] The great toe, as Professor Owen remarks [19], "which forms the fulcrum when standing or walking, is perhaps the most characteristic peculiarity in the human structure;" but in an embryo, about an inch in length, Prof. Wyman [20] found "that the great toe was shorter than the others; and, instead of being parallel to them, projected at an angle from the side of the foot, thus corresponding with the permanent condition of this part in the quadrumana." I will conclude with a quotation from Huxley [21] who after asking, does man originate in a different way from a dog, bird, frog or fish? says, "the reply is not doubtful for a moment; without question, the mode of origin, and the early stages of the development of man, are identical with those of the animals immediately below him in the scale: without a doubt in these respects, he is far nearer to apes than the apes are to the dog."

[16] Prof. Wyman in 'Proceedings of the American Academy of Sciences,' vol. iv. 1860, p. 17.
[17] Owen, 'Anatomy of Vertebrates,' vol. i. p. 533.
[18] 'Die Grosshirnwindungen des Menschen,' 1868, s. 95.
[19] 'Anatomy of Vertebrates,' vol. ii. p. 553.
[20] 'Proc. Soc. Nat. Hist.' Boston, 1863, vol. ix. p. 185.
[21] 'Man's Place in Nature,' p. 65.

RUDIMENTS.

This subject, though not intrinsically more important than the two last, will for several reasons be treated here more fully. [22] Not one of the higher animals can be named which does not bear some part in a rudimentary condition; and man forms no exception to the rule. Rudimentary organs must be distinguished from those that are nascent; though in some cases the distinction is not easy. The former are either absolutely useless, such as the mammæ of male quadrupeds, or the incisor teeth of ruminants which never cut through the gums; or they are of such slight service to their present possessors, that we can hardly suppose that they were developed under the conditions which now exist. Organs in this latter state are not strictly rudimentary, but they are tending in this direction. Nascent organs, on the other hand, though not fully developed, are of high service to their possessors, and are capable of further development. Rudimentary organs are eminently variable; and this is partly intelligible, as they are useless, or nearly useless, and consequently are no longer subjected to natural selection. They often become wholly suppressed. When this occurs, they are nevertheless liable to occasional reappearance through reversion—a circumstance well worthy of attention.

[22] I had written a rough copy of this chapter before reading a valuable paper, "Caratteri rudimentali in ordine all' origine dell' uomo" ('Annuario della Soc. d. Nat.,' Modena, 1867, p. 81), by G. Canestrini, to which paper I am considerably indebted. Häckel has given admirable discussions on this whole subject, under the title of Dysteleology, in his 'Generelle Morphologie' and 'Schöpfungsgeschichte.'

The chief agents in causing organs to become rudimentary seem to have been disuse at that period of life when the organ is chiefly used (and this is generally during maturity), and also inheritance at a corresponding period of life. The term "disuse" does not relate merely to the lessened action of muscles, but includes a diminished flow of blood to a part or organ, from being subjected to fewer alternations of pressure, or from becoming in any way less habitually active. Rudiments, however, may occur in one sex of those parts which are normally present in the other sex; and such rudiments, as we shall hereafter see, have often originated in a way distinct from those here referred to. In some cases, organs have been reduced by means of natural selection, from having become injurious to the species under changed habits of life. The process of reduction is probably often aided through the two principles of compensation and economy of growth; but the later stages of reduction, after disuse has done all that can fairly be attributed to it, and when the saving to be effected by the economy of growth would be very small [23], are difficult to understand. The final and complete suppression of a part, already useless and much reduced in size, in which case neither compensation nor economy can come into play, is perhaps intelligible by the aid of the hypothesis of pangenesis. But as the whole subject of rudimentary organs has been discussed and illustrated in my former works [24], I need here say no more on this head.

[23] Some good criticisms on this subject have been given by Messrs. Murie and Mivart, in 'Transact. Zoological Society,' 1869, vol. vii. p. 92.
[24] 'Variation of Animals and Plants under Domestication,' vol. ii pp. 317 and 397. See also 'Origin of Species,' 5th Edition p. 535.

Rudiments of various muscles have been observed in many parts of the human body; [25] and not a few muscles, which are regularly present in some of the lower animals can occasionally be detected in man in a greatly reduced condition. Every one must have noticed the power which many animals, especially horses, possess of moving or twitching their skin; and this is effected by the *panniculus carnosus*. Remnants of this muscle in an efficient state are found in various parts of our bodies; for instance, the muscle on the forehead, by which the eyebrows are raised. The *platysma myoides*, which is well developed on the neck, belongs to this system. Prof. Turner, of Edinburgh, has occasionally detected, as he informs me, muscular fasciculi in five different situations, namely in the *axillæ*, near the scapulae, etc., all of which must be referred to the system of the *panniculus*. He has also shown [26] that the *musculus sternalis* or *sternalis brutorum*, which is not an extension of the *rectus abdominalis*, but is closely allied to the *panniculus*, occurred in the proportion of about three per cent. in upwards of 600 bodies: he adds, that this muscle affords "an excellent illustration of the statement that occasional and rudimentary structures are especially liable to variation in arrangement."

[25] For instance, M. Richard ('Annales des Sciences Nat.,' 3rd series, Zoolog. 1852, tom. xviii. p. 13 describes and figures rudiments of what he calls the "muscle pédieux de la main," which he says is sometimes "infiniment petit." Another muscle, called "le tibial postérieur," is generally quite absent in the hand, but appears from time to time in a more or less rudimentary condition.
[26] Prof. W. Turner, 'Proceedings of the Royal Society of Edinburgh,' 1866-67, p. 65.

Some few persons have the power of contracting the superficial muscles on their scalps; and these muscles are in a variable and partially rudimentary condition. M. A. de Candolle has communicated to me a curious instance of the long-continued persistence or inheritance of this power, as well as of its unusual development. He knows a family, in which one member, the present head of the family, could, when a youth, pitch several heavy books from his head by the movement of the scalp alone; and he won wagers by performing this feat. His father, uncle, grandfather, and his three children possess the same power to the same unusual degree. This family became divided eight generations ago into two branches; so that the head of the above-mentioned branch is cousin in the seventh degree to the head of the other branch. This distant cousin resides in another part of France; and on being asked whether he possessed the same faculty, immediately exhibited his power. This case offers a good illustration how persistent may be the transmission of an absolutely useless faculty, probably derived from our remote semi-human progenitors; since many monkeys have, and frequently use the power, of largely moving their scalps up and down. [27]

[27] See my 'Expression of the Emotions in Man and Animals,' 1872, p. 144.

The extrinsic muscles which serve to move the external ear, and the intrinsic muscles which move the different parts, are in a rudimentary condition in man, and they all belong to the system of the *panniculus*; they are also variable in development, or at least in function. I have seen one man who could draw the whole ear forwards; other men can draw it upwards; another who could draw it backwards; [28] and from what one of these persons told me, it is probable that most of us, by often touching our ears, and thus directing our attention towards them, could recover some power of movement by repeated trials. The power of erecting and directing the shell of the ears to the various points of the compass, is no doubt of the highest service to many animals, as they thus perceive the direction of danger; but I have never heard, on sufficient evidence, of a man who possessed this power, the one which might be of use to him. The whole external shell may be considered a rudiment, together with the various folds and prominences (helix and anti-helix, tragus and anti-tragus, etc.) which in the lower animals strengthen and support the ear when erect, without adding much to its weight. Some authors, however, suppose that the cartilage of the shell serves to transmit vibrations to the acoustic nerve; but Mr. Toynbee [29], after collecting all the known evidence on this head, concludes that the external shell is of no distinct use. The ears of the chimpanzee and orang are curiously like those of man, and the proper muscles are likewise but very slightly developed. [30] I am also assured by the keepers in the Zoological Gardens that these animals never move or erect their ears; so that they are in an equally rudimentary condition with those of man, as far as function is concerned. Why these animals, as well as the progenitors of man, should have lost the power of erecting their ears, we cannot say. It may be, though I am not satisfied with this view, that owing to their arboreal habits and great strength they were but little exposed to danger, and so during a lengthened period moved their ears but little, and thus gradually lost the power of moving them. This would be a parallel case with that of those large and heavy birds, which, from inhabiting oceanic islands, have not been exposed to the attacks of beasts of prey, and have consequently lost the power of using their wings for flight. The inability to move the ears in man and several apes is, however, partly compensated by the freedom with which they can move the head in a horizontal plane, so as to catch sounds from all directions. It has been asserted that the ear of man alone possesses a lobule; but "a rudiment of it is found in the gorilla" [31]; and, as I hear from Prof. Preyer, it is not rarely absent in the negro.

[28] Canestrini quotes Hyrtl. ('Annuario della Soc. dei Naturalisti,' Modena, 1867, p. 97 to the same effect.
[29] 'The Diseases of the Ear,' by J. Toynbee, F.R.S., 1860, p. 12. A distinguished physiologist, Prof. Preyer, informs me that he had lately been experimenting on the function of the shell of the ear, and has come to nearly the same conclusion as that given here.
[30] Prof. A. Macalister, 'Annals and Magazine of Natural History,' vol. vii. 1871, p. 342.
[31] Mr. St. George Mivart, 'Elementary Anatomy,' 1873, p. 396.

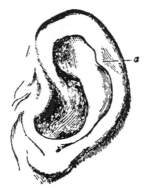

Fig. 2.—Human Ear, modelled and drawn by Mr. Woolner.
a. The projecting point.

The celebrated sculptor, Mr. Woolner, informs me of one little peculiarity in the external ear, which he has often observed both in men and women, and of which he perceived the full significance. His attention was first called to the subject whilst at work on his figure of Puck, to which he had given pointed ears. He was thus led to examine the ears of various monkeys, and subsequently more carefully those of man. The peculiarity consists in a little blunt point, projecting from the inwardly folded margin, or helix. When present, it is developed at birth, and, according to Prof. Ludwig Meyer, more frequently in man than in woman. Mr. Woolner made an exact model of one such case, and sent me the accompanying drawing. (Fig. 2). These points not only project inwards towards the centre of the ear, but often a little outwards from its plane, so as to be visible when the head is viewed from directly in front or behind. They are variable in size, and somewhat in position, standing either a little higher or lower; and they sometimes occur on one ear and not on the other. They are not confined to mankind, for I observed a case in one of the spider-monkeys (*Ateles beelzebuth*) in our Zoological Gardens; and Mr. E. Ray Lankester informs me of another case in a chimpanzee in the gardens at Hamburg. The helix obviously consists of the extreme margin of the ear folded inwards; and this folding appears to be in some manner connected with the whole external ear being permanently pressed backwards. In many monkeys, which do not stand high in the order, as baboons and some species of macacus [32], the upper portion of the ear is slightly pointed, and the margin is not at all folded inwards; but if the margin were to be thus folded, a slight point would necessarily project inwards towards the centre, and probably a little outwards from the plane of the ear; and this I believe to be their origin in many cases. On the other hand, Prof. L. Meyer, in an able paper recently published [33], maintains that the whole case is one of mere variability; and that the projections are not real ones, but are due to the internal cartilage on each side of the points not having been fully developed. I am quite ready to admit that this is the correct explanation in many instances, as in those figured by Prof. Meyer, in which there are several minute points, or the whole margin is sinuous. I have myself seen, through the kindness of Dr. L. Down, the ear of a microcephalous idiot, on which there is a projection on the outside of the helix, and not on the inward folded edge, so that this point can have no relation to a former apex of the ear. Nevertheless in some cases, my original view, that the points are vestiges of the tips of formerly erect and pointed ears, still seems to me probable. I think so from the frequency of their occurrence, and from the general correspondence in position with that of the tip of a pointed ear. In one case, of which a photograph has been sent me, the projection is so large, that supposing, in accordance with Prof. Meyer's view, the ear to be made perfect by the equal development of the cartilage throughout the whole extent of the margin, it would have covered fully one-third of the whole ear. Two cases have been communicated to me, one in North America, and the other in England, in which the upper margin is not at all folded inwards, but is pointed, so that it closely resembles the pointed ear of an ordinary quadruped in outline. In one of these cases, which was that of a young child, the father compared the ear with the drawing which I have given [34] of the ear of a monkey, the *Cynopithecus niger*, and says that their outlines are closely similar. If, in these two cases, the margin had been folded inwards in the normal manner, an inward projection must have been formed. I may add that in two other cases the outline still remains somewhat pointed, although the margin of the upper part of the ear is normally folded inwards—in one of them, however, very narrowly. The following woodcut (No. 3) is an accurate copy of a photograph of the fœtus of an orang (kindly sent me by Dr. Nitsche), in which it may be seen how different the pointed outline of the ear is at this period from its adult condition, when it bears a close general resemblance to that of man. It is evident that the folding over of the tip of such an ear, unless it changed greatly during its further development, would give rise to a point projecting inwards. On the whole, it still seems to me probable that the points in question are in some cases, both in man and apes, vestiges of a former condition.

[32] See also some remarks, and the drawings of the ears of the Lemuroidea, in Messrs. Murie and Mivart's excellent paper in 'Transactions of the Zoological Society,' vol. vii. 1869, pp. 6 and 90.
[33] 'Ueber das Darwin'sche Spitzohr,' Archiv fur Path. Anat. und Phys., 1871, p. 485.
[34] 'The Expression of the Emotions,' p. 136.

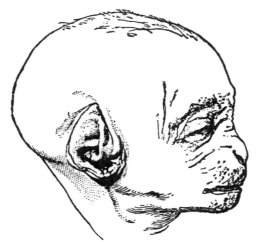

Fɪɢ. 3.—Fœtus of an Orang. Exact copy of a photograph, showing the form of the ear at this early age.

The nictitating membrane, or third eyelid, with its accessory muscles and other structures, is especially well developed in birds, and is of much functional importance to them, as it can be rapidly drawn across the whole eye-ball. It is found in some reptiles and amphibians, and in certain fishes, as in sharks. It is fairly well developed in the two lower divisions of the mammalian series, namely, in the monotremata and marsupials, and in some few of the higher mammals, as in the walrus. But in man, the quadrumana, and most other mammals, it exists, as is admitted by all anatomists, as a mere rudiment, called the semilunar fold. [35]

[35] Müller's 'Elements of Physiology,' Eng. translat. 1842, vol. ii. p. 1117. Owen, 'Anatomy of Vertebrates,' vol. iii. p. 260; ibid. on the Walrus, 'Proceedings of the Zoological Society,' November 8, 1854. See also R. Knox, 'Great Artists and Anatomists,' p. 106. This rudiment apparently is somewhat larger in Negroes and Australians than in Europeans, see Carl Vogt, 'Lectures on Man,' Eng. translat. p. 129.

The sense of smell is of the highest importance to the greater number of mammals— to some, as the ruminants, in warning them of danger; to others, as the Carnivora, in finding their prey; to others, again, as the wild boar, for both purposes combined. But the sense of smell is of extremely slight service, if any, even to the dark colored races of men, in whom it is much more highly developed than in the white and civilized races. [36] Nevertheless it does not warn them of danger, nor guide them to their food; nor does it prevent the Esquimaux from sleeping in the most fetid atmosphere, nor many savages from eating half-putrid meat. In Europeans the power differs greatly in different individuals, as I am assured by an eminent naturalist who possesses this sense highly developed, and who has attended to the subject. Those who believe in the principle of gradual evolution, will not readily admit that the sense of smell in its present state was originally acquired by man, as he now exists. He inherits the power in an enfeebled and so far rudimentary condition, from some early progenitor, to whom it was highly serviceable, and by whom it was continually used. In those animals which have this sense highly developed, such as dogs and horses, the recollection of persons and of places is strongly associated with their odor; and we can thus perhaps understand how it is, as Dr. Maudsley has truly remarked [37], that the sense of smell in man "is singularly effective in recalling vividly the ideas and images of forgotten scenes and places."

[36] The account given by Humboldt of the power of smell possessed by the natives of South America is well known, and has been confirmed by others. M. Houzeau ('Études sur les Facultés Mentales,' etc., tom. i. 1872, p. 91) asserts that he repeatedly made experiments, and proved that Negroes and Indians could recognize persons in the dark by their odor. Dr. W. Ogle has made some curious observations on the connection between the power of smell and the coloring matter of the mucous membrane of the olfactory region as well as of the skin of the body. I have, therefore, spoken in the text of the dark-colored races having a finer sense of smell than the white races. See his paper, 'Medico-Chirurgical Transactions,' London, vol. liii. 1870, p. 276.

[37] 'The Physiology and Pathology of Mind,' 2nd ed. 1868, p. 134.

Man differs conspicuously from all the other primates in being almost naked. But a few short straggling hairs are found over the greater part of the body in the man, and fine down on that of the woman. The different races differ much in hairiness; and in the individuals of the same race the hairs are highly variable, not only in abundance, but likewise in position: thus in some Europeans the shoulders are quite naked, whilst in others they bear thick tufts of hair. [38] There can be little doubt that the hairs thus scattered over the body are the rudiments of the uniform hairy coat of the lower animals. This view is rendered all the more probable, as it is known that fine, short, and pale-colored hairs on the limbs and other parts of the body, occasionally become developed into "thickset, long, and rather coarse dark hairs," when abnormally nourished near old-standing inflamed surfaces. [39]

[38] Eschricht, Ueber die Richtung der Haare am menschlichen Körper, Müller's 'Archiv für Anat. und Phys.' 1837, s. 47. I shall often have to refer to this very curious paper.
[39] Paget, 'Lectures on Surgical Pathology,' 1853, vol. i. p. 71.

I am informed by Sir James Paget that often several members of a family have a few hairs in their eyebrows much longer than the others; so that even this slight peculiarity seems to be inherited. These hairs, too, seem to have their representatives; for in the chimpanzee, and in certain species of Macacus, there are scattered hairs of considerable length rising from the naked skin above the eyes, and corresponding to our eyebrows; similar long hairs project from the hairy covering of the superciliary ridges in some baboons.

The fine wool-like hair, or so-called lanugo, with which the human fœtus during the sixth month is thickly covered, offers a more curious case. It is first developed, during the fifth month, on the eyebrows and face, and especially round the mouth, where it is much longer than that on the head. A moustache of this kind was observed by Eschricht [40] on a female fœtus; but this is not so surprising a circumstance as it may at first appear, for the two sexes generally resemble each other in all external characters during an early period of growth. The direction and arrangement of the hairs on all parts of the fœtal body are the same as in the adult, but are subject to much variability. The whole surface, including even the forehead and ears, is thus thickly clothed; but it is a significant fact that the palms of the hands and the soles of the feet are quite naked, like the inferior surfaces of all four extremities in most of the lower animals. As this can hardly be an accidental coincidence, the woolly covering of the fœtus probably represents the first permanent coat of hair in those mammals which are born hairy. Three or four cases have been recorded of persons born with their whole bodies and faces thickly covered with fine long hairs; and this strange condition is strongly inherited, and is correlated with an abnormal condition of the teeth. [41] Prof. Alex. Brandt informs me that he has compared the hair from the face of a man thus characterized, aged thirty-five, with the lanugo of a fœtus, and finds it quite similar in texture; therefore, as he remarks, the case may be attributed to an arrest of development in the hair, together with its continued growth. Many delicate children, as I have been assured by a surgeon to a hospital for children, have their backs covered by rather long silky hairs; and such cases probably come under the same head.

[40] Eschricht, ibid. s. 40, 47.
[41] See my 'Variation of Animals and Plants under Domestication,' vol. ii. p. 327. Prof. Alex. Brandt has recently sent me an additional case of a father and son, born in Russia, with these peculiarities. I have received drawings of both from Paris.

It appears as if the posterior molar or wisdom-teeth were tending to become rudimentary in the more civilized races of man. These teeth are rather smaller than the other molars, as is likewise the case with the corresponding teeth in the chimpanzee and orang; and they have only two separate fangs. They do not cut through the gums till about the seventeenth year, and I have been assured that they are much more liable to decay, and are earlier lost than the other teeth; but this is denied by some eminent dentists. They are also much more liable to vary, both in structure and in the period of their development, than the other teeth. [42] In the Melanian races, on the other hand, the wisdom-teeth are usually furnished with three separate fangs, and are generally sound; they also differ from the other molars in size, less than in the Caucasian races. [43] Prof. Schaaffhausen accounts for this difference between the races by "the posterior dental portion of the jaw being always shortened" in those that are civilized [44], and this shortening may, I presume, be attributed to civilized men habitually feeding on soft, cooked food, and thus using their jaws less. I am informed by Mr. Brace that it is becoming quite a common practice in the United States to remove some of the molar teeth of children, as the jaw does not grow large enough for the perfect development of the normal number. [45]

[42] Dr. Webb, 'Teeth in Man and the Anthropoid Apes,' as quoted by Dr. C. Carter Blake in Anthropological Review, July 1867, p. 299.
[43] Owen, 'Anatomy of Vertebrates,' vol. iii. pp. 320, 321, and 325.
[44] 'On the Primitive Form of the Skull,' Eng. translat., in 'Anthropological Review,' Oct. 1868, p. 426
[45] Prof. Montegazza writes to me from Florence, that he has lately been studying the last molar teeth in the different races of man, and has come to the same conclusion as that given in my text, viz., that in the higher or civilized races they are on the road towards atrophy or elimination.

With respect to the alimentary canal, I have met with an account of only a single rudiment, namely the vermiform appendage of the cæcum. The cæcum is a branch or diverticulum of the intestine, ending in a cul-de-sac, and is extremely long in many of the lower vegetable-feeding mammals. In the marsupial koala it is actually more than thrice as long as the whole body. [46] It is sometimes produced into a long gradually-tapering point, and is sometimes constricted in parts. It appears as if, in consequence of changed diet or habits, the cæcum had become much shortened in various animals, the vermiform appendage being left as a rudiment of the shortened part. That this appendage is a rudiment, we may infer from its small size, and from the evidence which Prof. Canestrini [47] has collected of its variability in man. It is occasionally quite absent, or again is largely developed. The passage is sometimes completely closed for half or two-thirds of its length, with the terminal part consisting of a flattened solid expansion. In the orang this appendage is long and convoluted: in man it arises from the end of the short cæcum, and is commonly from four to five inches in length, being only about the third of an inch in diameter. Not only is it useless, but it is sometimes the cause of death, of which fact I have lately heard two instances: this is due to small hard bodies, such as seeds, entering the passage, and causing inflammation. [48]

[46] Owen, 'Anatomy of Vertebrates,' vol. iii. pp. 416, 434, 441.
[47] 'Annuario della Soc. d. Nat.' Modena, 1867, p. 94.
[48] M. C. Martins ("De l'Unité Organique," in 'Revue des Deux Mondes,' June 15, 1862, p. 16) and Häckel ('Generelle Morphologie,' B. ii. s. 278), have both remarked on the singular fact of this rudiment sometimes causing death.

In some of the lower Quadrumana, in the Lemuridæ and Carnivora, as well as in many marsupials, there is a passage near the lower end of the humerus, called the supra-condyloid foramen, through which the great nerve of the fore limb and often the great artery pass. Now in the humerus of man, there is generally a trace of this passage, which is sometimes fairly well developed, being formed by a depending hook-like process of bone, completed by a band of ligament. Dr. Struthers, [49] who has closely attended to the subject, has now shown that this peculiarity is sometimes inherited, as it has occurred in a father, and in no less than four out of his seven children. When present, the great nerve invariably passes through it; and this clearly indicates that it is the homologue and rudiment of the supra-condyloid foramen of the lower animals. Prof. Turner estimates, as he informs me, that it occurs in about one per cent. of recent skeletons. But if the occasional development of this structure in man is, as seems probable, due to reversion, it is a return to a very ancient state of things, because in the higher Quadrumana it is absent.

[49] With respect to inheritance, see Dr. Struthers in the 'Lancet,' Feb. 15, 1873, and another important paper, ibid. Jan. 24, 1863, p. 83. Dr. Knox, as I am informed, was the first anatomist who drew attention to this peculiar structure in man; see his 'Great Artists and Anatomists,' p. 63. See also an important memoir on this process by Dr. Gruber, in the 'Bulletin de l'Acad. Imp. de St. Pétersbourg,' tom. xii. 1867, p. 448.

There is another foramen or perforation in the humerus, occasionally present in man, which may be called the inter-condyloid. This occurs, but not constantly, in various anthropoid and other apes, [50] and likewise in many of the lower animals. It is remarkable that this perforation seems to have been present in man much more frequently during ancient times than recently. Mr. Busk [51] has collected the following evidence on this head: Prof. Broca "noticed the perforation in four and a half per cent. of the arm-bones collected in the 'Cimetière du Sud,' at Paris; and in the Grotto of Orrony, the contents of which are referred to the Bronze period, as many as eight humeri out of thirty-two were perforated; but this extraordinary proportion, he thinks, might be due to the cavern having been a sort of 'family vault.' Again, M. Dupont found thirty per cent. of perforated bones in the caves of the Valley of the Lesse, belonging to the Reindeer period; whilst M. Leguay, in a sort of *dolmen* at Argenteuil, observed twenty-five per cent. to be perforated; and M. Pruner-Bey found twenty-six per cent. in the same condition in bones from Vauréal. Nor should it be left unnoticed that M. Pruner-Bey states that this condition is common in Guanche skeletons." It is an interesting fact that ancient races, in this and several other cases, more frequently present structures which resemble those of the lower animals than do the modern. One chief cause seems to be that the ancient races stand somewhat nearer in the long line of descent to their remote animal-like progenitors.

[50] Mr. St. George Mivart, 'Transactions Phil. Soc.' 1867, p. 310.
[51] "On the Caves of Gibraltar," 'Transactions of the International Congress of Prehistoric Archaeology,' Third Session, 1869, p. 159. Prof. Wyman has lately shown (Fourth Annual Report, Peabody Museum, 1871, p. 20), that this perforation is present in thirty-one per cent. of some human remains from ancient mounds in the Western United States, and in Florida. It frequently occurs in the negro.

In man, the *os coccyx*, together with certain other vertebræ hereafter to be described, though functionless as a tail, plainly represent this part in other vertebrate animals. At an early embryonic period it is free, and projects beyond the lower extremities; as may be seen in the drawing (Fig. 1.) of a human embryo. Even after birth it has been known, in certain rare and anomalous cases [52], to form a small external rudiment of a tail. The *os coccyx* is short, usually including only four vertebræ, all anchylosed together: and these are in a rudimentary condition, for they consist, with the exception of the basal one, of the centrum alone. [53] They are furnished with some small muscles; one of which, as I am informed by Prof. Turner, has been expressly described by Theile as a rudimentary repetition of the extensor of the tail, a muscle which is so largely developed in many mammals.

[52] Quatrefages has lately collected the evidence on this subject. 'Revue des Cours Scientifiques,' 1867-1868, p. 625. In 1840 Fleischmann exhibited a human fœtus bearing a free tail, which, as is not always the case, included vertebral bodies; and this tail was critically examined by the many anatomists present at the meeting of naturalists at Erlangen (see Marshall in Niederländischen Archiv fur Zoologie, December 1871).
[53] Owen, 'On the Nature of Limbs,' 1849, p. 114.

The spinal cord in man extends only as far downwards as the last dorsal or first lumbar vertebra; but a thread-like structure (the *filum terminale*) runs down the axis of the sacral part of the spinal canal, and even along the back of the coccygeal bones. The upper part of this filament, as Prof. Turner informs me, is undoubtedly homologous with the spinal cord; but the lower part apparently consists merely of the *pia mater*, or vascular investing membrane. Even in this case the *os coccyx* may be said to possess a vestige of so important a structure as the spinal cord, though no longer enclosed within a bony canal. The following fact, for which I am also indebted to Prof. Turner, shows how closely the *os coccyx* corresponds with the true tail in the lower animals: Luschka has recently discovered at the extremity of the coccygeal bones a very peculiar convoluted body, which is continuous with the middle sacral artery; and this discovery led Krause and Meyer to examine the tail of a monkey (Macacus), and of a cat, in both of which they found a similarly convoluted body, though not at the extremity.

The reproductive system offers various rudimentary structures; but these differ in one important respect from the foregoing cases. Here we are not concerned with the vestige of a part which does not belong to the species in an efficient state, but with a part efficient in the one sex, and represented in the other by a mere rudiment. Nevertheless, the occurrence of such rudiments is as difficult to explain, on the belief of the separate creation of each species, as in the foregoing cases. Hereafter I shall have to recur to these rudiments, and shall show that their presence generally depends merely on inheritance, that is, on parts acquired by one sex having been partially transmitted to the other. I will in this place only give some instances of such rudiments. It is well known that in the males of all mammals, including man, rudimentary mammæ exist. These in several instances have become well developed, and have yielded a copious supply of milk. Their essential identity in the two sexes is likewise shown by their occasional sympathetic enlargement in both during an attack of the measles. The *vesicula prostatica*, which has been observed in many male mammals, is now universally acknowledged to be the homologue of the female uterus, together with the connected passage. It is impossible to read Leuckart's able description of this organ, and his reasoning, without admitting the justness of his conclusion. This is especially clear in the case of those mammals in which the true female uterus bifurcates, for in the males of these the vesicula likewise bifurcates. [54] Some other rudimentary structures belonging to the reproductive system might have been here adduced. [55]

[54] Leuckart, in Todd's 'Cyclop. of Anatomy' 1849-52, vol. iv. p. 1415. In man this organ is only from three to six lines in length, but, like so many other rudimentary parts, it is variable in development as well as in other characters.
[55] See, on this subject, Owen, 'Anatomy of Vertebrates,' vol. iii. pp. 675, 676, 706.

The bearing of the three great classes of facts now given is unmistakable. But it would be superfluous fully to recapitulate the line of argument given in detail in my 'Origin of Species.' The homological construction of the whole frame in the members of the same class is intelligible, if we admit their descent from a common progenitor, together with their subsequent adaptation to diversified conditions. On any other view, the similarity of pattern between the hand of a man or monkey, the foot of a horse, the flipper of a seal, the wing of a bat, etc., is utterly inexplicable. [56] It is no scientific explanation to assert that they have all been formed on the same ideal plan. With respect to development, we can clearly understand, on the principle of variations supervening at a rather late embryonic period, and being inherited at a corresponding period, how it is that the embryos of wonderfully different forms should still retain, more or less perfectly, the structure of their common progenitor. No other explanation has ever been given of the marvellous fact that the embryos of a man, dog, seal, bat, reptile, etc., can at first hardly be distinguished from each other. In order to understand the existence of rudimentary organs, we have only to suppose that a former progenitor possessed the parts in question in a perfect state, and that under changed habits of life they became greatly reduced, either from simple disuse, or through the natural selection of those individuals which were least encumbered with a superfluous part, aided by the other means previously indicated.

[56] Prof. Bianconi, in a recently published work, illustrated by admirable engravings ('La Théorie Darwinienne et la création dite indépendante,' 1874), endeavours to show that homological structures, in the above and other cases, can be fully explained on mechanical principles, in accordance with their uses. No one has shown so well, how admirably such structures are adapted for their final purpose; and this adaptation can, as I believe, be explained through natural selection. In considering the wing of a bat, he brings forward (p. 218) what appears to me (to use Auguste Comte's words) a mere metaphysical principle, namely, the preservation "in its integrity of the mammalian nature of the animal." In only a few cases does he discuss rudiments, and then only those parts which are partially rudimentary, such as the little hoofs of the pig and ox, which do not touch the ground; these he shows clearly to be of service to the animal. It is unfortunate that he did not consider such cases as the minute teeth, which never cut through the jaw in the ox, or the mammæ of male quadrupeds, or the wings of certain beetles, existing under the soldered wing-covers, or the vestiges of the pistil and stamens in various flowers, and many other such cases. Although I greatly admire Prof. Bianconi's work, yet the belief now held by most naturalists seems to me left unshaken, that homological structures are inexplicable on the principle of mere adaptation.

Thus we can understand how it has come to pass that man and all other vertebrate animals have been constructed on the same general model, why they pass through the same early stages of development, and why they retain certain rudiments in common. Consequently we ought frankly to admit their community of descent: to take any other view, is to admit that our own structure, and that of all the animals around us, is a mere snare laid to entrap our judgment. This conclusion is greatly strengthened, if we look to the members of the whole animal series, and consider the evidence derived from their affinities or classification, their geographical distribution and geological succession. It is only our natural prejudice, and that arrogance which made our forefathers declare that they were descended from demi-gods, which leads us to demur to this conclusion. But the time will before long come, when it will be thought wonderful that naturalists, who were well acquainted with the comparative structure and development of man, and other mammals, should have believed that each was the work of a separate act of creation.

CHAPTER II.

ON THE MANNER OF DEVELOPMENT OF MAN FROM SOME LOWER FORM.

Variability of body and mind in man—Inheritance—Causes of variability—Laws of variation the same in man as in the lower animals—Direct action of the conditions of life—Effects of the increased use and disuse of parts—Arrested development—Reversion—Correlated variation—Rate of increase—Checks to increase—Natural selection—Man the most dominant animal in the world—Importance of his corporeal structure—The causes which have led to his becoming erect—Consequent changes of structure—Decrease in size of the canine teeth—Increased size and altered shape of the skull—Nakedness —Absence of a tail—Defenceless condition of man.

It is manifest that man is now subject to much variability. No two individuals of the same race are quite alike. We may compare millions of faces, and each will be distinct. There is an equally great amount of diversity in the proportions and dimensions of the various parts of the body; the length of the legs being one of the most variable points. [1] Although in some quarters of the world an elongated skull, and in other quarters a short skull prevails, yet there is great diversity of shape even within the limits of the same race, as with the aborigines of America and South Australia—the latter a race "probably as pure and homogeneous in blood, customs, and language as any in existence"—and even with the inhabitants of so confined an area as the Sandwich Islands. [2] An eminent dentist assures me that there is nearly as much diversity in the teeth as in the features. The chief arteries so frequently run in abnormal courses, that it has been found useful for surgical purposes to calculate from 1040 corpses how often each course prevails. [3] The muscles are eminently variable: thus those of the foot were found by Prof. Turner [4] not to be strictly alike in any two out of fifty bodies; and in some the deviations were considerable. He adds, that the power of performing the appropriate movements must have been modified in accordance with the several deviations. Mr. J. Wood has recorded [5] the occurrence of 295 muscular variations in thirty-six subjects, and in another set of the same number no less than 558 variations, those occurring on both sides of the body being only reckoned as one. In the last set, not one body out of the thirty-six was "found totally wanting in departures from the standard descriptions of the muscular system given in anatomical text books." A single body presented the extraordinary number of twenty-five distinct abnormalities. The same muscle sometimes varies in many ways: thus Prof. Macalister describes [6] no less than twenty distinct variations in the *palmaris accessorius*.

[1] 'Investigations in Military and Anthropological Statistics of American Soldiers,' by B.A. Gould, 1869, p. 256.
[2] With respect to the "Cranial forms of the American aborigines," see Dr. Aitken Meigs in 'Proc. Acad. Nat. Sci.' Philadelphia, May 1868. On the Australians, see Huxley, in Lyell's 'Antiquity of Man,' 1863, p. 87. On the Sandwich Islanders, Prof. J. Wyman, 'Observations on Crania,' Boston, 1868, p. 18.
[3] 'Anatomy of the Arteries,' by R. Quain. Preface, vol. i. 1844.
[4] 'Transactions of the Royal Society of Edinburgh,' vol. xxiv. pp. 175, 189.
[5] 'Proceedings Royal Society,' 1867, p. 544; also 1868, pp. 483, 524. There is a previous paper, 1866, p. 229.
[6] 'Proc. R. Irish Academy,' vol. x. 1868, p. 141.

The famous old anatomist, Wolff, [7] insists that the internal viscera are more variable than the external parts: *Nulla particula est quae non aliter et aliter in aliis se habeat hominibus.* He has even written a treatise on the choice of typical examples of the viscera for representation. A discussion on the beau-ideal of the liver, lungs, kidneys, etc., as of the human face divine, sounds strange in our ears.

[7] 'Act. Acad. St. Petersburg,' 1778, part ii. p. 217.

The variability or diversity of the mental faculties in men of the same race, not to mention the greater differences between the men of distinct races, is so notorious that not a word need here be said. So it is with the lower animals. All who have had charge of menageries admit this fact, and we see it plainly in our dogs and other domestic animals. Brehm especially insists that each individual monkey of those which he kept tame in Africa had its own peculiar disposition and temper: he mentions one baboon remarkable for its high intelligence; and the keepers in the Zoological Gardens pointed out to me a monkey, belonging to the New World division, equally remarkable for intelligence. Rengger, also, insists on the diversity in the various mental characters of the monkeys of the same species which he kept in Paraguay; and this diversity, as he adds, is partly innate, and partly the result of the manner in which they have been treated or educated.[8]

[8] Brehm, 'Thierleben,' B. i. ss. 58, 87. Rengger, 'Säugethiere von Paraguay,' s. 57.

I have elsewhere [9] so fully discussed the subject of Inheritance, that I need here add hardly anything. A greater number of facts have been collected with respect to the transmission of the most trifling, as well as of the most important characters in man, than in any of the lower animals; though the facts are copious enough with respect to the latter. So in regard to mental qualities, their transmission is manifest in our dogs, horses, and other domestic animals. Besides special tastes and habits, general intelligence, courage, bad and good temper, etc., are certainly transmitted. With man we see similar facts in almost every family; and we now know, through the admirable labors of Mr. Galton [10], that genius which implies a wonderfully complex combination of high faculties, tends to be inherited; and, on the other hand, it is too certain that insanity and deteriorated mental powers likewise run in families.

[9] 'Variation of Animals and Plants under Domestication,' vol. ii. chap. xii.
[10] 'Hereditary Genius: an Inquiry into its Laws and Consequences,' 1869.

With respect to the causes of variability, we are in all cases very ignorant; but we can see that in man as in the lower animals, they stand in some relation to the conditions to which each species has been exposed, during several generations. Domesticated animals vary more than those in a state of nature; and this is apparently due to the diversified and changing nature of the conditions to which they have been subjected. In this respect the different races of man resemble domesticated animals, and so do the individuals of the same race, when inhabiting a very wide area, like that of America. We see the influence of diversified conditions in the more civilized nations; for the members belonging to different grades of rank, and following different occupations, present a greater range of character than do the members of barbarous nations. But the uniformity of savages has often been exaggerated, and in some cases can hardly be said to exist. [11] It is, nevertheless, an error to speak of man, even if we look only to the conditions to which he has been exposed, as "far more domesticated" [12] than any other animal. Some savage races, such as the Australians, are not exposed to more diversified conditions than are many species which have a wide range. In another and much more important respect, man differs widely from any strictly domesticated animal; for his breeding has never long been controlled, either by methodical or unconscious selection. No race or body of men has been so completely subjugated by other men, as that certain individuals should be preserved, and thus unconsciously selected, from somehow excelling in utility to their masters. Nor have certain male and female individuals been intentionally picked out and matched, except in the well-known case of the Prussian grenadiers; and in this case man obeyed, as might have been expected, the law of methodical selection; for it is asserted that many tall men were reared in the villages inhabited by the grenadiers and their tall wives. In Sparta, also, a form of selection was followed, for it was enacted that all children should be examined shortly after birth; the well-formed and vigorous being preserved, the others left to perish. [13]

[11] Mr. Bates remarks ('The Naturalist on the Amazons,' 1863, vol. ii p. 159), with respect to the Indians of the same South American tribe, "no two of them were at all similar in the shape of the head; one man had an oval visage with fine features, and another was quite Mongolian in breadth and prominence of cheek, spread of nostrils, and obliquity of eyes."

[12] Blumenbach, 'Treatises on Anthropology.' Eng. translat., 1865, p. 205.

[13] Mitford's 'History of Greece,' vol. i. p. 282. It appears also from a passage in Xenophon's 'Memorabilia,' B. ii. 4 (to which my attention has been called by the Rev. J.N. Hoare), that it was a well recognized principle with the Greeks, that men ought to select their wives with a view to the health and vigor of their children. The Grecian poet, Theognis, who lived 550 B.C., clearly saw how important selection, if carefully applied, would be for the improvement of mankind. He saw, likewise, that wealth often checks the proper action of sexual selection. He thus writes:

> "With kine and horses, Kurnus! we proceed
> By reasonable rules, and choose a breed
> For profit and increase, at any price:
> Of a sound stock, without defect or vice.
> But, in the daily matches that we make,
> The price is everything: for money's sake,
> Men marry: women are in marriage given
> The churl or ruffian, that in wealth has thriven,
> May match his offspring with the proudest race:
> Thus everything is mix'd, noble and base!
> If then in outward manner, form, and mind,
> You find us a degraded, motley kind,
> Wonder no more, my friend! the cause is plain,
> And to lament the consequence is vain."
> (The Works of J. Hookham Frere, vol. ii. 1872, p. 334.)

If we consider all the races of man as forming a single species, his range is enormous; but some separate races, as the Americans and Polynesians, have very wide ranges. It is a well-known law that widely-ranging species are much more variable than species with restricted ranges; and the variability of man may with more truth be compared with that of widely-ranging species, than with that of domesticated animals.

Not only does variability appear to be induced in man and the lower animals by the same general causes, but in both the same parts of the body are affected in a closely analogous manner. This has been proved in such full detail by Godron and Quatrefages, that I need here only refer to their works. [14] Monstrosities, which graduate into slight variations, are likewise so similar in man and the lower animals, that the same classification and the same terms can be used for both, as has been shown by Isidore Geoffroy St.-Hilaire. [15] In my work on the variation of domestic animals, I have attempted to arrange in a rude fashion the laws of variation under the following heads:— The direct and definite action of changed conditions, as exhibited by all or nearly all the individuals of the same species, varying in the same manner under the same circumstances. The effects of the long-continued use or disuse of parts. The cohesion of homologous parts. The variability of multiple parts. Compensation of growth; but of this law I have found no good instance in the case of man. The effects of the mechanical pressure of one part on another; as of the pelvis on the cranium of the infant in the womb. Arrests of development, leading to the diminution or suppression of parts. The reappearance of long-lost characters through reversion. And lastly, correlated variation. All these so-called laws apply equally to man and the lower animals; and most of them even to plants. It would be superfluous here to discuss all of them [16]; but several are so important, that they must be treated at considerable length.

[14] Godron, 'De l'Espèce,' 1859, tom. ii. livre 3. Quatrefages, 'Unité de l'Espèce Humaine,' 1861. Also Lectures on Anthropology, given in the 'Revue des Cours Scientifiques,' 1866-1868.

[15] 'Hist. Gén. et Part. des Anomalies de l'Organisation,' in three volumes, tom. i. 1832.

[16] I have fully discussed these laws in my 'Variation of Animals and Plants under Domestication,' vol. ii. chap. xxii. and xxiii. M. J.P. Durand has lately (1868) published a valuable essay, 'De l'Influence des Milieux,' etc. He lays much stress, in the case of plants, on the nature of the soil.

THE DIRECT AND DEFINITE ACTION OF CHANGED CONDITIONS.

This is a most perplexing subject. It cannot be denied that changed conditions produce some, and occasionally a considerable effect, on organisms of all kinds; and it seems at first probable that if sufficient time were allowed this would be the invariable result. But I have failed to obtain clear evidence in favour of this conclusion; and valid reasons may be urged on the other side, at least as far as the innumerable structures are concerned, which are adapted for special ends. There can, however, be no doubt that changed conditions induce an almost indefinite amount of fluctuating variability, by which the whole organization is rendered in some degree plastic.

In the United States, above 1,000,000 soldiers, who served in the late war, were measured, and the States in which they were born and reared were recorded. [17] From this astonishing number of observations it is proved that local influences of some kind act directly on stature; and we further learn that "the State where the physical growth has in great measure taken place, and the State of birth, which indicates the ancestry, seem to exert a marked influence on the stature." For instance, it is established, "that residence in the Western States, during the years of growth, tends to produce increase of stature." On the other hand, it is certain that with sailors, their life delays growth, as shown "by the great difference between the statures of soldiers and sailors at the ages of seventeen and eighteen years." Mr. B.A. Gould endeavoured to ascertain the nature of the influences which thus act on stature; but he arrived only at negative results, namely that they did not relate to climate, the elevation of the land, soil, nor even "in any controlling degree" to the abundance or the need of the comforts of life. This latter conclusion is directly opposed to that arrived at by Villermé, from the statistics of the height of the conscripts in different parts of France. When we compare the differences in stature between the Polynesian chiefs and the lower orders within the same islands, or between the inhabitants of the fertile volcanic and low barren coral islands of the same ocean [18] or again between the Fuegians on the eastern and western shores of their country, where the means of subsistence are very different, it is scarcely possible to avoid the conclusion that better food and greater comfort do influence stature. But the preceding statements show how difficult it is to arrive at any precise result. Dr. Beddoe has lately proved that, with the inhabitants of Britain, residence in towns and certain occupations have a deteriorating influence on height; and he infers that the result is to a certain extent inherited, as is likewise the case in the United States. Dr. Beddoe further believes that wherever a "race attains its maximum of physical development, it rises highest in energy and moral vigor." [19]

[17] 'Investigations in Military and Anthrop. Statistics,' etc., 1869, by B.A. Gould, pp. 93, 107, 126, 131, 134.
[18] For the Polynesians, see Prichard's 'Physical History of Mankind,' vol. v. 1847, pp. 145, 283. Also Godron, 'De l'Espèce,' tom. ii. p. 289. There is also a remarkable difference in appearance between the closely-allied Hindoos inhabiting the Upper Ganges and Bengal: see Elphinstone's 'History of India,' vol. i. p. 324.
[19] 'Memoirs, Anthropological Society,' vol. iii. 1867-69, pp. 561, 565, 567.

Whether external conditions produce any other direct effect on man is not known. It might have been expected that differences of climate would have had a marked influence, inasmuch as the lungs and kidneys are brought into activity under a low temperature, and the liver and skin under a high one. [20] It was formerly thought that the color of the skin and the character of the hair were determined by light or heat; and although it can hardly be denied that some effect is thus produced, almost all observers now agree that the effect has been very small, even after exposure during many ages. But this subject will be more properly discussed when we treat of the different races of mankind. With our domestic animals there are grounds for believing that cold and damp directly affect the growth of the hair; but I have not met with any evidence on this head in the case of man.

[20] Dr. Brakenridge, 'Theory of Diathesis,' 'Medical Times,' June 19 and July 17, 1869.

EFFECTS OF THE INCREASED USE AND DISUSE OF PARTS.

It is well known that use strengthens the muscles in the individual, and complete disuse, or the destruction of the proper nerve, weakens them. When the eye is destroyed, the optic nerve often becomes atrophied. When an artery is tied, the lateral channels increase not only in diameter, but in the thickness and strength of their coats. When one kidney ceases to act from disease, the other increases in size, and does double work. Bones increase not only in thickness, but in length, from carrying a greater weight. [21] Different occupations, habitually followed, lead to changed proportions in various parts of the body. Thus it was ascertained by the United States Commission [22] that the legs of the sailors employed in the late war were longer by 0.217 of an inch than those of the soldiers, though the sailors were on an average shorter men; whilst their arms were shorter by 1.09 of an inch, and therefore, out of proportion, shorter in relation to their lesser height. This shortness of the arms is apparently due to their greater use, and is an unexpected result: but sailors chiefly use their arms in pulling, and not in supporting weights. With sailors, the girth of the neck and the depth of the instep are greater, whilst the circumference of the chest, waist, and hips is less, than in soldiers.

[21] I have given authorities for these several statements in my 'Variation of Animals and Plants under Domestication,' vol. ii. pp. 297-300. Dr. Jaeger, "Ueber das Längenwachsthum der Knochen," 'Jenaischen Zeitschrift,' B. v. Heft. i.
[22] 'Investigations,' etc., by B.A. Gould, 1869, p. 288.

Whether the several foregoing modifications would become hereditary, if the same habits of life were followed during many generations, is not known, but it is probable. Rengger [23] attributes the thin legs and thick arms of the Payaguas Indians to successive generations having passed nearly their whole lives in canoes, with their lower extremities motionless. Other writers have come to a similar conclusion in analogous cases. According to Cranz [24], who lived for a long time with the Esquimaux, "the natives believe that ingenuity and dexterity in seal-catching (their highest art and virtue) is hereditary; there is really something in it, for the son of a celebrated seal-catcher will distinguish himself, though he lost his father in childhood." But in this case it is mental aptitude, quite as much as bodily structure, which appears to be inherited. It is asserted that the hands of English laborers are at birth larger than those of the gentry. [25] From the correlation which exists, at least in some cases [26], between the development of the extremities and of the jaws, it is possible that in those classes which do not labor much with their hands and feet, the jaws would be reduced in size from this cause. That they are generally smaller in refined and civilized men than in hard-working men or savages, is certain. But with savages, as Mr. Herbert Spencer [27] has remarked, the greater use of the jaws in chewing coarse, uncooked food, would act in a direct manner on the masticatory muscles, and on the bones to which they are attached. In infants, long before birth, the skin on the soles of the feet is thicker than on any other part of the body; [28] and it can hardly be doubted that this is due to the inherited effects of pressure during a long series of generations.

[23] 'Säugethiere von Paraguay,' 1830, s. 4.
[24] 'History of Greenland,' Eng. translat., 1767, vol. i. p. 230.
[25] 'Intermarriage,' by Alex. Walker, 1838, p. 377.
[26] 'The Variation of Animals under Domestication,' vol. i. p. 173.
[27] 'Principles of Biology,' vol. i. p. 455.
[28] Paget, 'Lectures on Surgical Pathology,' vol. ii, 1853, p. 209.

It is familiar to every one that watchmakers and engravers are liable to be short-sighted, whilst men living much out of doors, and especially savages, are generally long-sighted. [29] Short-sight and long-sight certainly tend to be inherited. [30] The inferiority of Europeans, in comparison with savages, in eyesight and in the other senses, is no doubt the accumulated and transmitted effect of lessened use during many generations; for Rengger [31] states that he has repeatedly observed Europeans, who had been brought up and spent their whole lives with the wild Indians, who nevertheless did not equal them in the sharpness of their senses. The same naturalist observes that the cavities in the skull for the reception of the several sense-organs are larger in the American aborigines than in

Europeans; and this probably indicates a corresponding difference in the dimensions of the organs themselves. Blumenbach has also remarked on the large size of the nasal cavities in the skulls of the American aborigines, and connects this fact with their remarkably acute power of smell. The Mongolians of the plains of northern Asia, according to Pallas, have wonderfully perfect senses; and Prichard believes that the great breadth of their skulls across the zygomas follows from their highly-developed sense organs. [32]

[29] It is a singular and unexpected fact that sailors are inferior to landsmen in their mean distance of distinct vision. Dr. B.A. Gould ('Sanitary Memoirs of the War of the Rebellion,' 1869, p. 530), has proved this to be the case; and he accounts for it by the ordinary range of vision in sailors being "restricted to the length of the vessel and the height of the masts."

[30] 'The Variation of Animals under Domestication,' vol. i. p. 8.

[31] 'Säugethiere von Paraguay,' s. 8, 10. I have had good opportunities for observing the extraordinary power of eyesight in the Fuegians. See also Lawrence ('Lectures on Physiology,' etc., 1822, p. 404) on this same subject. M. Giraud-Teulon has recently collected ('Revue des Cours Scientifiques,' 1870, p. 625) a large and valuable body of evidence proving that the cause of short-sight, "*C'est le travail assidu, de prés.*"

[32] Prichard, 'Physical History of Mankind,' on the authority of Blumenbach, vol. i. 1851, p. 311; for the statement by Pallas, vol. iv. 1844, p. 407.

The Quechua Indians inhabit the lofty plateaux of Peru; and Alcide d'Orbigny states [33] that, from continually breathing a highly rarefied atmosphere, they have acquired chests and lungs of extraordinary dimensions. The cells, also, of the lungs are larger and more numerous than in Europeans. These observations have been doubted, but Mr. D. Forbes carefully measured many Aymaras, an allied race, living at the height of between 10,000 and 15,000 feet; and he informs me [34] that they differ conspicuously from the men of all other races seen by him in the circumference and length of their bodies. In his table of measurements, the stature of each man is taken at 1000, and the other measurements are reduced to this standard. It is here seen that the extended arms of the Aymaras are shorter than those of Europeans, and much shorter than those of Negroes. The legs are likewise shorter; and they present this remarkable peculiarity, that in every Aymara measured, the femur is actually shorter than the tibia. On an average, the length of the femur to that of the tibia is as 211 to 252; whilst in two Europeans, measured at the same time, the femora to the tibiae were as 244 to 230; and in three Negroes as 258 to 241. The humerus is likewise shorter relatively to the forearm. This shortening of that part of the limb which is nearest to the body, appears to be, as suggested to me by Mr. Forbes, a case of compensation in relation with the greatly increased length of the trunk. The Aymaras present some other singular points of structure, for instance, the very small projection of the heel.

[33] Quoted by Prichard, 'Researches into the Physical History of Mankind,' vol. v. p. 463.

[34] Mr. Forbes' valuable paper is now published in the 'Journal of the Ethnological Society of London,' new series, vol. ii. 1870, p.193.

These men are so thoroughly acclimatized to their cold and lofty abode, that when formerly carried down by the Spaniards to the low eastern plains, and when now tempted down by high wages to the gold-washings, they suffer a frightful rate of mortality. Nevertheless Mr. Forbes found a few pure families which had survived during two generations: and he observed that they still inherited their characteristic peculiarities. But it was manifest, even without measurement, that these peculiarities had all decreased; and on measurement, their bodies were found not to be so much elongated as those of the men on the high plateau; whilst their femora had become somewhat lengthened, as had their tibiae, although in a less degree. The actual measurements may be seen by consulting Mr. Forbes's memoir. From these observations, there can, I think, be no doubt that residence during many generations at a great elevation tends, both directly and indirectly, to induce inherited modifications in the proportions of the body. [35]

[35] Dr. Wilckens ('Landwirthschaft. Wochenblatt,' No. 10, 1869) has lately published an interesting essay showing how domestic animals, which live in mountainous regions, have their frames modified.

Although man may not have been much modified during the latter stages of his existence through the increased or decreased use of parts, the facts now given show that his liability in this respect has not been lost; and we positively know that the same law holds good with the lower animals. Consequently we may infer that when at a remote epoch the progenitors of man were in a transitional state, and were changing from quadrupeds into bipeds, natural selection would probably have been greatly aided by the inherited effects of the increased or diminished use of the different parts of the body.

ARRESTS OF DEVELOPMENT.

There is a difference between arrested development and arrested growth, for parts in the former state continue to grow whilst still retaining their early condition. Various monstrosities come under this head; and some, as a cleft palate, are known to be occasionally inherited. It will suffice for our purpose to refer to the arrested brain-development of microcephalous idiots, as described in Vogt's memoir. [36] Their skulls are smaller, and the convolutions of the brain are less complex than in normal men. The frontal sinus, or the projection over the eye-brows, is largely developed, and the jaws are prognathous to an "*effrayant*" degree; so that these idiots somewhat resemble the lower types of mankind. Their intelligence, and most of their mental faculties, are extremely feeble. They cannot acquire the power of speech, and are wholly incapable of prolonged attention, but are much given to imitation. They are strong and remarkably active, continually gambolling and jumping about, and making grimaces. They often ascend stairs on all-fours; and are curiously fond of climbing up furniture or trees. We are thus reminded of the delight shown by almost all boys in climbing trees; and this again reminds us how lambs and kids, originally alpine animals, delight to frisk on any hillock, however small. Idiots also resemble the lower animals in some other respects; thus several cases are recorded of their carefully smelling every mouthful of food before eating it. One idiot is described as often using his mouth in aid of his hands, whilst hunting for lice. They are often filthy in their habits, and have no sense of decency; and several cases have been published of their bodies being remarkably hairy. [37]

[36] 'Mémoire sur les Microcéphales,' 1867, pp. 50, 125, 169, 171, 184-198.

[37] Prof. Laycock sums up the character of brute-like idiots by calling them *theroid*; 'Journal of Mental Science,' July 1863. Dr. Scott ('The Deaf and Dumb,' 2nd ed. 1870, p. 10) has often observed the imbecile smelling their food. See, on this same subject, and on the hairiness of idiots, Dr. Maudsley, 'Body and Mind,' 1870, pp. 46-51. Pinel has also given a striking case of hairiness in an idiot.

REVERSION.

Many of the cases to be here given, might have been introduced under the last heading. When a structure is arrested in its development, but still continues growing, until it closely resembles a corresponding structure in some lower and adult member of the same group, it may in one sense be considered as a case of reversion. The lower members in a group give us some idea how the common progenitor was probably constructed; and it is hardly credible that a complex part, arrested at an early phase of embryonic development, should go on growing so as ultimately to perform its proper function, unless it had acquired such power during some earlier state of existence, when the present exceptional or arrested structure was normal. The simple brain of a microcephalous idiot, in as far as it resembles that of an ape, may in this sense be said to offer a case of reversion. [38] There are other cases which come more strictly under our present head of reversion. Certain structures, regularly occurring in the lower members of the group to which man belongs, occasionally make their appearance in him, though not found in the normal human embryo; or, if normally present in the human embryo, they become abnormally developed, although in a manner which is normal in the lower members of the group. These remarks will be rendered clearer by the following illustrations.

[38] In my 'Variation of Animals under Domestication' (vol. ii. p. 57), I attributed the not very rare cases of supernumerary mammæ in women to reversion. I was led to this as a probable conclusion, by the additional mammæ being generally placed symmetrically on the breast; and more especially from one case, in which a single efficient mamma occurred in the inguinal region of a woman, the daughter of another woman with supernumerary mammæ. But I now find (see, for instance, Prof. Preyer, 'Der Kampf um das Dasein,' 1869, s. 45) that *mammæ erraticae*, occur in other situations, as on the back, in the armpit, and on the thigh; the mammæ in this latter instance having given so much milk that the child was thus nourished. The probability that the additional mammæ are due to reversion is thus much weakened; nevertheless, it still seems to me probable, because two pairs are often found symmetrically on the breast; and of this I myself have received information in several cases. It is well known that some Lemurs normally have two pairs of mammæ on the breast. Five cases have been recorded of the presence of more than a pair of mammæ (of course rudimentary) in the male sex of mankind; see 'Journal of Anat. and Physiology,' 1872, p. 56, for a case given by Dr. Handyside, in which two brothers exhibited this peculiarity; see also a paper by Dr. Bartels, in 'Reichert's and du Bois-Reymond's Archiv.,' 1872, p. 304. In one of the cases alluded to by Dr. Bartels, a man bore five mammæ, one being medial and placed above the navel; Meckel von Hemsbach thinks that this latter case is illustrated by a medial mamma occurring in certain Cheiroptera. On the whole, we may well doubt if additional mammæ would ever have been developed in both sexes of mankind, had not his early progenitors been provided with more than a single pair.

In the above work (vol. ii. p. 12), I also attributed, though with much hesitation, the frequent cases of polydactylism in men and various animals to reversion. I was partly led to this through Prof. Owen's statement, that some of the Ichthyopterygia possess more than five digits, and therefore, as I supposed, had retained a primordial condition; but Prof. Gegenbaur ('Jenaischen Zeitschrift,' B. v. Heft 3, s. 341), disputes Owen's conclusion. On the other hand, according to the opinion lately advanced by Dr. Günther, on the paddle of Ceratodus, which is provided with articulated bony rays on both sides of a central chain of bones, there seems no great difficulty in admitting that six or more digits on one side, or on both sides, might reappear through reversion. I am informed by Dr. Zouteveen that there is a case on record of a man having twenty-four fingers and twenty-four toes! I was chiefly led to the conclusion that the presence of supernumerary digits might be due to reversion from the fact that such digits, not only are strongly inherited, but, as I then believed, had the power of regrowth after amputation, like the normal digits of the lower vertebrata. But I have explained in the second edition of my Variation under Domestication why I now place little reliance on the recorded cases of such regrowth. Nevertheless it deserves notice, inasmuch as arrested development and reversion are intimately related processes; that various structures in an embryonic or arrested condition, such as a cleft palate, bifid uterus, etc., are frequently accompanied by polydactylism. This has been strongly insisted on by Meckel and Isidore Geoffroy St.-Hilaire. But at present it is the safest course to give up altogether the idea that there is any relation between the development of supernumerary digits and reversion to some lowly organized progenitor of man.

In various mammals the uterus graduates from a double organ with two distinct orifices and two passages, as in the marsupials, into a single organ, which is in no way double except from having a slight internal fold, as in the higher apes and man. The rodents exhibit a perfect series of gradations between these two extreme states. In all mammals the uterus is developed from two simple primitive tubes, the inferior portions of which form the cornua; and it is in the words of Dr. Farre, "by the coalescence of the two cornua at their lower extremities that the body of the uterus is formed in man; while in those animals in which no middle portion or body exists, the cornua remain ununited. As the development of the uterus proceeds, the two cornua become gradually shorter, until at length they are lost, or, as it were, absorbed into the body of the uterus." The angles of the uterus are still produced into cornua, even in animals as high up in the scale as the lower apes and lemurs.

Now in women, anomalous cases are not very infrequent, in which the mature uterus is furnished with cornua, or is partially divided into two organs; and such cases, according to Owen, repeat "the grade of concentrative development," attained by certain rodents. Here perhaps we have an instance of a simple arrest of embryonic development, with subsequent growth and perfect functional development; for either side of the partially double uterus is capable of performing the proper office of gestation. In other and rarer cases, two distinct uterine cavities are formed, each having its proper orifice and passage. [39] No such stage is passed through during the ordinary development of the embryo; and it is difficult to believe, though perhaps not impossible, that the two simple, minute, primitive tubes should know how (if such an expression may be used) to grow into two distinct uteri, each with a well-constructed orifice and passage, and each furnished with numerous muscles, nerves, glands and vessels, if they had not formerly passed through a similar course of development, as in the case of existing marsupials. No one will pretend that so perfect a structure as the abnormal double uterus in woman could be the result of mere chance. But the principle of reversion, by which a long-lost structure is called back into existence, might serve as the guide for its full development, even after the lapse of an enormous interval of time.

[39] See Dr. A. Farre's well-known article in the 'Cyclopædia of Anatomy and Physiology,' vol. v. 1859, p. 642. Owen, 'Anatomy of Vertebrates,' vol. iii. 1868, p. 687. Professor Turner, in 'Edinburgh Medical Journal,' February, 1865.

Professor Canestrini, after discussing the foregoing and various analogous cases, arrives at the same conclusion as that just given. He adduces another instance, in the case of the malar bone, [40] which, in some of the Quadrumana and other mammals, normally consists of two portions. This is its condition in the human fœtus when two months old; and through arrested development, it sometimes remains thus in man when adult, more especially in the lower prognathous races. Hence Canestrini concludes that some ancient progenitor of man must have had this bone normally divided into two portions, which afterwards became fused together. In man the frontal bone consists of a single piece, but in the embryo, and in children, and in almost all the lower mammals, it consists of two pieces separated by a distinct suture. This suture occasionally persists more or less distinctly in man after maturity; and more frequently in ancient than in recent crania, especially, as Canestrini has observed, in those exhumed from the Drift, and belonging to the brachycephalic type. Here again he comes to the same conclusion as in the analogous case of the malar bones. In this, and other instances presently to be given, the cause of ancient races approaching the lower animals in certain characters more frequently than do the modern races, appears to be, that the latter stand at a somewhat greater distance in the long line of descent from their early semi-human progenitors.

[40] 'Annuario della Soc. dei Naturalisti,' Modena, 1867, p. 83. Prof. Canestrini gives extracts on this subject from various authorities. Laurillard remarks, that as he has found a complete similarity in the form, proportions, and connection of the two malar bones in several human subjects and in certain apes, he cannot consider this disposition of the parts as simply accidental. Another paper on this same anomaly has been published by Dr. Saviotti in the 'Gazzetta delle Cliniche,' Turin, 1871, where he says that traces of the division may be detected in about two per cent. of adult skulls; he also remarks that it more frequently occurs in prognathous skulls, not of the Aryan race, than in others. See also G. Delorenzi on the same subject; 'Tre nuovi casi d'anomalia dell' osso malare,' Torino, 1872. Also, E. Morselli, 'Sopra una rara anomalia dell' osso malare,' Modena, 1872. Still more recently Gruber has written a pamphlet on the division of this bone. I give these references because a reviewer, without any grounds or scruples, has thrown doubts on my statements.

Various other anomalies in man, more or less analogous to the foregoing, have been advanced by different authors, as cases of reversion; but these seem not a little doubtful, for we have to descend extremely low in the mammalian series, before we find such structures normally present. [41]

[41] A whole series of cases is given by Isidore Geoffroy St.-Hilaire, 'Hist. des

Anomalies,' tom, iii, p. 437. A reviewer ('Journal of Anatomy and Physiology,' 1871, p. 366) blames me much for not having discussed the numerous cases, which have been recorded, of various parts arrested in their development. He says that, according to my theory, "every transient condition of an organ, during its development, is not only a means to an end, but once was an end in itself." This does not seem to me necessarily to hold good. Why should not variations occur during an early period of development, having no relation to reversion; yet such variations might be preserved and accumulated, if in any way serviceable, for instance, in shortening and simplifying the course of development? And again, why should not injurious abnormalities, such as atrophied or hypertrophied parts, which have no relation to a former state of existence, occur at an early period, as well as during maturity?

In man, the canine teeth are perfectly efficient instruments for mastication. But their true canine character, as Owen [42] remarks, "is indicated by the conical form of the crown, which terminates in an obtuse point, is convex outward and flat or sub-concave within, at the base of which surface there is a feeble prominence. The conical form is best expressed in the Melanian races, especially the Australian. The canine is more deeply implanted, and by a stronger fang than the incisors." Nevertheless, this tooth no longer serves man as a special weapon for tearing his enemies or prey; it may, therefore, as far as its proper function is concerned, be considered as rudimentary. In every large collection of human skulls some may be found, as Häckel [43] observes, with the canine teeth projecting considerably beyond the others in the same manner as in the anthropomorphous apes, but in a less degree. In these cases, open spaces between the teeth in the one jaw are left for the reception of the canines of the opposite jaw. An inter-space of this kind in a Kaffir skull, figured by Wagner, is surprisingly wide. [44] Considering how few are the ancient skulls which have been examined, compared to recent skulls, it is an interesting fact that in at least three cases the canines project largely; and in the Naulette jaw they are spoken of as enormous. [45]

[42] 'Anatomy of Vertebrates,' vol. iii. 1868, p. 323.
[43] 'Generelle Morphologie,' 1866, B. ii. s. clv.
[44] Carl Vogt's 'Lectures on Man,' Eng. translat., 1864, p. 151.
[45] C. Carter Blake, on a jaw from La Naulette, 'Anthropological Review,' 1867, p. 295. Schaaffhausen, ibid. 1868, p. 426.

Of the anthropomorphous apes the males alone have their canines fully developed; but in the female gorilla, and in a less degree in the female gorilla, these teeth project considerably beyond the others; therefore the fact, of which I have been assured, that women sometimes have considerably projecting canines, is no serious objection to the belief that their occasional great development in man is a case of reversion to an ape-like progenitor. He who rejects with scorn the belief that the shape of his own canines, and their occasional great development in other men, are due to our early forefathers having been provided with these formidable weapons, will probably reveal, by sneering, the line of his descent. For though he no longer intends, nor has the power, to use these teeth as weapons, he will unconsciously retract his "snarling muscles" (thus named by Sir C. Bell) [46], so as to expose them ready for action, like a dog prepared to fight.

[46] The Anatomy of Expression, 1844, pp. 110, 131.

Many muscles are occasionally developed in man, which are proper to the Quadrumana or other mammals. Professor Vlacovich [47] examined forty male subjects, and found a muscle, called by him the ischio-pubic, in nineteen of them; in three others there was a ligament which represented this muscle; and in the remaining eighteen no trace of it. In only two out of thirty female subjects was this muscle developed on both sides, but in three others the rudimentary ligament was present. This muscle, therefore, appears to be much more common in the male than in the female sex; and on the belief in the descent of man from some lower form, the fact is intelligible; for it has been detected in several of the lower animals, and in all of these it serves exclusively to aid the male in the act of reproduction.

[47] Quoted by Prof. Canestrini in the 'Annuario della Soc. dei Naturalisti,' 1867, p. 90.

Mr. J. Wood, in his valuable series of papers [48], has minutely described a vast number of muscular variations in man, which resemble normal structures in the lower animals. The muscles which closely resemble those regularly present in our nearest allies, the Quadrumana, are too numerous to be here even specified. In a single male subject, having a strong bodily frame, and well-formed skull, no less than seven muscular variations were observed, all of which plainly represented muscles proper to various kinds of apes. This man, for instance, had on both sides of his neck a true and powerful "*levator claviculæ*," such as is found in all kinds of apes, and which is said to occur in about one out of sixty human subjects. [49] Again, this man had "a special abductor of the metatarsal bone of the fifth digit, such as Professor Huxley and Mr. Flower have shown to exist uniformly in the higher and lower apes." I will give only two additional cases; the *acromio-basilar* muscle is found in all mammals below man, and seems to be correlated with a quadrupedal gait, [50] and it occurs in about one out of sixty human subjects. In the lower extremities Mr. Bradley [51] found an *abductor ossis metatarsi quinti* in both feet of man; this muscle had not up to that time been recorded in mankind, but is always present in the anthropomorphous apes. The muscles of the hands and arms—parts which are so eminently characteristic of man—are extremely liable to vary, so as to resemble the corresponding muscles in the lower animals. [52] Such resemblances are either perfect or imperfect; yet in the latter case they are manifestly of a transitional nature. Certain variations are more common in man, and others in woman, without our being able to assign any reason. Mr. Wood, after describing numerous variations, makes the following pregnant remark. "Notable departures from the ordinary type of the muscular structures run in grooves or directions, which must be taken to indicate some unknown factor, of much importance to a comprehensive knowledge of general and scientific anatomy." [53]

[48] These papers deserve careful study by any one who desires to learn how frequently our muscles vary, and in varying come to resemble those of the Quadrumana. The following references relate to the few points touched on in my text: 'Proc. Royal Soc.' vol. xiv. 1865, pp. 379-384; vol. xv. 1866, pp. 241, 242; vol. xv. 1867, p. 544; vol. xvi. 1868, p. 524. I may here add that Dr. Murie and Mr. St. George Mivart have shown in their Memoir on the Lemuroidea ('Transactions, Zoological Society,' vol. vii. 1869, p. 96), how extraordinarily variable some of the muscles are in these animals, the lowest members of the Primates. Gradations, also, in the muscles

leading to structures found in animals still lower in the scale, are numerous in the Lemuroidea.

[49] See also Prof. Macalister in 'Proceedings, Royal Irish Academy,' vol. x. 1868, p. 124.

[50] Mr. Champneys in 'Journal of Anatomy and Physiology,' Nov. 1871, p. 178.

[51] Ibid. May 1872, p. 421.

[52] Prof. Macalister (ibid. p. 121) has tabulated his observations, and finds that muscular abnormalities are most frequent in the fore-arms, secondly, in the face, thirdly, in the foot, etc.

[53] The Rev. Dr. Haughton, after giving ('Proc. R. Irish Academy,' June 27, 1864, p. 715) a remarkable case of variation in the human *flexor pollicis longus*, adds, "This remarkable example shows that man may sometimes possess the arrangement of tendons of thumb and fingers characteristic of the macaque; but whether such a case should be regarded as a macaque passing upwards into a man, or a man passing downwards into a macaque, or as a congenital freak of nature, I cannot undertake to say." It is satisfactory to hear so capable an anatomist, and so embittered an opponent of evolutionism, admitting even the possibility of either of his first propositions. Prof. Macalister has also described ('Proceedings Royal Irish Academy,' vol. x. 1864, p. 138) variations in the *flexor pollicis longus*, remarkable from their relations to the same muscle in the Quadrumana.

That this unknown factor is reversion to a former state of existence may be admitted as in the highest degree probable. [54] It is quite incredible that a man should through mere accident abnormally resemble certain apes in no less than seven of his muscles, if there had been no genetic connection between them. On the other hand, if man is descended from some ape-like creature, no valid reason can be assigned why certain muscles should not suddenly reappear after an interval of many thousand generations, in the same manner as with horses, asses, and mules, dark-colored stripes suddenly reappear on the legs, and shoulders, after an interval of hundreds, or more probably of thousands of generations.

[54] Since the first edition of this book appeared, Mr. Wood has published another memoir in the Philosophical Transactions, 1870, p. 83, on the varieties of the muscles of the human neck, shoulder, and chest. He here shows how extremely variable these muscles are, and how often and how closely the variations resemble the normal muscles of the lower animals. He sums up by remarking, "It will be enough for my purpose if I have succeeded in showing the more important forms which, when occurring as varieties in the human subject, tend to exhibit in a sufficiently marked manner what may be considered as proofs and examples of the Darwinian principle of reversion, or law of inheritance, in this department of anatomical science."

These various cases of reversion are so closely related to those of rudimentary organs given in the first chapter, that many of them might have been indifferently introduced either there or here. Thus a human uterus furnished with cornua may be said to represent, in a rudimentary condition, the same organ in its normal state in certain mammals. Some parts which are rudimentary in man, as the *os coccyx* in both sexes, and the mammæ in the male sex, are always present; whilst others, such as the supracondyloid foramen, only occasionally appear, and therefore might have been introduced under the head of reversion. These several reversionary structures, as well as the strictly rudimentary ones, reveal the descent of man from some lower form in an unmistakable manner.

CORRELATED VARIATION.

In man, as in the lower animals, many structures are so intimately related, that when one part varies so does another, without our being able, in most cases, to assign any reason. We cannot say whether the one part governs the other, or whether both are governed by some earlier developed part. Various monstrosities, as I. Geoffroy repeatedly insists, are thus intimately connected. Homologous structures are particularly liable to change together, as we see on the opposite sides of the body, and in the upper and lower extremities. Meckel long ago remarked, that when the muscles of the arm depart from their proper type, they almost always imitate those of the leg; and so, conversely, with the muscles of the legs. The organs of sight and hearing, the teeth and hair, the color of the skin and of the hair, color and constitution, are more or less correlated. [55] Professor Schaaffhausen first drew attention to the relation apparently existing between a muscular frame and the strongly-pronounced supra-orbital ridges, which are so characteristic of the lower races of man.

[55] The authorities for these several statements are given in my 'Variation of Animals under Domestication,' vol. ii. pp. 320-335.

Besides the variations which can be grouped with more or less probability under the foregoing heads, there is a large class of variations which may be provisionally called spontaneous, for to our ignorance they appear to arise without any exciting cause. It can, however, be shown that such variations, whether consisting of slight individual differences, or of strongly-marked and abrupt deviations of structure, depend much more on the constitution of the organism than on the nature of the conditions to which it has been subjected. [56]

[56] This whole subject has been discussed in chap. xxiii. vol. ii. of my 'Variation of Animals and Plants under Domestication.'

RATE OF INCREASE.

Civilized populations have been known under favourable conditions, as in the United States, to double their numbers in twenty-five years; and, according to a calculation, by Euler, this might occur in a little over twelve years. [57] At the former rate, the present population of the United States (thirty millions), would in 657 years cover the whole terraqueous globe so thickly, that four men would have to stand on each square yard of surface. The primary or fundamental check to the continued increase of man is the difficulty of gaining subsistence, and of living in comfort. We may infer that this is the case from what we see, for instance, in the United States, where subsistence is easy, and there is plenty of room. If such means were suddenly doubled in Great Britain, our number would be quickly doubled. With civilized nations this primary check acts chiefly by restraining marriages. The greater death-rate of infants in the poorest classes is also very important; as well as the greater mortality, from various diseases, of the inhabitants of crowded and miserable houses, at all ages. The effects of severe epidemics and wars are soon counterbalanced, and more than counterbalanced, in nations placed under favourable conditions. Emigration also comes in aid as a temporary check, but, with the extremely poor classes, not to any great extent.

[57] See the ever memorable 'Essay on the Principle of Population,' by the Rev. T. Malthus, vol. i. 1826. pp. 6, 517.

There is reason to suspect, as Malthus has remarked, that the reproductive power is actually less in barbarous, than in civilized races. We know nothing positively on this head, for with savages no census has been taken; but from the concurrent testimony of missionaries, and of others who have long resided with such people, it appears that their families are usually small, and large ones rare. This may be partly accounted for, as it is believed, by the women suckling their infants during a long time; but it is highly probable that savages, who often suffer much hardship, and who do not obtain so much nutritious food as civilized men, would be actually less prolific. I have shown in a former work [58], that all our domesticated quadrupeds and birds, and all our cultivated plants, are more fertile than the corresponding species in a state of nature. It is no valid objection to this conclusion that animals suddenly supplied with an excess of food, or when grown very fat; and that most plants on sudden removal from very poor to very rich soil, are rendered more or less sterile. We might, therefore, expect that civilized men, who in one sense are highly domesticated, would be more prolific than wild men. It is also probable that the increased fertility of civilized nations would become, as with our domestic animals, an inherited character: it is at least known that with mankind a tendency to produce twins runs in families. [59]

[58] 'Variation of Animals and Plants under Domestication,' vol. ii. pp. 111-113, 163.
[59] Mr. Sedgwick, 'British and Foreign Medico-Chirurgical Review,' July 1863, p. 170.

Notwithstanding that savages appear to be less prolific than civilized people, they would no doubt rapidly increase if their numbers were not by some means rigidly kept down. The Santali, or hill-tribes of India, have recently afforded a good illustration of this fact; for, as shown by Mr. Hunter [60], they have increased at an extraordinary rate since vaccination has been introduced, other pestilences mitigated, and war sternly repressed. This increase, however, would not have been possible had not these rude people spread into the adjoining districts, and worked for hire. Savages almost always marry; yet there is some prudential restraint, for they do not commonly marry at the earliest possible age. The young men are often required to show that they can support a wife; and they generally have first to earn the price with which to purchase her from her parents. With savages the difficulty of obtaining subsistence occasionally limits their number in a much more direct manner than with civilized people, for all tribes periodically suffer from severe famines. At such times savages are forced to devour much bad food, and their health can hardly fail to be injured. Many accounts have been published of their protruding stomachs and emaciated limbs after and during famines. They are then, also, compelled to wander much, and, as I was assured in Australia, their infants perish in large numbers. As famines are periodical, depending chiefly on extreme seasons, all tribes must fluctuate in number. They cannot steadily and regularly increase, as there is no artificial increase in the supply of food. Savages, when hard pressed, encroach on each other's territories, and war is the result; but they are indeed almost always at war with their neighbours. They are liable to many accidents on land and water in their search for food; and in some countries they suffer much from the larger beasts of prey. Even in India, districts have been depopulated by the ravages of tigers.

[60] 'The Annals of Rural Bengal,' by W.W. Hunter, 1868, p. 259.

Malthus has discussed these several checks, but he does not lay stress enough on what is probably the most important of all, namely infanticide, especially of female infants, and the habit of procuring abortion. These practices now prevail in many quarters of the world; and infanticide seems formerly to have prevailed, as Mr. M'Lennan [61] has shown, on a still more extensive scale. These practices appear to have originated in savages recognising the difficulty, or rather the impossibility of supporting all the infants that are born. Licentiousness may also be added to the foregoing checks; but this does not follow from failing means of subsistence; though there is reason to believe that in some cases (as in Japan) it has been intentionally encouraged as a means of keeping down the population.

[61] 'Primitive Marriage,' 1865.

If we look back to an extremely remote epoch, before man had arrived at the dignity of manhood, he would have been guided more by instinct and less by reason than are the lowest savages at the present time. Our early semi-human progenitors would not have practised infanticide or polyandry; for the instincts of the lower animals are never so perverted [62] as to lead them regularly to destroy their own offspring, or to be quite devoid of jealousy. There would have been no prudential restraint from marriage, and the sexes would have freely united at an early age. Hence the progenitors of man would have tended to increase rapidly; but checks of some kind, either periodical or constant, must have kept down their numbers, even more severely than with existing savages. What the precise nature of these checks were, we cannot say, any more than with most other animals. We know that horses and cattle, which are not extremely prolific animals, when first turned loose in South America, increased at an enormous rate. The elephant, the slowest breeder of all known animals, would in a few thousand years stock the whole world. The increase of every species of monkey must be checked by some means; but not, as Brehm remarks, by the attacks of beasts of prey. No one will assume that the actual power of reproduction in the wild horses and cattle of America, was at first in any sensible degree increased; or that, as each district became fully stocked, this same power was diminished. No doubt, in this case, and in all others, many checks concur, and different checks under different circumstances; periodical dearths, depending on unfavorable seasons, being probably the most important of all. So it will have been with the early progenitors of man.

[62] A writer in the 'Spectator' (March 12, 1871, p. 320) comments as follows on this passage:—"Mr. Darwin finds himself compelled to reintroduce a new doctrine of the fall of man. He shows that the instincts of the higher animals are far nobler than the habits of savage races of men, and he finds himself, therefore, compelled to re-introduce,—in a form of the substantial orthodoxy of which he appears to be quite unconscious,—and to introduce as a scientific hypothesis the doctrine that man's gain

of *knowledge* was the cause of a temporary but long-enduring moral deterioration as indicated by the many foul customs, especially as to marriage, of savage tribes. What does the Jewish tradition of the moral degeneration of man through his snatching at a knowledge forbidden him by his highest instinct assert beyond this?"

NATURAL SELECTION.

We have now seen that man is variable in body and mind; and that the variations are induced, either directly or indirectly, by the same general causes, and obey the same general laws, as with the lower animals. Man has spread widely over the face of the earth, and must have been exposed, during his incessant migrations [63], to the most diversified conditions. The inhabitants of Tierra del Fuego, the Cape of Good Hope, and Tasmania in the one hemisphere, and of the arctic regions in the other, must have passed through many climates, and changed their habits many times, before they reached their present homes. [64] The early progenitors of man must also have tended, like all other animals, to have increased beyond their means of subsistence; they must, therefore, occasionally have been exposed to a struggle for existence, and consequently to the rigid law of natural selection. Beneficial variations of all kinds will thus, either occasionally or habitually, have been preserved and injurious ones eliminated. I do not refer to strongly-marked deviations of structure, which occur only at long intervals of time, but to mere individual differences. We know, for instance, that the muscles of our hands and feet, which determine our powers of movement, are liable, like those of the lower animals, [65] to incessant variability. If then the progenitors of man inhabiting any district, especially one undergoing some change in its conditions, were divided into two equal bodies, the one half which included all the individuals best adapted by their powers of movement for gaining subsistence, or for defending themselves, would on an average survive in greater numbers, and procreate more offspring than the other and less well endowed half.

[63] See some good remarks to this effect by W. Stanley Jevons, "A Deduction from Darwin's Theory," 'Nature,' 1869, p. 231.

[64] Latham, 'Man and his Migrations,' 1851, p. 135.

[65] Messrs. Murie and Mivart in their 'Anatomy of the Lemuroidea' ('Transact. Zoolog. Soc.' vol. vii. 1869, pp. 96-98) say, "some muscles are so irregular in their distribution that they cannot be well classed in any of the above groups." These muscles differ even on the opposite sides of the same individual.

Man in the rudest state in which he now exists is the most dominant animal that has ever appeared on this earth. He has spread more widely than any other highly organized form: and all others have yielded before him. He manifestly owes this immense superiority to his intellectual faculties, to his social habits, which lead him to aid and defend his fellows, and to his corporeal structure. The supreme importance of these characters has been proved by the final arbitrament of the battle for life. Through his powers of intellect, articulate language has been evolved; and on this his wonderful advancement has mainly depended. As Mr. Chauncey Wright remarks [66]: "a psychological analysis of the faculty of language shows, that even the smallest proficiency in it might require more brain power than the greatest proficiency in any other direction." He has invented and is able to use various weapons, tools, traps, etc., with which he defends himself, kills or catches prey, and otherwise obtains food. He has made rafts or canoes for fishing or crossing over to neighbouring fertile islands. He has discovered the art of making fire, by which hard and stringy roots can be rendered digestible, and poisonous roots or herbs innocuous. This discovery of fire, probably the greatest ever made by man, excepting language, dates from before the dawn of history. These several inventions, by which man in the rudest state has become so pre-eminent, are the direct results of the development of his powers of observation, memory, curiosity, imagination, and reason. I cannot, therefore, understand how it is that Mr. Wallace [67] maintains, that "natural selection could only have endowed the savage with a brain a little superior to that of an ape."

[66] Limits of Natural Selection, 'North American Review,' Oct. 1870, p. 295.
[67] 'Quarterly Review,' April 1869, p. 392. This subject is more fully discussed in Mr. Wallace's 'Contributions to the Theory of Natural Selection,' 1870, in which all the essays referred to in this work are re-published. The 'Essay on Man,' has been ably criticised by Prof. Claparede, one of the most distinguished zoologists in Europe, in an article published in the 'Bibliothèque Universelle,' June 1870. The remark quoted in my text will surprise every one who has read Mr. Wallace's celebrated paper on 'The Origin of Human Races Deduced from the Theory of Natural Selection,' originally published in the 'Anthropological Review,' May 1864, p. clviii. I cannot here resist quoting a most just remark by Sir J. Lubbock ('Prehistoric Times,' 1865, p. 479) in reference to this paper, namely, that Mr. Wallace, "with characteristic unselfishness, ascribes it (i.e. the idea of natural selection) unreservedly to Mr. Darwin, although, as is well known, he struck out the idea independently, and published it, though not with the same elaboration, at the same time."

Although the intellectual powers and social habits of man are of paramount importance to him, we must not underrate the importance of his bodily structure, to which subject the remainder of this chapter will be devoted; the development of the intellectual and social or moral faculties being discussed in a later chapter.

Even to hammer with precision is no easy matter, as every one who has tried to learn carpentry will admit. To throw a stone with as true an aim as a Fuegian in defending himself, or in killing birds, requires the most consummate perfection in the correlated action of the muscles of the hand, arm, and shoulder, and, further, a fine sense of touch. In throwing a stone or spear, and in many other actions, a man must stand firmly on his feet; and this again demands the perfect co-adaptation of numerous muscles. To chip a flint into the rudest tool, or to form a barbed spear or hook from a bone, demands the use of a perfect hand; for, as a most capable judge, Mr. Schoolcraft [68], remarks, the shaping fragments of stone into knives, lances, or arrow-heads, shows "extraordinary ability and long practice." This is to a great extent proved by the fact that primeval men practised a division of labor; each man did not manufacture his own flint tools or rude pottery, but certain individuals appear to have devoted themselves to such work, no doubt receiving in exchange the produce of the chase. Archaeologists are convinced that an enormous interval of time elapsed before our ancestors thought of grinding chipped flints into smooth tools. One can hardly doubt, that a man-like animal who possessed a hand and arm sufficiently perfect to throw a stone with precision, or to form a flint into a rude tool, could, with sufficient practice, as far as mechanical skill alone is concerned, make almost anything which a civilized man can make. The structure of the hand in this respect may be compared with that of the vocal organs, which in the apes are used for uttering various signal-cries, or, as in one genus, musical cadences; but in man the closely similar vocal organs have become adapted through the inherited effects of use for the utterance of articulate language.

[68] Quoted by Mr. Lawson Tait in his 'Law of Natural Selection,' 'Dublin Quarterly Journal of Medical Science,' Feb. 1869. Dr. Keller is likewise quoted to the same effect.

Turning now to the nearest allies of men, and therefore to the best representatives of our early progenitors, we find that the hands of the Quadrumana are constructed on the same general pattern as our own, but are far less perfectly adapted for diversified uses. Their hands do not serve for locomotion so well as the feet of a dog; as may be seen in such monkeys as the chimpanzee and orang, which walk on the outer margins of the palms, or on the knuckles. [69] Their hands, however, are admirably adapted for climbing trees. Monkeys seize thin branches or ropes, with the thumb on one side and the fingers and palm on the other, in the same manner as we do. They can thus also lift rather large objects, such as the neck of a bottle, to their mouths. Baboons turn over stones, and scratch up roots with their hands. They seize nuts, insects, or other small objects with the thumb in opposition to the fingers, and no doubt they thus extract eggs and young from the nests of birds. American monkeys beat the wild oranges on the branches until the rind is cracked, and then tear it off with the fingers of the two hands. In a wild state they break open hard fruits with stones. Other monkeys open mussel-shells with the two thumbs. With their fingers they pull out thorns and burs, and hunt for each other's parasites. They roll down stones, or throw them at their enemies: nevertheless, they are clumsy in these various actions, and, as I have myself seen, are quite unable to throw a stone with precision.

[69] Owen, 'Anatomy of Vertebrates,' vol. iii. p. 71.

It seems to me far from true that because "objects are grasped clumsily" by monkeys, "a much less specialized organ of prehension" would have served them [70] equally well with their present hands. On the contrary, I see no reason to doubt that more perfectly constructed hands would have been an advantage to them, provided that they were not thus rendered less fitted for climbing trees. We may suspect that a hand as perfect as that of man would have been disadvantageous for climbing; for the most arboreal monkeys in the world, namely, Ateles in America, Colobus in Africa, and Hylobates in Asia, are either thumbless, or their toes partially cohere, so that their limbs are converted into mere grasping hooks. [71]

[70] 'Quarterly Review,' April 1869, p. 392.
[71] In *Hylobates syndactylus*, as the name expresses, two of the toes regularly cohere; and this, as Mr. Blyth informs me, is occasionally the case with the toes of H. *agilis, lar*, and *leuciscus*. Colobus is strictly arboreal and extraordinarily active (Brehm, 'Thierleben,' B. i. s. 50), but whether a better climber than the species of the allied genera, I do not know. It deserves notice that the feet of the sloths, the most arboreal animals in the world, are wonderfully hook-like.

As soon as some ancient member in the great series of the Primates came to be less arboreal, owing to a change in its manner of procuring subsistence, or to some change in the surrounding conditions, its habitual manner of progression would have been modified: and thus it would have been rendered more strictly quadrupedal or bipedal. Baboons frequent hilly and rocky districts, and only from necessity climb high trees [72]; and they have acquired almost the gait of a dog. Man alone has become a biped; and we can, I think, partly see how he has come to assume his erect attitude, which forms one of his most conspicuous characters. Man could not have attained his present dominant position in the world without the use of his hands, which are so admirably adapted to act in obedience to his will. Sir C. Bell [73] insists that "the hand supplies all instruments, and by its correspondence with the intellect gives him universal dominion." But the hands and arms could hardly have become perfect enough to have manufactured weapons, or to have hurled stones and spears with a true aim, as long as they were habitually used for locomotion and for supporting the whole weight of the body, or, as before remarked, so long as they were especially fitted for climbing trees. Such rough treatment would also have blunted the sense of touch, on which their delicate use largely depends. From these causes alone it would have been an advantage to man to become a biped; but for many actions it is indispensable that the arms and whole upper part of the body should be free; and he must for this end stand firmly on his feet. To gain this great advantage, the feet have been rendered flat; and the great toe has been peculiarly modified, though this has entailed the almost complete loss of its power of prehension. It accords with the principle of the division of physiological labor, prevailing throughout the animal kingdom, that as the hands became perfected for prehension, the feet should have become perfected for support and locomotion. With some savages, however, the foot has not altogether lost its prehensile power, as shown by their manner of climbing trees, and of using them in other ways. [74]

[72] Brehm, 'Thierleben,' B. i. s. 80.
[73] 'The Hand,' etc., 'Bridgewater Treatise,' 1833, p. 38.
[74] Häckel has an excellent discussion on the steps by which man became a biped: 'Natürliche Schöpfungsgeschicte,' 1868, s. 507. Dr. Büchner ('Conferences sur la Théorie Darwinienne,' 1869, p. 135) has given good cases of the use of the foot as a prehensile organ by man; and has also written on the manner of progression of the higher apes, to which I allude in the following paragraph: see also Owen ('Anatomy of Vertebrates,' vol. iii. p. 71) on this latter subject.

If it be an advantage to man to stand firmly on his feet and to have his hands and arms free, of which, from his pre-eminent success in the battle of life there can be no doubt, then I can see no reason why it should not have been advantageous to the progenitors of man to have become more and more erect or bipedal. They would thus have been better able to defend themselves with stones or clubs, to attack their prey, or otherwise to obtain food. The best built individuals would in the long run have succeeded best, and have survived in larger numbers. If the gorilla and a few allied forms had become extinct, it might have been argued, with great force and apparent truth, that an animal could not have been gradually converted from a quadruped into a biped, as all the individuals in an intermediate condition would have been miserably ill-fitted for progression. But we know (and this is well worthy of reflection) that the anthropomorphous apes are now actually in an intermediate condition; and no one doubts that they are on the whole well adapted for their conditions of life. Thus the gorilla runs with a sidelong shambling gait, but more commonly progresses by resting on its bent hands. The long-armed apes occasionally use their arms like crutches, swinging their bodies forward between them, and some kinds of Hylobates, without having been taught, can walk or run upright with tolerable quickness; yet they move awkwardly, and much less securely than man. We see, in short, in existing monkeys a manner of progression intermediate between that of a quadruped and a biped; but, as an unprejudiced judge [75] insists, the anthropomorphous apes approach in structure more nearly to the bipedal than to the quadrupedal type.

[75] Prof. Broca, La Constitution des Vertèbres caudales; 'La Revue d'Anthropologie,' 1872, p. 26, (separate copy).

As the progenitors of man became more and more erect, with their hands and arms more and more modified for prehension and other purposes, with their feet and legs at the same time transformed for firm support and progression, endless other changes of structure would have become necessary. The pelvis would have to be broadened, the spine peculiarly curved, and the head fixed in an altered position, all which changes have been attained by man. Prof. Schaaffhausen [76] maintains that "the powerful mastoid processes of the human skull are the result of his erect position;" and these processes are absent in the orang, chimpanzee, etc., and are smaller in the gorilla than in man. Various other structures, which appear connected with man's erect position, might here have been added. It is very difficult to decide how far these correlated modifications are the result of natural selection, and how far of the inherited effects of the increased use of certain parts, or of the action of one part on another. No doubt these means of change often co-operate: thus when certain muscles, and the crests of bone to which they are attached, become enlarged by habitual use, this shows that certain actions are habitually performed and must be serviceable. Hence the individuals which performed them best, would tend to survive in greater numbers.

[76] 'On the Primitive Form of the Skull,' translated in 'Anthropological Review,' Oct. 1868, p. 428. Owen ('Anatomy of Vertebrates,' vol. ii. 1866, p. 551) on the mastoid processes in the higher apes.

The free use of the arms and hands, partly the cause and partly the result of man's erect position, appears to have led in an indirect manner to other modifications of structure. The early male forefathers of man were, as previously stated, probably furnished with great canine teeth; but as they gradually acquired the habit of using stones, clubs, or other weapons, for fighting with their enemies or rivals, they would use their jaws and teeth less and less. In this case, the jaws, together with the teeth, would become reduced in size, as we may feel almost sure from innumerable analogous cases. In a future chapter we shall meet with a closely parallel case, in the reduction or complete disappearance of the canine teeth in male ruminants, apparently in relation with the development of their horns; and in horses, in relation to their habit of fighting with their incisor teeth and hoofs.

In the adult male anthropomorphous apes, as Rütimeyer [77], and others, have insisted, it is the effect on the skull of the great development of the jaw-muscles that causes it to differ so greatly in many respects from that of man, and has given to these animals "a truly frightful physiognomy." Therefore, as the jaws and teeth in man's progenitors gradually become reduced in size, the adult skull would have come to resemble more and more that of existing man. As we shall hereafter see, a great reduction of the canine teeth in the males would almost certainly affect the teeth of the females through inheritance.

[77] 'Die Grenzen der Thierwelt, eine Betrachtung zu Darwin's Lehre,' 1868, s. 51.

As the various mental faculties gradually developed themselves the brain would almost certainly become larger. No one, I presume, doubts that the large proportion which the size of man's brain bears to his body, compared to the same proportion in the gorilla or orang, is closely connected with his higher mental powers. We meet with closely analogous facts with insects, for in ants the cerebral ganglia are of extraordinary dimensions, and in all the Hymenoptera these ganglia are many times larger than in the less intelligent orders, such as beetles. [78] On the other hand, no one supposes that the intellect of any two animals or of any two men can be accurately gauged by the cubic contents of their skulls. It is certain that there may be extraordinary mental activity with an extremely small absolute mass of nervous matter: thus the wonderfully diversified instincts, mental powers, and affections of ants are notorious, yet their cerebral ganglia are not so large as the quarter of a small pin's head. Under this point of view, the brain of an ant is one of the most marvellous atoms of matter in the world, perhaps more so than the brain of a man.

[78] Dujardin, 'Annales des Sciences Nat.' 3rd series, Zoolog., tom. xiv. 1850, p. 203. See also Mr. Lowne, 'Anatomy and Phys. of the *Musca vomitoria*,' 1870, p. 14. My son, Mr. F. Darwin, dissected for me the cerebral ganglia of the *Formica rufa*.

The belief that there exists in man some close relation between the size of the brain and the development of the intellectual faculties is supported by the comparison of the skulls of savage and civilized races, of ancient and modern people, and by the analogy of the whole vertebrate series. Dr. J. Barnard Davis has proved [79], by many careful measurements, that the mean internal capacity of the skull in Europeans is 92.3 cubic inches; in Americans 87.5; in Asiatics 87.1; and in Australians only 81.9 cubic inches. Professor Broca [80] found that the nineteenth century skulls from graves in Paris were larger than those from vaults of the twelfth century, in the proportion of 1484 to 1426; and that the increased size, as ascertained by measurements, was exclusively in the frontal part of the skull—the seat of the intellectual faculties. Prichard is persuaded that the present inhabitants of Britain have "much more capacious brain-cases" than the ancient inhabitants. Nevertheless, it must be admitted that some skulls of very high antiquity, such as the famous one of Neanderthal, are well developed and capacious. [81] With respect to the lower animals, M.E. Lartet [82], by comparing the crania of tertiary and recent mammals belonging to the same groups, has come to the remarkable conclusion that the brain is generally larger and the convolutions are more complex in the more recent forms. On the other hand, I have shown [83] that the brains of domestic rabbits are considerably reduced in bulk, in comparison with those of the wild rabbit or hare; and this may be attributed to their having been closely confined during many generations, so that they have exerted their intellect, instincts, senses and voluntary movements but little.

[79] 'Philosophical Transactions,' 1869, p. 513.

[80] 'Les Sélections,' M. P. Broca, 'Revue d'Anthropologies,' 1873; see also, as quoted in C. Vogt's 'Lectures on Man,' Engl. translat., 1864, pp. 88, 90. Prichard, 'Physical History of Mankind,' vol. i. 1838, p. 305.

[81] In the interesting article just referred to, Prof. Broca has well remarked, that in civilized nations, the average capacity of the skull must be lowered by the preservation of a considerable number of individuals, weak in mind and body, who would have been promptly eliminated in the savage state. On the other hand, with savages, the average includes only the more capable individuals, who have been able to survive under extremely hard conditions of life. Broca thus explains the otherwise inexplicable fact, that the mean capacity of the skull of the ancient Troglodytes of Lozère is greater than that of modern Frenchmen.

[82] 'Comptes-rendus des Sciences,' etc., June 1, 1868.

[83] The 'Variation of Animals and Plants under Domestication,' vol. i. pp. 124-129.

The gradually increasing weight of the brain and skull in man must have influenced the development of the supporting spinal column, more especially whilst he was becoming erect. As this change of position was being brought about, the internal pressure of the brain will also have influenced the form of the skull; for many facts show how easily the skull is thus affected. Ethnologists believe that it is modified by the kind of cradle in which infants sleep. Habitual spasms of the muscles, and a cicatrix from a severe burn, have permanently modified the facial bones. In young persons whose heads have become fixed either sideways or backwards, owing to disease, one of the two eyes has changed its position, and the shape of the skull has been altered apparently by the pressure of the brain in a new direction. [84] I have shown that with long-eared rabbits even so trifling a cause as the lopping forward of one ear drags forward almost every bone of the skull on that side; so that the bones on the opposite side no longer strictly correspond. Lastly, if any animal were to increase or diminish much in general size, without any change in its mental powers, or if the mental powers were to be much increased or diminished, without any great change in the size of the body, the shape of the skull would almost certainly be altered. I infer this from my observations on domestic rabbits, some kinds of which have become very much larger than the wild animal, whilst others have retained nearly the same size, but in both cases the brain has been much reduced relatively to the size of the body. Now I was at first much surprised on finding that in all these rabbits the skull had become elongated or dolichocephalic; for instance, of two skulls of nearly equal breadth, the one from a wild rabbit and the other from a large domestic kind, the former was 3.15 and the latter 4.3 inches in length. [85] One of the most marked distinctions in different races of men is that the skull in some is elongated, and in others rounded; and here the explanation suggested by the case of the rabbits may hold good; for Welcker finds that short "men incline more to brachycephaly, and tall men to dolichocephaly" [86]; and tall men may be compared with the larger and longer-bodied rabbits, all of which have elongated skulls or are dolichocephalic.

[84] Schaaffhausen gives from Blumenbach and Busch, the cases of the spasms and cicatrix, in 'Anthropological Review,' Oct. 1868, p. 420. Dr. Jarrold ('Anthropologia,' 1808, pp. 115, 116) adduces from Camper and from his own observations, cases of the modification of the skull from the head being fixed in an unnatural position. He

believes that in certain trades, such as that of a shoemaker, where the head is habitually held forward, the forehead becomes more rounded and prominent.

[85] 'Variation of Animals and Plants under Domestication,' vol. i. p. 117, on the elongation of the skull; p. 119, on the effect of the lopping of one ear.

[86] Quoted by Schaaffhausen, in 'Anthropological Review,' Oct. 1868, p. 419.

From these several facts we can understand, to a certain extent, the means by which the great size and more or less rounded form of the skull have been acquired by man; and these are characters eminently distinctive of him in comparison with the lower animals.

Another most conspicuous difference between man and the lower animals is the nakedness of his skin. Whales and porpoises (Cetacea), dugongs (Sirenia) and the hippopotamus are naked; and this may be advantageous to them for gliding through the water; nor would it be injurious to them from the loss of warmth, as the species, which inhabit the colder regions, are protected by a thick layer of blubber, serving the same purpose as the fur of seals and otters. Elephants and rhinoceroses are almost hairless; and as certain extinct species, which formerly lived under an Arctic climate, were covered with long wool or hair, it would almost appear as if the existing species of both genera had lost their hairy covering from exposure to heat. This appears the more probable, as the elephants in India which live on elevated and cool districts are more hairy [87] than those on the lowlands. May we then infer that man became divested of hair from having aboriginally inhabited some tropical land? That the hair is chiefly retained in the male sex on the chest and face, and in both sexes at the junction of all four limbs with the trunk, favours this inference—on the assumption that the hair was lost before man became erect; for the parts which now retain most hair would then have been most protected from the heat of the sun. The crown of the head, however, offers a curious exception, for at all times it must have been one of the most exposed parts, yet it is thickly clothed with hair. The fact, however, that the other members of the order of Primates, to which man belongs, although inhabiting various hot regions, are well clothed with hair, generally thickest on the upper surface [88], is opposed to the supposition that man became naked through the action of the sun. Mr. Belt believes [89] that within the tropics it is an advantage to man to be destitute of hair, as he is thus enabled to free himself of the multitude of ticks (acari) and other parasites, with which he is often infested, and which sometimes cause ulceration. But whether this evil is of sufficient magnitude to have led to the denudation of his body through natural selection, may be doubted, since none of the many quadrupeds inhabiting the tropics have, as far as I know, acquired any specialized means of relief. The view which seems to me the most probable is that man, or rather primarily woman, became divested of hair for ornamental purposes, as we shall see under Sexual Selection; and, according to this belief, it is not surprising that man should differ so greatly in hairiness from all other Primates, for characters, gained through sexual selection, often differ to an extraordinary degree in closely related forms.

[87] Owen, 'Anatomy of Vertebrates,' vol. iii. p. 619.

[88] Isidore Geoffroy St.-Hilaire remarks ('Histoire Nat. Générale,' tom. ii. 1859, pp. 215-217) on the head of man being covered with long hair; also on the upper surfaces of monkeys and of other mammals being more thickly clothed than the lower surfaces. This has likewise been observed by various authors. Prof. P. Gervais ('Histoire Nat. des Mammifères,' tom. i. 1854, p. 28), however, states that in the Gorilla the hair is thinner on the back, where it is partly rubbed off, than on the lower surface.

[89] The 'Naturalist in Nicaragua,' 1874, p. 209. As some confirmation of Mr. Belt's view, I may quote the following passage from Sir W. Denison ('Varieties of Vice-Regal Life,' vol. i. 1870, p. 440): "It is said to be a practice with the Australians, when the vermin get troublesome, to singe themselves."

According to a popular impression, the absence of a tail is eminently distinctive of man; but as those apes which come nearest to him are destitute of this organ, its disappearance does not relate exclusively to man. The tail often differs remarkably in length within the same genus: thus in some species of Macacus it is longer than the whole body, and is formed of twenty-four vertebræ; in others it consists of a scarcely visible stump, containing only three or four vertebræ. In some kinds of baboons there are twenty-five, whilst in the mandrill there are ten very small stunted caudal vertebræ, or, according to Cuvier [90], sometimes only five. The tail, whether it be long or short, almost always tapers towards the end; and this, I presume, results from the atrophy of the terminal muscles, together with their arteries and nerves, through disuse, leading to the atrophy of the terminal bones. But no explanation can at present be given of the great diversity which often occurs in its length. Here, however, we are more specially concerned with the complete external disappearance of the tail. Professor Broca has recently shown [91] that the tail in all quadrupeds consists of two portions, generally separated abruptly from each other; the basal portion consists of vertebræ, more or less perfectly channelled and furnished with apophyses like ordinary vertebræ; whereas those of the terminal portion are not channelled, are almost smooth, and scarcely resemble true vertebræ. A tail, though not externally visible, is really present in man and the anthropomorphous apes, and is constructed on exactly the same pattern in both. In the terminal portion the vertabræ, constituting the *os coccyx*, are quite rudimentary, being much reduced in size and number. In the basal portion, the vertebræ are likewise few, are united firmly together, and are arrested in development; but they have been rendered much broader and flatter than the corresponding vertebræ in the tails of other animals: they constitute what Broca calls the accessory sacral vertebræ. These are of functional importance by supporting certain internal parts and in other ways; and their modification is directly connected with the erect or semi-erect attitude of man and the anthropomorphous apes. This conclusion is the more trustworthy, as Broca formerly held a different view, which he has now abandoned. The modification, therefore, of the basal caudal vertebræ in man and the higher apes may have been effected, directly or indirectly, through natural selection.

[90] Mr. St. George Mivart, 'Proc. Zoolog. Soc.' 1865, pp. 562, 583. Dr. J.E. Gray, 'Cat. Brit. Mus.: 'Skeletons.' Owen, 'Anatomy of Vertebrates,' vol. ii. p. 517. Isidore Geoffroy, 'Hist. Nat. Gén.' tom. ii. p. 244.
[91] 'Revue d'Anthropologie,' 1872; 'La Constitution des Vertèbres caudales.'

But what are we to say about the rudimentary and variable vertebræ of the terminal portion of the tail, forming the *os coccyx*? A notion which has often been, and will no doubt again be ridiculed, namely, that friction has had something to do with the disappearance of the external portion of the tail, is not so ridiculous as it at first appears. Dr. Anderson [92] states that the extremely short tail of *Macacus brunneus* is formed of eleven vertebræ, including the imbedded basal ones. The extremity is tendinous and contains no vertebræ; this is succeeded by five rudimentary ones, so minute that together they are only one line and a half in length, and these are permanently bent to one side in the shape of a hook. The free part of the tail, only a little above an inch in length, includes only four more small vertebræ. This short tail is carried erect; but about a quarter of its total length is doubled on to itself to the left; and this terminal part, which includes the hook-like portion, serves "to fill up the interspace between the upper divergent portion of the callosities;" so that the animal sits on it, and thus renders it rough and callous. Dr. Anderson thus sums up his observations: "These facts seem to me to have only one explanation; this tail, from its short size, is in the monkey's way when it sits down, and frequently becomes placed under the animal while it is in this attitude; and from the circumstance that it does not extend beyond the extremity of the ischial tuberosities, it seems as if the tail originally had been bent round by the will of the animal, into the interspace between the callosities, to escape being pressed between them and the ground, and that in time the curvature became permanent, fitting in of itself when the organ happens to be sat upon." Under these circumstances it is not surprising that the surface of the tail should have been roughened and rendered callous, and Dr. Murie [93], who carefully observed this species in the Zoological Gardens, as well as three other closely allied forms with slightly longer tails, says that when the animal sits down, the tail "is necessarily thrust to one side of the buttocks; and whether long or short its root is consequently liable to be rubbed or chafed." As we now have evidence that mutilations occasionally produce an inherited effect [94], it is not very improbable that in short-tailed monkeys, the projecting part of the tail, being functionally useless, should after many generations have become rudimentary and distorted, from being continually rubbed and chafed. We see the projecting part in this condition in the *Macacus brunneus*, and absolutely aborted in the *M. ecaudatus* and in several of the higher apes. Finally, then, as far as we can judge, the tail has disappeared in man and the anthropomorphous apes, owing to the terminal portion having been injured by friction during a long lapse of time; the basal and embedded portion having been reduced and modified, so as to become suitable to the erect or semi-erect position.

[92] 'Proceedings Zoological Society,' 1872, p. 210.
[93] 'Proceedings Zoological Society,' 1872, p. 786.
[94] I allude to Dr. Brown-Séquard's observations on the transmitted effect of an operation causing epilepsy in guinea-pigs, and likewise more recently on the analogous effects of cutting the sympathetic nerve in the neck. I shall hereafter have occasion to refer to Mr. Salvin's interesting case of the apparently inherited effects of mot-mots biting off the barbs of their own tail-feathers. See also on the general subject 'Variation of Animals and Plants under Domestication,' vol. ii. pp. 22-24.

I have now endeavoured to show that some of the most distinctive characters of man have in all probability been acquired, either directly, or more commonly indirectly, through natural selection. We should bear in mind that modifications in structure or constitution which do not serve to adapt an organism to its habits of life, to the food which it consumes, or passively to the surrounding conditions, cannot have been thus acquired. We must not, however, be too confident in deciding what modifications are of service to each being: we should remember how little we know about the use of many parts, or what changes in the blood or tissues may serve to fit an organism for a new climate or new kinds of food. Nor must we forget the principle of correlation, by which, as Isidore Geoffroy has shown in the case of man, many strange deviations of structure are tied together. Independently of correlation, a change in one part often leads, through the increased or decreased use of other parts, to other changes of a quite unexpected nature. It is also well to reflect on such facts, as the wonderful growth of galls on plants caused by the poison of an insect, and on the remarkable changes of color in the plumage of parrots when fed on certain fishes, or inoculated with the poison of toads [95]; for we can thus see that the fluids of the system, if altered for some special purpose, might induce other changes. We should especially bear in mind that modifications acquired and continually used during past ages for some useful purpose, would probably become firmly fixed, and might be long inherited.

[95] The 'Variation of Animals and Plants under Domestication,' vol. ii. pp. 280, 282.

Thus a large yet undefined extension may safely be given to the direct and indirect results of natural selection; but I now admit, after reading the essay by Nägeli on plants, and the remarks by various authors with respect to animals, more especially those recently made by Professor Broca, that in the earlier editions of my 'Origin of Species' I perhaps attributed too much to the action of natural selection or the survival of the fittest. I have altered the fifth edition of the 'Origin' so as to confine my remarks to adaptive changes of structure; but I am convinced, from the light gained during even the last few years, that very many structures which now appear to us useless, will hereafter be proved to be useful, and will therefore come within the range of natural selection. Nevertheless, I did not formerly consider sufficiently the existence of structures, which, as far as we can at present judge, are neither beneficial nor injurious; and this I believe to be one of the greatest oversights as yet detected in my work. I may be permitted to say, as some excuse, that I had two distinct objects in view; firstly, to show that species had not been separately created, and secondly, that natural selection had been the chief agent of change, though largely aided by the inherited effects of habit, and slightly by the direct action of the surrounding conditions. I was not, however, able to annul the influence of my former belief, then almost universal, that each species had been purposely created; and this led to my tacit assumption that every detail of structure, excepting rudiments, was of some special, though unrecognized, service. Any one with this assumption in his mind would naturally extend too far the action of natural selection, either during past or present times. Some of those who admit the principle of evolution, but reject natural selection, seem to forget, when criticizing my book, that I had the above two objects in view; hence if I have erred in giving to natural selection great power, which I am very far from admitting, or in having exaggerated its power, which is in itself probable, I have at least, as I hope, done good service in aiding to overthrow the dogma of separate creations.

It is, as I can now see, probable that all organic beings, including man, possess peculiarities of structure, which neither are now, nor were formerly of any service to them, and which, therefore, are of no physiological importance. We know not what produces the numberless slight differences between the individuals of each species, for reversion only carries the problem a few steps backwards, but each peculiarity must have had its efficient cause. If these causes, whatever they may be, were to act more uniformly and energetically during a lengthened period (and against this no reason can be assigned), the result would probably be not a mere slight individual difference, but a well-marked and constant modification, though one of no physiological importance. Changed structures, which are in no way beneficial, cannot be kept uniform through natural selection, though the injurious will be thus eliminated. Uniformity of character would, however, naturally follow from the assumed uniformity of the exciting causes, and likewise from the free intercrossing of many individuals. During successive periods, the same organism might in this manner acquire successive modifications, which would be transmitted in a nearly uniform state as long as the exciting causes remained the same and there was free intercrossing. With respect to the exciting causes we can only say, as when speaking of so-called spontaneous variations, that they relate much more closely to the constitution of the varying organism, than to the nature of the conditions to which it has been subjected.

CONCLUSION.

In this chapter we have seen that as man at the present day is liable, like every other animal, to multiform individual differences or slight variations, so no doubt were the early progenitors of man; the variations being formerly induced by the same general causes, and governed by the same general and complex laws as at present. As all animals tend to multiply beyond their means of subsistence, so it must have been with the progenitors of man; and this would inevitably lead to a struggle for existence and to natural selection. The latter process would be greatly aided by the inherited effects of the increased use of parts, and these two processes would incessantly react on each other. It appears, also, as we shall hereafter see, that various unimportant characters have been acquired by man through sexual selection. An unexplained residuum of change must be left to the assumed uniform action of those unknown agencies, which occasionally induce strongly marked and abrupt deviations of structure in our domestic productions.

Judging from the habits of savages and of the greater number of the Quadrumana, primeval men, and even their ape-like progenitors, probably lived in society. With strictly social animals, natural selection sometimes acts on the individual, through the preservation of variations which are beneficial to the community. A community which includes a large number of well-endowed individuals increases in number, and is victorious over other less favoured ones; even although each separate member gains no advantage over the others of the same community. Associated insects have thus acquired many remarkable structures, which are of little or no service to the individual, such as the pollen-collecting apparatus, or the sting of the worker-bee, or the great jaws of soldier-ants. With the higher social animals, I am not aware that any structure has been modified solely for the good of the community, though some are of secondary service to it. For instance, the horns of ruminants and the great .canine teeth of baboons appear to have been acquired by the males as weapons for sexual strife, but they are used in defence of the herd or troop. In regard to certain mental powers the case, as we shall see in the fifth chapter, is wholly different; for these faculties have been chiefly, or even exclusively, gained for the benefit of the community, and the individuals thereof have at the same time gained an advantage indirectly.

It has often been objected to such views as the foregoing, that man is one of the most helpless and defenceless creatures in the world; and that during his early and less well-developed condition, he would have been still more helpless. The Duke of Argyll, for instance, insists [96] that "the human frame has diverged from the structure of brutes, in the direction of greater physical helplessness and weakness. That is to say, it is a divergence which of all others it is most impossible to ascribe to mere natural selection."

He adduces the naked and unprotected state of the body, the absence of great teeth or claws for defence, the small strength and speed of man, and his slight power of discovering food or of avoiding danger by smell. To these deficiencies there might be added one still more serious, namely, that he cannot climb quickly, and so escape from enemies. The loss of hair would not have been a great injury to the inhabitants of a warm country. For we know that the unclothed Fuegians can exist under a wretched climate. When we compare the defenceless state of man with that of apes, we must remember that the great canine teeth with which the latter are provided, are possessed in their full development by the males alone, and are chiefly used by them for fighting with their rivals; yet the females, which are not thus provided, manage to survive.

[96] 'Primeval Man,' 1869, p. 66.

In regard to bodily size or strength, we do not know whether man is descended from some small species, like the chimpanzee, or from one as powerful as the gorilla; and, therefore, we cannot say whether man has become larger and stronger, or smaller and weaker, than his ancestors. We should, however, bear in mind that an animal possessing great size, strength, and ferocity, and which, like the gorilla, could defend itself from all enemies, would not perhaps have become social: and this would most effectually have checked the acquirement of the higher mental qualities, such as sympathy and the love of his fellows. Hence it might have been an immense advantage to man to have sprung from some comparatively weak creature.

The small strength and speed of man, his want of natural weapons, etc., are more than counterbalanced, firstly, by his intellectual powers, through which he has formed for himself weapons, tools, etc., though still remaining in a barbarous state, and, secondly, by his social qualities which lead him to give and receive aid from his fellow-men. No country in the world abounds in a greater degree with dangerous beasts than Southern Africa; no country presents more fearful physical hardships than the Arctic regions; yet one of the puniest of races, that of the Bushmen, maintains itself in Southern Africa, as do the dwarfed Esquimaux in the Arctic regions. The ancestors of man were, no doubt, inferior in intellect, and probably in social disposition, to the lowest existing savages; but it is quite conceivable that they might have existed, or even flourished, if they had advanced in intellect, whilst gradually losing their brute-like powers, such as that of climbing trees, etc. But these ancestors would not have been exposed to any special danger, even if far more helpless and defenceless than any existing savages, had they inhabited some warm continent or large island, such as Australia, New Guinea, or Borneo, which is now the home of the orang. And natural selection arising from the competition of tribe with tribe, in some such large area as one of these, together with the inherited effects of habit, would, under favourable conditions, have sufficed to raise man to his present high position in the organic scale.

CHAPTER III.

COMPARISON OF THE MENTAL POWERS
OF MAN AND THE LOWER ANIMALS.

The difference in mental power between the highest ape and the lowest savage, immense—Certain instincts in common—The emotions—Curiosity—Imitation—Attention—Memory—Imagination—Reason—Progressive improvement —Tools and weapons used by animals—Abstraction, Self-consciousness—Language—Sense of beauty—Belief in God, spiritual agencies, superstitions.

We have seen in the last two chapters that man bears in his bodily structure clear traces of his descent from some lower form; but it may be urged that, as man differs so greatly in his mental power from all other animals, there must be some error in this conclusion. No doubt the difference in this respect is enormous, even if we compare the mind of one of the lowest savages, who has no words to express any number higher than four, and who uses hardly any abstract terms for common objects or for the affections [1], with that of the most highly organized ape. The difference would, no doubt, still remain immense, even if one of the higher apes had been improved or civilized as much as a dog has been in comparison with its parent-form, the wolf or jackal. The Fuegians rank amongst the lowest barbarians; but I was continually struck with surprise how closely the three natives on board H.M.S. "Beagle," who had lived some years in England, and could talk a little English, resembled us in disposition and in most of our mental faculties. If no organic being excepting man had possessed any mental power, or if his powers had been of a wholly different nature from those of the lower animals, then we should never have been able to convince ourselves that our high faculties had been gradually developed. But it can be shown that there is no fundamental difference of this kind. We must also admit that there is a much wider interval in mental power between one of the lowest fishes, as a lamprey or lancelet, and one of the higher apes, than between an ape and man; yet this interval is filled up by numberless gradations.

[1] See the evidence on those points, as given by Lubbock, 'Prehistoric Times,' p. 354, etc.

Nor is the difference slight in moral disposition between a barbarian, such as the man described by the old navigator Byron, who dashed his child on the rocks for dropping a basket of sea-urchins, and a Howard or Clarkson; and in intellect, between a savage who uses hardly any abstract terms, and a Newton or Shakspeare. Differences of this kind between the highest men of the highest races and the lowest savages, are connected by the finest gradations. Therefore it is possible that they might pass and be developed into each other.

My object in this chapter is to show that there is no fundamental difference between man and the higher mammals in their mental faculties. Each division of the subject might have been extended into a separate essay, but must here be treated briefly. As no classification of the mental powers has been universally accepted, I shall arrange my remarks in the order most convenient for my purpose; and will select those facts which have struck me most, with the hope that they may produce some effect on the reader.

With respect to animals very low in the scale, I shall give some additional facts under Sexual Selection, showing that their mental powers are much higher than might have been expected. The variability of the faculties in the individuals of the same species is an important point for us, and some few illustrations will here be given. But it would be superfluous to enter into many details on this head, for I have found on frequent enquiry, that it is the unanimous opinion of all those who have long attended to animals of many kinds, including birds, that the individuals differ greatly in every mental characteristic. In what manner the mental powers were first developed in the lowest organisms, is as hopeless an enquiry as how life itself first originated. These are problems for the distant future, if they are ever to be solved by man.

As man possesses the same senses as the lower animals, his fundamental intuitions must be the same. Man has also some few instincts in common, as that of self-preservation, sexual love, the love of the mother for her new-born offspring, the desire possessed by the latter to suck, and so forth. But man, perhaps, has somewhat fewer instincts than those possessed by the animals which come next to him in the series. The orang in the Eastern islands, and the chimpanzee in Africa, build platforms on which they sleep; and, as both species follow the same habit, it might be argued that this was due to instinct, but we cannot feel sure that it is not the result of both animals having similar wants, and possessing similar powers of reasoning. These apes, as we may assume, avoid the many poisonous fruits of the tropics, and man has no such knowledge: but as our domestic animals, when taken to foreign lands, and when first turned out in the spring, often eat poisonous herbs, which they afterwards avoid, we cannot feel sure that the apes do not learn from their own experience or from that of their parents what fruits to select. It is, however, certain, as we shall presently see, that apes have an instinctive dread of serpents, and probably of other dangerous animals.

The fewness and the comparative simplicity of the instincts in the higher animals are remarkable in contrast with those of the lower animals. Cuvier maintained that instinct and intelligence stand in an inverse ratio to each other; and some have thought that the intellectual faculties of the higher animals have been gradually developed from their instincts. But Pouchet, in an interesting essay [2], has shown that no such inverse ratio really exists. Those insects which possess the most wonderful instincts are certainly the most intelligent. In the vertebrate series, the least intelligent members, namely fishes and amphibians, do not possess complex instincts; and amongst mammals the animal most remarkable for its instincts, namely the beaver, is highly intelligent, as will be admitted by every one who has read Mr. Morgan's excellent work. [3]

[2] 'L'Instinct chez les Insectes,' 'Revue des Deux Mondes,' Feb. 1870, p. 690.
[3] 'The American Beaver and His Works,' 1868.

Although the first dawnings of intelligence, according to Mr. Herbert Spencer [4], have been developed through the multiplication and co-ordination of reflex actions, and although many of the simpler instincts graduate into reflex actions, and can hardly be distinguished from them, as in the case of young animals sucking, yet the more complex instincts seem to have originated independently of intelligence. I am, however, very far from wishing to deny that instinctive actions may lose their fixed and untaught character, and be replaced by others performed by the aid of the free will. On the other hand, some intelligent actions, after being performed during several generations, become converted into instincts and are inherited, as when birds on oceanic islands learn to avoid man.

These actions may then be said to be degraded in character, for they are no longer performed through reason or from experience. But the greater number of the more complex instincts appear to have been gained in a wholly different manner, through the natural selection of variations of simpler instinctive actions. Such variations appear to arise from the same unknown causes acting on the cerebral organization, which induce slight variations or individual differences in other parts of the body; and these variations, owing to our ignorance, are often said to arise spontaneously. We can, I think, come to no other conclusion with respect to the origin of the more complex instincts, when we reflect on the marvellous instincts of sterile worker-ants and bees, which leave no offspring to inherit the effects of experience and of modified habits.

[4] 'The Principles of Psychology,' 2nd edit., 1870, pp. 418-443.

Although, as we learn from the above-mentioned insects and the beaver, a high degree of intelligence is certainly compatible with complex instincts, and although actions, at first learnt voluntarily can soon through habit be performed with the quickness and certainty of a reflex action, yet it is not improbable that there is a certain amount of interference between the development of free intelligence and of instinct,—which latter implies some inherited modification of the brain. Little is known about the functions of the brain, but we can perceive that as the intellectual powers become highly developed, the various parts of the brain must be connected by very intricate channels of the freest intercommunication; and as a consequence each separate part would perhaps tend to be less well fitted to answer to particular sensations or associations in a definite and inherited—that is instinctive—manner. There seems even to exist some relation between a low degree of intelligence and a strong tendency to the formation of fixed, though not inherited habits; for as a sagacious physician remarked to me, persons who are slightly imbecile tend to act in everything by routine or habit; and they are rendered much happier if this is encouraged.

I have thought this digression worth giving, because we may easily underrate the mental powers of the higher animals, and especially of man, when we compare their actions founded on the memory of past events, on foresight, reason, and imagination, with exactly similar actions instinctively performed by the lower animals; in this latter case the capacity of performing such actions has been gained, step by step, through the variability of the mental organs and natural selection, without any conscious intelligence on the part of the animal during each successive generation. No doubt, as Mr. Wallace has argued [5], much of the intelligent work done by man is due to imitation and not to reason; but there is this great difference between his actions and many of those performed by the lower animals, namely, that man cannot, on his first trial, make, for instance, a stone hatchet or a canoe, through his power of imitation. He has to learn his work by practice; a beaver, on the other hand, can make its dam or canal, and a bird its nest, as well, or nearly as well, and a spider its wonderful web, quite as well [6], the first time it tries as when old and experienced.

[5] 'Contributions to the Theory of Natural Selection,' 1870, p. 212.
[6] For the evidence on this head, see Mr. J. Traherne Moggridge's most interesting work, 'Harvesting Ants and Trap-Door Spiders,' 1873, pp. 126, 128.

To return to our immediate subject: the lower animals, like man, manifestly feel pleasure and pain, happiness and misery. Happiness is never better exhibited than by young animals, such as puppies, kittens, lambs, etc., when playing together, like our own children. Even insects play together, as has been described by that excellent observer, P. Huber [7], who saw ants chasing and pretending to bite each other, like so many puppies.

[7] 'Recherches sur les Mœurs des Fourmis,' 1810, p. 173.

The fact that the lower animals are excited by the same emotions as ourselves is so well established, that it will not be necessary to weary the reader by many details. Terror acts in the same manner on them as on us, causing the muscles to tremble, the heart to palpitate, the sphincters to be relaxed, and the hair to stand on end. Suspicion, the offspring of fear, is eminently characteristic of most wild animals. It is, I think, impossible to read the account given by Sir E. Tennent, of the behaviour of the female elephants, used as decoys, without admitting that they intentionally practise deceit, and well know what they are about. Courage and timidity are extremely variable qualities in the individuals of the same species, as is plainly seen in our dogs. Some dogs and horses are ill-tempered, and easily turn sulky; others are good-tempered; and these qualities are certainly inherited. Every one knows how liable animals are to furious rage, and how plainly they show it. Many, and probably true, anecdotes have been published on the long-delayed and artful revenge of various animals. The accurate Rengger, and Brehm [8] state that the American and African monkeys which they kept tame, certainly revenged themselves. Sir Andrew Smith, a zoologist whose scrupulous accuracy was known to many persons, told me the following story of which he was himself an eye-witness; at the Cape of Good Hope an officer had often plagued a certain baboon, and the animal, seeing him approaching one Sunday for parade, poured water into a hole and hastily made some thick mud, which he skilfully dashed over the officer as he passed by, to the amusement of many bystanders. For long afterwards the baboon rejoiced and triumphed whenever he saw his victim.

[8] All the following statements, given on the authority of these two naturalists, are taken from Rengger's 'Naturgesch. der Säugethiere von Paraguay,' 1830, s. 41-57, and from Brehm's 'Thierleben,' B. i. s. 10-87.

The love of a dog for his master is notorious; as an old writer quaintly says [9], "A dog is the only thing on this earth that luvs you more than he luvs himself."

[9] Quoted by Dr. Lauder Lindsay, in his 'Physiology of Mind in the Lower Animals,' 'Journal of Mental Science,' April 1871, p. 38.

In the agony of death a dog has been known to caress his master, and every one has heard of the dog suffering under vivisection, who licked the hand of the operator; this man, unless the operation was fully justified by an increase of our knowledge, or unless he had a heart of stone, must have felt remorse to the last hour of his life.

As Whewell [10] has well asked, "who that reads the touching instances of maternal affection, related so often of the women of all nations, and of the females of all animals, can doubt that the principle of action is the same in the two cases?" We see maternal affection exhibited in the most trifling details; thus Rengger observed an American monkey (a Cebus) carefully driving away the flies which plagued her infant; and

Duvaucel saw a Hylobates washing the faces of her young ones in a stream. So intense is the grief of female monkeys for the loss of their young, that it invariably caused the death of certain kinds kept under confinement by Brehm in N. Africa. Orphan monkeys were always adopted and carefully guarded by the other monkeys, both males and females. One female baboon had so capacious a heart that she not only adopted young monkeys of other species, but stole young dogs and cats, which she continually carried about. Her kindness, however, did not go so far as to share her food with her adopted offspring, at which Brehm was surprised, as his monkeys always divided everything quite fairly with their own young ones. An adopted kitten scratched this affectionate baboon, who certainly had a fine intellect, for she was much astonished at being scratched, and immediately examined the kitten's feet, and without more ado bit off the claws. [11] In the Zoological Gardens, I heard from the keeper that an old baboon (*C. chacma*) had adopted a Rhesus monkey; but when a young drill and mandrill were placed in the cage, she seemed to perceive that these monkeys, though distinct species, were her nearer relatives, for she at once rejected the Rhesus and adopted both of them. The young Rhesus, as I saw, was greatly discontented at being thus rejected, and it would, like a naughty child, annoy and attack the young drill and mandrill whenever it could do so with safety; this conduct exciting great indignation in the old baboon. Monkeys will also, according to Brehm, defend their master when attacked by any one, as well as dogs to whom they are attached, from the attacks of other dogs. But we here trench on the subjects of sympathy and fidelity, to which I shall recur. Some of Brehm's monkeys took much delight in teasing a certain old dog whom they disliked, as well as other animals, in various ingenious ways.

[10] 'Bridgewater Treatise,' p. 263.
[11] A critic, without any grounds ('Quarterly Review,' July 1871, p. 72), disputes the possibility of this act as described by Brehm, for the sake of discrediting my work. Therefore I tried, and found that I could readily seize with my own teeth the sharp little claws of a kitten nearly five weeks old.

Most of the more complex emotions are common to the higher animals and ourselves. Every one has seen how jealous a dog is of his master's affection, if lavished on any other creature; and I have observed the same fact with monkeys. This shows that animals not only love, but have desire to be loved. Animals manifestly feel emulation. They love approbation or praise; and a dog carrying a basket for his master exhibits in a high degree self-complacency or pride. There can, I think, be no doubt that a dog feels shame, as distinct from fear, and something very like modesty when begging too often for food. A great dog scorns the snarling of a little dog, and this may be called magnanimity. Several observers have stated that monkeys certainly dislike being laughed at; and they sometimes invent imaginary offences. In the Zoological Gardens I saw a baboon who always got into a furious rage when his keeper took out a letter or book and read it aloud to him; and his rage was so violent that, as I witnessed on one occasion, he bit his own leg till the blood flowed. Dogs show what may be fairly called a sense of humour, as distinct from mere play; if a bit of stick or other such object be thrown to one, he will often carry it away for a short distance; and then squatting down with it on the ground close before him, will wait until his master comes quite close to take it away. The dog will then seize it and rush away in triumph, repeating the same manœuvre, and evidently enjoying the practical joke.

We will now turn to the more intellectual emotions and faculties, which are very important, as forming the basis for the development of the higher mental powers. Animals manifestly enjoy excitement, and suffer from ennui, as may be seen with dogs, and, according to Rengger, with monkeys. All animals feel *Wonder*, and many exhibit *Curiosity*. They sometimes suffer from this latter quality, as when the hunter plays antics and thus attracts them; I have witnessed this with deer, and so it is with the wary chamois, and with some kinds of wild-ducks. Brehm gives a curious account of the instinctive dread, which his monkeys exhibited, for snakes; but their curiosity was so great that they could not desist from occasionally satiating their horror in a most human fashion, by lifting up the lid of the box in which the snakes were kept. I was so much surprised at his account, that I took a stuffed and coiled-up snake into the monkey-house at the Zoological Gardens, and the excitement thus caused was one of the most curious spectacles which I ever beheld. Three species of Cercopithecus were the most alarmed; they dashed about their cages, and uttered sharp signal cries of danger, which were understood by the other monkeys. A few young monkeys and one old Anubis baboon alone took no notice of the snake. I then placed the stuffed specimen on the ground in one of the larger compartments. After a time all the monkeys collected round it in a large circle, and staring intently, presented a most ludicrous appearance. They became extremely nervous; so that when a wooden ball, with which they were familiar as a plaything, was accidentally moved in the straw, under which it was partly hidden, they all instantly started away. These monkeys behaved very differently when a dead fish, a mouse [12], a living turtle, and other new objects were placed in their cages; for though at first frightened, they soon approached, handled and examined them. I then placed a live snake in a paper bag, with the mouth loosely closed, in one of the larger compartments. One of the monkeys immediately approached, cautiously opened the bag a little, peeped in, and instantly dashed away. Then I witnessed what Brehm has described, for monkey after monkey, with head raised high and turned on one side, could not resist taking a momentary peep into the upright bag, at the dreadful object lying quietly at the bottom. It would almost appear as if monkeys had some notion of zoological affinities, for those kept by Brehm exhibited a strange, though mistaken, instinctive dread of innocent lizards and frogs. An orang, also, has been known to be much alarmed at the first sight of a turtle. [13]

[12] I have given a short account of their behaviour on this occasion in my 'Expression of the Emotions in Man and Animals,' p. 43.
[13] W.C.L. Martin, 'Natural History of Mammalia,' 1841, p. 405.

The principle of *Imitation* is strong in man, and especially, as I have myself observed, with savages. In certain morbid states of the brain this tendency is exaggerated to an extraordinary degree: some hemiplegic patients and others, at the commencement of inflammatory softening of the brain, unconsciously imitate every word which is uttered, whether in their own or in a foreign language, and every gesture or action which is performed near them. [14] Desor [15] has remarked that no animal voluntarily imitates an action performed by man, until in the ascending scale we come to monkeys, which are well known to be ridiculous mockers. Animals, however, sometimes imitate each other's actions: thus two species of wolves, which had been reared by dogs, learned to bark, as does sometimes the jackal [16], but whether this can be called voluntary imitation is another question. Birds imitate the songs of their parents, and sometimes of other birds;

and parrots are notorious imitators of any sound which they often hear. Dureau de la Malle gives an account [17] of a dog reared by a cat, who learnt to imitate the well-known action of a cat licking her paws, and thus washing her ears and face; this was also witnessed by the celebrated naturalist Audouin. I have received several confirmatory accounts; in one of these, a dog had not been suckled by a cat, but had been brought up with one, together with kittens, and had thus acquired the above habit, which he ever afterwards practised during his life of thirteen years. Dureau de la Malle's dog likewise learnt from the kittens to play with a ball by rolling it about with his fore paws, and springing on it. A correspondent assures me that a cat in his house used to put her paws into jugs of milk having too narrow a mouth for her head. A kitten of this cat soon learned the same trick, and practised it ever afterwards, whenever there was an opportunity.

[14] Dr. Bateman, 'On Aphasia,' 1870, p. 110.
[15] Quoted by Vogt, 'Mémoire sur les Microcéphales,' 1867, p. 168.
[16] The 'Variation of Animals and Plants under Domestication,' vol. i. p. 27.
[17] 'Annales des Sciences Nat.' (1st Series), tom. xxii. p. 397.

The parents of many animals, trusting to the principle of imitation in their young, and more especially to their instinctive or inherited tendencies, may be said to educate them. We see this when a cat brings a live mouse to her kittens; and Dureau de la Malle has given a curious account (in the paper above quoted) of his observations on hawks which taught their young dexterity, as well as judgment of distances, by first dropping through the air dead mice and sparrows, which the young generally failed to catch, and then bringing them live birds and letting them loose.

Hardly any faculty is more important for the intellectual progress of man than *Attention*. Animals clearly manifest this power, as when a cat watches by a hole and prepares to spring on its prey. Wild animals sometimes become so absorbed when thus engaged, that they may be easily approached. Mr. Bartlett has given me a curious proof how variable this faculty is in monkeys. A man who trains monkeys to act in plays, used to purchase common kinds from the Zoological Society at the price of five pounds for each; but he offered to give double the price, if he might keep three or four of them for a few days, in order to select one. When asked how he could possibly learn so soon, whether a particular monkey would turn out a good actor, he answered that it all depended on their power of attention. If when he was talking and explaining anything to a monkey, its attention was easily distracted, as by a fly on the wall or other trifling object, the case was hopeless. If he tried by punishment to make an inattentive monkey act, it turned sulky. On the other hand, a monkey which carefully attended to him could always be trained.

It is almost superfluous to state that animals have excellent *Memories* for persons and places. A baboon at the Cape of Good Hope, as I have been informed by Sir Andrew Smith, recognized him with joy after an absence of nine months. I had a dog who was savage and averse to all strangers, and I purposely tried his memory after an absence of five years and two days. I went near the stable where he lived, and shouted to him in my old manner; he showed no joy, but instantly followed me out walking, and obeyed me, exactly as if I had parted with him only half an hour before. A train of old associations, dormant during five years, had thus been instantaneously awakened in his mind. Even ants, as P. Huber [18] has clearly shown, recognized their fellow-ants belonging to the same community after a separation of four months. Animals can certainly by some means judge of the intervals of time between recurrent events.

[18] 'Les Mœurs des Fourmis,' 1810, p. 150.

The *Imagination* is one of the highest prerogatives of man. By this faculty he unites former images and ideas, independently of the will, and thus creates brilliant and novel results. A poet, as Jean Paul Richter remarks [19], "who must reflect whether he shall make a character say yes or no—to the devil with him; he is only a stupid corpse." Dreaming gives us the best notion of this power; as Jean Paul again says, "The dream is an involuntary art of poetry." The value of the products of our imagination depends of course on the number, accuracy, and clearness of our impressions, on our judgment and taste in selecting or rejecting the involuntary combinations, and to a certain extent on our power of voluntarily combining them. As dogs, cats, horses, and probably all the higher animals, even birds [20] have vivid dreams, and this is shown by their movements and the sounds uttered, we must admit that they possess some power of imagination. There must be something special, which causes dogs to howl in the night, and especially during moonlight, in that remarkable and melancholy manner called baying. All dogs do not do so; and, according to Houzeau [21], they do not then look at the moon, but at some fixed point near the horizon. Houzeau thinks that their imaginations are disturbed by the vague outlines of the surrounding objects, and conjure up before them fantastic images: if this be so, their feelings may almost be called superstitious.

[19] Quoted in Dr. Maudsley's 'Physiology and Pathology of Mind,' 1868, pp. 19, 220.
[20] Dr. Jerdon, 'Birds of India,' vol. i. 1862, p. xxi. Houzeau says that his parokeets and canary-birds dreamt: 'Etudes sur les Facultés Mentales,' tom. ii. p. 136.
[21] Facultés Mentales des Animaux 1872, tom. ii. p. 181.

Of all the faculties of the human mind, it will, I presume, be admitted that *Reason* stands at the summit. Only a few persons now dispute that animals possess some power of reasoning. Animals may constantly be seen to pause, deliberate, and resolve. It is a significant fact, that the more the habits of any particular animal are studied by a naturalist, the more he attributes to reason and the less to unlearnt instincts. [22] In future chapters we shall see that some animals extremely low in the scale apparently display a certain amount of reason. No doubt it is often difficult to distinguish between the power of reason and that of instinct. For instance, Dr. Hayes, in his work on 'The Open Polar Sea,' repeatedly remarks that his dogs, instead of continuing to draw the sledges in a compact body, diverged and separated when they came to thin ice, so that their weight might be more evenly distributed. This was often the first warning which the travellers received that the ice was becoming thin and dangerous. Now, did the dogs act thus from the experience of each individual, or from the example of the older and wiser dogs, or from an inherited habit, that is from instinct? This instinct, may possibly have arisen since the time, long ago, when dogs were first employed by the natives in drawing their sledges; or the Arctic wolves, the parent-stock of the Esquimaux dog, may have acquired an instinct impelling them not to attack their prey in a close pack, when on thin ice.

[22] Mr. L.H. Morgan's work on 'The American Beaver,' 1868, offers a good illustration of this remark. I cannot help thinking, however, that he goes too far in underrating

the power of instinct.

We can only judge by the circumstances under which actions are performed, whether they are due to instinct, or to reason, or to the mere association of ideas: this latter principle, however, is intimately connected with reason. A curious case has been given by Prof. Möbius [23], of a pike, separated by a plate of glass from an adjoining aquarium stocked with fish, and who often dashed himself with such violence against the glass in trying to catch the other fishes, that he was sometimes completely stunned. The pike went on thus for three months, but at last learnt caution, and ceased to do so. The plate of glass was then removed, but the pike would not attack these particular fishes, though he would devour others which were afterwards introduced; so strongly was the idea of a violent shock associated in his feeble mind with the attempt on his former neighbours. If a savage, who had never seen a large plate-glass window, were to dash himself even once against it, he would for a long time afterwards associate a shock with a window-frame; but very differently from the pike, he would probably reflect on the nature of the impediment, and be cautious under analogous circumstances. Now with monkeys, as we shall presently see, a painful or merely a disagreeable impression, from an action once performed, is sometimes sufficient to prevent the animal from repeating it. If we attribute this difference between the monkey and the pike solely to the association of ideas being so much stronger and more persistent in the one than the other, though the pike often received much the more severe injury, can we maintain in the case of man that a similar difference implies the possession of a fundamentally different mind?

[23] 'Die Bewegungen der Thiere,' etc., 1873, p. 11.

Houzeau relates [24] that, whilst crossing a wide and arid plain in Texas, his two dogs suffered greatly from thirst, and that between thirty and forty times they rushed down the hollows to search for water. These hollows were not valleys, and there were no trees in them, or any other difference in the vegetation, and as they were absolutely dry there could have been no smell of damp earth. The dogs behaved as if they knew that a dip in the ground offered them the best chance of finding water, and Houzeau has often witnessed the same behaviour in other animals.

[24] 'Etudes sur les Facultés Mentales des Animaux,' 1872, tom. ii. p. 265.

I have seen, as I daresay have others, that when a small object is thrown on the ground beyond the reach of one of the elephants in the Zoological Gardens, he blows through his trunk on the ground beyond the object, so that the current reflected on all sides may drive the object within his reach. Again a well-known ethnologist, Mr. Westropp, informs me that he observed in Vienna a bear deliberately making with his paw a current in some water, which was close to the bars of his cage, so as to draw a piece of floating bread within his reach. These actions of the elephant and bear can hardly be attributed to instinct or inherited habit, as they would be of little use to an animal in a state of nature. Now, what is the difference between such actions, when performed by an uncultivated man, and by one of the higher animals?

The savage and the dog have often found water at a low level, and the coincidence under such circumstances has become associated in their minds. A cultivated man would perhaps make some general proposition on the subject; but from all that we know of savages it is extremely doubtful whether they would do so, and a dog certainly would not. But a savage, as well as a dog, would search in the same way, though frequently disappointed; and in both it seems to be equally an act of reason, whether or not any general proposition on the subject is consciously placed before the mind. [25] The same would apply to the elephant and the bear making currents in the air or water. The savage would certainly neither know nor care by what law the desired movements were effected; yet his act would be guided by a rude process of reasoning, as surely as would a philosopher in his longest chain of deductions. There would no doubt be this difference between him and one of the higher animals, that he would take notice of much slighter circumstances and conditions, and would observe any connection between them after much less experience, and this would be of paramount importance. I kept a daily record of the actions of one of my infants, and when he was about eleven months old, and before he could speak a single word, I was continually struck with the greater quickness, with which all sorts of objects and sounds were associated together in his mind, compared with that of the most intelligent dogs I ever knew. But the higher animals differ in exactly the same way in this power of association from those low in the scale, such as the pike, as well as in that of drawing inferences and of observation.

[25] Prof. Huxley has analyzed with admirable clearness the mental steps by which a man, as well as a dog, arrives at a conclusion in a case analogous to that given in my text. See his article, 'Mr. Darwin's Critics,' in the 'Contemporary Review,' Nov. 1871, p. 462, and in his 'Critiques and Essays,' 1873, p. 279.

The promptings of reason, after very short experience, are well shown by the following actions of American monkeys, which stand low in their order. Rengger, a most careful observer, states that when he first gave eggs to his monkeys in Paraguay, they smashed them, and thus lost much of their contents; afterwards they gently hit one end against some hard body, and picked off the bits of shell with their fingers. After cutting themselves only *once* with any sharp tool, they would not touch it again, or would handle it with the greatest caution. Lumps of sugar were often given them wrapped up in paper; and Rengger sometimes put a live wasp in the paper, so that in hastily unfolding it they got stung; after this had *once* happened, they always first held the packet to their ears to detect any movement within. [26]

[26] Mr. Belt, in his most interesting work, 'The Naturalist in Nicaragua,' 1874, (p. 119,) likewise describes various actions of a tamed Cebus, which, I think, clearly show that this animal possessed some reasoning power.

The following cases relate to dogs. Mr. Colquhoun [27] winged two wild-ducks, which fell on the further side of a stream; his retriever tried to bring over both at once, but could not succeed; she then, though never before known to ruffle a feather, deliberately killed one, brought over the other, and returned for the dead bird. Col. Hutchinson relates that two partridges were shot at once, one being killed, the other wounded; the latter ran away, and was caught by the retriever, who on her return came across the dead bird; "she stopped, evidently greatly puzzled, and after one or two trials, finding she could not take it up without permitting the escape of the winged bird, she considered a moment, then deliberately murdered it by giving it a severe crunch, and afterwards brought away both together. This was the only known instance of her ever having wilfully injured any game." Here we have reason though not quite perfect, for the retriever might have brought the wounded bird first and then returned for the dead one, as in the case of the two wild-ducks. I give the above cases, as resting on the evidence of two independent witnesses, and because in both instances the retrievers, after deliberation, broke through a habit which is inherited by them (that of not killing the game retrieved), and because they show how strong their reasoning faculty must have been to overcome a fixed habit.

[27] 'The Moor and the Loch,' p. 45. Col. Hutchinson on 'Dog Breaking,' 1850, p. 46.

I will conclude by quoting a remark by the illustrious Humboldt. [28] "The muleteers in S. America say, 'I will not give you the mule whose step is easiest, but *la mas racional*,—the one that reasons best'"; and; as, he adds, "this popular expression, dictated by long experience, combats the system of animated machines, better perhaps than all the arguments of speculative philosophy." Nevertheless some writers even yet deny that the higher animals possess a trace of reason; and they endeavour to explain away, by what appears to be mere verbiage, [29] all such facts as those above given.

[28] 'Personal Narrative,' Eng. translat., vol. iii. p. 106.
[29] I am glad to find that so acute a reasoner as Mr. Leslie Stephen ('Darwinism and Divinity, Essays on Free Thinking,' 1873, p. 80), in speaking of the supposed impassable barrier between the minds of man and the lower animals, says, "The distinctions, indeed, which have been drawn, seem to us to rest upon no better foundation than a great many other metaphysical distinctions; that is, the assumption that because you can give two things different names, they must therefore have different natures. It is difficult to understand how anybody who has ever kept a dog, or seen an elephant, can have any doubt as to an animal's power of performing the essential processes of reasoning."

It has, I think, now been shown that man and the higher animals, especially the Primates, have some few instincts in common. All have the same senses, intuitions, and sensations,—similar passions, affections, and emotions, even the more complex ones, such as jealousy, suspicion, emulation, gratitude, and magnanimity; they practise deceit and are revengeful; they are sometimes susceptible to ridicule, and even have a sense of humour; they feel wonder and curiosity; they possess the same faculties of imitation, attention, deliberation, choice, memory, imagination, the association of ideas, and reason, though in very different degrees. The individuals of the same species graduate in intellect from absolute imbecility to high excellence. They are also liable to insanity, though far less often than in the case of man. [30] Nevertheless, many authors have insisted that man is divided by an insuperable barrier from all the lower animals in his mental faculties. I formerly made a collection of above a score of such aphorisms, but they are almost worthless, as their wide difference and number prove the difficulty, if not the impossibility, of the attempt. It has been asserted that man alone is capable of progressive improvement; that he alone makes use of tools or fire, domesticates other animals, or possesses property; that no animal has the power of abstraction, or of forming general concepts, is self-conscious and comprehends itself; that no animal employs language; that man alone has a sense of beauty, is liable to caprice, has the feeling of gratitude, mystery, etc.; believes in God, or is endowed with a conscience. I will hazard a few remarks on the more important and interesting of these points.

[30] See 'Madness in Animals,' by Dr. W. Lauder Lindsay, in 'Journal of Mental Science,' July 1871.

Archbishop Sumner formerly maintained [31] that man alone is capable of progressive improvement. That he is capable of incomparably greater and more rapid improvement than is any other animal, admits of no dispute; and this is mainly due to his power of speaking and handing down his acquired knowledge. With animals, looking first to the individual, every one who has had any experience in setting traps, knows that young animals can he caught much more easily than old ones; and they can be much more easily approached by an enemy. Even with respect to old animals, it is impossible to catch many in the same place and in the same kind of trap, or to destroy them by the same kind of poison; yet it is improbable that all should have partaken of the poison, and impossible that all should have been caught in a trap. They must learn caution by seeing their brethren caught or poisoned. In North America, where the fur-bearing animals have long been pursued, they exhibit, according to the unanimous testimony of all observers, an almost incredible amount of sagacity, caution and cunning; but trapping has been there so long carried on, that inheritance may possibly have come into play. I have received several accounts that when telegraphs are first set up in any district, many birds kill themselves by flying against the wires, but that in the course of a very few years they learn to avoid this danger, by seeing, as it would appear, their comrades killed. [32]

[31] Quoted by Sir C. Lyell, 'Antiquity of Man,' p. 497.
[32] For additional evidence, with details, see M. Houzeau, 'Etudes sur les Facultés Mentales des Animaux,' tom. ii. 1872, p. 147.

If we look to successive generations, or to the race, there is no doubt that birds and other animals gradually both acquire and lose caution in relation to man or other enemies [33]; and this caution is certainly in chief part an inherited habit or instinct, but in part the result of individual experience. A good observer, Leroy [34], states, that in districts where foxes are much hunted, the young, on first leaving their burrows, are incontestably much more wary than the old ones in districts where they are not much disturbed.

[33] See, with respect to birds on oceanic islands, my 'Journal of Researches during the Voyage of the "Beagle,"' 1845, p. 398. 'Origin of Species,' 5th ed. p. 260.
[34] 'Lettres Phil. sur l'Intelligence des Animaux,' nouvelle edit., 1802, p. 86.

Our domestic dogs are descended from wolves and jackals [35], and though they may not have gained in cunning, and may have lost in wariness and suspicion, yet they have progressed in certain moral qualities, such as in affection, trust-worthiness, temper, and probably in general intelligence. The common rat has conquered and beaten several other species throughout Europe, in parts of North America, New Zealand, and recently in Formosa, as well as on the mainland of China. Mr. Swinhoe [36], who describes these two latter cases, attributes the victory of the common rat over the large *Mus coninga* to its superior cunning; and this latter quality may probably be attributed to the habitual exercise of all its faculties in avoiding extirpation by man, as well as to nearly all the less cunning or weak-minded rats having been continuously destroyed by him. It is, however, possible that the success of the common rat may be due to its having possessed greater cunning than its fellow-species, before it became associated with man. To maintain, independently of any direct evidence, that no animal during the course of ages has progressed in intellect or other mental faculties, is to beg the question of the evolution of species. We have seen that, according to Lartet, existing mammals belonging to several orders have larger brains than their ancient tertiary prototypes.

[35] See the evidence on this head in chap. i. vol. i., 'On the Variation of Animals and Plants under Domestication.'
[36] 'Proceedings Zoological Society,' 1864, p. 186.

It has often been said that no animal uses any tool; but the chimpanzee in a state of nature cracks a native fruit, somewhat like a walnut, with a stone. [37] Rengger [38] easily taught an American monkey thus to break open hard palm-nuts; and afterwards of its own accord, it used stones to open other kinds of nuts, as well as boxes. It thus also removed the soft rind of fruit that had a disagreeable flavor. Another monkey was taught to open the lid of a large box with a stick, and afterwards it used the stick as a lever to move heavy bodies; and I have myself seen a young orang put a stick into a crevice, slip his hand to the other end, and use it in the proper manner as a lever. The tamed elephants in India are well known to break off branches of trees and use them to drive away the flies; and this same act has been observed in an elephant in a state of nature. [39] I have seen a young orang, when she thought she was going to be whipped, cover and protect herself with a blanket or straw. In these several cases stones and sticks were employed as implements; but they are likewise used as weapons. Brehm [40] states, on the authority of the well-known traveller Schimper, that in Abyssinia when the baboons belonging to one species (*C. gelada*) descend in troops from the mountains to plunder the fields, they sometimes encounter troops of another species (*C. hamadryas*), and then a fight ensues. The Geladas roll down great stones, which the Hamadryas try to avoid, and then both species, making a great uproar, rush furiously against each other. Brehm, when accompanying the Duke of Coburg-Gotha, aided in an attack with fire-arms on a troop of baboons in the pass of Mensa in Abyssinia. The baboons in return rolled so many stones down the mountain, some as large as a man's head, that the attackers had to beat a hasty retreat; and the pass was actually closed for a time against the caravan. It deserves notice that these baboons thus acted in concert. Mr. Wallace [41] on three occasions saw female orangs, accompanied by their young, "breaking off branches and the great spiny fruit of the Durian tree, with every appearance of rage; causing such a shower of missiles as effectually kept us from approaching too near the tree." As I have repeatedly seen, a chimpanzee will throw any object at hand at a person who offends him; and the before-mentioned baboon at the Cape of Good Hope prepared mud for the purpose.

[37] Savage and Wyman in 'Boston Journal of Natural History,' vol. iv. 1843-44, p. 383.
[38] 'Säugethiere von Paraguay,' 1830, s. 51-56.
[39] The Indian Field, March 4, 1871.
[40] 'Thierleben,' B. i. s. 79, 82.
[41] 'The Malay Archipelago,' vol. i. 1869, p. 87.

In the Zoological Gardens, a monkey, which had weak teeth, used to break open nuts with a stone; and I was assured by the keepers that after using the stone, he hid it in the straw, and would not let any other monkey touch it. Here, then, we have the idea of property; but this idea is common to every dog with a bone, and to most or all birds with their nests.

The Duke of Argyll [42] remarks, that the fashioning of an implement for a special purpose is absolutely peculiar to man; and he considers that this forms an immeasurable gulf between him and the brutes. This is no doubt a very important distinction; but there appears to me much truth in Sir J. Lubbock's suggestion [43], that when primeval man first used flint-stones for any purpose, he would have accidentally splintered them, and would then have used the sharp fragments. From this step it would be a small one to break the flints on purpose, and not a very wide step to fashion them rudely. This latter advance, however, may have taken long ages, if we may judge by the immense interval of time which elapsed before the men of the neolithic period took to grinding and polishing their stone tools. In breaking the flints, as Sir J. Lubbock likewise remarks, sparks would have been emitted, and in grinding them heat would have been evolved: thus the two usual methods of "obtaining fire may have originated." The nature of fire would have been known in the many volcanic regions where lava occasionally flows through forests. The anthropomorphous apes, guided probably by instinct, build for themselves temporary platforms; but as many instincts are largely controlled by reason, the simpler ones, such as this of building a platform, might readily pass into a voluntary and conscious act. The orang is known to cover itself at night with the leaves of the Pandanus; and Brehm states that one of his baboons used to protect itself from the heat of the sun by throwing a straw-mat over its head. In these several habits, we probably see the first steps towards some of the simpler arts, such as rude architecture and dress, as they arose amongst the early progenitors of man.

[42] 'Primeval Man,' 1869, pp. 145, 147.
[43] 'Prehistoric Times,' 1865, p. 473, etc.

ABSTRACTION, GENERAL CONCEPTIONS, SELF-CONSCIOUSNESS, MENTAL INDIVIDUALITY.

It would be very difficult for any one with even much more knowledge than I possess, to determine how far animals exhibit any traces of these high mental powers. This difficulty arises from the impossibility of judging what passes through the mind of an animal; and again, the fact that writers differ to a great extent in the meaning which they attribute to the above terms, causes a further difficulty. If one may judge from various articles which have been published lately, the greatest stress seems to be laid on the supposed entire absence in animals of the power of abstraction, or of forming general concepts. But when a dog sees another dog at a distance, it is often clear that he perceives that it is a dog in the abstract; for when he gets nearer his whole manner suddenly changes, if the other dog be a friend. A recent writer remarks, that in all such cases it is a pure assumption to assert that the mental act is not essentially of the same nature in the animal as in man. If either refers what he perceives with his senses to a mental concept, then so do both. [44] When I say to my terrier, in an eager voice (and I have made the trial many times), "Hi, hi, where is it?" she at once takes it as a sign that something is to be hunted, and generally first looks quickly all around, and then rushes into the nearest thicket, to scent for any game, but finding nothing, she looks up into any neighbouring tree for a squirrel. Now do not these actions clearly show that she had in her mind a general idea or concept that some animal is to be discovered and hunted?

[44] Mr. Hookham, in a letter to Prof. Max Müller, in the 'Birmingham News,' May 1873.

It may be freely admitted that no animal is self-conscious, if by this term it is implied, that he reflects on such points, as whence he comes or whither he will go, or what is life and death, and so forth. But how can we feel sure that an old dog with an excellent memory and some power of imagination, as shown by his dreams, never reflects on his past pleasures or pains in the chase? And this would be a form of self-consciousness. On the other hand, as Büchner [45] has remarked, how little can the hard-worked wife of a degraded Australian savage, who uses very few abstract words, and cannot count above four, exert her self-consciousness, or reflect on the nature of her own existence. It is generally admitted, that the higher animals possess memory, attention, association, and even some imagination and reason. If these powers, which differ much in different animals, are capable of improvement, there seems no great improbability in more complex faculties, such as the higher forms of abstraction, and self-consciousness, etc., having been evolved through the development and combination of the simpler ones. It has been urged against the views here maintained that it is impossible to say at what point in the ascending scale animals become capable of abstraction, etc.; but who can say at what age this occurs in our young children? We see at least that such powers are developed in children by imperceptible degrees.

[45] 'Conferences sur la Théorie Darwinienne,' French translat. 1869, p. 132.

That animals retain their mental individuality is unquestionable. When my voice awakened a train of old associations in the mind of the before-mentioned dog, he must have retained his mental individuality, although every atom of his brain had probably undergone change more than once during the interval of five years. This dog might have brought forward the argument lately advanced to crush all evolutionists, and said, "I abide amid all mental moods and all material changes...The teaching that atoms leave their impressions as legacies to other atoms falling into the places they have vacated is contradictory of the utterance of consciousness, and is therefore false; but it is the teaching necessitated by evolutionism, consequently the hypothesis is a false one." [46]

[46] The Rev. Dr. J. M'Cann, 'Anti-Darwinism,' 1869, p. 13.

LANGUAGE.

This faculty has justly been considered as one of the chief distinctions between man and the lower animals. But man, as a highly competent judge, Archbishop Whately remarks, "is not the only animal that can make use of language to express what is passing in his mind, and can understand, more or less, what is so expressed by another." [47] In Paraguay the *Cebus azaræ* when excited utters at least six distinct sounds, which excite in other monkeys similar emotions. [48] The movements of the features and gestures of monkeys are understood by us, and they partly understand ours, as Rengger and others declare. It is a more remarkable fact that the dog, since being domesticated, has learnt to bark [49] in at least four or five distinct tones. Although barking is a new art, no doubt the wild parent-species of the dog expressed their feelings by cries of various kinds. With the domesticated dog we have the bark of eagerness, as in the chase; that of anger, as well as growling; the yelp or howl of despair, as when shut up; the baying at night; the bark of joy, as when starting on a walk with his master; and the very distinct one of demand or supplication, as when wishing for a door or window to be opened. According to Houzeau, who paid particular attention to the subject, the domestic fowl utters at least a dozen significant sounds. [50]

[47] Quoted in 'Anthropological Review,' 1864, p. 158.
[48] Rengger, ibid. s. 45.
[49] See my 'Variation of Animals and Plants under Domestication,' vol. i. p. 27.
[50] 'Facultés Mentales des Animaux,' tom. ii. 1872, p. 346-349.

The habitual use of articulate language is, however, peculiar to man; but he uses, in common with the lower animals, inarticulate cries to express his meaning, aided by gestures and the movements of the muscles of the face. [51] This especially holds good with the more simple and vivid feelings, which are but little connected with our higher intelligence. Our cries of pain, fear, surprise, anger, together with their appropriate actions, and the murmur of a mother to her beloved child are more expressive than any words. That which distinguishes man from the lower animals is not the understanding of articulate sounds, for, as every one knows, dogs understand many words and sentences. In this respect they are at the same stage of development as infants, between the ages of ten and twelve months, who understand many words and short sentences, but cannot yet utter a single word. It is not the mere articulation which is our distinguishing character, for parrots and other birds possess this power. Nor is it the mere capacity of connecting definite sounds with definite ideas; for it is certain that some parrots, which have been taught to speak, connect unerringly words with things, and persons with events. [52] The lower animals differ from man solely in his almost infinitely larger power of associating together the most diversified sounds and ideas; and this obviously depends on the high development of his mental powers.

[51] See a discussion on this subject in Mr. E.B. Tylor's very interesting work, 'Researches into the Early History of Mankind,' 1865, chaps. ii. to iv.
[52] I have received several detailed accounts to this effect. Admiral Sir B.J. Sulivan, whom I know to be a careful observer, assures me that an African parrot, long kept in his father's house, invariably called certain persons of the household, as well as visitors, by their names. He said "good morning" to every one at breakfast, and "good night" to each as they left the room at night, and never reversed these salutations. To Sir B.J. Sulivan's father, he used to add to the "good morning" a short sentence, which was never once repeated after his father's death. He scolded violently a strange dog which came into the room through the open window; and he scolded another parrot (saying "you naughty polly") which had got out of its cage, and was eating apples on the kitchen table. See also, to the same effect, Houzeau on parrots, 'Facultés Mentales,' tom. ii. p. 309. Dr. A. Moschkau informs me that he knew a starling which never made a mistake in saying in German "good morning" to persons arriving, and "good bye, old fellow," to those departing. I could add several other such cases.

As Horne Tooke, one of the founders of the noble science of philology, observes, language is an art, like brewing or baking; but writing would have been a better simile. It certainly is not a true instinct, for every language has to be learnt. It differs, however, widely from all ordinary arts, for man has an instinctive tendency to speak, as we see in the babble of our young children; whilst no child has an instinctive tendency to brew, bake, or write. Moreover, no philologist now supposes that any language has been deliberately invented; it has been slowly and unconsciously developed by many steps. [53] The sounds uttered by birds offer in several respects the nearest analogy to language, for all the members of the same species utter the same instinctive cries expressive of their emotions; and all the kinds which sing, exert their power instinctively; but the actual song, and even the call-notes, are learnt from their parents or foster-parents. These sounds, as Daines Barrington [54] has proved, "are no more innate than language is in man." The first attempts to sing "may be compared to the imperfect endeavour in a child to babble." The young males continue practicing, or as the bird-catchers say, "recording," for ten or eleven months. Their first essays show hardly a rudiment of the future song; but as they grow older we can perceive what they are aiming at; and at last they are said "to sing their song round." Nestlings which have learnt the song of a distinct species, as with the canary-birds educated in the Tyrol, teach and transmit their new song to their offspring. The slight natural differences of song in the same species inhabiting different districts may be appositely compared, as Barrington remarks, "to provincial dialects"; and the songs of allied, though distinct species may be compared with the languages of distinct races of man. I have given the foregoing details to show that an instinctive tendency to acquire an art is not peculiar to man.

[53] See some good remarks on this head by Prof. Whitney, in his 'Oriental and Linguistic Studies,' 1873, p. 354. He observes that the desire of communication between man is the living force, which, in the development of language, "works both consciously and unconsciously; consciously as regards the immediate end to be attained; unconsciously as regards the further consequences of the act."
[54] Hon. Daines Barrington in 'Philosoph. Transactions,' 1773, p. 262. See also Dureau de la Malle, in 'Ann. des. Sc. Nat.' 3rd series, Zoolog., tom. x. p. 119.

With respect to the origin of articulate language, after having read on the one side the highly interesting works of Mr. Hensleigh Wedgwood, the Rev. F. Farrar, and Prof. Schleicher [55], and the celebrated lectures of Prof. Max Müller on the other side, I cannot doubt that language owes its origin to the imitation and modification of various natural sounds, the voices of other animals, and man's own instinctive cries, aided by signs and gestures. When we treat of sexual selection we shall see that primeval man, or rather some early progenitor of man, probably first used his voice in producing true musical cadences, that is in singing, as do some of the gibbon-apes at the present day; and we may conclude from a widely-spread analogy, that this power would have been especially exerted during the courtship of the sexes,—would have expressed various emotions, such as love, jealousy, triumph,—and would have served as a challenge to rivals. It is, therefore, probable that the imitation of musical cries by articulate sounds may have given rise to words expressive of various complex emotions. The strong tendency in our nearest allies, the monkeys, in microcephalous idiots [56], and in the barbarous races of mankind, to imitate whatever they hear deserves notice, as bearing on the subject of imitation. Since monkeys certainly understand much that is said to them by man, and when wild, utter signal-cries of danger to their fellows [57]; and since fowls

give distinct warnings for danger on the ground, or in the sky from hawks (both, as w
as a third cry, intelligible to dogs) [58], may not some unusually wise ape-like anima
have imitated the growl of a beast of prey, and thus told his fellow-monkeys the nature of
the expected danger? This would have been a first step in the formation of a language.

[55] 'On the Origin of Language,' by H. Wedgwood, 1866. 'Chapters on Language,' by
the Rev. F.W. Farrar, 1865. These works are most interesting. See also 'De la Phys.
et de Parole,' par Albert Lemoine, 1865, p. 190. The work on this subject, by the late
Prof. Aug. Schleicher, has been translated by Dr. Bikkers into English, under the title
of 'Darwinism tested by the Science of Language,' 1869.

[56] Vogt, 'Mémoire sur les Microcèphales,' 1867, p. 169. With respect to savages, I have
given some facts in my 'Journal of Researches,' etc., 1845, p. 206.

[57] See clear evidence on this head in the two works so often quoted, by Brehm and
Rengger.

[58] Houzeau gives a very curious account of his observations on this subject in his
'Facultés Mentales des Animaux,' tom. ii. p. 348.

As the voice was used more and more, the vocal organs would have been
strengthened and perfected through the principle of the inherited effects of use; and this
would have reacted on the power of speech. But the relation between the continued use of
language and the development of the brain, has no doubt been far more important. The
mental powers in some early progenitor of man must have been more highly developed
than in any existing ape, before even the most imperfect form of speech could have come
into use; but we may confidently believe that the continued use and advancement of this
power would have reacted on the mind itself, by enabling and encouraging it to carry on
long trains of thought. A complex train of thought can no more be carried on without the
aid of words, whether spoken or silent, than a long calculation without the use of figures
or algebra. It appears, also, that even an ordinary train of thought almost requires, or is
greatly facilitated by some form of language, for the dumb, deaf, and blind girl, Laura
Bridgman, was observed to use her fingers whilst dreaming. [59] Nevertheless, a long
succession of vivid and connected ideas may pass through the mind without the aid of
any form of language, as we may infer from the movements of dogs during their dreams.
We have, also, seen that animals are able to reason to a certain extent, manifestly without
the aid of language. The intimate connection between the brain, as it is now developed in
us, and the faculty of speech, is well shown by those curious cases of brain-disease in
which speech is specially affected, as when the power to remember substantives is lost,
whilst other words can be correctly used, or where substantives of a certain class, or all
except the initial letters of substantives and proper names are forgotten. [60] There is no
more improbability in the continued use of the mental and vocal organs leading to
inherited changes in their structure and functions, than in the case of hand-writing, which
depends partly on the form of the hand and partly on the disposition of the mind; and
handwriting is certainly inherited. [61]

[59] See remarks on this head by Dr. Maudsley, 'The Physiology and Pathology of Mind,'
2nd ed., 1868, p. 199.

[60] Many curious cases have been recorded. See, for instance, Dr. Bateman 'On
Aphasia,' 1870, pp. 27, 31, 53, 100, etc. Also, 'Inquiries Concerning the Intellectual
Powers,' by Dr. Abercrombie, 1838, p. 150.

in of Animals and Plants under Domestication,' vol. ii. p. 6.'

ers, more especially Prof. Max Müller [62], have lately insisted that the implies the power of forming general concepts; and that as no animals possess this power, an impassable barrier is formed between them and man. [63] With respect to animals, I have already endeavoured to show that they have this power, at least in a rude and incipient degree. As far as concerns infants of from ten to eleven months old, and deaf-mutes, it seems to me incredible, that they should be able to connect certain sounds with certain general ideas as quickly as they do, unless such ideas were already formed in their minds. The same remark may be extended to the more intelligent animals; as Mr. Leslie Stephen observes [64], "A dog frames a general concept of cats or sheep, and knows the corresponding words as well as a philosopher. And the capacity to understand is as good a proof of vocal intelligence, though in an inferior degree, as the capacity to speak."

[62] Lectures on 'Mr. Darwin's Philosophy of Language,' 1873.
[63] The judgment of a distinguished philologist, such as Prof. Whitney, will have far more weight on this point than anything that I can say. He remarks ('Oriental and Linguistic Studies,' 1873, p. 297), in speaking of Bleek's views: "Because on the grand scale language is the necessary auxiliary of thought, indispensable to the development of the power of thinking, to the distinctness and variety and complexity of cognitions to the full mastery of consciousness; therefore he would fain make thought absolutely impossible without speech, identifying the faculty with its instrument. He might just as reasonably assert that the human hand cannot act without a tool. With such a doctrine to start from, he cannot stop short of Max Müller's worst paradoxes, that an infant (in fans, not speaking) is not a human being, and that deaf-mutes do not become possessed of reason until they learn to twist their fingers into imitation of spoken words." Max Müller gives in italics ('Lectures on Mr. Darwin's Philosophy of Language,' 1873, third lecture) this aphorism: "There is no thought without words, as little as there are words without thought." What a strange definition must here be given to the word thought!
[64] 'Essays on Free Thinking,' etc., 1873, p. 82.

Why the organs now used for speech should have been originally perfected for this purpose, rather than any other organs, it is not difficult to see. Ants have considerable powers of intercommunication by means of their antennæ, as shown by Huber, who devotes a whole chapter to their language. We might have used our fingers as efficient instruments, for a person with practice can report to a deaf man every word of a speech rapidly delivered at a public meeting; but the loss of our hands, whilst thus employed, would have been a serious inconvenience. As all the higher mammals possess vocal organs, constructed on the same general plan as ours, and used as a means of communication, it was obviously probable that these same organs would be still further developed if the power of communication had to be improved; and this has been effected by the aid of adjoining and well adapted parts, namely the tongue and lips. [65] The fact of the higher apes not using their vocal organs for speech, no doubt depends on their intelligence not having been sufficiently advanced. The possession by them of organs, which with long-continued practice might have been used for speech, although not thus used, is paralleled by the case of many birds which possess organs fitted for singing, though they never sing. Thus, the nightingale and crow have vocal organs similarly constructed, these being used by the former for diversified song, and by the latter only for croaking. [66] If it be asked why apes have not had their intellects developed to the same degree as that of man, general causes only can be assigned in answer, and it is unreasonable to expect any thing more definite, considering our ignorance with respect to the successive stages of development through which each creature has passed.

[65] See some good remarks to this effect by Dr. Maudsley, 'The Physiology and Pathology of Mind,' 1868, p. 199.
[66] MacGillivray, 'Hist. of British Birds,' vol. ii. 1839, p. 29. An excellent observer, Mr. Blackwall, remarks that the magpie learns to pronounce single words, and even short sentences, more readily than almost any other British bird; yet, as he adds, after long and closely investigating its habits, he has never known it, in a state of nature, display any unusual capacity for imitation. 'Researches in Zoology,' 1834, p. 158.

The formation of different languages and of distinct species, and the proofs that both have been developed through a gradual process, are curiously parallel. [67] But we can trace the formation of many words further back than that of species, for we can perceive how they actually arose from the imitation of various sounds. We find in distinct languages striking homologies due to community of descent, and analogies due to a similar process of formation. The manner in which certain letters or sounds change when others change is very like correlated growth. We have in both cases the reduplication of parts, the effects of long-continued use, and so forth. The frequent presence of rudiments, both in languages and in species, is still more remarkable. The letter *m* in the word *am*, means *I*; so that in the expression *I am*, a superfluous and useless rudiment has been retained. In the spelling also of words, letters often remain as the rudiments of ancient forms of pronunciation. Languages, like organic beings, can be classed in groups under groups; and they can be classed either naturally according to descent, or artificially by other characters. Dominant languages and dialects spread widely, and lead to the gradual extinction of other tongues. A language, like a species, when once extinct, never, as Sir C. Lyell remarks, reappears. The same language never has two birth-places. Distinct languages may be crossed or blended together. [68] We see variability in every tongue, and new words are continually cropping up; but as there is a limit to the powers of the memory, single words, like whole languages, gradually become extinct. As Max Müller [69] has well remarked:—"A struggle for life is constantly going on amongst the words and grammatical forms in each language. The better, the shorter, the easier forms are constantly gaining the upper hand, and they owe their success to their own inherent virtue." To these more important causes of the survival of certain words, mere novelty and fashion may be added; for there is in the mind of man a strong love for slight changes in all things. The survival or preservation of certain favoured words in the struggle for existence is natural selection.

[67] See the very interesting parallelism between the development of species and languages, given by Sir C. Lyell in 'The Geological Evidences of the Antiquity of Man,' 1863, chap. xxiii.
[68] See remarks to this effect by the Rev. F.W. Farrar, in an interesting article, entitled 'Philology and Darwinism,' in 'Nature,' March 24th, 1870, p. 528.
[69] 'Nature,' January 6th, 1870, p. 257.

The perfectly regular and wonderfully complex construction of the languages of many barbarous nations has often been advanced as a proof, either of the divine origin of these languages, or of the high art and former civilization of their founders. Thus F. von Schlegel writes: "In those languages which appear to be at the lowest grade of intellectual culture, we frequently observe a very high and elaborate degree of art in their grammatical structure. This is especially the case with the Basque and the Lapponian, and many of the American languages." [70] But it is assuredly an error to speak of any language as an art, in the sense of its having been elaborately and methodically formed. Philologists now admit that conjugations, declensions, etc., originally existed as distinct words, since joined together; and as such words express the most obvious relations between objects and persons, it is not surprising that they should have been used by the men of most races during the earliest ages. With respect to perfection, the following illustration will best show how easily we may err: a Crinoid sometimes consists of no less than 150,000 pieces of shell [71], all arranged with perfect symmetry in radiating lines; but a naturalist does not consider an animal of this kind as more perfect than a bilateral one with comparatively few parts, and with none of these parts alike, excepting on the opposite sides of the body. He justly considers the differentiation and specialization of organs as the test of perfection. So with languages: the most symmetrical and complex ought not to be ranked above irregular, abbreviated, and bastardized languages, which have borrowed expressive words and useful forms of construction from various conquering, conquered, or immigrant races.

[70] Quoted by C.S. Wake, 'Chapters on Man,' 1868, p. 101.
[71] Buckland, 'Bridgewater Treatise,' p. 411.

From these few and imperfect remarks I conclude that the extremely complex and regular construction of many barbarous languages, is no proof that they owe their origin to a special act of creation. [72] Nor, as we have seen, does the faculty of articulate speech in itself offer any insuperable objection to the belief that man has been developed from some lower form.

[72] See some good remarks on the simplification of languages, by Sir J. Lubbock, 'Origin of Civilization,' 1870, p. 278.

SENSE OF BEAUTY.

This sense has been declared to be peculiar to man. I refer here only to the pleasure given by certain colors, forms, and sounds, and which may fairly be called a sense of the beautiful; with cultivated men such sensations are, however, intimately associated with complex ideas and trains of thought. When we behold a male bird elaborately displaying his graceful plumes or splendid colors before the female, whilst other birds, not thus decorated, make no such display, it is impossible to doubt that she admires the beauty of her male partner. As women everywhere deck themselves with these plumes, the beauty of such ornaments cannot be disputed. As we shall see later, the nests of humming-birds, and the playing passages of bower-birds are tastefully ornamented with gaily-colored objects; and this shows that they must receive some kind of pleasure from the sight of such things. With the great majority of animals, however, the taste for the beautiful is confined, as far as we can judge, to the attractions of the opposite sex. The sweet strains poured forth by many male birds during the season of love, are certainly admired by the females, of which fact evidence will hereafter be given. If female birds had been incapable of appreciating the beautiful colors, the ornaments, and voices of their male partners, all the labor and anxiety exhibited by the latter in displaying their charms before the females would have been thrown away; and this it is impossible to admit. Why certain bright colors should excite pleasure cannot, I presume, be explained, any more than why certain flavors and scents are agreeable; but habit has something to do with the result, for that which is at first unpleasant to our senses, ultimately becomes pleasant, and habits are inherited. With respect to sounds, Helmholtz has explained to a certain extent on physiological principles, why harmonies and certain cadences are agreeable. But besides this, sounds frequently recurring at irregular intervals are highly disagreeable, as every one will admit who has listened at night to the irregular flapping of a rope on board ship. The same principle seems to come into play with vision, as the eye prefers symmetry or figures with some regular recurrence. Patterns of this kind are employed by even the lowest savages as ornaments; and they have been developed through sexual selection for the adornment of some male animals. Whether we can or not give any reason for the pleasure thus derived from vision and hearing, yet man and many of the lower animals are alike pleased by the same colors, graceful shading and forms, and the same sounds.

The taste for the beautiful, at least as far as female beauty is concerned, is not of a special nature in the human mind; for it differs widely in the different races of man, and is not quite the same even in the different nations of the same race. Judging from the hideous ornaments, and the equally hideous music admired by most savages, it might be urged that their aesthetic faculty was not so highly developed as in certain animals, for instance, as in birds. Obviously no animal would be capable of admiring such scenes as the heavens at night, a beautiful landscape, or refined music; but such high tastes are acquired through culture, and depend on complex associations; they are not enjoyed by barbarians or by uneducated persons.

Many of the faculties, which have been of inestimable service to man for his progressive advancement, such as the powers of the imagination, wonder, curiosity, an undefined sense of beauty, a tendency to imitation, and the love of excitement or novelty, could hardly fail to lead to capricious changes of customs and fashions. I have alluded to this point, because a recent writer [73] has oddly fixed on Caprice "as one of the most remarkable and typical differences between savages and brutes." But not only can we partially understand how it is that man is from various conflicting influences rendered capricious, but that the lower animals are, as we shall hereafter see, likewise capricious in their affections, aversions, and sense of beauty. There is also reason to suspect that they love novelty, for its own sake.

[73] 'The Spectator,' Dec. 4th, 1869, p. 1430.

BELIEF IN GOD—RELIGION.

There is no evidence that man was aboriginally endowed with the ennobling belief in the existence of an Omnipotent God. On the contrary there is ample evidence, derived not from hasty travellers, but from men who have long resided with savages, that numerous races have existed, and still exist, who have no idea of one or more gods, and who have no words in their languages to express such an idea. [74] The question is of course wholly distinct from that higher one, whether there exists a Creator and Ruler of the universe; and this has been answered in the affirmative by some of the highest intellects that have ever existed.

[74] See an excellent article on this subject by the Rev. F.W. Farrar, in the 'Anthropological Review,' Aug. 1864, p. ccxvii. For further facts see Sir J. Lubbock, 'Prehistoric Times,' 2nd edit., 1869, p. 564; and especially the chapters on Religion in his 'Origin of Civilization,' 1870.

If, however, we include under the term "religion" the belief in unseen or spiritual agencies, the case is wholly different; for this belief seems to be universal with the less civilized races. Nor is it difficult to comprehend how it arose. As soon as the important faculties of the imagination, wonder, and curiosity, together with some power of reasoning, had become partially developed, man would naturally crave to understand what was passing around him, and would have vaguely speculated on his own existence. As Mr. M'Lennan [75] has remarked, "Some explanation of the phenomena of life, a man must feign for himself, and to judge from the universality of it, the simplest hypothesis, and the first to occur to men, seems to have been that natural phenomena are ascribable to the presence in animals, plants, and things, and in the forces of nature, of such spirits prompting to action as men are conscious they themselves possess." It is also probable, as Mr. Tylor has shown, that dreams may have first given rise to the notion of spirits; for savages do not readily distinguish between subjective and objective impressions. When a savage dreams, the figures which appear before him are believed to have come from a distance, and to stand over him; or "the soul of the dreamer goes out on its travels, and comes home with a remembrance of what it has seen." [76] But until the faculties of imagination, curiosity, reason, etc., had been fairly well developed in the mind of man, his dreams would not have led him to believe in spirits, any more than in the case of a dog.

[75] 'The Worship of Animals and Plants,' in the 'Fortnightly Review,' Oct. 1, 1869, p. 422.
[76] Tylor, 'Early History of Mankind,' 1865, p. 6. See also the three striking chapters on the 'Development of Religion,' in Lubbock's 'Origin of Civilization,' 1870. In a like manner Mr. Herbert Spencer, in his ingenious essay in the 'Fortnightly Review' (May 1st, 1870, p. 535), accounts for the earliest forms of religious belief throughout the world, by man being led through dreams, shadows, and other causes, to look at himself as a double essence, corporeal and spiritual. As the spiritual being is supposed to exist after death and to be powerful, it is propitiated by various gifts and ceremonies, and its aid invoked. He then further shows that names or nicknames given from some animal or other object, to the early progenitors or founders of a tribe, are supposed after a long interval to represent the real progenitor of the tribe; and such animal or object is then naturally believed still to exist as a spirit, is held sacred, and worshipped as a god. Nevertheless I cannot but suspect that there is a still earlier and ruder stage, when anything which manifests power or movement is thought to be endowed with some form of life, and with mental faculties analogous to our own.

The tendency in savages to imagine that natural objects and agencies are animated by spiritual or living essences, is perhaps illustrated by a little fact which I once noticed: my dog, a full-grown and very sensible animal, was lying on the lawn during a hot and still day; but at a little distance a slight breeze occasionally moved an open parasol, which would have been wholly disregarded by the dog, had any one stood near it. As it was, every time that the parasol slightly moved, the dog growled fiercely and barked. He must, I think, have reasoned to himself in a rapid and unconscious manner, that movement without any apparent cause indicated the presence of some strange living agent, and that no stranger had a right to be on his territory.

The belief in spiritual agencies would easily pass into the belief in the existence of one or more gods. For savages would naturally attribute to spirits the same passions, the same love of vengeance or simplest form of justice, and the same affections which they themselves feel. The Fuegians appear to be in this respect in an intermediate condition, for when the surgeon on board the "Beagle" shot some young ducklings as specimens, York Minster declared in the most solemn manner, "Oh, Mr. Bynoe, much rain, much snow, blow much"; and this was evidently a retributive punishment for wasting human food. So again he related how, when his brother killed a "wild man," storms long raged, much rain and snow fell. Yet we could never discover that the Fuegians believed in what we should call a God, or practised any religious rites; and Jemmy Button, with justifiable pride, stoutly maintained that there was no devil in his land. This latter assertion is the more remarkable, as with savages the belief in bad spirits is far more common than that in good ones.

The feeling of religious devotion is a highly complex one, consisting of love, complete submission to an exalted and mysterious superior, a strong sense of dependence [77], fear, reverence, gratitude, hope for the future, and perhaps other elements. No being could experience so complex an emotion until advanced in his intellectual and moral faculties to at least a moderately high level. Nevertheless, we see some distant approach to this state of mind in the deep love of a dog for his master, associated with complete submission, some fear, and perhaps other feelings. The behaviour of a dog when returning to his master after an absence, and, as I may add, of a monkey to his beloved keeper, is widely different from that towards their fellows. In the latter case the transports of joy appear to be somewhat less, and the sense of equality is shown in every action. Professor Braubach goes so far as to maintain that a dog looks on his master as on a god. [78]

[77] See an able article on the 'Physical Elements of Religion,' by Mr. L. Owen Pike, in 'Anthropological Review,' April 1870, p. lxiii.
[78] 'Religion, Moral, etc., der Darwin'schen Art-Lehre,' 1869, s. 53. It is said (Dr. W. Lauder Lindsay, 'Journal of Mental Science,' 1871, p. 43), that Bacon long ago, and the poet Burns, held the same notion.

The same high mental faculties which first led man to believe in unseen spiritual agencies, then in fetishism, polytheism, and ultimately in monotheism, would infallibly lead him, as long as his reasoning powers remained poorly developed, to various strange superstitions and customs. Many of these are terrible to think of—such as the sacrifice of human beings to a blood-loving god; the trial of innocent persons by the ordeal of poison or fire; witchcraft, etc.—yet it is well occasionally to reflect on these superstitions, for they show us what an infinite debt of gratitude we owe to the improvement of our reason, to science, and to our accumulated knowledge. As Sir J. Lubbock [79] has well observed, "it is not too much to say that the horrible dread of unknown evil hangs like a thick cloud over savage life, and embitters every pleasure." These miserable and indirect consequences of our highest faculties may be compared with the incidental and occasional mistakes of the instincts of the lower animals.

[79] 'Prehistoric Times,' 2nd edit., p. 571. In this work (p. 571) there will be found an excellent account of the many strange and capricious customs of savages.

CHAPTER IV.

COMPARISON OF THE MENTAL POWERS OF MAN AND THE LOWER ANIMALS—continued.

The moral sense—Fundamental proposition—The qualities of social animals—Origin of sociability—Struggle between opposed instincts—Man a social animal—The more enduring social instincts conquer other less persistent instincts—The social virtues alone regarded by savages—The self-regarding virtues acquired at a later stage of development—The importance of the judgment of the members of the same community on conduct—Transmission of moral tendencies—Summary.

I fully subscribe to the judgment of those writers [1] who maintain that of all the differences between man and the lower animals, the moral sense or conscience is by far the most important. This sense, as Mackintosh [2] remarks, "has a rightful supremacy over every other principle of human action"; it is summed up in that short but imperious word *ought*, so full of high significance. It is the most noble of all the attributes of man, leading him without a moment's hesitation to risk his life for that of a fellow-creature; or after due deliberation, impelled simply by the deep feeling of right or duty, to sacrifice it in some great cause. Immanuel Kant exclaims, "Duty! Wondrous thought, that workest neither by fond insinuation, flattery, nor by any threat, but merely by holding up thy naked law in the soul, and so extorting for thyself always reverence, if not always obedience; before whom all appetites are dumb, however secretly they rebel; whence thy original?" [3]

[1] See, for instance, on this subject, Quatrefages, 'Unité de l'Espèce Humaine,' 1861, p. 21, etc.
[2] 'Dissertation an Ethical Philosophy,' 1837, p. 231, etc.
[3] 'Metaphysics of Ethics,' translated by J.W. Semple, Edinburgh, 1836, p. 136.

This great question has been discussed by many writers [4] of consummate ability; and my sole excuse for touching on it, is the impossibility of here passing it over; and because, as far as I know, no one has approached it exclusively from the side of natural history. The investigation possesses, also, some independent interest, as an attempt to see how far the study of the lower animals throws light on one of the highest psychical faculties of man.

[4] Mr. Bain gives a list ('Mental and Moral Science,' 1868, pp. 543-725) of twenty-six British authors who have written on this subject, and whose names are familiar to every reader; to these, Mr. Bain's own name, and those of Mr. Lecky, Mr. Shadworth

Hodgson, Sir J. Lubbock, and others, might be added.

The following proposition seems to me in a high degree probable—namely, that any animal whatever, endowed with well-marked social instincts [5], the parental and filial affections being here included, would inevitably acquire a moral sense or conscience, as soon as its intellectual powers had become as well, or nearly as well developed, as in man. For, *firstly*, the social instincts lead an animal to take pleasure in the society of its fellows, to feel a certain amount of sympathy with them, and to perform various services for them. The services may be of a definite and evidently instinctive nature; or there may be only a wish and readiness, as with most of the higher social animals, to aid their fellows in certain general ways. But these feelings and services are by no means extended to all the individuals of the same species, only to those of the same association. *Secondly*, as soon as the mental faculties had become highly developed, images of all past actions and motives would be incessantly passing through the brain of each individual: and that feeling of dissatisfaction, or even misery, which invariably results, as we shall hereafter see, from any unsatisfied instinct, would arise, as often as it was perceived that the enduring and always present social instinct had yielded to some other instinct, at the time stronger, but neither enduring in its nature, nor leaving behind it a very vivid impression. It is clear that many instinctive desires, such as that of hunger, are in their nature of short duration; and after being satisfied, are not readily or vividly recalled. *Thirdly*, after the power of language had been acquired, and the wishes of the community could be expressed, the common opinion how each member ought to act for the public good, would naturally become in a paramount degree the guide to action. But it should be borne in mind that however great weight we may attribute to public opinion, our regard for the approbation and disapprobation of our fellows depends on sympathy, which, as we shall see, forms an essential part of the social instinct, and is indeed its foundation-stone. *Lastly*, habit in the individual would ultimately play a very important part in guiding the conduct of each member; for the social instinct, together with sympathy, is, like any other instinct, greatly strengthened by habit, and so consequently would be obedience to the wishes and judgment of the community. These several subordinate propositions must now be discussed, and some of them at considerable length.

[5] Sir B. Brodie, after observing that man is a social animal ('Psychological Enquiries,' 1854, p. 192), asks the pregnant question, "ought not this to settle the disputed question as to the existence of a moral sense?" Similar ideas have probably occurred to many persons, as they did long ago to Marcus Aurelius. Mr. J.S. Mill speaks, in his celebrated work, 'Utilitarianism,' (1864, pp. 45, 46), of the social feelings as a "powerful natural sentiment," and as "the natural basis of sentiment for utilitarian morality." Again he says, "Like the other acquired capacities above referred to, the moral faculty, if not a part of our nature, is a natural out-growth from it; capable, like them, in a certain small degree of springing up spontaneously." But in opposition to all this, he also remarks, "if, as in my own belief, the moral feelings are not innate, but acquired, they are not for that reason less natural." It is with hesitation that I venture to differ at all from so profound a thinker, but it can hardly be disputed that the social feelings are instinctive or innate in the lower animals; and why should they not be so in man? Mr. Bain (see, for instance, 'The Emotions and the Will,' 1865, p. 481) and others believe that the moral sense is acquired by each individual during his lifetime. On the general theory of evolution this is at least extremely improbable. The

ignoring of all transmitted mental qualities will, as it seems to me, be hereafter judged as a most serious blemish in the works of Mr. Mill.

It may be well first to premise that I do not wish to maintain that any strictly social animal, if its intellectual faculties were to become as active and as highly developed as in man, would acquire exactly the same moral sense as ours. In the same manner as various animals have some sense of beauty, though they admire widely-different objects, so they might have a sense of right and wrong, though led by it to follow widely different lines of conduct. If, for instance, to take an extreme case, men were reared under precisely the same conditions as hive-bees, there can hardly be a doubt that our unmarried females would, like the worker-bees, think it a sacred duty to kill their brothers, and mothers would strive to kill their fertile daughters; and no one would think of interfering. [6] Nevertheless, the bee, or any other social animal, would gain in our supposed case, as it appears to me, some feeling of right or wrong, or a conscience. For each individual would have an inward sense of possessing certain stronger or more enduring instincts, and others less strong or enduring; so that there would often be a struggle as to which impulse should be followed; and satisfaction, dissatisfaction, or even misery would be felt, as past impressions were compared during their incessant passage through the mind. In this case an inward monitor would tell the animal that it would have been better to have followed the one impulse rather than the other. The one course ought to have been followed, and the other ought not; the one would have been right and the other wrong; but to these terms I shall recur.

[6] Mr. H. Sidgwick remarks, in an able discussion on this subject (the 'Academy,' June 15, 1872, p. 231), "a superior bee, we may feel sure, would aspire to a milder solution of the population question." Judging, however, from the habits of many or most savages, man solves the problem by female infanticide, polyandry and promiscuous intercourse; therefore it may well be doubted whether it would be by a milder method. Miss Cobbe, in commenting ('Darwinism in Morals,' 'Theological Review,' April 1872, pp. 188-191) on the same illustration, says, the *principles* of social duty would be thus reversed; and by this, I presume, she means that the fulfilment of a social duty would tend to the injury of individuals; but she overlooks the fact, which she would doubtless admit, that the instincts of the bee have been acquired for the good of the community. She goes so far as to say that if the theory of ethics advocated in this chapter were ever generally accepted, "I cannot but believe that in the hour of their triumph would be sounded the knell of the virtue of mankind!" It is to be hoped that the belief in the permanence of virtue on this earth is not held by many persons on so weak a tenure.

SOCIABILITY.

Animals of many kinds are social; we find even distinct species living together; for example, some American monkeys; and united flocks of rooks, jackdaws, and starlings. Man shows the same feeling in his strong love for the dog, which the dog returns with interest. Every one must have noticed how miserable horses, dogs, sheep, etc., are when separated from their companions, and what strong mutual affection the two former kinds, at least, show on their reunion. It is curious to speculate on the feelings of a dog, who will rest peacefully for hours in a room with his master or any of the family, without the least notice being taken of him; but if left for a short time by himself, barks or howls dismally. We will confine our attention to the higher social animals; and pass over insects, although some of these are social, and aid one another in many important ways. The most common mutual service in the higher animals is to warn one another of danger by means of the united senses of all. Every sportsman knows, as Dr. Jaeger remarks [7], how difficult it is to approach animals in a herd or troop. Wild horses and cattle do not, I believe, make any danger-signal; but the attitude of any one of them who first discovers an enemy, warns the others. Rabbits stamp loudly on the ground with their hind-feet as a signal: sheep and chamois do the same with their forefeet, uttering likewise a whistle. Many birds, and some mammals, post sentinels, which in the case of seals are said [8] generally to be the females. The leader of a troop of monkeys acts as the sentinel, and utters cries expressive both of danger and of safety. [9] Social animals perform many little services for each other: horses nibble, and cows lick each other, on any spot which itches: monkeys search each other for external parasites; and Brehm states that after a troop of the *Cercopithecus griseo-viridis* has rushed through a thorny brake, each monkey stretches itself on a branch, and another monkey sitting by, "conscientiously" examines its fur, and extracts every thorn or burr.

[7] 'Die Darwin'sche Théorie,' s. 101.
[8] Mr. R. Brown in 'Proc. Zoolog. Soc.' 1868, p. 409.
[9] Brehm, 'Thierleben,' B. i. 1864, s. 52, 79. For the case of the monkeys extracting thorns from each other, see s. 54. With respect to the Hamadryas turning over stones, the fact is given (s. 76), on the evidence of Alvarez, whose observations Brehm thinks quite trustworthy. For the cases of the old male baboons attacking the dogs, see s. 79; and with respect to the eagle, s. 56.

Animals also render more important services to one another: thus wolves and some other beasts of prey hunt in packs, and aid one another in attacking their victims. Pelicans fish in concert. The Hamadryas baboons turn over stones to find insects, etc.; and when they come to a large one, as many as can stand round, turn it over together and share the booty. Social animals mutually defend each other. Bull bisons in N. America, when there is danger, drive the cows and calves into the middle of the herd, whilst they defend the outside. I shall also in a future chapter give an account of two young wild bulls at Chillingham attacking an old one in concert, and of two stallions together trying to drive away a third stallion from a troop of mares. In Abyssinia, Brehm encountered a great troop of baboons who were crossing a valley: some had already ascended the opposite mountain, and some were still in the valley; the latter were attacked by the dogs, but the old males immediately hurried down from the rocks, and with mouths widely opened, roared so fearfully, that the dogs quickly drew back. They were again encouraged to the attack; but by this time all the baboons had reascended the heights, excepting a young one, about six months old, who, loudly calling for aid, climbed on a block of rock, and was surrounded. Now one of the largest males, a true hero, came down again from the mountain, slowly went to the young one, coaxed him, and triumphantly led him away— the dogs being too much astonished to make an attack. I cannot resist giving another scene which was witnessed by this same naturalist; an eagle seized a young Cercopithecus, which, by clinging to a branch, was not at once carried off; it cried loudly for assistance, upon which the other members of the troop, with much uproar, rushed to the rescue, surrounded the eagle, and pulled out so many feathers, that he no longer thought of his prey, but only how to escape. This eagle, as Brehm remarks, assuredly would never again attack a single monkey of a troop. [10]

[10] Mr. Belt gives the case of a spider-monkey (Ateles) in Nicaragua, which was heard screaming for nearly two hours in the forest, and was found with an eagle perched close by it. The bird apparently feared to attack as long as it remained face to face; and Mr. Belt believes, from what he has seen of the habits of these monkeys, that they protect themselves from eagles by keeping two or three together. 'The Naturalist in Nicaragua,' 1874, p. 118.

It is certain that associated animals have a feeling of love for each other, which is not felt by non-social adult animals. How far in most cases they actually sympathize in the pains and pleasures of others, is more doubtful, especially with respect to pleasures. Mr. Buxton, however, who had excellent means of observation [11], states that his macaws, which lived free in Norfolk, took "an extravagant interest" in a pair with a nest; and whenever the female left it, she was surrounded by a troop "screaming horrible acclamations in her honour." It is often difficult to judge whether animals have any feeling for the sufferings of others of their kind. Who can say what cows feel, when they surround and stare intently on a dying or dead companion; apparently, however, as Houzeau remarks, they feel no pity. That animals sometimes are far from feeling any sympathy is too certain; for they will expel a wounded animal from the herd, or gore or worry it to death. This is almost the blackest fact in natural history, unless, indeed, the explanation which has been suggested is true, that their instinct or reason leads them to expel an injured companion, lest beasts of prey, including man, should be tempted to follow the troop. In this case their conduct is not much worse than that of the North American Indians, who leave their feeble comrades to perish on the plains; or the Fijians, who, when their parents get old, or fall ill, bury them alive. [12]

[11] 'Annals and Magazine of Natural History,' November 1868, p. 382.
[12] Sir J. Lubbock, 'Prehistoric Times,' 2nd ed., p. 446.

Many animals, however, certainly sympathize with each other's distress or danger. This is the case even with birds. Captain Stansbury [13] found on a salt lake in Utah an old and completely blind pelican, which was very fat, and must have been well fed for a long time by his companions. Mr. Blyth, as he informs me, saw Indian crows feeding two or three of their companions which were blind; and I have heard of an analogous case with the domestic cock. We may, if we choose, call these actions instinctive; but such cases are much too rare for the development of any special instinct. [14] I have myself seen a dog, who never passed a cat who lay sick in a basket, and was a great friend of his, without giving her a few licks with his tongue, the surest sign of kind feeling in a dog.

[13] As quoted by Mr. L.H. Morgan, 'The American Beaver,' 1868, p. 272. Capt. Stansbury also gives an interesting account of the manner in which a very young pelican, carried away by a strong stream, was guided and encouraged in its attempts to reach the shore by half a dozen old birds.
[14] As Mr. Bain states, "effective aid to a sufferer springs from sympathy proper:" 'Mental and Moral Science,' 1868, p. 245.

It must be called sympathy that leads a courageous dog to fly at any one who strikes his master, as he certainly will. I saw a person pretending to beat a lady, who had a very timid little dog on her lap, and the trial had never been made before; the little creature instantly jumped away, but after the pretended beating was over, it was really pathetic to see how perseveringly he tried to lick his mistress's face, and comfort her. Brehm [15] states that when a baboon in confinement was pursued to be punished, the others tried to protect him. It must have been sympathy in the cases above given which led the baboons and Cercopitheci to defend their young comrades from the dogs and the eagle. I will give only one other instance of sympathetic and heroic conduct, in the case of a little American monkey. Several years ago a keeper at the Zoological Gardens showed me some deep and scarcely healed wounds on the nape of his own neck, inflicted on him, whilst kneeling on the floor, by a fierce baboon. The little American monkey, who was a warm friend of this keeper, lived in the same large compartment, and was dreadfully afraid of the great baboon. Nevertheless, as soon as he saw his friend in peril, he rushed to the rescue, and by screams and bites so distracted the baboon that the man was able to escape, after, as the surgeon thought, running great risk of his life.

[15] 'Thierleben,' B. i. s. 85.

Besides love and sympathy, animals exhibit other qualities connected with the social instincts, which in us would be called moral; and I agree with Agassiz [16] that dogs possess something very like a conscience.

[16] 'De l'Espèce et de la Classe,' 1869, p. 97.

Dogs possess some power of self-command, and this does not appear to be wholly the result of fear. As Braubach [17] remarks, they will refrain from stealing food in the absence of their master. They have long been accepted as the very type of fidelity and obedience. But the elephant is likewise very faithful to his driver or keeper, and probably considers him as the leader of the herd. Dr. Hooker informs me that an elephant, which he was riding in India, became so deeply bogged that he remained stuck fast until the next day, when he was extricated by men with ropes. Under such circumstances elephants will seize with their trunks any object, dead or alive, to place under their knees, to prevent their sinking deeper in the mud; and the driver was dreadfully afraid lest the animal should have seized Dr. Hooker and crushed him to death. But the driver himself, as Dr. Hooker was assured, ran no risk. This forbearance under an emergency so dreadful for a heavy animal, is a wonderful proof of noble fidelity. [18]

[17] 'Die Darwin'sche Art-Lehre,' 1869, s. 54.
[18] See also Hooker's 'Himalayan Journals,' vol. ii. 1854, p. 333.

All animals living in a body, which defend themselves or attack their enemies in concert, must indeed be in some degree faithful to one another; and those that follow a leader must be in some degree obedient. When the baboons in Abyssinia [19] plunder a garden, they silently follow their leader; and if an imprudent young animal makes a noise, he receives a slap from the others to teach him silence and obedience. Mr. Galton, who has had excellent opportunities for observing the half-wild cattle in S. Africa, says [20], that they cannot endure even a momentary separation from the herd. They are essentially slavish, and accept the common determination, seeking no better lot than to be led by any one ox who has enough self-reliance to accept the position. The men who break in these animals for harness, watch assiduously for those who, by grazing apart, show a self-reliant disposition, and these they train as fore-oxen. Mr. Galton adds that such animals are rare and valuable; and if many were born they would soon be eliminated, as lions are always on the look-out for the individuals which wander from the herd.

[19] Brehm, 'Thierleben,' B. i. s. 76.
[20] See his extremely interesting paper on 'Gregariousness in Cattle, and in Man,' 'Macmillan's Magazine,' Feb. 1871, p. 353.

With respect to the impulse which leads certain animals to associate together, and to aid one another in many ways, we may infer that in most cases they are impelled by the same sense of satisfaction or pleasure which they experience in performing other instinctive actions; or by the same sense of dissatisfaction as when other instinctive actions are checked. We see this in innumerable instances, and it is illustrated in a striking manner by the acquired instincts of our domesticated animals; thus a young shepherd-dog delights in driving and running round a flock of sheep, but not in worrying them; a young fox-hound delights in hunting a fox, whilst some other kinds of dogs, as I have witnessed, utterly disregard foxes. What a strong feeling of inward satisfaction must impel a bird, so full of activity, to brood day after day over her eggs. Migratory birds are quite miserable if stopped from migrating; perhaps they enjoy starting on their long flight; but it is hard to believe that the poor pinioned goose, described by Audubon, which started on foot at the proper time for its journey of probably more than a thousand miles, could have felt any joy in doing so. Some instincts are determined solely by painful feelings, as by fear, which leads to self-preservation, and is in some cases directed towards special enemies. No one, I presume, can analyze the sensations of pleasure or pain. In many instances, however, it is probable that instincts are persistently followed from the mere force of inheritance, without the stimulus of either pleasure or pain. A young pointer, when it first scents game, apparently cannot help pointing. A squirrel in a cage who pats the nuts which it cannot eat, as if to bury them in the ground, can hardly be thought to act thus, either from pleasure or pain. Hence the common assumption that men must be impelled to every action by experiencing some pleasure or pain may be erroneous. Although a habit may be blindly and implicitly followed, independently of any pleasure or pain felt at the moment, yet if it be forcibly and abruptly checked, a vague sense of dissatisfaction is generally experienced.

It has often been assumed that animals were in the first place rendered social, and that they feel as a consequence uncomfortable when separated from each other, and comfortable whilst together; but it is a more probable view that these sensations were first developed, in order that those animals which would profit by living in society, should be induced to live together, in the same manner as the sense of hunger and the pleasure of eating were, no doubt, first acquired in order to induce animals to eat. The feeling of pleasure from society is probably an extension of the parental or filial affections, since the social instinct seems to be developed by the young remaining for a long time with their parents; and this extension may be attributed in part to habit, but chiefly to natural selection. With those animals which were benefited by living in close association, the individuals which took the greatest pleasure in society would best escape various dangers, whilst those that cared least for their comrades, and lived solitary, would perish in greater numbers. With respect to the origin of the parental and filial affections, which apparently lie at the base of the social instincts, we know not the steps by which they have been gained; but we may infer that it has been to a large extent through natural selection. So it has almost certainly been with the unusual and opposite feeling of hatred between the nearest relations, as with the worker-bees which kill their brother drones, and with the queen-bees which kill their daughter-queens; the desire to destroy their nearest relations having been in this case of service to the community. Parental affection, or some feeling which replaces it, has been developed in certain animals extremely low in the scale, for example, in star-fishes and spiders. It is also occasionally present in a few members alone in a whole group of animals, as in the genus Forficula, or earwigs.

The all-important emotion of sympathy is distinct from that of love. A mother may passionately love her sleeping and passive infant, but she can hardly at such times be said to feel sympathy for it. The love of a man for his dog is distinct from sympathy, and so is that of a dog for his master. Adam Smith formerly argued, as has Mr. Bain recently, that the basis of sympathy lies in our strong retentiveness of former states of pain or pleasure. Hence, "the sight of another person enduring hunger, cold, fatigue, revives in us some recollection of these states, which are painful even in idea." We are thus impelled to relieve the sufferings of another, in order that our own painful feelings may be at the same time relieved. In like manner we are led to participate in the pleasures of others. [21] But I cannot see how this view explains the fact that sympathy is excited, in an immeasurably stronger degree, by a beloved, than by an indifferent person. The mere sight of suffering, independently of love, would suffice to call up in us vivid recollections and associations. The explanation may lie in the fact that, with all animals, sympathy is directed solely towards the members of the same community, and therefore towards known, and more or less beloved members, but not to all the individuals of the same species. This fact is not more surprising than that the fears of many animals should be directed against special enemies. Species which are not social, such as lions and tigers, no doubt feel sympathy for the suffering of their own young, but not for that of any other animal. With mankind, selfishness, experience, and imitation, probably add, as Mr. Bain has shown, to the power of sympathy; for we are led by the hope of receiving good in return to perform acts of sympathetic kindness to others; and sympathy is much strengthened by habit. In however complex a manner this feeling may have originated, as it is one of high importance to all those animals which aid and defend one another, it will have been increased through natural selection; for those communities, which included the greatest number of the most sympathetic members, would flourish best, and rear the greatest number of offspring.

[21] See the first and striking chapter in Adam Smith's 'Theory of Moral Sentiments.' Also 'Mr. Bain's Mental and Moral Science,' 1868, pp. 244, and 275-282. Mr. Bain states, that, "sympathy is, indirectly, a source of pleasure to the sympathiser"; and he accounts for this through reciprocity. He remarks that "the person benefited, or others in his stead, may make up, by sympathy and good offices returned, for all the sacrifice." But if, as appears to be the case, sympathy is strictly an instinct, its exercise would give direct pleasure, in the same manner as the exercise, as before remarked, of almost every other instinct.

It is, however, impossible to decide in many cases whether certain social instincts have been acquired through natural selection, or are the indirect result of other instincts and faculties, such as sympathy, reason, experience, and a tendency to imitation; or again, whether they are simply the result of long-continued habit. So remarkable an instinct as the placing sentinels to warn the community of danger, can hardly have been the indirect result of any of these faculties; it must, therefore, have been directly acquired. On the other hand, the habit followed by the males of some social animals of defending the community, and of attacking their enemies or their prey in concert, may perhaps have originated from mutual sympathy; but courage, and in most cases strength, must have been previously acquired, probably through natural selection.

Of the various instincts and habits, some are much stronger than others; that is, some either give more pleasure in their performance, and more distress in their prevention, than others; or, which is probably quite as important, they are, through inheritance, more persistently followed, without exciting any special feeling of pleasure or pain. We are ourselves conscious that some habits are much more difficult to cure or change than others. Hence a struggle may often be observed in animals between different instincts, or between an instinct and some habitual disposition; as when a dog rushes after a hare, is rebuked, pauses, hesitates, pursues again, or returns ashamed to his master; or as between the love of a female dog for her young puppies and for her master,—for she may be seen to slink away to them, as if half ashamed of not accompanying her master. But the most curious instance known to me of one instinct getting the better of another, is the migratory instinct conquering the maternal instinct. The former is wonderfully strong; a confined bird will at the proper season beat her breast against the wires of her cage, until it is bare and bloody. It causes young salmon to leap out of the fresh water, in which they could continue to exist, and thus unintentionally to commit suicide. Every one knows how strong the maternal instinct is, leading even timid birds to face great danger, though with hesitation, and in opposition to the instinct of self-preservation. Nevertheless, the migratory instinct is so powerful, that late in the autumn swallows, house-martins, and swifts frequently desert their tender young, leaving them to perish miserably in their nests. [22]

[22] This fact, the Rev. L. Jenyns states (see his edition of 'White's Nat. Hist. of Selborne,' 1853, p. 204) was first recorded by the illustrious Jenner, in 'Phil. Transact.' 1824, and has since been confirmed by several observers, especially by Mr. Blackwall. This latter careful observer examined, late in the autumn, during two years, thirty-six nests; he found that twelve contained young dead birds, five contained eggs on the point of being hatched, and three, eggs not nearly hatched. Many birds, not yet old enough for a prolonged flight, are likewise deserted and left behind. See Blackwall, 'Researches in Zoology,' 1834, pp. 108, 118. For some additional evidence, although this is not wanted, see Leroy, 'Lettres Phil.' 1802, p. 217. For Swifts, Gould's 'Introduction to the Birds of Great Britain,' 1823, p. 5. Similar cases have been observed in Canada by Mr. Adams; 'Pop. Science Review,' July 1873, p. 283.

We can perceive that an instinctive impulse, if it be in any way more beneficial to a species than some other or opposed instinct, would be rendered the more potent of the two through natural selection; for the individuals which had it most strongly developed would survive in larger numbers. Whether this is the case with the migratory in comparison with the maternal instinct, may be doubted. The great persistence, or steady action of the former at certain seasons of the year during the whole day, may give it for a time paramount force.

MAN A SOCIAL ANIMAL.

Every one will admit that man is a social being. We see this in his dislike of solitude, and in his wish for society beyond that of his own family. Solitary confinement is one of the severest punishments which can be inflicted. Some authors suppose that man primevally lived in single families; but at the present day, though single families, or only two or three together, roam the solitudes of some savage lands, they always, as far as I can discover, hold friendly relations with other families inhabiting the same district. Such families occasionally meet in council, and unite for their common defence. It is no argument against savage man being a social animal, that the tribes inhabiting adjacent districts are almost always at war with each other; for the social instincts never extend to all the individuals of the same species. Judging from the analogy of the majority of the Quadrumana, it is probable that the early ape-like progenitors of man were likewise social; but this is not of much importance for us. Although man, as he now exists, has few special instincts, having lost any which his early progenitors may have possessed, this is no reason why he should not have retained from an extremely remote period some degree of instinctive love and sympathy for his fellows. We are indeed all conscious that we do possess such sympathetic feelings [23]; but our consciousness does not tell us whether they are instinctive, having originated long ago in the same manner as with the lower animals, or whether they have been acquired by each of us during our early years. As man is a social animal, it is almost certain that he would inherit a tendency to be faithful to his comrades, and obedient to the leader of his tribe; for these qualities are common to most social animals. He would consequently possess some capacity for self-command. He would from an inherited tendency be willing to defend, in concert with others, his fellow-men; and would be ready to aid them in any way, which did not too greatly interfere with his own welfare or his own strong desires.

[23] Hume remarks ('An Enquiry Concerning the Principles of Morals,' edit. of 1751, p. 132), "There seems a necessity for confessing that the happiness and misery of others are not spectacles altogether indifferent to us, but that the view of the former...communicates a secret joy; the appearance of the latter... throws a melancholy damp over the imagination."

The social animals which stand at the bottom of the scale are guided almost exclusively, and those which stand higher in the scale are largely guided, by special instincts in the aid which they give to the members of the same community; but they are likewise in part impelled by mutual love and sympathy, assisted apparently by some amount of reason. Although man, as just remarked, has no special instincts to tell him how to aid his fellow-men, he still has the impulse, and with his improved intellectual faculties would naturally be much guided in this respect by reason and experience. Instinctive sympathy would also cause him to value highly the approbation of his fellows; for, as Mr. Bain has clearly shown [24], the love of praise and the strong feeling of glory, and the still stronger horror of scorn and infamy, "are due to the workings of sympathy." Consequently man would be influenced in the highest degree by the wishes, approbation, and blame of his fellow-men, as expressed by their gestures and language. Thus the social instincts, which must have been acquired by man in a very rude state, and probably even by his early ape-like progenitors, still give the impulse to some of his best actions; but his actions are in a higher degree determined by the expressed wishes and judgment of his fellow-men, and unfortunately very often by his own strong selfish desires. But as love, sympathy and self-command become strengthened by habit, and as the power of reasoning becomes clearer, so that man can value justly the judgments of his fellows, he will feel himself impelled, apart from any transitory pleasure or pain, to certain lines of conduct. He might then declare—not that any barbarian or uncultivated man could thus think—I am the supreme judge of my own conduct, and in the words of Kant, I will not in my own person violate the dignity of humanity.

[24] 'Mental and Moral Science,' 1868, p. 254.

THE MORE ENDURING SOCIAL INSTINCTS CONQUER THE LESS PERSISTENT INSTINCTS.

We have not, however, as yet considered the main point, on which, from our present point of view, the whole question of the moral sense turns. Why should a man feel that he ought to obey one instinctive desire rather than another? Why is he bitterly regretful, if he has yielded to a strong sense of self-preservation, and has not risked his life to save that of a fellow-creature? or why does he regret having stolen food from hunger?

It is evident in the first place, that with mankind the instinctive impulses have different degrees of strength; a savage will risk his own life to save that of a member of the same community, but will be wholly indifferent about a stranger: a young and timid mother urged by the maternal instinct will, without a moment's hesitation, run the greatest danger for her own infant, but not for a mere fellow-creature. Nevertheless many a civilized man, or even boy, who never before risked his life for another, but full of courage and sympathy, has disregarded the instinct of self-preservation, and plunged at once into a torrent to save a drowning man, though a stranger. In this case man is impelled by the same instinctive motive, which made the heroic little American monkey, formerly described, save his keeper, by attacking the great and dreaded baboon. Such actions as the above appear to be the simple result of the greater strength of the social or maternal instincts rather than that of any other instinct or motive; for they are performed too instantaneously for reflection, or for pleasure or pain to be felt at the time; though, if prevented by any cause, distress or even misery might be felt. In a timid man, on the other hand, the instinct of self-preservation might be so strong, that he would be unable to force himself to run any such risk, perhaps not even for his own child.

I am aware that some persons maintain that actions performed impulsively, as in the above cases, do not come under the dominion of the moral sense, and cannot be called moral. They confine this term to actions done deliberately, after a victory over opposing desires, or when prompted by some exalted motive. But it appears scarcely possible to draw any clear line of distinction of this kind. [25] As far as exalted motives are concerned, many instances have been recorded of savages, destitute of any feeling of general benevolence towards mankind, and not guided by any religious motive, who have deliberately sacrificed their lives as prisoners [26], rather than betray their comrades; and surely their conduct ought to be considered as moral. As far as deliberation, and the victory over opposing motives are concerned, animals may be seen doubting between opposed instincts, in rescuing their offspring or comrades from danger; yet their actions, though done for the good of others, are not called moral. Moreover, anything performed very often by us, will at last be done without deliberation or hesitation, and can then hardly be distinguished from an instinct; yet surely no one will pretend that such an action ceases to be moral. On the contrary, we all feel that an act cannot be considered as perfect, or as performed in the most noble manner, unless it be done impulsively, without deliberation or effort, in the same manner as by a man in whom the requisite qualities are innate. He who is forced to overcome his fear or want of sympathy before he acts, deserves, however, in one way higher credit than the man whose innate disposition leads him to a good act without effort. As we cannot distinguish between motives, we rank all actions of a certain class as moral, if performed by a moral being. A moral being is one who is capable of comparing his past and future actions or motives, and of approving or disapproving of them. We have no reason to suppose that any of the lower animals have this capacity; therefore, when a Newfoundland dog drags a child out of the water, or a monkey faces danger to rescue its comrade, or takes charge of an orphan monkey, we do not call its conduct moral. But in the case of man, who alone can with certainty be ranked as a moral being, actions of a certain class are called moral, whether performed deliberately, after a struggle with opposing motives, or impulsively through instinct, or from the effects of slowly-gained habit.

[25] I refer here to the distinction between what has been called *material* and *formal* morality. I am glad to find that Professor Huxley ('Critiques and Addresses,' 1873, p. 287) takes the same view on this subject as I do. Mr. Leslie Stephen remarks ('Essays on Freethinking and Plain Speaking,' 1873, p. 83), "the metaphysical distinction, between material and formal morality is as irrelevant as other such distinctions."

[26] I have given one such case, namely of three Patagonian Indians who preferred being shot, one after the other, to betraying the plans of their companions in war ('Journal of Researches,' 1845, p. 103).

But to return to our more immediate subject. Although some instincts are more powerful than others, and thus lead to corresponding actions, yet it is untenable, that in man the social instincts (including the love of praise and fear of blame) possess greater strength, or have, through long habit, acquired greater strength than the instincts of self-preservation, hunger, lust, vengeance, etc. Why then does man regret, even though trying to banish such regret, that he has followed the one natural impulse rather than the other; and why does he further feel that he ought to regret his conduct? Man in this respect differs profoundly from the lower animals. Nevertheless we can, I think, see with some degree of clearness the reason of this difference.

Man, from the activity of his mental faculties, cannot avoid reflection: past impressions and images are incessantly and clearly passing through his mind. Now with those animals which live permanently in a body, the social instincts are ever present and persistent. Such animals are always ready to utter the danger-signal, to defend the community, and to give aid to their fellows in accordance with their habits; they feel at all times, without the stimulus of any special passion or desire, some degree of love and sympathy for them; they are unhappy if long separated from them, and always happy to be again in their company. So it is with ourselves. Even when we are quite alone, how often do we think with pleasure or pain of what others think of us,—of their imagined approbation or disapprobation; and this all follows from sympathy, a fundamental element of the social instincts. A man who possessed no trace of such instincts would be an unnatural monster. On the other hand, the desire to satisfy hunger, or any passion such as vengeance, is in its nature temporary, and can for a time be fully satisfied. Nor is it easy, perhaps hardly possible, to call up with complete vividness the feeling, for instance, of hunger; nor indeed, as has often been remarked, of any suffering. The instinct of self-preservation is not felt except in the presence of danger; and many a coward has thought himself brave until he has met his enemy face to face. The wish for another man's property is perhaps as persistent a desire as any that can be named; but even in this case the satisfaction of actual possession is generally a weaker feeling than the desire: many a thief, if not a habitual one, after success has wondered why he stole some article. [27]

[27] Enmity or hatred seems also to be a highly persistent feeling, perhaps more so than any other that can be named. Envy is defined as hatred of another for some excellence or success; and Bacon insists (Essay ix.), "Of all other affections envy is the most importune and continual." Dogs are very apt to hate both strange men and strange dogs, especially if they live near at hand, but do not belong to the same family, tribe, or clan; this feeling would thus seem to be innate, and is certainly a most persistent one. It seems to be the complement and converse of the true social instinct. From what we hear of savages, it would appear that something of the same kind holds good with them. If this be so, it would be a small step in any one to transfer such feelings to any member of the same tribe if he had done him an injury and had become his enemy. Nor is it probable that the primitive conscience would reproach a man for injuring his enemy; rather it would reproach him, if he had not revenged himself. To do good in return for evil, to love your enemy, is a height of morality to which it may be doubted whether the social instincts would, by themselves, have ever led us. It is necessary that these instincts, together with sympathy, should have been highly cultivated and extended by the aid of reason, instruction, and the love or fear of God, before any such golden rule would ever be thought of and obeyed.

A man cannot prevent past impressions often repassing through his mind; he will thus be driven to make a comparison between the impressions of past hunger, vengeance satisfied, or danger shunned at other men's cost, with the almost ever-present instinct of sympathy, and with his early knowledge of what others consider as praiseworthy or blamable. This knowledge cannot be banished from his mind, and from instinctive sympathy is esteemed of great moment. He will then feel as if he had been baulked in following a present instinct or habit, and this with all animals causes dissatisfaction, or even misery.

The above case of the swallow affords an illustration, though of a reversed nature, of a temporary though for the time strongly persistent instinct conquering another instinct, which is usually dominant over all others. At the proper season these birds seem all day long to be impressed with the desire to migrate; their habits change; they become restless, are noisy and congregate in flocks. Whilst the mother-bird is feeding, or brooding over her nestlings, the maternal instinct is probably stronger than the migratory; but the instinct which is the more persistent gains the victory, and at last, at a moment when her young ones are not in sight, she takes flight and deserts them. When arrived at the end of her long journey, and the migratory instinct has ceased to act, what an agony of remorse the bird would feel, if, from being endowed with great mental activity, she could not prevent the image constantly passing through her mind, of her young ones perishing in the bleak north from cold and hunger.

At the moment of action, man will no doubt be apt to follow the stronger impulse; and though this may occasionally prompt him to the noblest deeds, it will more commonly lead him to gratify his own desires at the expense of other men. But after their gratification when past and weaker impressions are judged by the ever-enduring social instinct, and by his deep regard for the good opinion of his fellows, retribution will surely come. He will then feel remorse, repentance, regret, or shame; this latter feeling, however, relates almost exclusively to the judgment of others. He will consequently resolve more or less firmly to act differently for the future; and this is conscience; for conscience looks backwards, and serves as a guide for the future.

The nature and strength of the feelings which we call regret, shame, repentance or remorse, depend apparently not only on the strength of the violated instinct, but partly on the strength of the temptation, and often still more on the judgment of our fellows. How far each man values the appreciation of others, depends on the strength of his innate or acquired feeling of sympathy; and on his own capacity for reasoning out the remote consequences of his acts. Another element is most important, although not necessary, the reverence or fear of the Gods, or Spirits believed in by each man: and this applies especially in cases of remorse. Several critics have objected that though some slight regret or repentance may be explained by the view advocated in this chapter, it is impossible thus to account for the soul-shaking feeling of remorse. But I can see little force in this objection. My critics do not define what they mean by remorse, and I can find no definition implying more than an overwhelming sense of repentance. Remorse seems to bear the same relation to repentance, as rage does to anger, or agony to pain. It is far from strange that an instinct so strong and so generally admired, as maternal love, should, if disobeyed, lead to the deepest misery, as soon as the impression of the past cause of disobedience is weakened. Even when an action is opposed to no special instinct, merely to know that our friends and equals despise us for it is enough to cause great misery. Who can doubt that the refusal to fight a duel through fear has caused many men an agony of shame? Many a Hindoo, it is said, has been stirred to the bottom of his soul by having partaken of unclean food. Here is another case of what must, I think, be called remorse. Dr. Landor acted as a magistrate in West Australia, and relates [28], that a native on his farm, after losing one of his wives from disease, came and said that, "he was going to a distant tribe to spear a woman, to satisfy his sense of duty to his wife. I told him that if he did so, I would send him to prison for life. He remained about the farm for some months, but got exceedingly thin, and complained that he could not rest or eat, that his wife's spirit was haunting him, because he had not taken a life for hers. I was inexorable, and assured him that nothing should save him if he did." Nevertheless the man disappeared for more than a year, and then returned in high condition; and his other wife told Dr. Landor that her husband had taken the life of a woman belonging to a distant tribe; but it was impossible to obtain legal evidence of the act. The breach of a rule held sacred by the tribe, will thus, as it seems, give rise to the deepest feelings,—and this quite apart from the social instincts, excepting in so far as the rule is grounded on the judgment of the community. How so many strange superstitions have arisen throughout the world we know not; nor can we tell how some real and great crimes, such as incest, have come to be held in an abhorrence (which is not however quite universal) by the lowest savages. It is even doubtful whether in some tribes incest would be looked on with greater horror, than would the marriage of a man with a woman bearing the same name, though not a relation. "To violate this law is a crime which the Australians hold in the greatest abhorrence, in this agreeing exactly with certain tribes of North America. When the question is put in either district, is it worse to kill a girl of a foreign tribe, or to marry a girl of one's own, an answer just opposite to ours would be given without hesitation." [29] We may, therefore, reject the belief, lately insisted on by some writers, that the abhorrence of incest is due to our possessing a special God-implanted conscience. On the whole it is intelligible, that a man urged by so powerful a sentiment as remorse, though arising as above explained, should be led to act in a manner, which he has been taught to believe serves as an expiation, such as delivering himself up to justice.

[28] 'Insanity in Relation to Law,' Ontario, United States, 1871, p. 1.
[29] E.B. Tylor, in 'Contemporary Review,' April 1873, p. 707.

Man prompted by his conscience, will through long habit acquire such perfect self-command, that his desires and passions will at last yield instantly and without a struggle to his social sympathies and instincts, including his feeling for the judgment of his fellows. The still hungry, or the still revengeful man will not think of stealing food, or of wreaking his vengeance. It is possible, or as we shall hereafter see, even probable, that the habit of self-command may, like other habits, be inherited. Thus at last man comes to feel, through acquired and perhaps inherited habit, that it is best for him to obey his more persistent impulses. The imperious word *ought* seems merely to imply the consciousness of the existence of a rule of conduct, however it may have originated. Formerly it must have been often vehemently urged that an insulted gentleman *ought* to fight a duel. We even say that a pointer *ought* to point, and a retriever to retrieve game. If they fail to do so, they fail in their duty and act wrongly.

If any desire or instinct leading to an action opposed to the good of others still appears, when recalled to mind, as strong as, or stronger than, the social instinct, a man will feel no keen regret at having followed it; but he will be conscious that if his conduct were known to his fellows, it would meet with their disapprobation; and few are so destitute of sympathy as not to feel discomfort when this is realised. If he has no such sympathy, and if his desires leading to bad actions are at the time strong, and when recalled are not over-mastered by the persistent social instincts, and the judgment of others, then he is essentially a bad man [30] gives many curious cases of the worst criminals, who apparently have been entirely destitute of conscience.); and the sole restraining motive left is the fear of punishment, and the conviction that in the long run it would be best for his own selfish interests to regard the good of others rather than his own.

[30] Dr. Prosper Despine, in his Psychologie Naturelle, 1868 (tom. i. p. 243; tom. ii. p. 169

It is obvious that every one may with an easy conscience gratify his own desires, if they do not interfere with his social instincts, that is with the good of others; but in order to be quite free from self-reproach, or at least of anxiety, it is almost necessary for him to avoid the disapprobation, whether reasonable or not, of his fellow-men. Nor must he break through the fixed habits of his life, especially if these are supported by reason; for if he does, he will assuredly feel dissatisfaction. He must likewise avoid the reprobation of the one God or gods in whom, according to his knowledge or superstition, he may believe; but in this case the additional fear of divine punishment often supervenes.

THE STRICTLY SOCIAL VIRTUES AT FIRST ALONE REGARDED.

The above view of the origin and nature of the moral sense, which tells us what we ought to do, and of the conscience which reproves us if we disobey it, accords well with what we see of the early and undeveloped condition of this faculty in mankind. The virtues which must be practised, at least generally, by rude men, so that they may associate in a body, are those which are still recognized as the most important. But they are practised almost exclusively in relation to the men of the same tribe; and their opposites are not regarded as crimes in relation to the men of other tribes. No tribe could hold together if murder, robbery, treachery, etc., were common; consequently such crimes within the limits of the same tribe "are branded with everlasting infamy" [31]; but excite no such sentiment beyond these limits. A North-American Indian is well pleased with himself, and is honoured by others, when he scalps a man of another tribe; and a Dyak cuts off the head of an unoffending person, and dries it as a trophy. The murder of infants has prevailed on the largest scale throughout the world [32], and has met with no reproach; but infanticide, especially of females, has been thought to be good for the tribe, or at least not injurious. Suicide during former times was not generally considered as a crime [33], but rather, from the courage displayed, as an honourable act; and it is still practised by some semi-civilized and savage nations without reproach, for it does not obviously concern others of the tribe. It has been recorded that an Indian Thug conscientiously regretted that he had not robbed and strangled as many travellers as did his father before him. In a rude state of civilization the robbery of strangers is, indeed, generally considered as honourable.

[31] See an able article in the 'North British Review,' 1867, p. 395. See also Mr. W. Bagehot's articles on the Importance of Obedience and Coherence to Primitive Man, in the 'Fortnightly Review,' 1867, p. 529, and 1868, p. 457, etc.

[32] The fullest account which I have met with is by Dr. Gerland, in his 'Ueber den Aussterben der Naturvölker,' 1868; but I shall have to recur to the subject of infanticide in a future chapter.

[33] See the very interesting discussion on suicide in Lecky's 'History of European Morals,' vol. i. 1869, p. 223. With respect to savages, Mr. Winwood Reade informs me that the negroes of West Africa often commit suicide. It is well known how common it was amongst the miserable aborigines of South America after the Spanish conquest. For New Zealand, see the voyage of the "Novara," and for the Aleutian Islands, Müller, as quoted by Houzeau, 'Les Facultés Mentales,' etc., tom. ii. p. 136.

Slavery, although in some ways beneficial during ancient times [34], is a great crime; yet it was not so regarded until quite recently, even by the most civilized nations. And this was especially the case, because the slaves belonged in general to a race different from that of their masters. As barbarians do not regard the opinion of their women, wives are commonly treated like slaves. Most savages are utterly indifferent to the sufferings of strangers, or even delight in witnessing them. It is well known that the women and children of the North-American Indians aided in torturing their enemies. Some savages take a horrid pleasure in cruelty to animals [35], and humanity is an unknown virtue. Nevertheless, besides the family affections, kindness is common, especially during sickness, between the members of the same tribe, and is sometimes extended beyond these limits. Mungo Park's touching account of the kindness of the negro women of the interior to him is well known. Many instances could be given of the noble fidelity of savages towards each other, but not to strangers; common experience justifies the maxim of the Spaniard, "Never, never trust an Indian." There cannot be fidelity without truth; and this fundamental virtue is not rare between the members of the same tribe: thus Mungo Park heard the negro women teaching their young children to love the truth. This, again, is one of the virtues which becomes so deeply rooted in the mind, that it is sometimes practised by savages, even at a high cost, towards strangers; but to lie to your enemy has rarely been thought a sin, as the history of modern diplomacy too plainly shows. As soon as a tribe has a recognized leader, disobedience becomes a crime, and even abject submission is looked at as a sacred virtue.

[34] See Mr. Bagehot, 'Physics and Politics,' 1872, p. 72.
[35] See, for instance, Mr. Hamilton's account of the Kaffirs, 'Anthropological Review,' 1870, p. xv.

As during rude times no man can be useful or faithful to his tribe without courage, this quality has universally been placed in the highest rank; and although in civilized countries a good yet timid man may be far more useful to the community than a brave one, we cannot help instinctively honoring the latter above a coward, however benevolent. Prudence, on the other hand, which does not concern the welfare of others, though a very useful virtue, has never been highly esteemed. As no man can practise the virtues necessary for the welfare of his tribe without self-sacrifice, self-command, and the power of endurance, these qualities have been at all times highly and most justly valued. The American savage voluntarily submits to the most horrid tortures without a groan, to prove and strengthen his fortitude and courage; and we cannot help admiring him, or even an Indian Fakir, who, from a foolish religious motive, swings suspended by a hook buried in his flesh.

The other so-called self-regarding virtues, which do not obviously, though they may really, affect the welfare of the tribe, have never been esteemed by savages, though now highly appreciated by civilized nations. The greatest intemperance is no reproach with savages. Utter licentiousness, and unnatural crimes, prevail to an astounding extent. [36] As soon, however, as marriage, whether polygamous, or monogamous, becomes common, jealousy will lead to the inculcation of female virtue; and this, being honoured, will tend to spread to the unmarried females. How slowly it spreads to the male sex, we see at the present day. Chastity eminently requires self-command; therefore it has been honoured from a very early period in the moral history of civilized man. As a consequence of this, the senseless practice of celibacy has been ranked from a remote period as a virtue. [37] The hatred of indecency, which appears to us so natural as to be thought innate, and which is so valuable an aid to chastity, is a modern virtue, appertaining exclusively, as Sir G. Staunton remarks [38], to civilized life. This is shown by the ancient religious rites of various nations, by the drawings on the walls of Pompeii, and by the practices of many savages.

[36] Mr. M'Lennan has given ('Primitive Marriage,' 1865, p. 176) a good collection of facts on this head.
[37] Lecky, 'History of European Morals,' vol. i. 1869, p. 109.
[38] 'Embassy to China,' vol. ii. p. 348.

We have now seen that actions are regarded by savages, and were probably so regarded by primeval man, as good or bad, solely as they obviously affect the welfare of the tribe,—not that of the species, nor that of an individual member of the tribe. This conclusion agrees well with the belief that the so-called moral sense is aboriginally derived from the social instincts, for both relate at first exclusively to the community.

The chief causes of the low morality of savages, as judged by our standard, are, firstly, the confinement of sympathy to the same tribe. Secondly, powers of reasoning insufficient to recognize the bearing of many virtues, especially of the self-regarding virtues, on the general welfare of the tribe. Savages, for instance, fail to trace the multiplied evils consequent on a want of temperance, chastity, etc. And, thirdly, weak power of self-command; for this power has not been strengthened through long-continued, perhaps inherited, habit, instruction and religion.

I have entered into the above details on the immorality of savages [39], because some authors have recently taken a high view of their moral nature, or have attributed most of their crimes to mistaken benevolence. [40] These authors appear to rest their conclusion on savages possessing those virtues which are serviceable, or even necessary, for the existence of the family and of the tribe,—qualities which they undoubtedly do possess, and often in a high degree.

[39] See on this subject copious evidence in Chap. vii. of Sir J. Lubbock, 'Origin of Civilization,' 1870.
[40] For instance Lecky, 'History of European Morals,' vol. i. p. 124.

CONCLUDING REMARKS.

It was assumed formerly by philosophers of the derivative [41] school of morals that the foundation of morality lay in a form of Selfishness; but more recently the "Greatest happiness principle" has been brought prominently forward. It is, however, more correct to speak of the latter principle as the standard, and not as the motive of conduct. Nevertheless, all the authors whose works I have consulted, with a few exceptions [42], write as if there must be a distinct motive for every action, and that this must be associated with some pleasure or displeasure. But man seems often to act impulsively, that is from instinct or long habit, without any consciousness of pleasure, in the same manner as does probably a bee or ant, when it blindly follows its instincts. Under circumstances of extreme peril, as during a fire, when a man endeavours to save a fellow-creature without a moment's hesitation, he can hardly feel pleasure; and still less has he time to reflect on the dissatisfaction which he might subsequently experience if he did not make the attempt. Should he afterwards reflect over his own conduct, he would feel that there lies within him an impulsive power widely different from a search after pleasure or happiness; and this seems to be the deeply planted social instinct.

[41] This term is used in an able article in the 'Westminster Review,' Oct. 1869, p. 498. For the "Greatest happiness principle," see J.S. Mill, 'Utilitarianism,' p. 17.
[42] Mill recognizes ('System of Logic,' vol. ii. p. 422) in the clearest manner, that actions may be performed through habit without the anticipation of pleasure. Mr. H. Sidgwick also, in his Essay on Pleasure and Desire ('The Contemporary Review,' April 1872, p. 671), remarks: "To sum up, in contravention of the doctrine that our

conscious active impulses are always directed towards the production of agreeable sensations in ourselves, I would maintain that we find everywhere in consciousness extra-regarding impulse, directed towards something that is not pleasure; that in many cases the impulse is so far incompatible with the self-regarding that the two do not easily co-exist in the same moment of consciousness." A dim feeling that our impulses do not by any means always arise from any contemporaneous or anticipated pleasure, has, I cannot but think, been one chief cause of the acceptance of the intuitive theory of morality, and of the rejection of the utilitarian or "Greatest happiness" theory. With respect to the latter theory the standard and the motive of conduct have no doubt often been confused, but they are really in some degree blended.

In the case of the lower animals it seems much more appropriate to speak of their social instincts, as having been developed for the general good rather than for the general happiness of the species. The term, general good, may be defined as the rearing of the greatest number of individuals in full vigor and health, with all their faculties perfect, under the conditions to which they are subjected. As the social instincts both of man and the lower animals have no doubt been developed by nearly the same steps, it would be advisable, if found practicable, to use the same definition in both cases, and to take as the standard of morality, the general good or welfare of the community, rather than the general happiness; but this definition would perhaps require some limitation on account of political ethics.

When a man risks his life to save that of a fellow-creature, it seems also more correct to say that he acts for the general good, rather than for the general happiness of mankind. No doubt the welfare and the happiness of the individual usually coincide; and a contented, happy tribe will flourish better than one that is discontented and unhappy. We have seen that even at an early period in the history of man, the expressed wishes of the community will have naturally influenced to a large extent the conduct of each member; and as all wish for happiness, the "greatest happiness principle" will have become a most important secondary guide and object; the social instinct, however, together with sympathy (which leads to our regarding the approbation and disapprobation of others), having served as the primary impulse and guide. Thus the reproach is removed of laying the foundation of the noblest part of our nature in the base principle of selfishness; unless, indeed, the satisfaction which every animal feels, when it follows its proper instincts, and the dissatisfaction felt when prevented, be called selfish.

The wishes and opinions of the members of the same community, expressed at first orally, but later by writing also, either form the sole guides of our conduct, or greatly reinforce the social instincts; such opinions, however, have sometimes a tendency directly opposed to these instincts. This latter fact is well exemplified by the *Law of Honour*, that is, the law of the opinion of our equals, and not of all our countrymen. The breach of this law, even when the breach is known to be strictly accordant with true morality, has caused many a man more agony than a real crime. We recognize the same influence in the burning sense of shame which most of us have felt, even after the interval of years, when calling to mind some accidental breach of a trifling, though fixed, rule of etiquette. The judgment of the community will generally be guided by some rude experience of what is best in the long run for all the members; but this judgment will not rarely err from ignorance and weak powers of reasoning. Hence the strangest customs and superstitions, in complete opposition to the true welfare and happiness of mankind, have become all-powerful throughout the world. We see this in the horror felt by a Hindoo who breaks his caste, and in many other such cases. It would be difficult to distinguish between the remorse felt by a Hindoo who has yielded to the temptation of eating unclean food, from that felt after committing a theft; but the former would probably be the more severe.

How so many absurd rules of conduct, as well as so many absurd religious beliefs, have originated, we do not know; nor how it is that they have become, in all quarters of the world, so deeply impressed on the mind of men; but it is worthy of remark that a belief constantly inculcated during the early years of life, whilst the brain is impressible, appears to acquire almost the nature of an instinct; and the very essence of an instinct is that it is followed independently of reason. Neither can we say why certain admirable virtues, such as the love of truth, are much more highly appreciated by some savage tribes than by others [43]; nor, again, why similar differences prevail even amongst highly civilized nations. Knowing how firmly fixed many strange customs and superstitions have become, we need feel no surprise that the self-regarding virtues, supported as they are by reason, should now appear to us so natural as to be thought innate, although they were not valued by man in his early condition.

[43] Good instances are given by Mr. Wallace in 'Scientific Opinion,' Sept. 15, 1869; and more fully in his 'Contributions to the Theory of Natural Selection,' 1870, p. 353.

Not withstanding many sources of doubt, man can generally and readily distinguish between the higher and lower moral rules. The higher are founded on the social instincts, and relate to the welfare of others. They are supported by the approbation of our fellow-men and by reason. The lower rules, though some of them when implying self-sacrifice hardly deserve to be called lower, relate chiefly to self, and arise from public opinion, matured by experience and cultivation; for they are not practised by rude tribes.

As man advances in civilization, and small tribes are united into larger communities, the simplest reason would tell each individual that he ought to extend his social instincts and sympathies to all the members of the same nation, though personally unknown to him. This point being once reached, there is only an artificial barrier to prevent his sympathies extending to the men of all nations and races. If, indeed, such men are separated from him by great differences in appearance or habits, experience unfortunately shows us how long it is, before we look at them as our fellow-creatures. Sympathy beyond the confines of man, that is, humanity to the lower animals, seems to be one of the latest moral acquisitions. It is apparently unfelt by savages, except towards their pets. How little the old Romans knew of it is shown by their abhorrent gladiatorial exhibitions. The very idea of humanity, as far as I could observe, was new to most of the Gauchos of the Pampas. This virtue, one of the noblest with which man is endowed, seems to arise incidentally from our sympathies becoming more tender and more widely diffused, until they are extended to all sentient beings. As soon as this virtue is honoured and practised by some few men, it spreads through instruction and example to the young, and eventually becomes incorporated in public opinion.

The highest possible stage in moral culture is when we recognize that we ought to control our thoughts, and "not even in inmost thought to think again the sins that made the past so pleasant to us." [44] Whatever makes any bad action familiar to the mind, renders its performance by so much the easier. As Marcus Aurelius long ago said, "Such as are thy habitual thoughts, such also will be the character of thy mind; for the soul is dyed by the thoughts." [45]

[44] Tennyson, Idylls of the King, p. 244.
[45] 'The Thoughts of the Emperor M. Aurelius Antoninus,' English translation, 2nd edit., 1869. p. 112. Marcus Aurelius ws born A.D. 121.

Our great philosopher, Herbert Spencer, has recently explained his views on the moral sense. He says [46], "I believe that the experiences of utility organized and consolidated through all past generations of the human race, have been producing corresponding modifications, which, by continued transmission and accumulation, have become in us certain faculties of moral intuition—certain emotions responding to right and wrong conduct, which have no apparent basis in the individual experiences of utility." There is not the least inherent improbability, as it seems to me, in virtuous tendencies being more or less strongly inherited; for, not to mention the various dispositions and habits transmitted by many of our domestic animals to their offspring, I have heard of authentic cases in which a desire to steal and a tendency to lie appeared to run in families of the upper ranks; and as stealing is a rare crime in the wealthy classes, we can hardly account by accidental coincidence for the tendency occurring in two or three members of the same family. If bad tendencies are transmitted, it is probable that good ones are likewise transmitted. That the state of the body by affecting the brain, has great influence on the moral tendencies is known to most of those who have suffered from chronic derangements of the digestion or liver. The same fact is likewise shown by the "perversion or destruction of the moral sense being often one of the earliest symptoms of mental derangement" [47]; and insanity is notoriously often inherited. Except through the principle of the transmission of moral tendencies, we cannot understand the differences believed to exist in this respect between the various races of mankind.

[46] Letter to Mr. Mill in Bain's 'Mental and Moral Science,' 1868, p. 722.
[47] Maudsley, 'Body and Mind,' 1870, p. 60.

Even the partial transmission of virtuous tendencies would be an immense assistance to the primary impulse derived directly and indirectly from the social instincts. Admitting for a moment that virtuous tendencies are inherited, it appears probable, at least in such cases as chastity, temperance, humanity to animals, etc., that they become first impressed on the mental organization through habit, instruction and example, continued during several generations in the same family, and in a quite subordinate degree, or not at all, by the individuals possessing such virtues having succeeded best in the struggle for life. My chief source of doubt with respect to any such inheritance, is that senseless customs, superstitions, and tastes, such as the horror of a Hindoo for unclean food, ought on the same principle to be transmitted. I have not met with any evidence in support of the transmission of superstitious customs or senseless habits, although in itself it is perhaps not less probable than that animals should acquire inherited tastes for certain kinds of food or fear of certain foes.

Finally the social instincts, which no doubt were acquired by man as by the lower animals for the good of the community, will from the first have given to him some wish to aid his fellows, some feeling of sympathy, and have compelled him to regard their approbation and disapprobation. Such impulses will have served him at a very early period as a rude rule of right and wrong. But as man gradually advanced in intellectual power, and was enabled to trace the more remote consequences of his actions; as he acquired sufficient knowledge to reject baneful customs and superstitions; as he regarded more and more, not only the welfare, but the happiness of his fellow-men; as from habit, following on beneficial experience, instruction and example, his sympathies became more tender and widely diffused, extending to men of all races, to the imbecile, maimed, and other useless members of society, and finally to the lower animals,—so would the standard of his morality rise higher and higher. And it is admitted by moralists of the derivative school and by some intuitionists, that the standard of morality has risen since an early period in the history of man. [48]

[48] A writer in the 'North British Review' (July 1869, p. 531), well capable of forming a sound judgment, expresses himself strongly in favour of this conclusion. Mr. Lecky ('History of Morals,' vol. i. p. 143) seems to a certain extent to coincide therein.

As a struggle may sometimes be seen going on between the various instincts of the lower animals, it is not surprising that there should be a struggle in man between his social instincts, with their derived virtues, and his lower, though momentarily stronger impulses or desires. This, as Mr. Galton [49] has remarked, is all the less surprising, as man has emerged from a state of barbarism within a comparatively recent period. After having yielded to some temptation we feel a sense of dissatisfaction, shame, repentance, or remorse, analogous to the feelings caused by other powerful instincts or desires, when left unsatisfied or baulked. We compare the weakened impression of a past temptation with the ever present social instincts, or with habits, gained in early youth and strengthened during our whole lives, until they have become almost as strong as instincts. If with the temptation still before us we do not yield, it is because either the social instinct or some custom is at the moment predominant, or because we have learnt that it will appear to us hereafter the stronger, when compared with the weakened impression of the temptation, and we realise that its violation would cause us suffering. Looking to future generations, there is no cause to fear that the social instincts will grow weaker, and we may expect that virtuous habits will grow stronger, becoming perhaps fixed by inheritance. In this case the struggle between our higher and lower impulses will be less severe, and virtue will be triumphant.

[49] See his remarkable work on 'Hereditary Genius,' 1869, p. 349. The Duke of Argyll ('Primeval Man,' 1869, p. 188) has some good remarks on the contest in man's nature between right and wrong.

SUMMARY OF THE LAST TWO CHAPTERS.

There can be no doubt that the difference between the mind of the lowest man and that of the highest animal is immense. An anthropomorphous ape, if he could take a dispassionate view of his own case, would admit that though he could form an artful plan to plunder a garden—though he could use stones for fighting or for breaking open nuts, yet that the thought of fashioning a stone into a tool was quite beyond his scope. Still less, as he would admit, could he follow out a train of metaphysical reasoning, or solve a mathematical problem, or reflect on God, or admire a grand natural scene. Some apes, however, would probably declare that they could and did admire the beauty of the colored skin and fur of their partners in marriage. They would admit, that though they could make other apes understand by cries some of their perceptions and simpler wants, the notion of expressing definite ideas by definite sounds had never crossed their minds. They might insist that they were ready to aid their fellow-apes of the same troop in many ways, to risk their lives for them, and to take charge of their orphans; but they would be forced to acknowledge that disinterested love for all living creatures, the most noble attribute of man, was quite beyond their comprehension.

Nevertheless the difference in mind between man and the higher animals, great as it is, certainly is one of degree and not of kind. We have seen that the senses and intuitions, the various emotions and faculties, such as love, memory, attention, curiosity, imitation, reason, etc., of which man boasts, may be found in an incipient, or even sometimes in a well-developed condition, in the lower animals. They are also capable of some inherited improvement, as we see in the domestic dog compared with the wolf or jackal. If it could be proved that certain high mental powers, such as the formation of general concepts, self-consciousness, etc., were absolutely peculiar to man, which seems extremely doubtful, it is not improbable that these qualities are merely the incidental results of other highly-advanced intellectual faculties; and these again mainly the result of the continued use of a perfect language. At what age does the new-born infant possess the power of abstraction, or become self-conscious, and reflect on its own existence? We cannot answer; nor can we answer in regard to the ascending organic scale. The half-art, half-instinct of language still bears the stamp of its gradual evolution. The ennobling belief in God is not universal with man; and the belief in spiritual agencies naturally follows from other mental powers. The moral sense perhaps affords the best and highest distinction between man and the lower animals; but I need say nothing on this head, as I have so lately endeavoured to show that the social instincts,—the prime principle of man's moral constitution [50]—with the aid of active intellectual powers and the effects of habit, naturally lead to the golden rule, "As ye would that men should do to you, do ye to them likewise;" and this lies at the foundation of morality.

[50] 'The Thoughts of Marcus Aurelius,' etc., p. 139.

In the next chapter I shall make some few remarks on the probable steps and means by which the several mental and moral faculties of man have been gradually evolved. That such evolution is at least possible, ought not to be denied, for we daily see these faculties developing in every infant; and we may trace a perfect gradation from the mind of an utter idiot, lower than that of an animal low in the scale, to the mind of a Newton.

CHAPTER V.

ON THE DEVELOPMENT OF THE INTELLECTUAL AND MORAL FACULTIES DURING PRIMEVAL AND CIVILIZED TIMES.

Advancement of the intellectual powers through natural selection—Importance of imitation—Social and moral faculties—Their development within the limits of the same tribe—Natural selection as affecting civilized nations—Evidence that civilized nations were once barbarous.

The subjects to be discussed in this chapter are of the highest interest, but are treated by me in an imperfect and fragmentary manner. Mr. Wallace, in an admirable paper before referred to [1], argues that man, after he had partially acquired those intellectual and moral faculties which distinguish him from the lower animals, would have been but little liable to bodily modifications through natural selection or any other means. For man is enabled through his mental faculties "to keep with an unchanged body in harmony with the changing universe." He has great power of adapting his habits to new conditions of life. He invents weapons, tools, and various stratagems to procure food and to defend himself. When he migrates into a colder climate he uses clothes, builds sheds, and makes fires; and by the aid of fire cooks food otherwise indigestible. He aids his fellow-men in many ways, and anticipates future events. Even at a remote period he practised some division of labor.

[1] Anthropological Review, May 1864, p. clviii.

The lower animals, on the other hand, must have their bodily structure modified in order to survive under greatly changed conditions. They must be rendered stronger, or acquire more effective teeth or claws, for defence against new enemies; or they must be reduced in size, so as to escape detection and danger. When they migrate into a colder climate, they must become clothed with thicker fur, or have their constitutions altered. If they fail to be thus modified, they will cease to exist.

The case, however, is widely different, as Mr. Wallace has with justice insisted, in relation to the intellectual and moral faculties of man. These faculties are variable; and we have every reason to believe that the variations tend to be inherited. Therefore, if they were formerly of high importance to primeval man and to his ape-like progenitors, they would have been perfected or advanced through natural selection. Of the high importance of the intellectual faculties there can be no doubt, for man mainly owes to them his predominant position in the world. We can see, that in the rudest state of society, the individuals who were the most sagacious, who invented and used the best weapons or traps, and who were best able to defend themselves, would rear the greatest number of offspring. The tribes, which included the largest number of men thus endowed, would increase in number and supplant other tribes. Numbers depend primarily on the means of subsistence, and this depends partly on the physical nature of the country, but in a much higher degree on the arts which are there practised. As a tribe increases and is victorious, it is often still further increased by the absorption of other tribes. [2] The stature and strength of the men of a tribe are likewise of some importance for its success, and these depend in part on the nature and amount of the food which can be obtained. In Europe the men of the Bronze period were supplanted by a race more powerful, and, judging from their sword-handles, with larger hands [3]; but their success was probably still more due to their superiority in the arts.

[2] After a time the members or tribes which are absorbed into another tribe assume, as Sir Henry Maine remarks ('Ancient Law,' 1861, p. 131), that they are the co-descendants of the same ancestors.
[3] Morlot, 'Soc. Vaud. Sc. Nat.' 1860, p. 294.

All that we know about savages, or may infer from their traditions and from old monuments, the history of which is quite forgotten by the present inhabitants, show that from the remotest times successful tribes have supplanted other tribes. Relics of extinct or forgotten tribes have been discovered throughout the civilized regions of the earth, on the wild plains of America, and on the isolated islands in the Pacific Ocean. At the present day civilized nations are everywhere supplanting barbarous nations, excepting where the climate opposes a deadly barrier; and they succeed mainly, though not exclusively, through their arts, which are the products of the intellect. It is, therefore, highly probable that with mankind the intellectual faculties have been mainly and gradually perfected through natural selection; and this conclusion is sufficient for our purpose. Undoubtedly it would be interesting to trace the development of each separate faculty from the state in which it exists in the lower animals to that in which it exists in man; but neither my ability nor knowledge permits the attempt.

It deserves notice that, as soon as the progenitors of man became social (and this probably occurred at a very early period), the principle of imitation, and reason, and experience would have increased, and much modified the intellectual powers in a way, of which we see only traces in the lower animals. Apes are much given to imitation, as are the lowest savages; and the simple fact previously referred to, that after a time no animal can be caught in the same place by the same sort of trap, shows that animals learn by experience, and imitate the caution of others. Now, if some one man in a tribe, more sagacious than the others, invented a new snare or weapon, or other means of attack or defence, the plainest self-interest, without the assistance of much reasoning power, would prompt the other members to imitate him; and all would thus profit. The habitual practice of each new art must likewise in some slight degree strengthen the intellect. If the new invention were an important one, the tribe would increase in number, spread, and supplant other tribes. In a tribe thus rendered more numerous there would always be a rather greater chance of the birth of other superior and inventive members. If such men left children to inherit their mental superiority, the chance of the birth of still more ingenious members would be somewhat better, and in a very small tribe decidedly better. Even if they left no children, the tribe would still include their blood-relations; and it has been ascertained by agriculturists [4] that by preserving and breeding from the family of an animal, which when slaughtered was found to be valuable, the desired character has been obtained.

[4] I have given instances in my Variation of Animals under Domestication, vol. ii. p. 196.

Turning now to the social and moral faculties. In order that primeval men, or the ape-like progenitors of man, should become social, they must have acquired the same instinctive feelings, which impel other animals to live in a body; and they no doubt exhibited the same general disposition. They would have felt uneasy when separated from their comrades, for whom they would have felt some degree of love; they would have warned each other of danger, and have given mutual aid in attack or defence. All this implies some degree of sympathy, fidelity, and courage. Such social qualities, the paramount importance of which to the lower animals is disputed by no one, were no doubt acquired by the progenitors of man in a similar manner, namely, through natural selection, aided by inherited habit. When two tribes of primeval man, living in the same country, came into competition, if (other circumstances being equal) the one tribe included a great number of courageous, sympathetic and faithful members, who were always ready to warn each other of danger, to aid and defend each other, this tribe would succeed better and conquer the other. Let it be borne in mind how all-important in the never-ceasing wars of savages, fidelity and courage must be. The advantage which disciplined soldiers have over undisciplined hordes follows chiefly from the confidence which each man feels in his comrades. Obedience, as Mr. Bagehot has well shown [5], is of the highest value, for any form of government is better than none. Selfish and contentious people will not cohere, and without coherence nothing can be effected. A tribe rich in the above qualities would spread and be victorious over other tribes: but in the course of time it would, judging from all past history, be in its turn overcome by some other tribe still more highly endowed. Thus the social and moral qualities would tend slowly to advance and be diffused throughout the world.

[5] See a remarkable series of articles on 'Physics and Politics,' in the 'Fortnightly Review,' Nov. 1867; April 1, 1868; July 1, 1869, since separately published.

But it may be asked, how within the limits of the same tribe did a large number of members first become endowed with these social and moral qualities, and how was the standard of excellence raised? It is extremely doubtful whether the offspring of the more sympathetic and benevolent parents, or of those who were the most faithful to their comrades, would be reared in greater numbers than the children of selfish and treacherous parents belonging to the same tribe. He who was ready to sacrifice his life, as many a savage has been, rather than betray his comrades, would often leave no offspring to inherit his noble nature. The bravest men, who were always willing to come to the front in war, and who freely risked their lives for others, would on an average perish in larger numbers than other men. Therefore, it hardly seems probable, that the number of men gifted with such virtues, or that the standard of their excellence, could be increased through natural selection, that is, by the survival of the fittest; for we are not here speaking of one tribe being victorious over another.

Although the circumstances, leading to an increase in the number of those thus endowed within the same tribe, are too complex to be clearly followed out, we can trace some of the probable steps. In the first place, as the reasoning powers and foresight of the members became improved, each man would soon learn that if he aided his fellow-men, he would commonly receive aid in return. From this low motive he might acquire the habit of aiding his fellows; and the habit of performing benevolent actions certainly strengthens the feeling of sympathy which gives the first impulse to benevolent actions. Habits, moreover, followed during many generations probably tend to be inherited.

But another and much more powerful stimulus to the development of the social virtues, is afforded by the praise and the blame of our fellow-men. To the instinct of sympathy, as we have already seen, it is primarily due, that we habitually bestow both praise and blame on others, whilst we love the former and dread the latter when applied to ourselves; and this instinct no doubt was originally acquired, like all the other social instincts, through natural selection. At how early a period the progenitors of man in the course of their development, became capable of feeling and being impelled by, the praise or blame of their fellow-creatures, we cannot of course say. But it appears that even dogs appreciate encouragement, praise, and blame. The rudest savages feel the sentiment of glory, as they clearly show by preserving the trophies of their prowess, by their habit of excessive boasting, and even by the extreme care which they take of their personal appearance and decorations; for unless they regarded the opinion of their comrades, such habits would be senseless.

They certainly feel shame at the breach of some of their lesser rules, and apparently remorse, as shown by the case of the Australian who grew thin and could not rest from having delayed to murder some other woman, so as to propitiate his dead wife's spirit. Though I have not met with any other recorded case, it is scarcely credible that a savage, who will sacrifice his life rather than betray his tribe, or one who will deliver himself up as a prisoner rather than break his parole [6], would not feel remorse in his inmost soul, if he had failed in a duty, which he held sacred.

[6] Mr. Wallace gives cases in his 'Contributions to the Theory of Natural Selection,' 1870, p. 354.

We may therefore conclude that primeval man, at a very remote period, was influenced by the praise and blame of his fellows. It is obvious, that the members of the same tribe would approve of conduct which appeared to them to be for the general good, and would reprobate that which appeared evil. To do good unto others—to do unto others as ye would they should do unto you—is the foundation-stone of morality. It is, therefore, hardly possible to exaggerate the importance during rude times of the love of praise and the dread of blame. A man who was not impelled by any deep, instinctive feeling, to sacrifice his life for the good of others, yet was roused to such actions by a sense of glory, would by his example excite the same wish for glory in other men, and would strengthen by exercise the noble feeling of admiration. He might thus do far more good to his tribe than by begetting offspring with a tendency to inherit his own high character.

With increased experience and reason, man perceives the more remote consequences of his actions, and the self-regarding virtues, such as temperance, chastity, etc., which during early times are, as we have before seen, utterly disregarded, come to be highly esteemed or even held sacred. I need not, however, repeat what I have said on this head in the fourth chapter. Ultimately our moral sense or conscience becomes a highly complex sentiment—originating in the social instincts, largely guided by the approbation of our fellow-men, ruled by reason, self-interest, and in later times by deep religious feelings, and confirmed by instruction and habit.

It must not be forgotten that although a high standard of morality gives but a slight or no advantage to each individual man and his children over the other men of the same tribe, yet that an increase in the number of well-endowed men and an advancement in the standard of morality will certainly give an immense advantage to one tribe over another. A tribe including many members who, from possessing in a high degree the spirit of patriotism, fidelity, obedience, courage, and sympathy, were always ready to aid one another, and to sacrifice themselves for the common good, would be victorious over most other tribes; and this would be natural selection. At all times throughout the world tribes have supplanted other tribes; and as morality is one important element in their success, the standard of morality and the number of well-endowed men will thus everywhere tend to rise and increase.

It is, however, very difficult to form any judgment why one particular tribe and not another has been successful and has risen in the scale of civilization. Many savages are in the same condition as when first discovered several centuries ago. As Mr. Bagehot has remarked, we are apt to look at progress as normal in human society; but history refutes this. The ancients did not even entertain the idea, nor do the Oriental nations at the present day. According to another high authority, Sir Henry Maine [7], "the greatest part of mankind has never shown a particle of desire that its civil institutions should be improved." Progress seems to depend on many concurrent favourable conditions, far too complex to be followed out. But it has often been remarked, that a cool climate, from leading to industry and to the various arts, has been highly favourable thereto. The Esquimaux, pressed by hard necessity, have succeeded in many ingenious inventions, but their climate has been too severe for continued progress. Nomadic habits, whether over wide plains, or through the dense forests of the tropics, or along the shores of the sea, have in every case been highly detrimental. Whilst observing the barbarous inhabitants of Tierra del Fuego, it struck me that the possession of some property, a fixed abode, and the union of many families under a chief, were the indispensable requisites for civilization. Such habits almost necessitate the cultivation of the ground; and the first steps in cultivation would probably result, as I have elsewhere shown [8], from some such accident as the seeds of a fruit-tree falling on a heap of refuse, and producing an unusually fine variety. The problem, however, of the first advance of savages towards civilization is at present much too difficult to be solved.

[7] 'Ancient Law,' 1861, p. 22. For Mr. Bagehot's remarks, 'Fortnightly Review,' April 1, 1868, p. 452.
[8] 'The Variation of Animals and Plants under Domestication,' vol. i. p. 309.

NATURAL SELECTION AS AFFECTING CIVILIZED NATIONS.

I have hitherto only considered the advancement of man from a semi-human condition to that of the modern savage. But some remarks on the action of natural selection on civilized nations may be worth adding. This subject has been ably discussed by Mr. W.R. Greg [9], and previously by Mr. Wallace and Mr. Galton. [10] Most of my remarks are taken from these three authors. With savages, the weak in body or mind are soon eliminated; and those that survive commonly exhibit a vigorous state of health. We civilized men, on the other hand, do our utmost to check the process of elimination; we build asylums for the imbecile, the maimed, and the sick; we institute poor-laws; and our medical men exert their utmost skill to save the life of every one to the last moment. There is reason to believe that vaccination has preserved thousands, who from a weak constitution would formerly have succumbed to small-pox. Thus the weak members of civilized societies propagate their kind. No one who has attended to the breeding of domestic animals will doubt that this must be highly injurious to the race of man. It is surprising how soon a want of care, or care wrongly directed, leads to the degeneration of a domestic race; but excepting in the case of man himself, hardly any one is so ignorant as to allow his worst animals to breed.

[9] 'Fraser's Magazine,' Sept. 1868, p. 353. This article seems to have struck many persons, and has given rise to two remarkable essays and a rejoinder in the 'Spectator,' Oct. 3rd and 17th, 1868. It has also been discussed in the 'Quarterly Journal of Science,' 1869, p. 152, and by Mr. Lawson Tait in the 'Dublin Quarterly Journal of Medical Science,' Feb. 1869, and by Mr. E. Ray Lankester in his 'Comparative Longevity,' 1870, p. 128. Similar views appeared previously in the 'Australasian,' July 13, 1867. I have borrowed ideas from several of these writers.
[10] For Mr. Wallace, see 'Anthropological Review,' as before cited. Mr. Galton in 'Macmillan's Magazine,' Aug. 1865, p. 318; also his great work, 'Hereditary Genius,' 1870.

The aid which we feel impelled to give to the helpless is mainly an incidental result of the instinct of sympathy, which was originally acquired as part of the social instincts, but subsequently rendered, in the manner previously indicated, more tender and more widely diffused. Nor could we check our sympathy, even at the urging of hard reason, without deterioration in the noblest part of our nature. The surgeon may harden himself whilst performing an operation, for he knows that he is acting for the good of his patient; but if we were intentionally to neglect the weak and helpless, it could only be for a contingent benefit, with an overwhelming present evil. We must therefore bear the undoubtedly bad effects of the weak surviving and propagating their kind; but there appears to be at least one check in steady action, namely that the weaker and inferior members of society do not marry so freely as the sound; and this check might be indefinitely increased by the weak in body or mind refraining from marriage, though this is more to be hoped for than expected.

In every country in which a large standing army is kept up, the finest young men are taken by the conscription or are enlisted. They are thus exposed to early death during war, are often tempted into vice, and are prevented from marrying during the prime of life. On the other hand the shorter and feebler men, with poor constitutions, are left at home, and consequently have a much better chance of marrying and propagating their kind. [11]

[11] Prof. H. Fick ('Einfluss der Naturwissenschaft auf das Recht,' June 1872) has some good remarks on this head, and on other such points.

Man accumulates property and bequeaths it to his children, so that the children of the rich have an advantage over the poor in the race for success, independently of bodily or mental superiority. On the other hand, the children of parents who are short-lived, and are therefore on an average deficient in health and vigor, come into their property sooner than other children, and will be likely to marry earlier, and leave a larger number of offspring to inherit their inferior constitutions. But the inheritance of property by itself is very far from an evil; for without the accumulation of capital the arts could not progress; and it is chiefly through their power that the civilized races have extended, and are now everywhere extending their range, so as to take the place of the lower races. Nor does the moderate accumulation of wealth interfere with the process of selection. When a poor man becomes moderately rich, his children enter trades or professions in which there is struggle enough, so that the able in body and mind succeed best. The presence of a body of well-instructed men, who have not to labor for their daily bread, is important to a degree which cannot be over-estimated; as all high intellectual work is carried on by them, and on such work, material progress of all kinds mainly depends, not to mention other and higher advantages. No doubt wealth when very great tends to convert men into useless drones, but their number is never large; and some degree of elimination here occurs, for we daily see rich men, who happen to be fools or profligate, squandering away their wealth.

Primogeniture with entailed estates is a more direct evil, though it may formerly have been a great advantage by the creation of a dominant class, and any government is better than none. Most eldest sons, though they may be weak in body or mind, marry, whilst the younger sons, however superior in these respects, do not so generally marry. Nor can worthless eldest sons with entailed estates squander their wealth. But here, as elsewhere, the relations of civilized life are so complex that some compensatory checks intervene. The men who are rich through primogeniture are able to select generation after generation the more beautiful and charming women; and these must generally be healthy in body and active in mind. The evil consequences, such as they may be, of the continued preservation of the same line of descent, without any selection, are checked by men of rank always wishing to increase their wealth and power; and this they effect by marrying heiresses. But the daughters of parents who have produced single children, are themselves, as Mr. Galton [12] has shown, apt to be sterile; and thus noble families are continually cut off in the direct line, and their wealth flows into some side channel; but unfortunately this channel is not determined by superiority of any kind.

[12] 'Hereditary Genius,' 1870, pp. 132-140.

Although civilization thus checks in many ways the action of natural selection, it apparently favours the better development of the body, by means of good food and the freedom from occasional hardships. This may be inferred from civilized men having been found, wherever compared, to be physically stronger than savages. [13] They appear also to have equal powers of endurance, as has been proved in many adventurous expeditions. Even the great luxury of the rich can be but little detrimental; for the expectation of life of our aristocracy, at all ages and of both sexes, is very little inferior to that of healthy English lives in the lower classes. [14]

[13] Quatrefages, 'Revue des Cours Scientifiques,' 1867-68, p. 659.
[14] See the fifth and sixth columns, compiled from good authorities, in the table given in Mr. E.R. Lankester's 'Comparative Longevity,' 1870, p. 115.

We will now look to the intellectual faculties. If in each grade of society the members were divided into two equal bodies, the one including the intellectually superior and the other the inferior, there can be little doubt that the former would succeed best in all occupations, and rear a greater number of children. Even in the lowest walks of life, skill and ability must be of some advantage; though in many occupations, owing to the great division of labor, a very small one. Hence in civilized nations there will be some tendency to an increase both in the number and in the standard of the intellectually able. But I do not wish to assert that this tendency may not be more than counterbalanced in other ways, as by the multiplication of the reckless and improvident; but even to such as these, ability must be some advantage.

It has often been objected to views like the foregoing, that the most eminent men who have ever lived have left no offspring to inherit their great intellect. Mr. Galton says, "I regret I am unable to solve the simple question whether, and how far, men and women who are prodigies of genius are infertile. I have, however, shown that men of eminence are by no means so." [15] Great lawgivers, the founders of beneficent religions, great philosophers and discoverers in science, aid the progress of mankind in a far higher degree by their works than by leaving a numerous progeny. In the case of corporeal structures, it is the selection of the slightly better-endowed and the elimination of the slightly less well-endowed individuals, and not the preservation of strongly-marked and rare anomalies, that leads to the advancement of a species. [16] So it will be with the intellectual faculties, since the somewhat abler men in each grade of society succeed rather better than the less able, and consequently increase in number, if not otherwise prevented. When in any nation the standard of intellect and the number of intellectual men have increased, we may expect from the law of the deviation from an average, that prodigies of genius will, as shown by Mr. Galton, appear somewhat more frequently than before.

[15] 'Hereditary Genius,' 1870, p. 330.
[16] 'Origin of Species' (fifth edition, 1869), p. 104.

In regard to the moral qualities, some elimination of the worst dispositions is always in progress even in the most civilized nations. Malefactors are executed, or imprisoned for long periods, so that they cannot freely transmit their bad qualities. Melancholic and insane persons are confined, or commit suicide. Violent and quarrelsome men often come to a bloody end. The restless who will not follow any steady occupation—and this relic of barbarism is a great check to civilization [17]—emigrate to newly-settled countries; where they prove useful pioneers. Intemperance is so highly destructive, that the expectation of life of the intemperate, at the age of thirty for instance, is only 13.8 years; whilst for the rural laborers of England at the same age it is 40.59 years. [18] Profligate women bear few children, and profligate men rarely marry; both suffer from disease. In the breeding of domestic animals, the elimination of those individuals, though few in number, which are in any marked manner inferior, is by no means an unimportant element towards success. This especially holds good with injurious characters which tend to reappear through reversion, such as blackness in sheep; and with mankind some of the worst dispositions, which occasionally without any assignable cause make their appearance in families, may perhaps be reversions to a savage state, from which we are not removed by very many generations. This view seems indeed recognized in the common expression that such men are the black sheep of the family.

[17] 'Hereditary Genius,' 1870, p. 347.
[18] E. Ray Lankester, 'Comparative Longevity,' 1870, p. 115. The table of the intemperate is from Neison's 'Vital Statistics.' In regard to profligacy, see Dr. Farr, 'Influence of Marriage on Mortality,' 'Nat. Assoc. for the Promotion of Social Science,' 1858.

With civilized nations, as far as an advanced standard of morality, and an increased number of fairly good men are concerned, natural selection apparently effects but little; though the fundamental social instincts were originally thus gained. But I have already said enough, whilst treating of the lower races, on the causes which lead to the advance of morality, namely, the approbation of our fellow-men—the strengthening of our sympathies by habit—example and imitation—reason—experience, and even self-interest—instruction during youth, and religious feelings.

A most important obstacle in civilized countries to an increase in the number of men of a superior class has been strongly insisted on by Mr. Greg and Mr. Galton [19], namely, the fact that the very poor and reckless, who are often degraded by vice, almost invariably marry early, whilst the careful and frugal, who are generally otherwise virtuous, marry late in life, so that they may be able to support themselves and their children in comfort. Those who marry early produce within a given period not only a greater number of generations, but, as shown by Dr. Duncan [20], they produce many more children. The children, moreover, that are borne by mothers during the prime of life are heavier and larger, and therefore probably more vigorous, than those born at other periods. Thus the reckless, degraded, and often vicious members of society, tend to increase at a quicker rate than the provident and generally virtuous members. Or as Mr. Greg puts the case: "The careless, squalid, unaspiring Irishman multiplies like rabbits: the frugal, foreseeing, self-respecting, ambitious Scot, stern in his morality, spiritual in his faith, sagacious and disciplined in his intelligence, passes his best years in struggle and in celibacy, marries late, and leaves few behind him. Given a land originally peopled by a thousand Saxons and a thousand Celts—and in a dozen generations five-sixths of the population would be Celts, but five-sixths of the property, of the power, of the intellect, would belong to the one-sixth of Saxons that remained. In the eternal 'struggle for existence,' it would be the inferior and *less* favoured race that had prevailed—and prevailed by virtue not of its good qualities but of its faults."

[19] 'Fraser's Magazine,' Sept. 1868, p. 353. 'Macmillan's Magazine,' Aug. 1865, p. 318. The Rev. F.W. Farrar ('Fraser's Magazine,' Aug. 1870, p. 264) takes a different view.
[20] 'On the Laws of the Fertility of Women,' in 'Transactions of the Royal Society,' Edinburgh, vol. xxiv. p. 287; now published separately under the title of 'Fecundity, Fertility, and Sterility,' 1871. See, also, Mr. Galton, 'Hereditary Genius,' pp. 352-357, for observations to the above effect.

There are, however, some checks to this downward tendency. We have seen that the intemperate suffer from a high rate of mortality, and the extremely profligate leave few offspring. The poorest classes crowd into towns, and it has been proved by Dr. Stark from the statistics of ten years in Scotland [21], that at all ages the death-rate is higher in towns than in rural districts, "and during the first five years of life the town death-rate is almost exactly double that of the rural districts." As these returns include both the rich and the poor, no doubt more than twice the number of births would be requisite to keep up the number of the very poor inhabitants in the towns, relatively to those in the country. With women, marriage at too early an age is highly injurious; for it has been found in France that, "Twice as many wives under twenty die in the year, as died out of the same number of the unmarried." The mortality, also, of husbands under twenty is "excessively high" [22], but what the cause of this may be, seems doubtful. Lastly, if the men who prudently delay marrying until they can bring up their families in comfort, were to select, as they often do, women in the prime of life, the rate of increase in the better class would be only slightly lessened.

[21] 'Tenth Annual Report of Births, Deaths, etc., in Scotland,' 1867, p. xxix.
[22] These quotations are taken from our highest authority on such questions, namely, Dr. Farr, in his paper 'On the Influence of Marriage on the Mortality of the French People,' read before the Nat. Assoc. for the Promotion of Social Science, 1858.

It was established from an enormous body of statistics, taken during 1853, that the unmarried men throughout France, between the ages of twenty and eighty, die in a much larger proportion than the married: for instance, out of every 1000 unmarried men, between the ages of twenty and thirty, 11.3 annually died, whilst of the married, only 6.5 died. [23] A similar law was proved to hold good, during the years 1863 and 1864, with the entire population above the age of twenty in Scotland: for instance, out of every 1000 unmarried men, between the ages of twenty and thirty, 14.97 annually died, whilst of the married only 7.24 died, that is less than half. [24] Dr. Stark remarks on this,

"Bachelorhood is more destructive to life than the most unwholesome trades, or than residence in an unwholesome house or district where there has never been the most distant attempt at sanitary improvement." He considers that the lessened mortality is the direct result of "marriage, and the more regular domestic habits which attend that state." He admits, however, that the intemperate, profligate, and criminal classes, whose duration of life is low, do not commonly marry; and it must likewise be admitted that men with a weak constitution, ill health, or any great infirmity in body or mind, will often not wish to marry, or will be rejected. Dr. Stark seems to have come to the conclusion that marriage in itself is a main cause of prolonged life, from finding that aged married men still have a considerable advantage in this respect over the unmarried of the same advanced age; but every one must have known instances of men, who with weak health during youth did not marry, and yet have survived to old age, though remaining weak, and therefore always with a lessened chance of life or of marrying. There is another remarkable circumstance which seems to support Dr. Stark's conclusion, namely, that widows and widowers in France suffer in comparison with the married a very heavy rate of mortality; but Dr. Farr attributes this to the poverty and evil habits consequent on the disruption of the family, and to grief. On the whole we may conclude with Dr. Farr that the lesser mortality of married than of unmarried men, which seems to be a general law, "is mainly due to the constant elimination of imperfect types, and to the skilful selection of the finest individuals out of each successive generation;" the selection relating only to the marriage state, and acting on all corporeal, intellectual, and moral qualities. [25] We may, therefore, infer that sound and good men who out of prudence remain for a time unmarried, do not suffer a high rate of mortality.

[23] Dr. Farr, ibid. The quotations given below are extracted from the same striking paper.

[24] I have taken the mean of the quinquennial means, given in 'The Tenth Annual Report of Births, Deaths, etc., in Scotland,' 1867. The quotation from Dr. Stark is copied from an article in the 'Daily News,' Oct. 17, 1868, which Dr. Farr considers very carefully written.

[25] Dr. Duncan remarks ('Fecundity, Fertility, etc.' 1871, p. 334) on this subject: "At every age the healthy and beautiful go over from the unmarried side to the married, leaving the unmarried columns crowded with the sickly and unfortunate."

If the various checks specified in the two last paragraphs, and perhaps others as yet unknown, do not prevent the reckless, the vicious and otherwise inferior members of society from increasing at a quicker rate than the better class of men, the nation will retrograde, as has too often occurred in the history of the world. We must remember that progress is no invariable rule. It is very difficult to say why one civilized nation rises, becomes more powerful, and spreads more widely, than another; or why the same nation progresses more quickly at one time than at another. We can only say that it depends on an increase in the actual number of the population, on the number of men endowed with high intellectual and moral faculties, as well as on their standard of excellence. Corporeal structure appears to have little influence, except so far as vigor of body leads to vigor of mind.

It has been urged by several writers that as high intellectual powers are advantageous to a nation, the old Greeks, who stood some grades higher in intellect than any race that has ever existed [26], ought, if the power of natural selection were real, to have risen still higher in the scale, increased in number, and stocked the whole of Europe. Here we have the tacit assumption, so often made with respect to corporeal structures, that there is some innate tendency towards continued development in mind and body. But development of all kinds depends on many concurrent favourable circumstances. Natural selection acts only tentatively. Individuals and races may have acquired certain indisputable advantages, and yet have perished from failing in other characters. The Greeks may have retrograded from a want of coherence between the many small states, from the small size of their whole country, from the practice of slavery, or from extreme sensuality; for they did not succumb until "they were enervated and corrupt to the very core." [27] The western nations of Europe, who now so immeasurably surpass their former savage progenitors, and stand at the summit of civilization, owe little or none of their superiority to direct inheritance from the old Greeks, though they owe much to the written works of that wonderful people.

[26] See the ingenious and original argument on this subject by Mr. Galton, 'Hereditary Genius,' pp. 340-342.

[27] Mr. Greg, 'Fraser's Magazine,' Sept. 1868, p. 357.

Who can positively say why the Spanish nation, so dominant at one time, has been distanced in the race. The awakening of the nations of Europe from the dark ages is a still more perplexing problem. At that early period, as Mr. Galton has remarked, almost all the men of a gentle nature, those given to meditation or culture of the mind, had no refuge except in the bosom of a Church which demanded celibacy [28]; and this could hardly fail to have had a deteriorating influence on each successive generation. During this same period the Holy Inquisition selected with extreme care the freest and boldest men in order to burn or imprison them. In Spain alone some of the best men—those who doubted and questioned, and without doubting there can be no progress—were eliminated during three centuries at the rate of a thousand a year. The evil which the Catholic Church has thus effected is incalculable, though no doubt counterbalanced to a certain, perhaps to a large, extent in other ways; nevertheless, Europe has progressed at an unparallèled rate.

[28] 'Hereditary Genius,' 1870, pp. 357-359. The Rev. F.W. Farrar ('Fraser's Magazine,' Aug. 1870, p. 257) advances arguments on the other side. Sir C. Lyell had already ('Principles of Geology,' vol. ii. 1868, p. 489), in a striking passage called attention to the evil influence of the Holy Inquisition in having, through selection, lowered the general standard of intelligence in Europe.

The remarkable success of the English as colonists, compared to other European nations, has been ascribed to their "daring and persistent energy"; a result which is well illustrated by comparing the progress of the Canadians of English and French extraction; but who can say how the English gained their energy? There is apparently much truth in the belief that the wonderful progress of the United States, as well as the character of the people, are the results of natural selection; for the more energetic, restless, and courageous men from all parts of Europe have emigrated during the last ten or twelve generations to that great country, and have there succeeded best. [29] Looking to the distant future, I do not think that the Rev. Mr. Zincke takes an exaggerated view when he says [30]: "All other series of events—as that which resulted in the culture of mind in Greece, and that which resulted in the empire of Rome—only appear to have purpose and value when viewed in connection with, or rather as subsidiary to...the great stream of Anglo-Saxon emigration to the west." Obscure as is the problem of the advance of civilization, we can at least see that a nation which produced during a lengthened period the greatest number of highly intellectual, energetic, brave, patriotic, and benevolent men, would generally prevail over less favoured nations.

[29] Mr. Galton, 'Macmillan's Magazine,' August 1865, p. 325. See also, 'Nature,' 'On Darwinism and National Life,' Dec. 1869, p. 184.
[30] 'Last Winter in the United States,' 1868, p. 29.

Natural selection follows from the struggle for existence; and this from a rapid rate of increase. It is impossible not to regret bitterly, but whether wisely is another question, the rate at which man tends to increase; for this leads in barbarous tribes to infanticide and many other evils, and in civilized nations to abject poverty, celibacy, and to the late marriages of the prudent. But as man suffers from the same physical evils as the lower animals, he has no right to expect an immunity from the evils consequent on the struggle for existence. Had he not been subjected during primeval times to natural selection, assuredly he would never have attained to his present rank. Since we see in many parts of the world enormous areas of the most fertile land capable of supporting numerous happy homes, but peopled only by a few wandering savages, it might be argued that the struggle for existence had not been sufficiently severe to force man upwards to his highest standard. Judging from all that we know of man and the lower animals, there has always been sufficient variability in their intellectual and moral faculties, for a steady advance through natural selection. No doubt such advance demands many favourable concurrent circumstances; but it may well be doubted whether the most favourable would have sufficed, had not the rate of increase been rapid, and the consequent struggle for existence extremely severe. It even appears from what we see, for instance, in parts of S. America, that a people which may be called civilized, such as the Spanish settlers, is liable to become indolent and to retrograde, when the conditions of life are very easy. With highly civilized nations continued progress depends in a subordinate degree on natural selection; for such nations do not supplant and exterminate one another as do savage tribes. Nevertheless the more intelligent members within the same community will succeed better in the long run than the inferior, and leave a more numerous progeny, and this is a form of natural selection. The more efficient causes of progress seem to consist of a good education during youth whilst the brain is impressible, and of a high standard of excellence, inculcated by the ablest and best men, embodied in the laws, customs and traditions of the nation, and enforced by public opinion. It should, however, be borne in mind, that the enforcement of public opinion depends on our appreciation of the approbation and disapprobation of others; and this appreciation is founded on our sympathy, which it can hardly be doubted was originally developed through natural selection as one of the most important elements of the social instincts. [31]

[31] I am much indebted to Mr. John Morley for some good criticisms on this subject: see, also Broca, 'Les Selections,' 'Revue d'Anthropologie,' 1872.

ON THE EVIDENCE THAT ALL CIVILIZED
NATIONS WERE ONCE BARBAROUS.

The present subject has been treated in so full and admirable a manner by Sir J. Lubbock [32], Mr. Tylor, Mr. M'Lennan, and others, that I need here give only the briefest summary of their results. The arguments recently advanced by the Duke of Argyll [33] and formerly by Archbishop Whately, in favour of the belief that man came into the world as a civilized being, and that all savages have since undergone degradation, seem to me weak in comparison with those advanced on the other side. Many nations, no doubt, have fallen away in civilization, and some may have lapsed into utter barbarism, though on this latter head I have met with no evidence. The Fuegians were probably compelled by other conquering hordes to settle in their inhospitable country, and they may have become in consequence somewhat more degraded; but it would be difficult to prove that they have fallen much below the Botocudos, who inhabit the finest parts of Brazil.

[32] 'On the Origin of Civilization,' 'Proceedings of the Ethnological Society,' Nov. 26, 1867.
[33] 'Primeval Man,' 1869.

The evidence that all civilized nations are the descendants of barbarians, consists, on the one side, of clear traces of their former low condition in still-existing customs, beliefs, language, etc.; and on the other side, of proofs that savages are independently able to raise themselves a few steps in the scale of civilization, and have actually thus risen. The evidence on the first head is extremely curious, but cannot be here given: I refer to such cases as that of the art of enumeration, which, as Mr. Tylor clearly shows by reference to the words still used in some places, originated in counting the fingers, first of one hand and then of the other, and lastly of the toes. We have traces of this in our own decimal system, and in the Roman numerals, where, after the V, which is supposed to be an abbreviated picture of a human hand, we pass on to VI, etc., when the other hand no doubt was used. So again, "when we speak of three-score and ten, we are counting by the vigesimal system, each score thus ideally made, standing for 20—for 'one man' as a Mexican or Carib would put it." [34] According to a large and increasing school of philologists, every language bears the marks of its slow and gradual evolution. So it is with the art of writing, for letters are rudiments of pictorial representations. It is hardly possible to read Mr. M'Lennan's work [35] and not admit that almost all civilized nations still retain traces of such rude habits as the forcible capture of wives. What ancient nation, as the same author asks, can be named that was originally monogamous? The primitive idea of justice, as shown by the law of battle and other customs of which vestiges still remain, was likewise most rude. Many existing superstitions are the remnants of former false religious beliefs. The highest form of religion—the grand idea of God hating sin and loving righteousness—was unknown during primeval times.

[34] 'Royal Institution of Great Britain,' March 15, 1867. Also, 'Researches into the Early History of Mankind,' 1865.
[35] 'Primitive Marriage,' 1865. See, likewise, an excellent article, evidently by the same author, in the 'North British Review,' July 1869. Also, Mr. L.H. Morgan, 'A

Conjectural Solution of the Origin of the Class, System of Relationship,' in 'Proc. American Acad. of Sciences,' vol. vii. Feb. 1868. Prof. Schaaffhausen ('Anthropolog. Review,' Oct. 1869, p. 373) remarks on "the vestiges of human sacrifices found both in Homer and the Old Testament."

Turning to the other kind of evidence: Sir J. Lubbock has shown that some savages have recently improved a little in some of their simpler arts. From the extremely curious account which he gives of the weapons, tools, and arts, in use amongst savages in various parts of the world, it cannot be doubted that these have nearly all been independent discoveries, excepting perhaps the art of making fire. [36] The Australian boomerang is a good instance of one such independent discovery. The Tahitians when first visited had advanced in many respects beyond the inhabitants of most of the other Polynesian islands. There are no just grounds for the belief that the high culture of the native Peruvians and Mexicans was derived from abroad [37]; many native plants were there cultivated, and a few native animals domesticated. We should bear in mind that, judging from the small influence of most missionaries, a wandering crew from some semi-civilized land, if washed to the shores of America, would not have produced any marked effect on the natives, unless they had already become somewhat advanced. Looking to a very remote period in the history of the world, we find, to use Sir J. Lubbock's well-known terms, a paleolithic and neolithic period; and no one will pretend that the art of grinding rough flint tools was a borrowed one. In all parts of Europe, as far east as Greece, in Palestine, India, Japan, New Zealand, and Africa, including Egypt, flint tools have been discovered in abundance; and of their use the existing inhabitants retain no tradition. There is also indirect evidence of their former use by the Chinese and ancient Jews. Hence there can hardly be a doubt that the inhabitants of these countries, which include nearly the whole civilized world, were once in a barbarous condition. To believe that man was aboriginally civilized and then suffered utter degradation in so many regions, is to take a pitiably low view of human nature. It is apparently a truer and more cheerful view that progress has been much more general than retrogression; that man has risen, though by slow and interrupted steps, from a lowly condition to the highest standard as yet attained by him in knowledge, morals and religion.

[36] Sir J. Lubbock, 'Prehistoric Times,' 2nd edit. 1869, chaps. xv. and xvi. *et passim.* See also the excellent 9th Chapter in Tylor's 'Early History of Mankind,' 2nd edit., 1870.

[37] Dr. F. Müller has made some good remarks to this effect in the 'Reise der Novara: Anthropolog. Theil,' Abtheil. iii. 1868, s. 127.

CHAPTER VI.

ON THE AFFINITIES AND GENEALOGY OF MAN.

Position of man in the animal series—The natural system genealogical—Adaptive characters of slight value—Various small points of resemblance between man and the Quadrumana—Rank of man in the natural system—Birthplace and antiquity of man—Absence of fossil connecting links—Lower stages in the genealogy of man, as inferred, firstly from his affinities and secondly from his structure—Early androgynous condition of the Vertebrata—Conclusion.

Even if it be granted that the difference between man and his nearest allies is as great in corporeal structure as some naturalists maintain, and although we must grant that the difference between them is immense in mental power, yet the facts given in the earlier chapters appear to declare, in the plainest manner, that man is descended from some lower form, notwithstanding that connecting-links have not hitherto been discovered.

Man is liable to numerous, slight, and diversified variations, which are induced by the same general causes, are governed and transmitted in accordance with the same general laws, as in the lower animals. Man has multiplied so rapidly, that he has necessarily been exposed to struggle for existence, and consequently to natural selection. He has given rise to many races, some of which differ so much from each other, that they have often been ranked by naturalists as distinct species. His body is constructed on the same homological plan as that of other mammals. He passes through the same phases of embryological development. He retains many rudimentary and useless structures, which no doubt were once serviceable. Characters occasionally make their re-appearance in him, which we have reason to believe were possessed by his early progenitors. If the origin of man had been wholly different from that of all other animals, these various appearances would be mere empty deceptions; but such an admission is incredible. These appearances, on the other hand, are intelligible, at least to a large extent, if man is the co-descendant with other mammals of some unknown and lower form.

Some naturalists, from being deeply impressed with the mental and spiritual powers of man, have divided the whole organic world into three kingdoms, the Human, the Animal, and the Vegetable, thus giving to man a separate kingdom. [1] Spiritual powers cannot be compared or classed by the naturalist: but he may endeavour to show, as I have done, that the mental faculties of man and the lower animals do not differ in kind, although immensely in degree. A difference in degree, however great, does not justify us in placing man in a distinct kingdom, as will perhaps be best illustrated by comparing the mental powers of two insects, namely, a coccus or scale-insect and an ant, which undoubtedly belong to the same class. The difference is here greater than, though of a somewhat different kind from, that between man and the highest mammal. The female coccus, whilst young, attaches itself by its proboscis to a plant; sucks the sap, but never moves again; is fertilized and lays eggs; and this is its whole history. On the other hand, to describe the habits and mental powers of worker-ants, would require, as Pierre Huber has shown, a large volume; I may, however, briefly specify a few points. Ants certainly communicate information to each other, and several unite for the same work, or for games of play. They recognize their fellow-ants after months of absence, and feel sympathy for each other. They build great edifices, keep them clean, close the doors in the evening, and post sentries. They make roads as well as tunnels under rivers, and temporary bridges over them, by clinging together. They collect food for the community, and when an object, too large for entrance, is brought to the nest, they enlarge the door, and afterwards build it up again. They store up seeds, of which they prevent the germination, and which, if damp, are brought up to the surface to dry. They keep aphides and other insects as milch-cows. They go out to battle in regular bands, and freely sacrifice their lives for the common weal. They emigrate according to a preconcerted plan. They capture slaves. They move the eggs of their aphides, as well as their own eggs and cocoons, into warm parts of the nest, in order that they may be quickly hatched; and endless similar facts could be given. [2] On the whole, the difference in mental power between an ant and a coccus is immense; yet no one has ever dreamed of placing these insects in distinct classes, much less in distinct kingdoms. No doubt the difference is bridged over by other insects; and this is not the case with man and the higher apes. But we have every reason to believe that the breaks in the series are simply the results of many forms having become extinct.

[1] Isidore Geoffroy St.-Hilaire gives a detailed account of the position assigned to man by various naturalists in their classifications: 'Hist. Nat. Gen.' tom. ii. 1859, pp. 170-189.

[2] Some of the most interesting facts ever published on the habits of ants are given by Mr. Belt, in his 'Naturalist in Nicaragua,' 1874. See also Mr. Moggridge's admirable work, 'Harvesting Ants,' etc., 1873, also 'L'Instinct chez les Insectes,' by M. George Pouchet, 'Revue des Deux Mondes,' Feb. 1870, p. 682.

Professor Owen, relying chiefly on the structure of the brain, has divided the mammalian series into four sub-classes. One of these he devotes to man; in another he places both the marsupials and the Monotremata; so that he makes man as distinct from all other mammals as are these two latter groups conjoined. This view has not been accepted, as far as I am aware, by any naturalist capable of forming an independent judgment, and therefore need not here be further considered.

We can understand why a classification founded on any single character or organ—even an organ so wonderfully complex and important as the brain—or on the high development of the mental faculties, is almost sure to prove unsatisfactory. This principle has indeed been tried with hymenopterous insects; but when thus classed by their habits or instincts, the arrangement proved thoroughly artificial. [3] Classifications may, of course, be based on any character whatever, as on size, color, or the element inhabited; but naturalists have long felt a profound conviction that there is a natural system. This system, it is now generally admitted, must be, as far as possible, genealogical in arrangement,—that is, the co-descendants of the same form must be kept together in one group, apart from the co-descendants of any other form; but if the parent-forms are related, so will be their descendants, and the two groups together will form a larger group. The amount of difference between the several groups—that is the amount of modification which each has undergone—is expressed by such terms as genera, families, orders, and classes. As we have no record of the lines of descent, the pedigree can be discovered only by observing the degrees of resemblance between the beings which are to be classed. For this object numerous points of resemblance are of much more importance than the amount of similarity or dissimilarity in a few points. If two languages were found to resemble each other in a multitude of words and points of construction, they would be universally recognized as having sprung from a common source, notwithstanding that they differed greatly in some few words or points of construction. But with organic beings the points of resemblance must not consist of adaptations to similar habits of life: two animals may, for instance, have had their whole frames modified for living in the water, and yet they will not be brought any nearer to each other in the natural system. Hence we can see how it is that resemblances in several unimportant structures, in useless and rudimentary organs, or not now functionally active, or in an embryological condition, are by far the most serviceable for classification; for they can hardly be due to adaptations within a late period; and thus they reveal the old lines of descent or of true affinity.

[3] Westwood, 'Modern Classification of Insects,' vol. ii. 1840, p. 87.

We can further see why a great amount of modification in some one character ought not to lead us to separate widely any two organisms. A part which already differs much from the same part in other allied forms has already, according to the theory of evolution, varied much; consequently it would (as long as the organism remained exposed to the same exciting conditions) be liable to further variations of the same kind; and these, if beneficial, would be preserved, and thus be continually augmented. In many cases the continued development of a part, for instance, of the beak of a bird, or of the teeth of a mammal, would not aid the species in gaining its food, or for any other object; but with man we can see no definite limit to the continued development of the brain and mental faculties, as far as advantage is concerned. Therefore in determining the position of man in the natural or genealogical system, the extreme development of his brain ought not to outweigh a multitude of resemblances in other less important or quite unimportant points.

The greater number of naturalists who have taken into consideration the whole structure of man, including his mental faculties, have followed Blumenbach and Cuvier, and have placed man in a separate Order, under the title of the Bimana, and therefore on an equality with the orders of the Quadrumana, Carnivora, etc. Recently many of our best naturalists have recurred to the view first propounded by Linnæus, so remarkable for his sagacity, and have placed man in the same Order with the Quadrumana, under the title of the Primates. The justice of this conclusion will be admitted: for in the first place, we must bear in mind the comparative insignificance for classification of the great development of the brain in man, and that the strongly-marked differences between the skulls of man and the Quadrumana (lately insisted upon by Bischoff, Aeby, and others) apparently follow from their differently developed brains. In the second place, we must remember that nearly all the other and more important differences between man and the Quadrumana are manifestly adaptive in their nature, and relate chiefly to the erect position of man; such as the structure of his hand, foot, and pelvis, the curvature of his spine, and the position of his head. The family of Seals offers a good illustration of the small importance of adaptive characters for classification. These animals differ from all other Carnivora in the form of their bodies and in the structure of their limbs, far more than does man from the higher apes; yet in most systems, from that of Cuvier to the most recent one by Mr. Flower [4], seals are ranked as a mere family in the Order of the Carnivora. If man had not been his own classifier, he would never have thought of founding a separate order for his own reception.

[4] 'Proceedings Zoological Society,' 1863, p. 4.

It would be beyond my limits, and quite beyond my knowledge, even to name the innumerable points of structure in which man agrees with the other Primates. Our great anatomist and philosopher, Prof. Huxley, has fully discussed this subject, [5] and concludes that man in all parts of his organization differs less from the higher apes, than these do from the lower members of the same group. Consequently there "is no justification for placing man in a distinct order."

[5] 'Evidence as to Man's Place in Nature,' 1863, p. 70, *et passim*.

In an early part of this work I brought forward various facts, showing how closely man agrees in constitution with the higher mammals; and this agreement must depend on our close similarity in minute structure and chemical composition. I gave, as instances, our liability to the same diseases, and to the attacks of allied parasites; our tastes in common for the same stimulants, and the similar effects produced by them, as well as by various drugs, and other such facts.

As small unimportant points of resemblance between man and the Quadrumana are not commonly noticed in systematic works, and as, when numerous, they clearly reveal our relationship, I will specify a few such points. The relative position of our features is manifestly the same; and the various emotions are displayed by nearly similar movements of the muscles and skin, chiefly above the eyebrows and round the mouth. Some few expressions are, indeed, almost the same, as in the weeping of certain kinds of monkeys and in the laughing noise made by others, during which the corners of the mouth are drawn backwards, and the lower eyelids wrinkled. The external ears are curiously alike. In man the nose is much more prominent than in most monkeys; but we may trace the commencement of an aquiline curvature in the nose of the Hoolock Gibbon; and this in the *Semnopithecus nasica* is carried to a ridiculous extreme.

The faces of many monkeys are ornamented with beards, whiskers, or moustaches. The hair on the head grows to a great length in some species of Semnopithecus [6]; and in the Bonnet monkey (*Macacus radiatus*) it radiates from a point on the crown, with a parting down the middle. It is commonly said that the forehead gives to man his noble and intellectual appearance; but the thick hair on the head of the Bonnet monkey terminates downwards abruptly, and is succeeded by hair so short and fine that at a little distance the forehead, with the exception of the eyebrows, appears quite naked. It has been erroneously asserted that eyebrows are not present in any monkey. In the species just named the degree of nakedness of the forehead differs in different individuals; and Eschricht states [7] that in our children the limit between the hairy scalp and the naked forehead is sometimes not well defined; so that here we seem to have a trifling case of reversion to a progenitor, in whom the forehead had not as yet become quite naked.

[6] Isidore Geoffroy St.-Hilaire, 'Hist. Nat. Gen.' tom. ii. 1859, p. 217.
[7] 'Ueber die Richtung der Haare,' etc., Müller's 'Archiv fur Anat. und Phys.' 1837, s. 51.

It is well known that the hair on our arms tends to converge from above and below to a point at the elbow. This curious arrangement, so unlike that in most of the lower mammals, is common to the gorilla, chimpanzee, orang, some species of Hylobates, and even to some few American monkeys. But in *Hylobates agilis* the hair on the fore-arm is directed downwards or towards the wrist in the ordinary manner; and in *H. lar* it is nearly erect, with only a very slight forward inclination; so that in this latter species it is in a transitional state. It can hardly be doubted that with most mammals the thickness of the hair on the back and its direction, is adapted to throw off the rain; even the transverse hairs on the fore-legs of a dog may serve for this end when he is coiled up asleep. Mr. Wallace, who has carefully studied the habits of the orang, remarks that the convergence of the hair towards the elbow on the arms of the orang may be explained as serving to throw off the rain, for this animal during rainy weather sits with its arms bent, and with the hands clasped round a branch or over its head. According to Livingstone, the gorilla also "sits in pelting rain with his hands over his head." [8] If the above explanation is correct, as seems probable, the direction of the hair on our own arms offers a curious record of our former state; for no one supposes that it is now of any use in throwing off the rain; nor, in our present erect condition, is it properly directed for this purpose.

[8] Quoted by Reade, 'The African Sketch Book,' vol. i. 1873, p. 152.

It would, however, be rash to trust too much to the principle of adaptation in regard to the direction of the hair in man or his early progenitors; for it is impossible to study the figures given by Eschricht of the arrangement of the hair on the human fœtus (this being the same as in the adult) and not agree with this excellent observer that other and more complex causes have intervened. The points of convergence seem to stand in some relation to those points in the embryo which are last closed in during development. There appears, also, to exist some relation between the arrangement of the hair on the limbs, and the course of the medullary arteries. [9]

[9] On the hair in Hylobates, see 'Natural History of Mammals,' by C.L. Martin, 1841, p. 415. Also, Isidore Geoffroy on the American monkeys and other kinds, 'Hist. Nat. Gen.' vol. ii. 1859, pp. 216, 243. Eschricht, ibid. s. 46, 55, 61. Owen, 'Anatomy of Vertebrates,' vol. iii. p. 619. Wallace, 'Contributions to the Theory of Natural Selection,' 1870, p. 344.

It must not be supposed that the resemblances between man and certain apes in the above and in many other points—such as in having a naked forehead, long tresses on the head, etc.,—are all necessarily the result of unbroken inheritance from a common progenitor, or of subsequent reversion. Many of these resemblances are more probably due to analogous variation, which follows, as I have elsewhere attempted to show [10], from co-descended organisms having a similar constitution, and having been acted on by like causes inducing similar modifications. With respect to the similar direction of the hair on the fore-arms of man and certain monkeys, as this character is common to almost all the anthropomorphous apes, it may probably be attributed to inheritance; but this is not certain, as some very distinct American monkeys are thus characterized.

[10] 'Origin of Species,' 5th edit. 1869, p.194. 'The Variation of Animals and Plants under Domestication,' vol. ii. 1868, p. 348.

Although, as we have now seen, man has no just right to form a separate Order for his own reception, he may perhaps claim a distinct Sub-order or Family. Prof. Huxley, in his last work [11], divides the primates into three Sub-orders; namely, the Anthropidæ with man alone, the Simiadæ including monkeys of all kinds, and the Lemuridæ with the diversified genera of lemurs. As far as differences in certain important points of structure are concerned, man may no doubt rightly claim the rank of a Sub-order; and this rank is too low, if we look chiefly to his mental faculties. Nevertheless, from a genealogical point of view it appears that this rank is too high, and that man ought to form merely a Family, or possibly even only a Sub-family. If we imagine three lines of descent proceeding from a common stock, it is quite conceivable that two of them might after the lapse of ages be so slightly changed as still to remain as species of the same genus, whilst the third line might become so greatly modified as to deserve to rank as a distinct Sub-family, Family, or even Order. But in this case it is almost certain that the third line would still retain through inheritance numerous small points of resemblance with the other two. Here, then, would occur the difficulty, at present insoluble, how much weight we ought to assign in our classifications to strongly-marked differences in some few points,—that is, to the amount of modification undergone; and how much to close resemblance in numerous unimportant points, as indicating the lines of descent or genealogy. To attach much weight to the few but strong differences is the most obvious and perhaps the safest course, though it appears more correct to pay great attention to the many small resemblances, as giving a truly natural classification.

[11] 'An Introduction to the Classification of Animals,' 1869, p. 99.

In forming a judgment on this head with reference to man, we must glance at the classification of the Simiadæ. This family is divided by almost all naturalists into the Catarhine group, or Old World monkeys, all of which are characterized (as their name expresses) by the peculiar structure of their nostrils, and by having four premolars in each jaw; and into the Platyrhine group or New World monkeys (including two very distinct sub-groups), all of which are characterized by differently constructed nostrils, and by having six premolars in each jaw. Some other small differences might be mentioned. Now man unquestionably belongs in his dentition, in the structure of his nostrils, and some other respects, to the Catarhine or Old World division; nor does he resemble the Platyrhines more closely than the Catarhines in any characters, excepting in a few of not much importance and apparently of an adaptive nature. It is therefore against all probability that some New World species should have formerly varied and produced a man-like creature, with all the distinctive characters proper to the Old World division; losing at the same time all its own distinctive characters. There can, consequently, hardly be a doubt that man is an off-shoot from the Old World Simian stem; and that under a genealogical point of view he must be classed with the Catarhine division. [12]

[12] This is nearly the same classification as that provisionally adopted by Mr. St. George Mivart, ('Transactions, Philosophical Society," 1867, p. 300), who, after separating the Lemuridæ, divides the remainder of the Primates into the Hominidæ, the Simiadæ which answer to the Catarhines, the Cebidæ, and the Hapalidæ,—these two latter groups answering to the Platyrhines. Mr. Mivart still abides by the same view; see 'Nature,' 1871, p. 481.

The anthropomorphous apes, namely the gorilla, chimpanzee, orang, and hylobates, are by most naturalists separated from the other Old World monkeys, as a distinct sub-group. I am aware that Gratiolet, relying on the structure of the brain, does not admit the existence of this sub-group, and no doubt it is a broken one. Thus the orang, as Mr. St. G. Mivart remarks, "is one of the most peculiar and aberrant forms to be found in the Order." [13] The remaining non-anthropomorphous Old World monkeys, are again divided by some naturalists into two or three smaller sub-groups; the genus Semnopithecus, with its peculiar sacculated stomach, being the type of one sub-group. But it appears from M. Gaudry's wonderful discoveries in Attica, that during the Miocene period a form existed there, which connected Semnopithecus and Macacus; and this probably illustrates the manner in which the other and higher groups were once blended together.

[13] 'Transactions, Zoolog. Soc.' vol. vi. 1867, p. 214.

If the anthropomorphous apes be admitted to form a natural sub-group, then as man agrees with them, not only in all those characters which he possesses in common with the whole Catarhine group, but in other peculiar characters, such as the absence of a tail and of callosities, and in general appearance, we may infer that some ancient member of the anthropomorphous sub-group gave birth to man. It is not probable that, through the law of analogous variation, a member of one of the other lower sub-groups should have given rise to a man-like creature, resembling the higher anthropomorphous apes in so many respects. No doubt man, in comparison with most of his allies, has undergone an extraordinary amount of modification, chiefly in consequence of the great development of his brain and his erect position; nevertheless, we should bear in mind that he "is but one of several exceptional forms of Primates." [14]

[14] Mr. St. G. Mivart, 'Transactions of the Philosophical Society,' 1867, p. 410.

Every naturalist, who believes in the principle of evolution, will grant that the two main divisions of the Simiadæ, namely the Catarhine and Platyrhine monkeys, with their sub-groups, have all proceeded from some one extremely ancient progenitor. The early descendants of this progenitor, before they had diverged to any considerable extent from each other, would still have formed a single natural group; but some of the species or incipient genera would have already begun to indicate by their diverging characters the future distinctive marks of the Catarhine and Platyrhine divisions. Hence the members of this supposed ancient group would not have been so uniform in their dentition, or in the structure of their nostrils, as are the existing Catarhine monkeys in one way and the Platyrhines in another way, but would have resembled in this respect the allied Lemuridæ, which differ greatly from each other in the form of their muzzles [15], and to an extraordinary degree in their dentition.

[15] Messrs. Murie and Mivart on the Lemuroidea, 'Transactions, Zoological Society,' vol. vii, 1869, p. 5.

The Catarhine and Platyrhine monkeys agree in a multitude of characters, as is shown by their unquestionably belonging to one and the same Order. The many characters which they possess in common can hardly have been independently acquired by so many distinct species; so that these characters must have been inherited. But a naturalist would undoubtedly have ranked as an ape or a monkey, an ancient form which possessed many characters common to the Catarhine and Platyrhine monkeys, other characters in an intermediate condition, and some few, perhaps, distinct from those now found in either group. And as man from a genealogical point of view belongs to the Catarhine or Old World stock, we must conclude, however much the conclusion may revolt our pride, that our early progenitors would have been properly thus designated. [16] But we must not fall into the error of supposing that the early progenitor of the whole Simian stock, including man, was identical with, or even closely resembled, any existing ape or monkey.

[16] Häckel has come to this same conclusion. See 'Ueber die Entstehung des Menschengeschlechts,' in Virchow's 'Sammlung. gemein. wissen. Vortrage,' 1868, s. 61. Also his 'Natürliche Schöpfungsgeschicte,' 1868, in which he gives in detail his views on the genealogy of man.

ON THE BIRTHPLACE AND ANTIQUITY OF MAN.

We are naturally led to enquire, where was the birthplace of man at that stage of descent when our progenitors diverged from the Catarhine stock? The fact that they belonged to this stock clearly shows that they inhabited the Old World; but not Australia nor any oceanic island, as we may infer from the laws of geographical distribution. In each great region of the world the living mammals are closely related to the extinct species of the same region. It is therefore probable that Africa was formerly inhabited by extinct apes closely allied to the gorilla and chimpanzee; and as these two species are now man's nearest allies, it is somewhat more probable that our early progenitors lived on the African continent than elsewhere. But it is useless to speculate on this subject; for two or three anthropomorphous apes, one the Dryopithecus [17] of Lartet, nearly as large as a man, and closely allied to Hylobates, existed in Europe during the Miocene age; and since so remote a period the earth has certainly undergone many great revolutions, and there has been ample time for migration on the largest scale.

[17] Dr. C. Forsyth Major, 'Sur les Singes fossiles trouvés en Italie:' 'Soc. Ital. des Sc. Nat.' tom. xv. 1872.

At the period and place, whenever and wherever it was, when man first lost his hairy covering, he probably inhabited a hot country; a circumstance favourable for the frugiferous diet on which, judging from analogy, he subsisted. We are far from knowing how long ago it was when man first diverged from the Catarhine stock; but it may have occurred at an epoch as remote as the Eocene period; for that the higher apes had diverged from the lower apes as early as the Upper Miocene period is shown by the existence of the Dryopithecus. We are also quite ignorant at how rapid a rate organisms, whether high or low in the scale, may be modified under favourable circumstances; we know, however, that some have retained the same form during an enormous lapse of time. From what we see going on under domestication, we learn that some of the co-descendants of the same species may be not at all, some a little, and some greatly changed, all within the same period. Thus it may have been with man, who has undergone a great amount of modification in certain characters in comparison with the higher apes.

The great break in the organic chain between man and his nearest allies, which cannot be bridged over by any extinct or living species, has often been advanced as a grave objection to the belief that man is descended from some lower form; but this objection will not appear of much weight to those who, from general reasons, believe in the general principle of evolution. Breaks often occur in all parts of the series, some being wide, sharp and defined, others less so in various degrees; as between the orang and its nearest allies—between the Tarsius and the other Lemuridæ—between the elephant, and in a more striking manner between the Ornithorhynchus or Echidna, and all other mammals. But these breaks depend merely on the number of related forms which have become extinct. At some future period, not very distant as measured by centuries, the civilized races of man will almost certainly exterminate, and replace, the savage races throughout the world. At the same time the anthropomorphous apes, as Professor Schaaffhausen has remarked [18], will no doubt be exterminated. The break between man and his nearest allies will then be wider, for it will intervene between man in a more civilized state, as we may hope, even than the Caucasian, and some ape as low as a baboon, instead of as now between the negro or Australian and the gorilla.

[18] 'Anthropological Review,' April 1867, p. 236.

With respect to the absence of fossil remains, serving to connect man with his ape-like progenitors, no one will lay much stress on this fact who reads Sir C. Lyell's discussion [19], where he shows that in all the vertebrate classes the discovery of fossil remains has been a very slow and fortuitous process. Nor should it be forgotten that those regions which are the most likely to afford remains connecting man with some extinct ape-like creature, have not as yet been searched by geologists.

[19] 'Elements of Geology,' 1865, pp. 583-585. 'Antiquity of Man,' 1863, p. 145.

LOWER STAGES IN THE GENEALOGY OF MAN.

We have seen that man appears to have diverged from the Catarhine or Old World division of the Simiadæ, after these had diverged from the New World division. We will now endeavour to follow the remote traces of his genealogy, trusting principally to the mutual affinities between the various classes and orders, with some slight reference to the periods, as far as ascertained, of their successive appearance on the earth. The Lemuridæ stand below and near to the Simiadæ, and constitute a very distinct family of the primates, or, according to Häckel and others, a distinct Order. This group is diversified and broken to an extraordinary degree, and includes many aberrant forms. It has, therefore, probably suffered much extinction. Most of the remnants survive on islands, such as Madagascar and the Malayan archipelago, where they have not been exposed to so severe a competition as they would have been on well-stocked continents. This group likewise presents many gradations, leading, as Huxley remarks, [20] "insensibly from the crown and summit of the animal creation down to creatures from which there is but a step, as it seems, to the lowest, smallest, and least intelligent of the placental mammalia." From these various considerations it is probable that the Simiadæ were originally developed from the progenitors of the existing Lemuridæ; and these in their turn from forms standing very low in the mammalian series.

[20] 'Man's Place in Nature,' p. 105.

The Marsupials stand in many important characters below the placental mammals. They appeared at an earlier geological period, and their range was formerly much more extensive than at present. Hence the Placentata are generally supposed to have been derived from the Implacentata or Marsupials; not, however, from forms closely resembling the existing Marsupials, but from their early progenitors. The Monotremata are plainly allied to the Marsupials, forming a third and still lower division in the great mammalian series. They are represented at the present day solely by the Ornithorhynchus and Echidna; and these two forms may be safely considered as relics of a much larger group, representatives of which have been preserved in Australia through some favourable concurrence of circumstances. The Monotremata are eminently interesting, as leading in several important points of structure towards the class of reptiles.

In attempting to trace the genealogy of the Mammalia, and therefore of man, lower down in the series, we become involved in greater and greater obscurity; but as a most capable judge, Mr. Parker, has remarked, we have good reason to believe, that no true bird or reptile intervenes in the direct line of descent. He who wishes to see what ingenuity and knowledge can effect, may consult Prof. Häckel's works. [21] I will content myself with a few general remarks. Every evolutionist will admit that the five great vertebrate classes, namely, mammals, birds, reptiles, amphibians, and fishes, are descended from some one prototype; for they have much in common, especially during their embryonic state. As the class of fishes is the most lowly organised, and appeared before the others, we may conclude that all the members of the vertebrate kingdom are derived from some fishlike animal. The belief that animals so distinct as a monkey, an elephant, a humming-bird, a snake, a frog, and a fish, etc., could all have sprung from the same parents, will appear monstrous to those who have not attended to the recent progress of natural history. For this belief implies the former existence of links binding closely together all these forms, now so utterly unlike.

[21] Elaborate tables are given in his 'Generelle Morphologie' (B. ii. s. cliii. and s. 425); and with more especial reference to man in his 'Natürliche Schöpfungsgeschichte,' 1868. Prof. Huxley, in reviewing this latter work ('The Academy,' 1869, p. 42) says, that he considers the phylum or lines of descent of the Vertebrata to be admirably discussed by Häckel, although he differs on some points. He expresses, also, his high estimate of the general tenor and spirit of the whole work.

Nevertheless, it is certain that groups of animals have existed, or do now exist, which serve to connect several of the great vertebrate classes more or less closely. We have seen that the Ornithorhynchus graduates towards reptiles; and Prof. Huxley has discovered, and is confirmed by Mr. Cope and others, that the Dinosaurians are in many important characters intermediate between certain reptiles and certain birds—the birds referred to being the ostrich-tribe (itself evidently a widely-diffused remnant of a larger group) and the Archeopteryx, that strange Secondary bird, with a long lizard-like tail. Again, according to Prof. Owen [22], the Ichthyosaurians—great sea-lizards furnished with paddles—present many affinities with fishes, or rather, according to Huxley, with amphibians; a class which, including in its highest division frogs and toads, is plainly allied to the Ganoid fishes. These latter fishes swarmed during the earlier geological periods, and were constructed on what is called a generalized type, that is, they presented diversified affinities with other groups of organisms. The Lepidosiren is also so closely allied to amphibians and fishes, that naturalists long disputed in which of these two classes to rank it; it, and also some few Ganoid fishes, have been preserved from utter extinction by inhabiting rivers, which are harbors of refuge, and are related to the great waters of the ocean in the same way that islands are to continents.

[22] 'Palæontology' 1860, p. 199.

Lastly, one single member of the immense and diversified class of fishes, namely, the lancelet or amphioxus, is so different from all other fishes, that Häckel maintains that it ought to form a distinct class in the vertebrate kingdom. This fish is remarkable for its negative characters; it can hardly be said to possess a brain, vertebral column, or heart, etc.; so that it was classed by the older naturalists amongst the worms. Many years ago Prof. Goodsir perceived that the lancelet presented some affinities with the Ascidians, which are invertebrate, hermaphrodite, marine creatures permanently attached to a support. They hardly appear like animals, and consist of a simple, tough, leathery sack, with two small projecting orifices. They belong to the Mulluscoida of Huxley—a lower division of the great kingdom of the Mollusca; but they have recently been placed by some naturalists amongst the Vermes or worms. Their larvæ somewhat resemble tadpoles in shape [23], and have the power of swimming freely about. Mr. Kovalevsky [24] has lately observed that the larvæ of Ascidians are related to the Vertebrata, in their manner of development, in the relative position of the nervous system, and in possessing a structure closely like the *chorda dorsalis* of vertebrate animals; and in this he has been since confirmed by Prof. Kupffer. M. Kovalevsky writes to me from Naples, that he has now carried these observations yet further, and should his results be well established, the whole will form a discovery of the very greatest value. Thus, if we may rely on embryology, ever the safest guide in classification, it seems that we have at last gained a clue to the source whence the Vertebrata were derived. [25] We should then be justified in believing that at an extremely remote period a group of animals existed, resembling in many respects the larvæ of our present Ascidians, which diverged into two great branches—the one retrograding in development and producing the present class of Ascidians, the other rising to the crown and summit of the animal kingdom by giving birth to the Vertebrata.

[23] At the Falkland Islands I had the satisfaction of seeing, in April, 1833, and therefore some years before any other naturalist, the locomotive larvæ of a compound Ascidian, closely allied to Synoicum, but apparently generically distinct from it. The tail was about five times as long as the oblong head, and terminated in a very fine filament. It was, as sketched by me under a simple microscope, plainly divided by transverse opaque partitions, which I presume represent the great cells figured by Kovalevsky. At an early stage of development the tail was closely coiled round the head of the larva.

[24] 'Memoires de l'Acad. des Sciences de St. Pétersbourg,' tom. x. No. 15, 1866.

[25] But I am bound to add that some competent judges dispute this conclusion; for instance, M. Giard, in a series of papers in the 'Archives de Zoologie Expérimentale,' for 1872. Nevertheless, this naturalist remarks, p. 281, "L'organisation de la larve ascidienne en dehors de toute hypothèse et de toute Théorie, nous montre comment la nature peut produire la disposition fondamentale du type vertébré (l'existence d'une corde dorsale) chez un invertébré par la seule condition vitale de l'adaptation, et cette simple possibilité du passage supprime l'abîme entre les deux sous-règnes, encore bien qu'en ignore par où le passage s'est fait en réalité."

We have thus far endeavoured rudely to trace the genealogy of the Vertebrata by the aid of their mutual affinities. We will now look to man as he exists; and we shall, I think, be able partially to restore the structure of our early progenitors, during successive periods, but not in due order of time. This, can be effected by means of the rudiments which man still retains, by the characters which occasionally make their appearance in him through reversion, and by the aid of the principles of morphology and embryology. The various facts, to which I shall here allude, have been given in the previous chapters.

The early progenitors of man must have been once covered with hair, both sexes having beards; their ears were probably pointed, and capable of movement; and their bodies were provided with a tail, having the proper muscles. Their limbs and bodies were also acted on by many muscles which now only occasionally reappear, but are normally present in the Quadrumana. At this or some earlier period, the great artery and nerve of the humerus ran through a supra-condyloid foramen. The intestine gave forth a much larger diverticulum or cæcum than that now existing. The foot was then prehensile, judging from the condition of the great toe in the fœtus; and our progenitors, no doubt, were arboreal in their habits, and frequented some warm, forest-clad land. The males had great canine teeth, which served them as formidable weapons. At a much earlier period the uterus was double; the excreta were voided through a cloaca; and the eye was protected by a third eyelid or nictitating membrane. At a still earlier period the progenitors of man must have been aquatic in their habits; for morphology plainly tells us that our lungs consist of a modified swim-bladder, which once served as a float. The clefts on the neck in the embryo of man show where the branchiæ once existed. In the lunar or weekly recurrent periods of some of our functions we apparently still retain traces of our primordial birthplace, a shore washed by the tides. At about this same early period the true kidneys were replaced by the corpora wolffiana. The heart existed as a simple pulsating vessel; and the chorda dorsalis took the place of a vertebral column. These early ancestors of man, thus seen in the dim recesses of time, must have been as simply, or even still more simply organised than the lancelet or amphioxus.

There is one other point deserving a fuller notice. It has long been known that in the vertebrate kingdom one sex bears rudiments of various accessory parts, appertaining to the reproductive system, which properly belong to the opposite sex; and it has now been ascertained that at a very early embryonic period both sexes possess true male and female glands. Hence some remote progenitor of the whole vertebrate kingdom appears to have been hermaphrodite or androgynous. [26] But here we encounter a singular difficulty. In the mammalian class the males possess rudiments of a uterus with the adjacent passage, in their vesiculæ prostaticæ; they bear also rudiments of mammæ, and some male Marsupials have traces of a marsupial sack. [27] Other analogous facts could be added. Are we, then, to suppose that some extremely ancient mammal continued androgynous, after it had acquired the chief distinctions of its class, and therefore after it had diverged from the lower classes of the vertebrate kingdom? This seems very improbable, for we have to look to fishes, the lowest of all the classes, to find any still existent androgynous forms. [28] That various accessory parts, proper to each sex, are found in a rudimentary condition in the opposite sex, may be explained by such organs having been gradually acquired by the one sex, and then transmitted in a more or less imperfect state to the other. When we treat of sexual selection, we shall meet with innumerable instances of this form of transmission,—as in the case of the spurs, plumes, and brilliant colors, acquired for battle or ornament by male birds, and inherited by the females in an imperfect or rudimentary condition.

[26] This is the conclusion of Prof. Gegenbaur, one of the highest authorities in comparative anatomy: see 'Grundzüge der vergleich. Anat.' 1870, s. 876. The result has been arrived at chiefly from the study of the Amphibia; but it appears from the researches of Waldeyer (as quoted in 'Journal of Anat. and Phys.' 1869, p. 161), that the sexual organs of even "the higher vertebrata are, in their early condition, hermaphrodite." Similar views have long been held by some authors, though until recently without a firm basis.

[27] The male Thylacinus offers the best instance. Owen, 'Anatomy of Vertebrates,' vol. iii. p. 771.

[28] Hermaphroditism has been observed in several species of Serranus, as well as in

some other fishes, where it is either normal and symmetrical, or abnormal and unilateral. Dr. Zouteveen has given me references on this subject, more especially to a paper by Prof. Halbertsma, in the 'Transact. of the Dutch Acad. of Sciences,' vol. xvi. Dr. Günther doubts the fact, but it has now been recorded by too many good observers to be any longer disputed. Dr. M. Lessona writes to me, that he has verified the observations made by Cavolini on Serranus. Prof. Ercolani has recently shown ('Accad. delle Scienze,' Bologna, Dec. 28, 1871) that eels are androgynous.

The possession by male mammals of functionally imperfect mammary organs is, in some respects, especially curious. The Monotremata have the proper milk-secreting glands with orifices, but no nipples; and as these animals stand at the very base of the mammalian series, it is probable that the progenitors of the class also had milk-secreting glands, but no nipples. This conclusion is supported by what is known of their manner of development; for Professor Turner informs me, on the authority of Kölliker and Langer, that in the embryo the mammary glands can be distinctly traced before the nipples are in the least visible; and the development of successive parts in the individual generally represents and accords with the development of successive beings in the same line of descent. The Marsupials differ from the Monotremata by possessing nipples; so that probably these organs were first acquired by the Marsupials, after they had diverged from, and risen above, the Monotremata, and were then transmitted to the placental mammals. [29] No one will suppose that the marsupials still remained androgynous, after they had approximately acquired their present structure. How then are we to account for male mammals possessing mammæ? It is possible that they were first developed in the females and then transferred to the males, but from what follows this is hardly probable.

[29] Prof. Gegenbaur has shown ('Jenaische Zeitschrift,' Bd. vii. p. 212) that two distinct types of nipples prevail throughout the several mammalian orders, but that it is quite intelligible how both could have been derived from the nipples of the Marsupials, and the latter from those of the Monotremata. See, also, a memoir by Dr. Max Huss, on the mammary glands, ibid. B. viii. p. 176.

It may be suggested, as another view, that long after the progenitors of the whole mammalian class had ceased to be androgynous, both sexes yielded milk, and thus nourished their young; and in the case of the Marsupials, that both sexes carried their young in marsupial sacks. This will not appear altogether improbable, if we reflect that the males of existing syngnathous fishes receive the eggs of the females in their abdominal pouches, hatch them, and afterwards, as some believe, nourish the young; [30]—that certain other male fishes hatch the eggs within their mouths or branchial cavities;—that certain male toads take the chaplets of eggs from the females, and wind them round their own thighs, keeping them there until the tadpoles are born;—that certain male birds undertake the whole duty of incubation, and that male pigeons, as well as the females, feed their nestlings with a secretion from their crops. But the above suggestion first occurred to me from mammary glands of male mammals being so much more perfectly developed than the rudiments of the other accessory reproductive parts, which are found in the one sex though proper to the other. The mammary glands and nipples, as they exist in male mammals, can indeed hardly be called rudimentary; they are merely not fully developed, and not functionally active. They are sympathetically affected under the influence of certain diseases, like the same organs in the female. They often secrete a few drops of milk at birth and at puberty: this latter fact occurred in the curious case, before referred to, where a young man possessed two pairs of mammæ. In man and some other male mammals these organs have been known occasionally to become so well developed during maturity as to yield a fair supply of milk. Now if we suppose that during a former prolonged period male mammals aided the females in nursing their offspring, [31] and that afterwards from some cause (as from the production of a smaller number of young) the males ceased to give this aid, disuse of the organs during maturity would lead to their becoming inactive; and from two well-known principles of inheritance, this state of inactivity would probably be transmitted to the males at the corresponding age of maturity. But at an earlier age these organs would be left unaffected, so that they would be almost equally well developed in the young of both sexes.

[30] Mr. Lockwood believes (as quoted in 'Quart. Journal of Science,' April 1868, p. 269), from what he has observed of the development of Hippocampus, that the walls of the abdominal pouch of the male in some way afford nourishment. On male fishes hatching the ova in their mouths, see a very interesting paper by Prof. Wyman, in 'Proc. Boston Soc. of Nat. Hist.' Sept. 15, 1857; also Prof. Turner, in 'Journal of Anatomy and Physiology,' Nov. 1, 1866, p. 78. Dr. Günther has likewise described similar cases.

[31] Mlle. C. Royer has suggested a similar view in her 'Origine de l'Homme,' etc., 1870.

CONCLUSION.

Von Baer has defined advancement or progress in the organic scale better than any one else, as resting on the amount of differentiation and specialization of the several parts of a being,—when arrived at maturity, as I should be inclined to add. Now as organisms have become slowly adapted to diversified lines of life by means of natural selection, their parts will have become more and more differentiated and specialized for various functions from the advantage gained by the division of physiological labor. The same part appears often to have been modified first for one purpose, and then long afterwards for some other and quite distinct purpose; and thus all the parts are rendered more and more complex. But each organism still retains the general type of structure of the progenitor from which it was aboriginally derived. In accordance with this view it seems, if we turn to geological evidence, that organization on the whole has advanced throughout the world by slow and interrupted steps. In the great kingdom of the Vertebrata it has culminated in man. It must not, however, be supposed that groups of organic beings are always supplanted, and disappear as soon as they have given birth to other and more perfect groups. The latter, though victorious over their predecessors, may not have become better adapted for all places in the economy of nature. Some old forms appear to have survived from inhabiting protected sites, where they have not been exposed to very severe competition; and these often aid us in constructing our genealogies, by giving us a fair idea of former and lost populations. But we must not fall into the error of looking at the existing members of any lowly-organised group as perfect representatives of their ancient predecessors.

The most ancient progenitors in the kingdom of the Vertebrata, at which we are able to obtain an obscure glance, apparently consisted of a group of marine animals [32], resembling the larvæ of existing Ascidians. These animals probably gave rise to a group of fishes, as lowly organised as the lancelet; and from these the Ganoids, and other fishes like the Lepidosiren, must have been developed. From such fish a very small advance would carry us on to the Amphibians. We have seen that birds and reptiles were once intimately connected together; and the Monotremata now connect mammals with reptiles in a slight degree. But no one can at present say by what line of descent the three higher and related classes, namely, mammals, birds, and reptiles, were derived from the two lower vertebrate classes, namely, amphibians and fishes. In the class of mammals the steps are not difficult to conceive which led from the ancient Monotremata to the ancient Marsupials; and from these to the early progenitors of the placental mammals. We may thus ascend to the Lemuridæ; and the interval is not very wide from these to the Simiadæ. The Simiadæ then branched off into two great stems, the New World and Old World monkeys; and from the latter, at a remote period, Man, the wonder and glory of the Universe, proceeded.

[32] The inhabitants of the seashore must be greatly affected by the tides; animals living either about the *mean* high-water mark, or about the mean low-water mark, pass through a complete cycle of tidal changes in a fortnight. Consequently, their food supply will undergo marked changes week by week. The vital functions of such animals, living under these conditions for many generations, can hardly fail to run their course in regular weekly periods. Now it is a mysterious fact that in the higher and now terrestrial Vertebrata, as well as in other classes, many normal and abnormal processes have one or more whole weeks as their periods; this would be rendered intelligible if the Vertebrata are descended from an animal allied to the existing tidal Ascidians. Many instances of such periodic processes might be given, as the gestation of mammals, the duration of fevers, etc. The hatching of eggs affords also a good example, for, according to Mr. Bartlett ('Land and Water,' Jan. 7, 1871), the eggs of the pigeon are hatched in two weeks; those of the fowl in three; those of the duck in four; those of the goose in five; and those of the ostrich in seven weeks. As far as we can judge, a recurrent period, if approximately of the right duration for any process or function, would not, when once gained, be liable to change; consequently it might be thus transmitted through almost any number of generations. But if the function changed, the period would have to change, and would be apt to change almost abruptly by a whole week. This conclusion, if sound, is highly remarkable; for the period of gestation in each mammal, and the hatching of each bird's eggs, and many other vital processes, thus betray to us the primordial birthplace of these animals.

Thus we have given to man a pedigree of prodigious length, but not, it may be said, of noble quality. The world, it has often been remarked, appears as if it had long been preparing for the advent of man: and this, in one sense is strictly true, for he owes his birth to a long line of progenitors. If any single link in this chain had never existed, man would not have been exactly what he now is. Unless we wilfully close our eyes, we may, with our present knowledge, approximately recognize our parentage; nor need we feel ashamed of it. The most humble organism is something much higher than the inorganic dust under our feet; and no one with an unbiased mind can study any living creature, however humble, without being struck with enthusiasm at its marvellous structure and properties.

CHAPTER VII.

ON THE RACES OF MAN.

The nature and value of specific characters—Application to the races of man—Arguments in favour of, and opposed to, ranking the so-called races of man as district species—Sub-species—Monogenists and polygenists—Convergence of character—Numerous points of resemblance in body and mind between the most distinct races of man—The state of man when he first spread over the earth—Each race not descended from a single pair—The extinction of races—The formation of races—The effects of crossing—Slight influence of the direct action of the conditions of life—Slight or no influence of natural selection—Sexual selection.

It is not my intention here to describe the several so-called races of men; but I am about to enquire what is the value of the differences between them under a classificatory point of view, and how they have originated. In determining whether two or more allied forms ought to be ranked as species or varieties, naturalists are practically guided by the following considerations; namely, the amount of difference between them, and whether such differences relate to few or many points of structure, and whether they are of physiological importance; but more especially whether they are constant. Constancy of character is what is chiefly valued and sought for by naturalists. Whenever it can be shown, or rendered probable, that the forms in question have remained distinct for a long period, this becomes an argument of much weight in favour of treating them as species. Even a slight degree of sterility between any two forms when first crossed, or in their offspring, is generally considered as a decisive test of their specific distinctness; and their continued persistence without blending within the same area, is usually accepted as sufficient evidence, either of some degree of mutual sterility, or in the case of animals of some mutual repugnance to pairing.

Independently of fusion from intercrossing, the complete absence, in a well-investigated region, of varieties linking together any two closely-allied forms, is probably the most important of all the criterions of their specific distinctness; and this is a somewhat different consideration from mere constancy of character, for two forms may be highly variable and yet not yield intermediate varieties. Geographical distribution is often brought into play unconsciously and sometimes consciously; so that forms living in two widely separated areas, in which most of the other inhabitants are specifically distinct, are themselves usually looked at as distinct; but in truth this affords no aid in distinguishing geographical races from so-called good or true species.

Now let us apply these generally-admitted principles to the races of man, viewing him in the same spirit as a naturalist would any other animal. In regard to the amount of difference between the races, we must make some allowance for our nice powers of discrimination gained by the long habit of observing ourselves. In India, as Elphinstone remarks, although a newly-arrived European cannot at first distinguish the various native races, yet they soon appear to him extremely dissimilar [1]; and the Hindoo cannot at first perceive any difference between the several European nations. Even the most distinct races of man are much more like each other in form than would at first be supposed; certain negro tribes must be excepted, whilst others, as Dr. Rohlfs writes to me, and as I have myself seen, have Caucasian features. This general similarity is well shown by the

French photographs in the Collection Anthropologique du Museum de Paris of the men belonging to various races, the greater number of which might pass for Europeans, as many persons to whom I have shown them have remarked. Nevertheless, these men, if seen alive, would undoubtedly appear very distinct, so that we are clearly much influenced in our judgment by the mere color of the skin and hair, by slight differences in the features, and by expression.

[1] 'History of India,' 1841, vol. i. p. 323. Father Ripa makes exactly the same remark with respect to the Chinese.

There is, however, no doubt that the various races, when carefully compared and measured, differ much from each other,—as in the texture of the hair, the relative proportions of all parts of the body [2], the capacity of the lungs, the form and capacity of the skull, and even in the convolutions of the brain. [3] But it would be an endless task to specify the numerous points of difference. The races differ also in constitution, in acclimatization and in liability to certain diseases. Their mental characteristics are likewise very distinct; chiefly as it would appear in their emotional, but partly in their intellectual faculties. Every one who has had the opportunity of comparison, must have been struck with the contrast between the taciturn, even morose, aborigines of S. America and the light-hearted, talkative negroes. There is a nearly similar contrast between the Malays and the Papuans [4], who live under the same physical conditions, and are separated from each other only by a narrow space of sea.

[2] A vast number of measurements of Whites, Blacks, and Indians, are given in the 'Investigations in the Military and Anthropolog. Statistics of American Soldiers,' by B.A. Gould, 1869, pp. 298-358; 'On the capacity of the lungs,' p. 471. See also the numerous and valuable tables, by Dr. Weisbach, from the observations of Dr. Scherzer and Dr. Schwarz, in the 'Reise der Novara: Anthropolog. Theil,' 1867.

[3] See, for instance, Mr. Marshall's account of the brain of a Bushwoman, in 'Philosophical Transactions,' 1864, p. 519.

[4] Wallace, 'The Malay Archipelago,' vol. ii. 1869, p. 178.

We will first consider the arguments which may be advanced in favour of classing the races of man as distinct species, and then the arguments on the other side. If a naturalist, who had never before seen a Negro, Hottentot, Australian, or Mongolian, were to compare them, he would at once perceive that they differed in a multitude of characters, some of slight and some of considerable importance. On enquiry he would find that they were adapted to live under widely different climates, and that they differed somewhat in bodily constitution and mental disposition. If he were then told that hundreds of similar specimens could be brought from the same countries, he would assuredly declare that they were as good species as many to which he had been in the habit of affixing specific names. This conclusion would be greatly strengthened as soon as he had ascertained that these forms had all retained the same character for many centuries; and that negroes, apparently identical with existing negroes, had lived at least 4000 years ago. [5] He would also hear, on the authority of an excellent observer, Dr. Lund [6], that the human skulls found in the caves of Brazil, entombed with many extinct mammals, belonged to the same type as that now prevailing throughout the American Continent.

[5] With respect to the figures in the famous Egyptian caves of Abou-Simbel, M. Pouchet says ('The Plurality of the Human Races,' Eng. translat., 1864, p. 50), that he was far from finding recognizable representations of the dozen or more nations which some authors believe that they can recognize. Even some of the most strongly-marked races cannot be identified with that degree of unanimity which might have been expected from what has been written on the subject. Thus Messrs. Nott and Gliddon ('Types of Mankind,' p. 148), state that Rameses II., or the Great, has features superbly European; whereas Knox, another firm believer in the specific distinctness of the races of man ('Races of Man,' 1850, p. 201), speaking of young Memnon (the same as Rameses II., as I am informed by Mr. Birch), insists in the strongest manner that he is identical in character with the Jews of Antwerp. Again, when I looked at the statue of Amunoph III., I agreed with two officers of the establishment, both competent judges, that he had a strongly-marked negro type of features; but Messrs. Nott and Gliddon (ibid. p. 146, fig. 53), describe him as a hybrid, but not of "negro intermixture."

[6] As quoted by Nott and Gliddon, 'Types of Mankind,' 1854, p. 439. They give also corroborative evidence; but C. Vogt thinks that the subject requires further investigation.

Our naturalist would then perhaps turn to geographical distribution, and he would probably declare that those forms must be distinct species, which differ not only in appearance, but are fitted for hot, as well as damp or dry countries, and for the Artic regions. He might appeal to the fact that no species in the group next to man—namely, the Quadrumana, can resist a low temperature, or any considerable change of climate; and that the species which come nearest to man have never been reared to maturity, even under the temperate climate of Europe. He would be deeply impressed with the fact, first noticed by Agassiz [7], that the different races of man are distributed over the world in the same zoological provinces, as those inhabited by undoubtedly distinct species and genera of mammals. This is manifestly the case with the Australian, Mongolian, and Negro races of man; in a less well-marked manner with the Hottentots; but plainly with the Papuans and Malays, who are separated, as Mr. Wallace has shown, by nearly the same line which divides the great Malayan and Australian zoological provinces. The Aborigines of America range throughout the Continent; and this at first appears opposed to the above rule, for most of the productions of the Southern and Northern halves differ widely: yet some few living forms, as the opossum, range from the one into the other, as did formerly some of the gigantic Edentata. The Esquimaux, like other Arctic animals, extend round the whole polar regions. It should be observed that the amount of difference between the mammals of the several zoological provinces does not correspond with the degree of separation between the latter; so that it can hardly be considered as an anomaly that the Negro differs more, and the American much less from the other races of man, than do the mammals of the African and American continents from the mammals of the other provinces. Man, it may be added, does not appear to have aboriginally inhabited any oceanic island; and in this respect, he resembles the other members of his class.

[7] 'Diversity of Origin of the Human Races,' in the 'Christian Examiner,' July 1850.

In determining whether the supposed varieties of the same kind of domestic animal should be ranked as such, or as specifically distinct, that is, whether any of them are descended from distinct wild species, every naturalist would lay much stress on the fact of their external parasites being specifically distinct. All the more stress would be laid on this fact, as it would be an exceptional one; for I am informed by Mr. Denny that the most different kinds of dogs, fowls, and pigeons, in England, are infested by the same species of Pediculi or lice. Now Mr. A. Murray has carefully examined the Pediculi collected in different countries from the different races of man [8]; and he finds that they differ, not only in color, but in the structure of their claws and limbs. In every case in which many specimens were obtained the differences were constant. The surgeon of a whaling ship in the Pacific assured me that when the Pediculi, with which some Sandwich Islanders on board swarmed, strayed on to the bodies of the English sailors, they died in the course of three or four days. These Pediculi were darker colored, and appeared different from those proper to the natives of Chiloe in South America, of which he gave me specimens. These, again, appeared larger and much softer than European lice. Mr. Murray procured four kinds from Africa, namely, from the Negroes of the Eastern and Western coasts, from the Hottentots and Kaffirs; two kinds from the natives of Australia; two from North and two from South America. In these latter cases it may be presumed that the Pediculi came from natives inhabiting different districts. With insects slight structural differences, if constant, are generally esteemed of specific value: and the fact of the races of man being infested by parasites, which appear to be specifically distinct, might fairly be urged as an argument that the races themselves ought to be classed as distinct species.

[8] 'Transactions of the Royal Society of Edinburgh,' vol. xxii, 1861, p. 567.

Our supposed naturalist having proceeded thus far in his investigation, would next enquire whether the races of men, when crossed, were in any degree sterile. He might consult the work [9] of Professor Broca, a cautious and philosophical observer, and in this he would find good evidence that some races were quite fertile together, but evidence of an opposite nature in regard to other races. Thus it has been asserted that the native women of Australia and Tasmania rarely produce children to European men; the evidence, however, on this head has now been shown to be almost valueless. The half-castes are killed by the pure blacks: and an account has lately been published of eleven half-caste youths murdered and burnt at the same time, whose remains were found by the police. [10] Again, it has often been said that when mulattoes intermarry, they produce few children; on the other hand, Dr. Bachman, of Charleston [11], positively asserts that he has known mulatto families which have intermarried for several generations, and have continued on an average as fertile as either pure whites or pure blacks. Enquiries formerly made by Sir C. Lyell on this subject led him, as he informs me, to the same conclusion. [12] In the United States the census for the year 1854 included, according to Dr. Bachman, 405,751 mulattoes; and this number, considering all the circumstances of the case, seems small; but it may partly be accounted for by the degraded and anomalous position of the class, and by the profligacy of the women. A certain amount of absorption of mulattoes into negroes must always be in progress; and this would lead to an apparent diminution of the former. The inferior vitality of mulattoes is spoken of in a trustworthy work [13] as a well-known phenomenon; and this, although a different consideration from their lessened fertility, may perhaps be advanced as a proof of the specific distinctness of the parent races. No doubt both animal and vegetable hybrids, when produced from extremely distinct species, are liable to premature death; but the parents of mulattoes cannot be put under the category of extremely distinct species. The common Mule, so notorious for long life and vigor, and yet so sterile, shows how little necessary connection there is in hybrids between lessened fertility and vitality; other analogous cases could be cited.

[9] 'On the Phenomena of Hybridity in the Genus Homo,' Eng. translat., 1864.
[10] See the interesting letter by Mr. T.A. Murray, in the 'Anthropological Review,' April 1868, p. liii. In this letter Count Strzelecki's statement that Australian women who have borne children to a white man, are afterwards sterile with their own race, is disproved. M. A. de Quatrefages has also collected (Revue des Cours Scientifiques, March, 1869, p. 239), much evidence that Australians and Europeans are not sterile when crossed.
[11] 'An Examination of Prof. Agassiz's Sketch of the Nat. Provinces of the Animal World,' Charleston, 1855, p. 44.
[12] Dr. Rohlfs writes to me that he found the mixed races in the Great Sahara, derived from Arabs, Berbers, and Negroes of three tribes, extraordinarily fertile. On the other hand, Mr. Winwood Reade informs me that the Negroes on the Gold Coast, though admiring white men and mulattoes, have a maxim that mulattoes should not intermarry, as the children are few and sickly. This belief, as Mr. Reade remarks, deserves attention, as white men have visited and resided on the Gold Coast for four hundred years, so that the natives have had ample time to gain knowledge through experience.
[13] 'Military and Anthropological Statistics of American Soldiers,' by B.A. Gould, 1869, p. 319.

Even if it should hereafter be proved that all the races of men were perfectly fertile together, he who was inclined from other reasons to rank them as distinct species, might with justice argue that fertility and sterility are not safe criterions of specific distinctness. We know that these qualities are easily affected by changed conditions of life, or by close inter-breeding, and that they are governed by highly complex laws, for instance, that of the unequal fertility of converse crosses between the same two species. With forms which must be ranked as undoubted species, a perfect series exists from those which are absolutely sterile when crossed, to those which are almost or completely fertile. The degrees of sterility do not coincide strictly with the degrees of difference between the parents in external structure or habits of life. Man in many respects may be compared with those animals which have long been domesticated, and a large body of evidence can be advanced in favour of the Pallasian doctrine, [14] that domestication tends to eliminate the sterility which is so general a result of the crossing of species in a state of nature. From these several considerations, it may be justly urged that the perfect fertility of the intercrossed races of man, if established, would not absolutely preclude us from ranking them as distinct species.

[14] The 'Variation of Animals and Plants under Domestication,' vol. ii. p. 109. I may here remind the reader that the sterility of species when crossed is not a specially-acquired quality, but, like the incapacity of certain trees to be grafted together, is incidental on other acquired differences. The nature of these differences is unknown, but they relate more especially to the reproductive system, and much less so to external structure or to ordinary differences in constitution. One important element in

the sterility of crossed species apparently lies in one or both having been long habituated to fixed conditions; for we know that changed conditions have a special influence on the reproductive system, and we have good reason to believe (as before remarked) that the fluctuating conditions of domestication tend to eliminate that sterility which is so general with species, in a natural state, when crossed. It has elsewhere been shown by me (ibid. vol. ii. p. 185, and 'Origin of Species,' 5th edit. p. 317), that the sterility of crossed species has not been acquired through natural selection: we can see that when two forms have already been rendered very sterile, it is scarcely possible that their sterility should be augmented by the preservation or survival of the more and more sterile individuals; for, as the sterility increases, fewer and fewer offspring will be produced from which to breed, and at last only single individuals will be produced at the rarest intervals. But there is even a higher grade of sterility than this. Both Gartner and Kölreuter have proved that in genera of plants, including many species, a series can be formed from species which, when crossed, yield fewer and fewer seeds, to species which never produce a single seed, but yet are affected by the pollen of the other species, as shown by the swelling of the germen. It is here manifestly impossible to select the more sterile individuals, which have already ceased to yield seeds; so that the acme of sterility, when the germen alone is affected, cannot have been gained through selection. This acme, and no doubt the other grades of sterility, are the incidental results of certain unknown differences in the constitution of the reproductive system of the species which are crossed.

Independently of fertility, the characters presented by the offspring from a cross have been thought to indicate whether or not the parent-forms ought to be ranked as species or varieties; but after carefully studying the evidence, I have come to the conclusion that no general rules of this kind can be trusted. The ordinary result of a cross is the production of a blended or intermediate form; but in certain cases some of the offspring take closely after one parent-form, and some after the other. This is especially apt to occur when the parents differ in characters which first appeared as sudden variations or monstrosities. [15] I refer to this point, because Dr. Rohlfs informs me that he has frequently seen in Africa the offspring of negroes crossed with members of other races, either completely black or completely white, or rarely piebald. On the other hand, it is notorious that in America mulattoes commonly present an intermediate appearance.

[15] 'The Variation of Animals,' etc., vol. ii. p. 92.

We have now seen that a naturalist might feel himself fully justified in ranking the races of man as distinct species; for he has found that they are distinguished by many differences in structure and constitution, some being of importance. These differences have, also, remained nearly constant for very long periods of time. Our naturalist will have been in some degree influenced by the enormous range of man, which is a great anomaly in the class of mammals, if mankind be viewed as a single species. He will have been struck with the distribution of the several so-called races, which accords with that of other undoubtedly distinct species of mammals. Finally, he might urge that the mutual fertility of all the races has not as yet been fully proved, and even if proved would not be an absolute proof of their specific identity.

On the other side of the question, if our supposed naturalist were to enquire whether the forms of man keep distinct like ordinary species, when mingled together in large numbers in the same country, he would immediately discover that this was by no means the case. In Brazil he would behold an immense mongrel population of Negroes and Portuguese; in Chiloe, and other parts of South America, he would behold the whole population consisting of Indians and Spaniards blended in various degrees. [16] In many parts of the same continent he would meet with the most complex crosses between Negroes, Indians, and Europeans; and judging from the vegetable kingdom, such triple crosses afford the severest test of the mutual fertility of the parent forms. In one island of the Pacific he would find a small population of mingled Polynesian and English blood; and in the Fiji Archipelago a population of Polynesian and Negritos crossed in all degrees. Many analogous cases could be added; for instance, in Africa. Hence the races of man are not sufficiently distinct to inhabit the same country without fusion; and the absence of fusion affords the usual and best test of specific distinctness.

[16] M. de Quatrefages has given ('Anthropological Review,' Jan. 1869, p. 22), an interesting account of the success and energy of the Paulistas in Brazil, who are a much crossed race of Portuguese and Indians, with a mixture of the blood of other races.

Our naturalist would likewise be much disturbed as soon as he perceived that the distinctive characters of all the races were highly variable. This fact strikes every one on first beholding the negro slaves in Brazil, who have been imported from all parts of Africa. The same remark holds good with the Polynesians, and with many other races. It may be doubted whether any character can be named which is distinctive of a race and is constant. Savages, even within the limits of the same tribe, are not nearly so uniform in character, as has been often asserted. Hottentot women offer certain peculiarities, more strongly marked than those occurring in any other race, but these are known not to be of constant occurrence. In the several American tribes, color and hairiness differ considerably; as does color to a certain degree, and the shape of the features greatly, in the Negroes of Africa. The shape of the skull varies much in some races [17]; and so it is with every other character. Now all naturalists have learnt by dearly bought experience, how rash it is to attempt to define species by the aid of inconstant characters.

[17] For instance, with the aborigines of America and Australia, Prof. Huxley says ('Transact. Internat. Congress of Prehist. Arch.' 1868, p. 105), that the skulls of many South Germans and Swiss are "as short and as broad as those of the Tartars," etc.

But the most weighty of all the arguments against treating the races of man as distinct species, is that they graduate into each other, independently in many cases, as far as we can judge, of their having intercrossed. Man has been studied more carefully than any other animal, and yet there is the greatest possible diversity amongst capable judges whether he should be classed as a single species or race, or as two (Virey), as three (Jacquinot), as four (Kant), five (Blumenbach), six (Buffon), seven (Hunter), eight (Agassiz), eleven (Pickering), fifteen (Bory St. Vincent), sixteen (Desmoulins), twenty-two (Morton), sixty (Crawfurd), or as sixty-three, according to Burke. [18] This diversity of judgment does not prove that the races ought not to be ranked as species, but it shows that they graduate into each other, and that it is hardly possible to discover clear distinctive characters between them.

[18] See a good discussion on this subject in Waitz, 'Introduction to Anthropology,' Eng. translat., 1863, pp. 198-208, 227. I have taken some of the above statements from H. Tuttle's 'Origin and Antiquity of Physical Man,' Boston, 1866, p. 35.

Every naturalist who has had the misfortune to undertake the description of a group of highly varying organisms, has encountered cases (I speak after experience) precisely like that of man; and if of a cautious disposition, he will end by uniting all the forms which graduate into each other, under a single species; for he will say to himself that he has no right to give names to objects which he cannot define. Cases of this kind occur in the Order which includes man, namely in certain genera of monkeys; whilst in other genera, as in Cercopithecus, most of the species can be determined with certainty. In the American genus Cebus, the various forms are ranked by some naturalists as species, by others as mere geographical races. Now if numerous specimens of Cebus were collected from all parts of South America, and those forms which at present appear to be specifically distinct, were found to graduate into each other by close steps, they would usually be ranked as mere varieties or races; and this course has been followed by most naturalists with respect to the races of man. Nevertheless, it must be confessed that there are forms, at least in the vegetable kingdom [19], which we cannot avoid naming as species, but which are connected together by numberless gradations, independently of intercrossing.

[19] Prof. Nägeli has carefully described several striking cases in his 'Botanische Mittheilungen,' B. ii. 1866, ss. 294-369. Prof. Asa Gray has made analogous remarks on some intermediate forms in the Compositæ of N. America.

Some naturalists have lately employed the term "sub-species" to designate forms which possess many of the characteristics of true species, but which hardly deserve so high a rank. Now if we reflect on the weighty arguments above given, for raising the races of man to the dignity of species, and the insuperable difficulties on the other side in defining them, it seems that the term "sub-species" might here be used with propriety. But from long habit the term "race" will perhaps always be employed. The choice of terms is only so far important in that it is desirable to use, as far as possible, the same terms for the same degrees of difference. Unfortunately this can rarely be done: for the larger genera generally include closely-allied forms, which can be distinguished only with much difficulty, whilst the smaller genera within the same family include forms that are perfectly distinct; yet all must be ranked equally as species. So again, species within the same large genus by no means resemble each other to the same degree: on the contrary, some of them can generally be arranged in little groups round other species, like satellites round planets. [20]

[20] 'Origin of Species,' 5th edit. p. 68.

The question whether mankind consists of one or several species has of late years been much discussed by anthropologists, who are divided into the two schools of monogenists and polygenists. Those who do not admit the principle of evolution, must look at species as separate creations, or in some manner as distinct entities; and they must decide what forms of man they will consider as species by the analogy of the method commonly pursued in ranking other organic beings as species. But it is a hopeless endeavour to decide this point, until some definition of the term "species" is generally accepted; and the definition must not include an indeterminate element such as an act of creation. We might as well attempt without any definition to decide whether a certain number of houses should be called a village, town, or city. We have a practical illustration of the difficulty in the never-ending doubts whether many closely-allied mammals, birds, insects, and plants, which represent each other respectively in North America and Europe, should be ranked as species or geographical races; and the like holds true of the productions of many islands situated at some little distance from the nearest continent.

Those naturalists, on the other hand, who admit the principle of evolution, and this is now admitted by the majority of rising men, will feel no doubt that all the races of man are descended from a single primitive stock; whether or not they may think fit to designate the races as distinct species, for the sake of expressing their amount of difference. [21] With our domestic animals the question whether the various races have arisen from one or more species is somewhat different. Although it may be admitted that all the races, as well as all the natural species within the same genus, have sprung from the same primitive stock, yet it is a fit subject for discussion, whether all the domestic races of the dog, for instance, have acquired their present amount of difference since some one species was first domesticated by man; or whether they owe some of their characters to inheritance from distinct species, which had already been differentiated in a state of nature. With man no such question can arise, for he cannot be said to have been domesticated at any particular period.

[21] See Prof. Huxley to this effect in the 'Fortnightly Review,' 1865, p. 275.

During an early stage in the divergence of the races of man from a common stock, the differences between the races and their number must have been small; consequently as far as their distinguishing characters are concerned, they then had less claim to rank as distinct species than the existing so-called races. Nevertheless, so arbitrary is the term of species, that such early races would perhaps have been ranked by some naturalists as distinct species, if their differences, although extremely slight, had been more constant than they are at present, and had not graduated into each other.

It is however possible, though far from probable, that the early progenitors of man might formerly have diverged much in character, until they became more unlike each other than any now existing races; but that subsequently, as suggested by Vogt [22], they converged in character. When man selects the offspring of two distinct species for the same object, he sometimes induces a considerable amount of convergence, as far as general appearance is concerned. This is the case, as shown by von Nathusius [23], with the improved breeds of the pig, which are descended from two distinct species; and in a less marked manner with the improved breeds of cattle. A great anatomist, Gratiolet, maintains that the anthropomorphous apes do not form a natural sub-group; but that the orang is a highly developed gibbon or semnopithecus, the chimpanzee a highly developed macacus, and the gorilla a highly developed mandrill. If this conclusion, which rests almost exclusively on brain-characters, be admitted, we should have a case of convergence at least in external characters, for the anthropomorphous apes are certainly more like each other in many points, than they are to other apes. All analogical resemblances, as of a whale to a fish, may indeed be said to be cases of convergence; but this term has never been applied to superficial and adaptive resemblances. It would, however, be extremely rash to attribute to convergence close similarity of character in many points of structure amongst the modified descendants of widely distinct beings. The form of a crystal is determined solely by the molecular forces, and it is not surprising that dissimilar substances should sometimes assume the same form; but with organic beings we should bear in mind that the form of each depends on an infinity of complex relations, namely on variations, due to causes far too intricate to be followed,—on the nature of the variations preserved, these depending on the physical conditions, and still more on the surrounding organisms which compete with each,—and lastly, on inheritance (in itself a fluctuating element) from innumerable progenitors, all of which have had their forms determined through equally complex relations. It appears incredible that the modified descendants of two organisms, if these differed from each other in a marked manner, should ever afterwards converge so closely as to lead to a near approach to identity throughout their whole organization. In the case of the convergent races of pigs above referred to, evidence of their descent from two primitive stocks is, according to von Nathusius, still plainly retained, in certain bones of their skulls. If the races of man had descended, as is supposed by some naturalists, from two or more species, which differed from each other as much, or nearly as much, as does the orang from the gorilla, it can hardly be doubted that marked differences in the structure of certain bones would still be discoverable in man as he now exists.

[22] 'Lectures on Man,' Eng. translat., 1864, p. 468.
[23] 'Die Racen des Schweines,' 1860, s. 46. 'Vorstudien für Geschichte,' etc., Schweinescähdel, 1864, s. 104. With respect to cattle, see M. de Quatrefages, 'Unité de l'Espèce Humaine,' 1861, p. 119.

Although the existing races of man differ in many respects, as in color, hair, shape of skull, proportions of the body, etc., yet if their whole structure be taken into consideration they are found to resemble each other closely in a multitude of points. Many of these are of so unimportant or of so singular a nature, that it is extremely improbable that they should have been independently acquired by aboriginally distinct species or races. The same remark holds good with equal or greater force with respect to the numerous points of mental similarity between the most distinct races of man. The American aborigines, Negroes and Europeans are as different from each other in mind as any three races that can be named; yet I was incessantly struck, whilst living with the Feugians on board the "Beagle," with the many little traits of character, showing how similar their minds were to ours; and so it was with a full-blooded negro with whom I happened once to be intimate.

He who will read Mr. Tylor's and Sir J. Lubbock's interesting works [24] can hardly fail to be deeply impressed with the close similarity between the men of all races in tastes, dispositions and habits. This is shown by the pleasure which they all take in dancing, rude music, acting, painting, tattooing, and otherwise decorating themselves; in their mutual comprehension of gesture-language, by the same expression in their features, and by the same inarticulate cries, when excited by the same emotions. This similarity, or rather identity, is striking, when contrasted with the different expressions and cries made by distinct species of monkeys. There is good evidence that the art of shooting with bows and arrows has not been handed down from any common progenitor of mankind, yet as Westropp and Nilsson have remarked [25], the stone arrow-heads, brought from the most distant parts of the world, and manufactured at the most remote periods, are almost identical; and this fact can only be accounted for by the various races having similar inventive or mental powers. The same observation has been made by archaeologists [26] with respect to certain widely-prevalent ornaments, such as zig-zags, etc.; and with respect to various simple beliefs and customs, such as the burying of the dead under megalithic structures. I remember observing in South America [27], that there, as in so many other parts of the world, men have generally chosen the summits of lofty hills, to throw up piles of stones, either as a record of some remarkable event, or for burying their dead.

[24] Tylor's 'Early History of Mankind,' 1865: with respect to gesture-language, see p. 54. Lubbock's 'Prehistoric Times,' 2nd edit. 1869.
[25] 'On Analogous Forms of Implements,' in 'Memoirs of Anthropological Society' by H.M. Westropp. 'The Primitive Inhabitants of Scandinavia,' Eng. translat., edited by Sir J. Lubbock, 1868, p. 104.
[26] Westropp 'On Cromlechs,' etc., 'Journal of Ethnological Soc.' as given in 'Scientific Opinion,' June 2nd, 1869, p. 3.
[27] 'Journal of Researches: Voyage of the "Beagle,"' p. 46.

Now when naturalists observe a close agreement in numerous small details of habits, tastes, and dispositions between two or more domestic races, or between nearly-allied natural forms, they use this fact as an argument that they are descended from a common progenitor who was thus endowed; and consequently that all should be classed under the same species. The same argument may be applied with much force to the races of man.

As it is improbable that the numerous and unimportant points of resemblance between the several races of man in bodily structure and mental faculties (I do not here refer to similar customs) should all have been independently acquired, they must have been inherited from progenitors who had these same characters. We thus gain some insight into the early state of man, before he had spread step by step over the face of the earth. The spreading of man to regions widely separated by the sea, no doubt, preceded any great amount of divergence of character in the several races; for otherwise we should sometimes meet with the same race in distinct continents; and this is never the case. Sir J. Lubbock, after comparing the arts now practised by savages in all parts of the world, specifies those which man could not have known, when he first wandered from his original birthplace; for if once learnt they would never have been forgotten. [28] He thus shows that "the spear, which is but a development of the knife-point, and the club, which is but a long hammer, are the only things left." He admits, however, that the art of making fire probably had been already discovered, for it is common to all the races now existing, and was known to the ancient cave-inhabitants of Europe. Perhaps the art of making rude canoes or rafts was likewise known; but as man existed at a remote epoch, when the land in many places stood at a very different level to what it does now, he would have been able, without the aid of canoes, to have spread widely. Sir J. Lubbock further remarks how improbable it is that our earliest ancestors could have "counted as high as ten, considering that so many races now in existence cannot get beyond four." Nevertheless, at this early period, the intellectual and social faculties of man could hardly have been inferior in any extreme degree to those possessed at present by the lowest savages; otherwise primeval man could not have been so eminently successful in the struggle for life, as proved by his early and wide diffusion.

[28] 'Prehistoric Times,' 1869, p. 574.

From the fundamental differences between certain languages, some philologists have inferred that when man first became widely diffused, he was not a speaking animal; but it may be suspected that languages, far less perfect than any now spoken, aided by gestures, might have been used, and yet have left no traces on subsequent and more highly-developed tongues. Without the use of some language, however imperfect, it appears doubtful whether man's intellect could have risen to the standard implied by his dominant position at an early period.

Whether primeval man, when he possessed but few arts, and those of the rudest kind, and when his power of language was extremely imperfect, would have deserved to be called man, must depend on the definition which we employ. In a series of forms graduating insensibly from some ape-like creature to man as he now exists, it would be impossible to fix on any definite point where the term "man" ought to be used. But this is a matter of very little importance. So again, it is almost a matter of indifference whether the so-called races of man are thus designated, or are ranked as species or sub-species; but the latter term appears the more appropriate. Finally, we may conclude that when the principle of evolution is generally accepted, as it surely will be before long, the dispute between the monogenists and the polygenists will die a silent and unobserved death.

One other question ought not to be passed over without notice, namely, whether, as is sometimes assumed, each sub-species or race of man has sprung from a single pair of progenitors. With our domestic animals a new race can readily be formed by carefully matching the varying offspring from a single pair, or even from a single individual possessing some new character; but most of our races have been formed, not intentionally from a selected pair, but unconsciously by the preservation of many individuals which have varied, however slightly, in some useful or desired manner. If in one country stronger and heavier horses, and in another country lighter and fleeter ones, were habitually preferred, we may feel sure that two distinct sub-breeds would be produced in the course of time, without any one pair having been separated and bred from, in either country. Many races have been thus formed, and their manner of formation is closely analogous to that of natural species. We know, also, that the horses taken to the Falkland Islands have, during successive generations, become smaller and weaker, whilst those which have run wild on the Pampas have acquired larger and coarser heads; and such changes are manifestly due, not to any one pair, but to all the individuals having been subjected to the same conditions, aided, perhaps, by the principle of reversion. The new sub-breeds in such cases are not descended from any single pair, but from many individuals which have varied in different degrees, but in the same general manner; and we may conclude that the races of man have been similarly produced, the modifications being either the direct result of exposure to different conditions, or the indirect result of some form of selection. But to this latter subject we shall presently return.

ON THE EXTINCTION OF THE RACES OF MAN.

The partial or complete extinction of many races and sub-races of man is historically known. Humboldt saw in South America a parrot which was the sole living creature that could speak a word of the language of a lost tribe. Ancient monuments and stone implements found in all parts of the world, about which no tradition has been preserved by the present inhabitants, indicate much extinction. Some small and broken tribes, remnants of former races, still survive in isolated and generally mountainous districts. In Europe the ancient races were all, according to Shaaffhausen [29], "lower in the scale than the rudest living savages"; they must therefore have differed, to a certain extent, from any existing race. The remains described by Professor Broca from Les Eyzies, though they unfortunately appear to have belonged to a single family, indicate a race with a most singular combination of low or simious, and of high characteristics. This race is "entirely different from any other, ancient or modern, that we have heard of." [30] It differed, therefore, from the quaternary race of the caverns of Belgium.

[29] Translation in 'Anthropological Review,' Oct. 1868, p. 431.
[30] 'Transactions, International Congress of Prehistoric Archaeology' 1868, pp. 172-175. See also Broca (tr.) in 'Anthropological Review,' Oct. 1868, p. 410.

Man can long resist conditions which appear extremely unfavorable for his existence. [31] He has long lived in the extreme regions of the North, with no wood for his canoes or implements, and with only blubber as fuel, and melted snow as drink. In the southern extremity of America the Fuegians survive without the protection of clothes, or of any building worthy to be called a hovel. In South Africa the aborigines wander over arid plains, where dangerous beasts abound. Man can withstand the deadly influence of the Terai at the foot of the Himalaya, and the pestilential shores of tropical Africa.

[31] Dr. Gerland, 'Ueber das Aussterben der Naturvölker,' 1868, s. 82.

Extinction follows chiefly from the competition of tribe with tribe, and race with race. Various checks are always in action, serving to keep down the numbers of each savage tribe,—such as periodical famines, nomadic habits and the consequent deaths of infants, prolonged suckling, wars, accidents, sickness, licentiousness, the stealing of women, infanticide, and especially lessened fertility. If any one of these checks increases in power, even slightly, the tribe thus affected tends to decrease; and when of two adjoining tribes one becomes less numerous and less powerful than the other, the contest is soon settled by war, slaughter, cannibalism, slavery, and absorption. Even when a weaker tribe is not thus abruptly swept away, if it once begins to decrease, it generally goes on decreasing until it becomes extinct. [32]

[32] Gerland (ibid. s. 12) gives facts in support of this statement.

When civilized nations come into contact with barbarians the struggle is short, except where a deadly climate gives its aid to the native race. Of the causes which lead to the victory of civilized nations, some are plain and simple, others complex and obscure. We can see that the cultivation of the land will be fatal in many ways to savages, for they cannot, or will not, change their habits. New diseases and vices have in some cases proved highly destructive; and it appears that a new disease often causes much death, until those who are most susceptible to its destructive influence are gradually weeded out [33]; and so it may be with the evil effects from spirituous liquors, as well as with the unconquerably strong taste for them shown by so many savages. It further appears, mysterious as is the fact, that the first meeting of distinct and separated people generates disease. [34] Mr. Sproat, who in Vancouver Island closely attended to the subject of extinction, believed that changed habits of life, consequent on the advent of Europeans, induces much ill health. He lays, also, great stress on the apparently trifling cause that the natives become "bewildered and dull by the new life around them; they lose the motives for exertion, and get no new ones in their place." [35]

[33] See remarks to this effect in Sir H. Holland's 'Medical Notes and Reflections,' 1839, p. 390.

[34] I have collected ('Journal of Researches: Voyage of the "Beagle,"' p. 435) a good many cases bearing on this subject; see also Gerland, ibid. s. 8. Poeppig speaks of the "breath of civilization as poisonous to savages."

[35] Sproat, 'Scenes and Studies of Savage Life,' 1868, p. 284.

The grade of their civilization seems to be a most important element in the success of competing nations. A few centuries ago Europe feared the inroads of Eastern barbarians; now any such fear would be ridiculous. It is a more curious fact, as Mr. Bagehot has remarked, that savages did not formerly waste away before the classical nations, as they now do before modern civilized nations; had they done so, the old moralists would have mused over the event; but there is no lament in any writer of that period over the perishing barbarians. [36] The most potent of all the causes of extinction, appears in many cases to be lessened fertility and ill-health, especially amongst the children, arising from changed conditions of life, notwithstanding that the new conditions may not be injurious in themselves. I am much indebted to Mr. H.H. Howorth for having called my attention to this subject, and for having given me information respecting it. I have collected the following cases.

[36] Bagehot, 'Physics and Politics,' 'Fortnightly Review,' April 1, 1868, p. 455.

When Tasmania was first colonized the natives were roughly estimated by some at 7000 and by others at 20,000. Their number was soon greatly reduced, chiefly by fighting with the English and with each other. After the famous hunt by all the colonists, when the remaining natives delivered themselves up to the government, they consisted only of 120 individuals [37], who were in 1832 transported to Flinders Island. This island, situated between Tasmania and Australia, is forty miles long, and from twelve to eighteen miles broad: it seems healthy, and the natives were well treated. Nevertheless, they suffered greatly in health. In 1834 they consisted (Bonwick, p. 250) of forty-seven adult males, forty-eight adult females, and sixteen children, or in all of 111 souls. In 1835 only one hundred were left. As they continued rapidly to decrease, and as they themselves thought that they should not perish so quickly elsewhere, they were removed in 1847 to Oyster Cove in the southern part of Tasmania. They then consisted (Dec. 20th, 1847) of fourteen men, twenty-two women and ten children. [38] But the change of site did no good. Disease and death still pursued them, and in 1864 one man (who died in 1869), and three elderly women alone survived. The infertility of the women is even a more remarkable fact than the liability of all to ill-health and death. At the time when only nine women were left at Oyster Cove, they told Mr. Bonwick (p. 386), that only two had ever borne children: and these two had together produced only three children!

[37] All the statements here given are taken from 'The Last of the Tasmanians,' by J.

Bonwick, 1870.

[38] This is the statement of the Governor of Tasmania, Sir W. Denison, 'Varieties of Vice-Regal Life,' 1870, vol. i. p. 67.

With respect to the cause of this extraordinary state of things, Dr. Story remarks that death followed the attempts to civilize the natives. "If left to themselves to roam as they were wont and undisturbed, they would have reared more children, and there would have been less mortality." Another careful observer of the natives, Mr. Davis, remarks, "The births have been few and the deaths numerous. This may have been in a great measure owing to their change of living and food; but more so to their banishment from the mainland of Van Diemen's Land, and consequent depression of spirits" (Bonwick, pp. 388, 390).

Similar facts have been observed in two widely different parts of Australia. The celebrated explorer, Mr. Gregory, told Mr. Bonwick, that in Queensland "the want of reproduction was being already felt with the blacks, even in the most recently settled parts, and that decay would set in." Of thirteen aborigines from Shark's Bay who visited Murchison River, twelve died of consumption within three months. [39]

[39] For these cases, see Bonwick's 'Daily Life of the Tasmanians,' 1870, p. 90: and the 'Last of the Tasmanians,' 1870, p. 386.

The decrease of the Maories of New Zealand has been carefully investigated by Mr. Fenton, in an admirable Report, from which all the following statements, with one exception, are taken. [40] The decrease in number since 1830 is admitted by every one, including the natives themselves, and is still steadily progressing. Although it has hitherto been found impossible to take an actual census of the natives, their numbers were carefully estimated by residents in many districts. The result seems trustworthy, and shows that during the fourteen years, previous to 1858, the decrease was 19.42 per cent. Some of the tribes, thus carefully examined, lived above a hundred miles apart, some on the coast, some inland; and their means of subsistence and habits differed to a certain extent (p. 28). The total number in 1858 was believed to be 53,700, and in 1872, after a second interval of fourteen years, another census was taken, and the number is given as only 36,359, showing a decrease of 32.29 per cent! [41] Mr. Fenton, after showing in detail the insufficiency of the various causes, usually assigned in explanation of this extraordinary decrease, such as new diseases, the profligacy of the women, drunkenness, wars, etc., concludes on weighty grounds that it depends chiefly on the unproductiveness of the women, and on the extraordinary mortality of the young children (pp. 31, 34). In proof of this he shows (p. 33) that in 1844 there was one non-adult for every 2.57 adults; whereas in 1858 there was only one non-adult for every 3.27 adults. The mortality of the adults is also great. He adduces as a further cause of the decrease the inequality of the sexes; for fewer females are born than males. To this latter point, depending perhaps on a widely distinct cause, I shall return in a future chapter. Mr. Fenton contrasts with astonishment the decrease in New Zealand with the increase in Ireland; countries not very dissimilar in climate, and where the inhabitants now follow nearly similar habits. The Maories themselves (p. 35) "attribute their decadence, in some measure, to the introduction of new food and clothing, and the attendant change of habits"; and it will be seen, when we consider the influence of changed conditions on fertility, that they are probably right. The diminution began between the years 1830 and 1840; and Mr. Fenton shows (p. 40) that about 1830, the art of manufacturing putrid corn (maize), by long steeping in water, was discovered and largely practised; and this proves that a change of habits was beginning amongst the natives, even when New Zealand was only thinly inhabited by Europeans. When I visited the Bay of Islands in 1835, the dress and food of the inhabitants had already been much modified: they raised potatoes, maize, and other agricultural produce, and exchanged them for English manufactured goods and tobacco.

[40] 'Observations on the Aboriginal Inhabitants of New Zealand,' published by the Government, 1859.
[41] 'New Zealand,' by Alex. Kennedy, 1873, p. 47.

It is evident from many statements in the life of Bishop Patteson, [42] that the Melanesians of the New Hebrides and neighbouring archipelagoes, suffered to an extraordinary degree in health, and perished in large numbers, when they were removed to New Zealand, Norfolk Island, and other salubrious places, in order to be educated as missionaries.

[42] 'Life of J.C. Patteson,' by C.M. Younge, 1874; see more especially vol. i. p. 530.

The decrease of the native population of the Sandwich Islands is as notorious as that of New Zealand. It has been roughly estimated by those best capable of judging, that when Cook discovered the Islands in 1779, the population amounted to about 300,000. According to a loose census in 1823, the numbers then were 142,050. In 1832, and at several subsequent periods, an accurate census was officially taken, but I have been able to obtain only the following returns:

YEAR.	NATIVE POPULATION. (Except during 1832 and 1836, when the few foreigners in the islands were included.)	Annual rate of decrease per cent., assuming it to have been uniform between the successive censuses; these censuses being taken at irregular intervals.
1832	130,313	
		4.46
1836	108,579	
		2.47
1853	71,019	
		0.81
1860	67,084	
		2.18
1866	58,765	
		2.17
1872	51,531	

We here see that in the interval of forty years, between 1832 and 1872, the population has decreased no less than sixty-eight per cent.! This has been attributed by most writers to the profligacy of the women, to former bloody wars, and to the severe labor imposed on conquered tribes and to newly introduced diseases, which have been on several occasions extremely destructive. No doubt these and other such causes have been highly efficient, and may account for the extraordinary rate of decrease between the years

1832 and 1836; but the most potent of all the causes seems to be lessened fertility. According to Dr. Ruschenberger of the U.S. Navy, who visited these islands between 1835 and 1837, in one district of Hawaii, only twenty-five men out of 1134, and in another district only ten out of 637, had a family with as many as three children. Of eighty married women, only thirty-nine had ever borne children; and "the official report gives an average of half a child to each married couple in the whole island." This is almost exactly the same average as with the Tasmanians at Oyster Cove. Jarves, who published his History in 1843, says that "families who have three children are freed from all taxes; those having more, are rewarded by gifts of land and other encouragements." This unparallèled enactment by the government well shows how infertile the race had become. The Rev. A. Bishop stated in the Hawaiian 'Spectator' in 1839, that a large proportion of the children die at early ages, and Bishop Staley informs me that this is still the case, just as in New Zealand. This has been attributed to the neglect of the children by the women, but it is probably in large part due to innate weakness of constitution in the children, in relation to the lessened fertility of their parents. There is, moreover, a further resemblance to the case of New Zealand, in the fact that there is a large excess of male over female births: the census of 1872 gives 31,650 males to 25,247 females of all ages, that is 125.36 males for every 100 females; whereas in all civilized countries the females exceed the males. No doubt the profligacy of the women may in part account for their small fertility; but their changed habits of life is a much more probable cause, and which will at the same time account for the increased mortality, especially of the children. The islands were visited by Cook in 1779, Vancouver in 1794, and often subsequently by whalers. In 1819 missionaries arrived, and found that idolatry had been already abolished, and other changes effected by the king. After this period there was a rapid change in almost all the habits of life of the natives, and they soon became "the most civilized of the Pacific Islanders." One of my informants, Mr. Coan, who was born on the islands, remarks that the natives have undergone a greater change in their habits of life in the course of fifty years than Englishmen during a thousand years. From information received from Bishop Staley, it does not appear that the poorer classes have ever much changed their diet, although many new kinds of fruit have been introduced, and the sugar-cane is in universal use. Owing, however, to their passion for imitating Europeans, they altered their manner of dressing at an early period, and the use of alcoholic drinks became very general. Although these changes appear inconsiderable, I can well believe, from what is known with respect to animals, that they might suffice to lessen the fertility of the natives. [43]

[43] The foregoing statements are taken chiefly from the following works: Jarves' 'History of the Hawaiian Islands,' 1843, pp. 400-407. Cheever, 'Life in the Sandwich Islands,' 1851, p. 277. Ruschenberger is quoted by Bonwick, 'Last of the Tasmanians,' 1870, p. 378. Bishop is quoted by Sir E. Belcher, 'Voyage Round the World,' 1843, vol. i. p. 272. I owe the census of the several years to the kindness of Mr. Coan, at the request of Dr. Youmans of New York; and in most cases I have compared the Youmans figures with those given in several of the above-named works. I have omitted the census for 1850, as I have seen two widely different numbers given.

Lastly, Mr. Macnamara states [44] that the low and degraded inhabitants of the Andaman Islands, on the eastern side of the Gulf of Bengal, are "eminently susceptible to any change of climate: in fact, take them away from their island homes, and they are almost certain to die, and that independently of diet or extraneous influences." He further states that the inhabitants of the Valley of Nepal, which is extremely hot in summer, and also the various hill-tribes of India, suffer from dysentery and fever when on the plains; and they die if they attempt to pass the whole year there.

[44] 'The Indian Medical Gazette,' Nov. 1, 1871, p. 240.

We thus see that many of the wilder races of man are apt to suffer much in health when subjected to changed conditions or habits of life, and not exclusively from being transported to a new climate. Mere alterations in habits, which do not appear injurious in themselves, seem to have this same effect; and in several cases the children are particularly liable to suffer. It has often been said, as Mr. Macnamara remarks, that man can resist with impunity the greatest diversities of climate and other changes; but this is true only of the civilized races. Man in his wild condition seems to be in this respect almost as susceptible as his nearest allies, the anthropoid apes, which have never yet survived long, when removed from their native country.

Lessened fertility from changed conditions, as in the case of the Tasmanians, Maories, Sandwich Islanders, and apparently the Australians, is still more interesting than their liability to ill-health and death; for even a slight degree of infertility, combined with those other causes which tend to check the increase of every population, would sooner or later lead to extinction. The diminution of fertility may be explained in some cases by the profligacy of the women (as until lately with the Tahitians), but Mr. Fenton has shown that this explanation by no means suffices with the New Zealanders, nor does it with the Tasmanians.

In the paper above quoted, Mr. Macnamara gives reasons for believing that the inhabitants of districts subject to malaria are apt to be sterile; but this cannot apply in several of the above cases. Some writers have suggested that the aborigines of islands have suffered in fertility and health from long continued inter-breeding; but in the above cases infertility has coincided too closely with the arrival of Europeans for us to admit this explanation. Nor have we at present any reason to believe that man is highly sensitive to the evil effects of inter-breeding, especially in areas so large as New Zealand, and the Sandwich archipelago with its diversified stations. On the contrary, it is known that the present inhabitants of Norfolk Island are nearly all cousins or near relations, as are the Todas in India, and the inhabitants of some of the Western Islands of Scotland; and yet they seem not to have suffered in fertility. [45]

[45] On the close relationship of the Norfolk Islanders, Sir W. Denison, 'Varieties of Vice-Regal Life,' vol. i. 1870, p. 410. For the Todas, see Col. Marshall's work 1873, p. 110. For the Western Islands of Scotland, Dr. Mitchell, 'Edinburgh Medical Journal,' March to June, 1865.

A much more probable view is suggested by the analogy of the lower animals. The reproductive system can be shown to be susceptible to an extraordinary degree (though why we know not) to changed conditions of life; and this susceptibility leads both to beneficial and to evil results. A large collection of facts on this subject is given in chap. xviii. of vol. ii. of my 'Variation of Animals and Plants under Domestication,' I can here give only the briefest abstract; and every one interested in the subject may consult the above work. Very slight changes increase the health, vigor, and fertility of most or all organic beings, whilst other changes are known to render a large number of animals sterile. One of the most familiar cases, is that of tamed elephants not breeding in India; though they often breed in Ava, where the females are allowed to roam about the forests to some extent, and are thus placed under more natural conditions. The case of various American monkeys, both sexes of which have been kept for many years together in their own countries, and yet have very rarely or never bred, is a more apposite instance, because of their relationship to man. It is remarkable how slight a change in the conditions often induces sterility in a wild animal when captured; and this is the more strange as all our domesticated animals have become more fertile than they were in a state of nature; and some of them can resist the most unnatural conditions with undiminished fertility. [46] Certain groups of animals are much more liable than others to be affected by captivity; and generally all the species of the same group are affected in the same manner. But sometimes a single species in a group is rendered sterile, whilst the others are not so; on the other hand, a single species may retain its fertility whilst most of the others fail to breed. The males and females of some species when confined, or when allowed to live almost, but not quite free, in their native country, never unite; others thus circumstanced frequently unite but never produce offspring; others again produce some offspring, but fewer than in a state of nature; and as bearing on the above cases of man, it is important to remark that the young are apt to be weak and sickly, or malformed, and to perish at an early age.

[46] For the evidence on this head, see 'Variation of Animals,' etc., vol. ii. p. 111.

Seeing how general is this law of the susceptibility of the reproductive system to changed conditions of life, and that it holds good with our nearest allies, the Quadrumana, I can hardly doubt that it applies to man in his primeval state. Hence if savages of any race are induced suddenly to change their habits of life, they become more or less sterile, and their young offspring suffer in health, in the same manner and from the same cause, as do the elephant and hunting-leopard in India, many monkeys in America, and a host of animals of all kinds, on removal from their natural conditions.

We can see why it is that aborigines, who have long inhabited islands, and who must have been long exposed to nearly uniform conditions, should be specially affected by any change in their habits, as seems to be the case. Civilized races can certainly resist changes of all kinds far better than savages; and in this respect they resemble domesticated animals, for though the latter sometimes suffer in health (for instance European dogs in India), yet they are rarely rendered sterile, though a few such instances have been recorded. [47] The immunity of civilized races and domesticated animals is probably due to their having been subjected to a greater extent, and therefore having grown somewhat more accustomed, to diversified or varying conditions, than the majority of wild animals; and to their having formerly immigrated or been carried from country to country, and to different families or sub-races having inter-crossed. It appears that a cross with civilized races at once gives to an aboriginal race an immunity from the evil consequences of changed conditions. Thus the crossed offspring from the Tahitians and English, when settled in Pitcairn Island, increased so rapidly that the island was soon overstocked; and in June 1856 they were removed to Norfolk Island. They then consisted of 60 married persons and 134 children, making a total of 194. Here they likewise increased so rapidly, that although sixteen of them returned to Pitcairn Island in 1859, they numbered in

January 1868, 300 souls; the males and females being in exactly equal numbers. What a contrast does this case present with that of the Tasmanians; the Norfolk Islanders *increased* in only twelve and a half years from 194 to 300; whereas the Tasmanians *decreased* during fifteen years from 120 to 46, of which latter number only ten were children. [48]

[47] 'Variation of Animals,' etc., vol. ii. p. 16.
[48] These details are taken from 'The Mutineers of the "Bounty,"' by Lady Belcher, 1870; and from 'Pitcairn Island,' ordered to be printed by the House of Commons, May 29, 1863. The following statements about the Sandwich Islanders are from the 'Honolulu Gazette,' and from Mr. Coan.

So again in the interval between the census of 1866 and 1872 the natives of full blood in the Sandwich Islands decreased by 8081, whilst the half-castes, who are believed to be healthier, increased by 847; but I do not know whether the latter number includes the offspring from the half-castes, or only the half-castes of the first generation.

The cases which I have here given all relate to aborigines, who have been subjected to new conditions as the result of the immigration of civilized men. But sterility and ill-health would probably follow, if savages were compelled by any cause, such as the inroad of a conquering tribe, to desert their homes and to change their habits. It is an interesting circumstance that the chief check to wild animals becoming domesticated, which implies the power of their breeding freely when first captured, and one chief check to wild men, when brought into contact with civilization, surviving to form a civilized race, is the same, namely, sterility from changed conditions of life.

Finally, although the gradual decrease and ultimate extinction of the races of man is a highly complex problem, depending on many causes which differ in different places and at different times; it is the same problem as that presented by the extinction of one of the higher animals—of the fossil horse, for instance, which disappeared from South America, soon afterwards to be replaced, within the same districts, by countless troops of the Spanish horse. The New Zealander seems conscious of this parallelism, for he compares his future fate with that of the native rat now almost exterminated by the European rat. Though the difficulty is great to our imagination, and really great, if we wish to ascertain the precise causes and their manner of action, it ought not to be so to our reason, as long as we keep steadily in mind that the increase of each species and each race is constantly checked in various ways; so that if any new check, even a slight one, be superadded, the race will surely decrease in number; and decreasing numbers will sooner or later lead to extinction; the end, in most cases, being promptly determined by the inroads of conquering tribes.

ON THE FORMATION OF THE RACES OF MAN.

In some cases the crossing of distinct races has led to the formation of a new race. The singular fact that the Europeans and Hindoos, who belong to the same Aryan stock, and speak a language fundamentally the same, differ widely in appearance, whilst Europeans differ but little from Jews, who belong to the Semitic stock, and speak quite another language, has been accounted for by Broca [49], through certain Aryan branches having been largely crossed by indigenous tribes during their wide diffusion. When two races in close contact cross, the first result is a heterogeneous mixture: thus Mr. Hunter, in describing the Santali or hill-tribes of India, says that hundreds of imperceptible gradations may be traced "from the black, squat tribes of the mountains to the tall olive-colored Brahman, with his intellectual brow, calm eyes, and high but narrow head"; so that it is necessary in courts of justice to ask the witnesses whether they are Santalis or Hindoos. [50] Whether a heterogeneous people, such as the inhabitants of some of the Polynesian islands, formed by the crossing of two distinct races, with few or no pure members left, would ever become homogeneous, is not known from direct evidence. But as with our domesticated animals, a cross-breed can certainly be fixed and made uniform by careful selection [51] in the course of a few generations, we may infer that the free intercrossing of a heterogeneous mixture during a long descent would supply the place of selection, and overcome any tendency to reversion; so that the crossed race would ultimately become homogeneous, though it might not partake in an equal degree of the characters of the two parent-races.

[49] 'On Anthropology,' translation, 'Anthropological Review,' Jan. 1868, p. 38.
[50] 'The Annals of Rural Bengal,' 1868, p. 134.
[51] 'The Variation of Animals and Plants under Domestication,' vol. ii. p. 95.

Of all the differences between the races of man, the color of the skin is the most conspicuous and one of the best marked. It was formerly thought that differences of this kind could be accounted for by long exposure to different climates; but Pallas first showed that this is not tenable, and he has since been followed by almost all anthropologists. [52] This view has been rejected chiefly because the distribution of the variously colored races, most of whom must have long inhabited their present homes, does not coincide with corresponding differences of climate. Some little weight may be given to such cases as that of the Dutch families, who, as we hear on excellent authority [53], have not undergone the least change of color after residing for three centuries in South Africa. An argument on the same side may likewise be drawn from the uniform appearance in various parts of the world of gipsies and Jews, though the uniformity of the latter has been somewhat exaggerated. [54] A very damp or a very dry atmosphere has been supposed to be more influential in modifying the color of the skin than mere heat; but as D'Orbigny in South America, and Livingstone in Africa, arrived at diametrically opposite conclusions with respect to dampness and dryness, any conclusion on this head must be considered as very doubtful. [55]

[52] Pallas, 'Act. Acad. St. Petersburg,' 1780, part ii. p. 69. He was followed by Rudolphi, in his 'Beyträge zur Anthropologie,' 1812. An excellent summary of the evidence is given by Godron, 'De l'Espèce,' 1859, vol. ii. p. 246, etc.
[53] Sir Andrew Smith, as quoted by Knox, 'Races of Man,' 1850, p. 473.
[54] See De Quatrefages on this head, 'Revue des Cours Scientifiques,' Oct. 17, 1868, p. 731.
[55] Livingstone's 'Travels and Researches in S. Africa,' 1857, pp. 338, 339. D'Orbigny, as quoted by Godron, 'De l'Espèce,' vol. ii. p. 266.

Various facts, which I have given elsewhere, prove that the color of the skin and hair is sometimes correlated in a surprising manner with a complete immunity from the action of certain vegetable poisons, and from the attacks of certain parasites. Hence it occurred to me, that negroes and other dark races might have acquired their dark tints by the darker individuals escaping from the deadly influence of the miasma of their native countries, during a long series of generations.

I afterwards found that this same idea had long ago occurred to Dr. Wells. [56] It has long been known that negroes, and even mulattoes, are almost completely exempt from the yellow-fever, so destructive in tropical America. [57] They likewise escape to a large extent the fatal intermittent fevers, that prevail along at least 2600 miles of the shores of Africa, and which annually cause one-fifth of the white settlers to die, and another fifth to return home invalided. [58] This immunity in the negro seems to be partly inherent, depending on some unknown peculiarity of constitution, and partly the result of acclimatization. Pouchet [59] states that the negro regiments recruited near the Soudan, and borrowed from the Viceroy of Egypt for the Mexican war, escaped the yellow-fever almost equally with the negroes originally brought from various parts of Africa and accustomed to the climate of the West Indies. That acclimatization plays a part, is shown by the many cases in which negroes have become somewhat liable to tropical fevers, after having resided for some time in a colder climate. [60] The nature of the climate under which the white races have long resided, likewise has some influence on them; for during the fearful epidemic of yellow fever in Demerara during 1837, Dr. Blair found that the death-rate of the immigrants was proportional to the latitude of the country whence they had come. With the negro the immunity, as far as it is the result of acclimatization, implies exposure during a prodigious length of time; for the aborigines of tropical America who have resided there from time immemorial, are not exempt from yellow fever; and the Rev. H.B. Tristram states, that there are districts in Northern Africa which the native inhabitants are compelled annually to leave, though the negroes can remain with safety.

[56] See a paper read before the Royal Soc. in 1813, and published in his Essays in 1818. I have given an account of Dr. Wells' views in the Historical Sketch (p. xvi.) to my 'Origin of Species.' Various cases of color correlated with constitutional peculiarities are given in my 'Variation of Animals and Plants under Domestication,' vol. ii. pp. 227, 335.
[57] See, for instance, Nott and Gliddon, 'Types of Mankind,' p. 68.
[58] Major Tulloch, in a paper read before the Statistical Society, April 20, 1840, and given in the 'Athenaeum,' 1840, p. 353.
[59] 'The Plurality of the Human Race' (translat.), 1864, p. 60.
[60] Quatrefages, 'Unite de l'Espèce Humaine,' 1861, p. 205. Waitz, 'Introduction to Anthropology,' translat., vol. i. 1863, p. 124. Livingstone gives analogous cases in his 'Travels.'

That the immunity of the negro is in any degree correlated with the color of his skin is a mere conjecture: it may be correlated with some difference in his blood, nervous system, or other tissues. Nevertheless, from the facts above alluded to, and from some connection apparently existing between complexion and a tendency to consumption, the conjecture seemed to me not improbable. Consequently I endeavoured, with but little success [61], to ascertain how far it holds good. The late Dr. Daniell, who had long lived on the West Coast of Africa, told me that he did not believe in any such relation. He was himself unusually fair, and had withstood the climate in a wonderful manner. When he first arrived as a boy on the coast, an old and experienced negro chief predicted from his appearance that this would prove the case. Dr. Nicholson, of Antigua, after having attended to this subject, writes to me that dark-colored Europeans escape the yellow fever more than those that are light-colored. Mr. J.M. Harris altogether denies that Europeans with dark hair withstand a hot climate better than other men: on the contrary, experience has taught him in making a selection of men for service on the coast of Africa, to choose those with red hair. [62] As far, therefore, as these slight indications go, there seems no foundation for the hypothesis, that blackness has resulted from the darker and darker individuals having survived better during long exposure to fever-generating miasma.

[61] In the spring of 1862 I obtained permission from the Director-General of the Medical department of the Army, to transmit to the surgeons of the various regiments on foreign service a blank table, with the following appended remarks, but I have received no returns. "As several well-marked cases have been recorded with our domestic animals of a relation between the color of the dermal appendages and the constitution; and it being notorious that there is some limited degree of relation between the color of the races of man and the climate inhabited by them; the following investigation seems worth consideration. Namely, whether there is any relation in Europeans between the color of their hair, and their liability to the diseases of tropical countries. If the surgeons of the several regiments, when stationed in unhealthy tropical districts, would be so good as first to count, as a standard of comparison, how many men, in the force whence the sick are drawn, have dark and light-colored hair, and hair of intermediate or doubtful tints; and if a similar account were kept by the same medical gentlemen, of all the men who suffered from malarious and yellow fevers, or from dysentery, it would soon be apparent, after some thousand cases had been tabulated, whether there exists any relation between the color of the hair and constitutional liability to tropical diseases. Perhaps no such relation would be discovered, but the investigation is well worth making. In case any positive result were obtained, it might be of some practical use in selecting men for any particular service. Theoretically the result would be of high interest, as indicating one means by which a race of men inhabiting from a remote period an unhealthy tropical climate, might have become dark-colored by the better preservation of dark-haired or dark-complexioned individuals during a long succession of generations."

[62] 'Anthropological Review,' Jan. 1866, p. xxi. Dr. Sharpe also says, with respect to India ('Man a Special Creation,' 1873, p. 118), "that it has been noticed by some medical officers that Europeans with light hair and florid complexions suffer less from diseases of tropical countries than persons with dark hair and sallow complexions; and, so far as I know, there appear to be good grounds for this remark." On the other hand, Mr. Heddle, of Sierra Leone, "who has had more clerks killed under him than any other man," by the climate of the West African Coast (W. Reade, 'African Sketch Book,' vol. ii. p. 522), holds a directly opposite view, as does Capt. Burton.

Dr. Sharpe remarks [63], that a tropical sun, which burns and blisters a white skin, does not injure a black one at all; and, as he adds, this is not due to habit in the individual, for children only six or eight months old are often carried about naked, and are not affected. I have been assured by a medical man, that some years ago during each summer, but not during the winter, his hands became marked with light brown patches, like, although larger than freckles, and that these patches were never affected by sun-burning, whilst the white parts of his skin have on several occasions been much inflamed and blistered. With the lower animals there is, also, a constitutional difference in liability to the action of the sun between those parts of the skin clothed with white hair and other parts. [64] Whether the saving of the skin from being thus burnt is of sufficient importance to account for a dark tint having been gradually acquired by man through natural selection, I am unable to judge. If it be so, we should have to assume that the natives of tropical America have lived there for a much shorter time than the Negroes in Africa, or the Papuans in the southern parts of the Malay archipelago, just as the lighter-colored Hindoos have resided in India for a shorter time than the darker aborigines of the central and southern parts of the peninsula.

[63] 'Man a Special Creation,' 1873, p. 119.
[64] 'Variation of Animals and Plants under Domestication,' vol. ii. pp. 336, 337.

Although with our present knowledge we cannot account for the differences of color in the races of man, through any advantage thus gained, or from the direct action of climate; yet we must not quite ignore the latter agency, for there is good reason to believe that some inherited effect is thus produced. [65]

[65] See, for instance, Quatrefages ('Revue des Cours Scientifiques,' Oct. 10, 1868, p. 724) on the effects of residence in Abyssinia and Arabia, and other analogous cases. Dr. Rolle ('Der Mensch, seine Abstammung,' etc., 1865, s. 99) states, on the authority of Khanikof, that the greater number of German families settled in Georgia, have acquired in the course of two generations dark hair and eyes. Mr. D. Forbes informs me that the Quichuas in the Andes vary greatly in color, according to the position of the valleys inhabited by them.

We have seen in the second chapter that the conditions of life affect the development of the bodily frame in a direct manner, and that the effects are transmitted. Thus, as is generally admitted, the European settlers in the United States undergo a slight but extraordinary rapid change of appearance. Their bodies and limbs become elongated; and I hear from Col. Bernys that during the late war in the United States, good evidence was afforded of this fact by the ridiculous appearance presented by the German regiments, when dressed in ready-made clothes manufactured for the American market, and which were much too long for the men in every way. There is, also, a considerable body of evidence showing that in the Southern States the house-slaves of the third generation present a markedly different appearance from the field-slaves. [66]

[66] Harlan, 'Medical Researches,' p. 532. Quatrefages ('Unité de l'Espèce Humaine,' 1861, p. 128) has collected much evidence on this head.

If, however, we look to the races of man as distributed over the world, we must infer that their characteristic differences cannot be accounted for by the direct action of different conditions of life, even after exposure to them for an enormous period of time. The Esquimaux live exclusively on animal food; they are clothed in thick fur, and are exposed to intense cold and to prolonged darkness; yet they do not differ in any extreme degree from the inhabitants of Southern China, who live entirely on vegetable food, and are exposed almost naked to a hot, glaring climate. The unclothed Fuegians live on the marine productions of their inhospitable shores; the Botocudos of Brazil wander about the hot forests of the interior and live chiefly on vegetable productions; yet these tribes resemble each other so closely that the Fuegians on board the "Beagle" were mistaken by some Brazilians for Botocudos. The Botocudos again, as well as the other inhabitants of tropical America, are wholly different from the Negroes who inhabit the opposite shores of the Atlantic, are exposed to a nearly similar climate, and follow nearly the same habits of life.

Nor can the differences between the races of man be accounted for by the inherited effects of the increased or decreased use of parts, except to a quite insignificant degree. Men who habitually live in canoes, may have their legs somewhat stunted; those who inhabit lofty regions may have their chests enlarged; and those who constantly use certain sense-organs may have the cavities in which they are lodged somewhat increased in size, and their features consequently a little modified. With civilized nations, the reduced size of the jaws from lessened use—the habitual play of different muscles serving to express different emotions—and the increased size of the brain from greater intellectual activity, have together produced a considerable effect on their general appearance when compared with savages. [67] Increased bodily stature, without any corresponding increase in the size of the brain, may (judging from the previously adduced case of rabbits), have given to some races an elongated skull of the dolichocephalic type.

[67] See Prof. Schaaffhausen, translat., in 'Anthropological Review,' Oct. 1868, p. 429.

Lastly, the little-understood principle of correlated development has sometimes come into action, as in the case of great muscular development and strongly projecting supra-orbital ridges. The color of the skin and hair are plainly correlated, as is the texture of the hair with its color in the Mandans of North America. [68] The color also of the skin, and the odor emitted by it, are likewise in some manner connected. With the breeds of sheep the number of hairs within a given space and the number of excretory pores are related. [69] If we may judge from the analogy of our domesticated animals, many modifications of structure in man probably come under this principle of correlated development.

[68] Mr. Catlin states ('N. American Indians,' 3rd ed., 1842, vol. i. p. 49) that in the whole tribe of the Mandans, about one in ten or twelve of the members, of all ages and both sexes, have bright silvery grey hair, which is hereditary. Now this hair is as coarse and harsh as that of a horse's mane, whilst the hair of other colors is fine and soft.
[69] On the odor of the skin, Godron, 'Sur l'Espèce,' tom. ii. p. 217. On the pores in the skin, Dr. Wilckens, 'Die Aufgaben der Landwirth. Zootechnik,' 1869, s. 7.

We have now seen that the external characteristic differences between the races of man cannot be accounted for in a satisfactory manner by the direct action of the conditions of life, nor by the effects of the continued use of parts, nor through the principle of correlation. We are therefore led to enquire whether slight individual differences, to which man is eminently liable, may not have been preserved and augmented during a long series of generations through natural selection. But here we are at once met by the objection that beneficial variations alone can be thus preserved; and as far as we are enabled to judge, although always liable to err on this head, none of the differences between the races of man are of any direct or special service to him. The intellectual and moral or social faculties must of course be excepted from this remark.

The great variability of all the external differences between the races of man, likewise indicates that they cannot be of much importance; for if important, they would long ago have been either fixed and preserved, or eliminated. In this respect man resembles those forms, called by naturalists protean or polymorphic, which have remained extremely variable, owing, as it seems, to such variations being of an indifferent nature, and to their having thus escaped the action of natural selection.

We have thus far been baffled in all our attempts to account for the differences between the races of man; but there remains one important agency, namely Sexual Selection, which appears to have acted powerfully on man, as on many other animals. I do not intend to assert that sexual selection will account for all the differences between the races. An unexplained residuum is left, about which we can only say, in our ignorance, that as individuals are continually born with, for instance, heads a little rounder or narrower, and with noses a little longer or shorter, such slight differences might become fixed and uniform, if the unknown agencies which induced them were to act in a more constant manner, aided by long-continued intercrossing. Such variations come under the provisional class, alluded to in our second chapter, which for want of a better term are often called spontaneous. Nor do I pretend that the effects of sexual selection can be indicated with scientific precision; but it can be shown that it would be an inexplicable fact if man had not been modified by this agency, which appears to have acted powerfully on innumerable animals. It can further be shown that the differences between the races of man, as in color, hairiness, form of features, etc., are of a kind which might have been expected to come under the influence of sexual selection. But in order to treat this subject properly, I have found it necessary to pass the whole animal kingdom in review. I have therefore devoted to it the Second Part of this work. At the close I shall return to man, and, after attempting to show how far he has been modified through sexual selection, will give a brief summary of the chapters in this First Part.

NOTE ON THE RESEMBLANCES AND DIFFERENCES IN THE STRUCTURE AND THE DEVELOPMENT OF THE BRAIN IN MAN AND APES BY PROFESSOR HUXLEY, F.R.S.

The controversy respecting the nature and the extent of the differences in the structure of the brain in man and the apes, which arose some fifteen years ago, has not yet come to an end, though the subject matter of the dispute is, at present, totally different from what it was formerly. It was originally asserted and re-asserted, with singular pertinacity, that the brain of all the apes, even the highest, differs from that of man, in the absence of such conspicuous structures as the posterior lobes of the cerebral hemispheres, with the posterior cornu of the lateral ventricle and the *hippocampus minor*, contained in those lobes, which are so obvious in man.

But the truth that the three structures in question are as well developed in apes' as in human brains, or even better; and that it is characteristic of all the *Primates* (if we exclude the Lemurs) to have these parts well developed, stands at present on as secure a basis as any proposition in comparative anatomy. Moreover, it is admitted by every one of the long series of anatomists who, of late years, have paid special attention to the arrangement of the complicated sulci and gyri which appear upon the surface of the cerebral hemispheres in man and the higher apes, that they are disposed after the very same pattern in him, as in them. Every principal gyrus and sulcus of a chimpanzee's brain is clearly represented in that of a man, so that the terminology which applies to the one answers for the other. On this point there is no difference of opinion. Some years since, Professor Bischoff published a memoir [70] on the cerebral convolutions of man and apes; and as the purpose of my learned colleague was certainly not to diminish the value of the differences between apes and men in this respect, I am glad to make a citation from him.

[70] 'Die Grosshirn-Windungen des Menschen;' 'Abhandlungen der K. Bayerischen Akademie,' B. x. 1868.

"That the apes, and especially the orang, chimpanzee and gorilla, come very close to man in their organization, much nearer than to any other animal, is a well known fact, disputed by nobody. Looking at the matter from the point of view of organization alone, no one probably would ever have disputed the view of Linnæus, that man should be placed, merely as a peculiar species, at the head of the mammalia and of those apes. Both show, in all their organs, so close an affinity, that the most exact anatomical investigation is needed in order to demonstrate those differences which really exist. So it is with the brains. The brains of man, the orang, the chimpanzee, the gorilla, in spite of all the important differences which they present, come very close to one another" (loc. cit. p. 101).

There remains, then, no dispute as to the resemblance in fundamental characters, between the ape's brain and man's: nor any as to the wonderfully close similarity between the chimpanzee, orang and man, in even the details of the arrangement of the gyri and sulci of the cerebral hemispheres. Nor, turning to the differences between the brains of the highest apes and that of man, is there any serious question as to the nature and extent of these differences. It is admitted that the man's cerebral hemispheres are absolutely and relatively larger than those of the orang and chimpanzee; that his frontal lobes are less excavated by the upward protrusion of the roof of the orbits; that his gyri and sulci are, as a rule, less symmetrically disposed, and present a greater number of secondary plications. And it is admitted that, as a rule, in man, the temporo-occipital or "external perpendicular" fissure, which is usually so strongly marked a feature of the ape's brain is but faintly marked. But it is also clear, that none of these differences constitutes a sharp demarcation between the man's and the ape's brain. In respect to the external perpendicular fissure of Gratiolet, in the human brain for instance, Professor Turner remarks: [71]

[71] 'Convolutions of the Human Cerebrum Topographically Considered,' 1866, p. 12.

"In some brains it appears simply as an indentation of the margin of the hemisphere, but, in others, it extends for some distance more or less transversely outwards. I saw it in the right hemisphere of a female brain pass more than two inches outwards; and on another specimen, also the right hemisphere, it proceeded for four-tenths of an inch outwards, and then extended downwards, as far as the lower margin of the outer surface of the hemisphere. The imperfect definition of this fissure in the majority of human brains, as compared with its remarkable distinctness in the brain of most Quadrumana, is owing to the presence, in the former, of certain superficial, well marked, secondary convolutions which bridge it over and connect the parietal with the occipital lobe. The closer the first of these bridging gyri lies to the longitudinal fissure, the shorter is the external parieto-occipital fissure" (loc. cit. p. 12).

The obliteration of the external perpendicular fissure of Gratiolet, therefore, is not a constant character of the human brain. On the other hand, its full development is not a constant character of the higher ape's brain. For, in the chimpanzee, the more or less extensive obliteration of the external perpendicular sulcus by "bridging convolutions," on one side or the other, has been noted over and over again by Prof. Rolleston, Mr. Marshall, M. Broca and Professor Turner. At the conclusion of a special paper on this subject the latter writes: [72]

[72] Notes more especially on the bridging convolutions in the Brain of the Chimpanzee, 'Proceedings of the Royal Society of Edinburgh,' 1865-6.

"The three specimens of the brain of a chimpanzee, just described, prove, that the generalization which Gratiolet has attempted to draw of the complete absence of the first connecting convolution and the concealment of the second, as essentially characteristic features in the brain of this animal, is by no means universally applicable. In only one specimen did the brain, in these particulars, follow the law which Gratiolet has expressed. As regards the presence of the superior bridging convolution, I am inclined to think that it has existed in one hemisphere, at least, in a majority of the brains of this animal which have, up to this time, been figured or described. The superficial position of the second bridging convolution is evidently less frequent, and has as yet, I believe, only been seen in the brain (A) recorded in this communication. The asymmetrical arrangement in the convolutions of the two hemispheres, which previous observers have referred to in their descriptions, is also well illustrated in these specimens" (pp. 8, 9).

Even were the presence of the temporo-occipital, or external perpendicular, sulcus, a mark of distinction between the higher apes and man, the value of such a distinctive character would be rendered very doubtful by the structure of the brain in the Platyrhine apes. In fact, while the temporo-occipital is one of the most constant of sulci in the Catarhine, or Old World, apes, it is never very strongly developed in the New World apes; it is absent in the smaller Platyrrhini; rudimentary in *Pithecia* [73]; and more or less obliterated by bridging convolutions in *Ateles*.

[73] Flower, 'On the Anatomy of *Pithecia Monachus*,' 'Proceedings of the Zoological Society,' 1862.

A character which is thus variable within the limits of a single group can have no great taxonomic value.

It is further established, that the degree of asymmetry of the convolution of the two sides in the human brain is subject to much individual variation; and that, in those individuals of the Bushman race who have been examined, the gyri and sulci of the two hemispheres are considerably less complicated and more symmetrical than in the European brain, while, in some individuals of the chimpanzee, their complexity and asymmetry become notable. This is particularly the case in the brain of a young male chimpanzee figured by M. Broca. ('L'ordre des Primates,' p. 165, fig. 11.)

Again, as respects the question of absolute size, it is established that the difference between the largest and the smallest healthy human brain is greater than the difference between the smallest healthy human brain and the largest chimpanzee's or orang's brain.

Moreover, there is one circumstance in which the orang's and chimpanzee's brains resemble man's, but in which they differ from the lower apes, and that is the presence of two corpora candicantia—the *Cynomorpha* having but one.

In view of these facts I do not hesitate in this year 1874, to repeat and insist upon the proposition which I enunciated in 1863: [74]

[74] 'Man's Place in Nature,' p. 102.

"So far as cerebral structure goes, therefore, it is clear that man differs less from the chimpanzee or the orang, than these do even from the monkeys, and that the difference between the brain of the chimpanzee and of man is almost insignificant when compared with that between the chimpanzee brain and that of a Lemur."

In the paper to which I have referred, Professor Bischoff does not deny the second part of this statement, but he first makes the irrelevant remark that it is not wonderful if the brains of an orang and a Lemur are very different; and secondly, goes on to assert that, "If we successively compare the brain of a man with that of an orang; the brain of this with that of a chimpanzee; of this with that of a gorilla, and so on of a *Hylobates, Semnopithecus, Cynocephalus, Cercopithecus, Macacus, Cebus, Callithrix, Lemur, Stenops, Hapale,* we shall not meet with a greater, or even as great a, break in the degree of development of the convolutions, as we find between the brain of a man and that of an orang or chimpanzee."

To which I reply, firstly, that whether this assertion be true or false, it has nothing whatever to do with the proposition enunciated in 'Man's Place in Nature,' which refers not to the development of the convolutions alone, but to the structure of the whole brain. If Professor Bischoff had taken the trouble to refer to p. 96 of the work he criticises, in fact, he would have found the following passage: "And it is a remarkable circumstance that though, so far as our present knowledge extends, there IS one true structural break in the series of forms of Simian brains, this hiatus does not lie between man and the manlike apes, but between the lower and the lowest Simians, or in other words, between the Old and New World apes and monkeys and the Lemurs. Every Lemur which has yet been examined, in fact, has its cerebellum partially visible from above; and its posterior lobe, with the contained posterior cornu and hippocampus minor, more or less rudimentary. Every marmoset, American monkey, Old World monkey, baboon or manlike ape, on the contrary, has its cerebellum entirely hidden, posteriorly, by the cerebral lobes, and possesses a large posterior cornu with a well-developed hippocampus minor."

This statement was a strictly accurate account of what was known when it was made; and it does not appear to me to be more than apparently weakened by the subsequent discovery of the relatively small development of the posterior lobes in the Siamang and in the Howling monkey. Notwithstanding the exceptional brevity of the posterior lobes in these two species, no one will pretend that their brains, in the slightest degree, approach those of the Lemurs. And if, instead of putting *Hapale* out of its natural place, as Professor Bischoff most unaccountably does, we write the series of animals he has chosen to mention as follows: *Homo, Pithecus, Troglodytes, Hylobates, Semnopithecus, Cynocephalus, Cercopithecus, Macacus, Cebus, Callithrix, Hapale, Lemur, Stenops,* I venture to reaffirm that the great break in this series lies between *Hapale* and *Lemur,* and that this break is considerably greater than that between any other two terms of that series. Professor Bischoff ignores the fact that long before he wrote, Gratiolet had suggested the separation of the Lemurs from the other Primates on the very ground of the difference in their cerebral characters; and that Professor Flower had made the following observations in the course of his description of the brain of the Javan Loris: [75]

[75] 'Transactions of the Zoological Society,' vol. v. 1862.

"And it is especially remarkable that, in the development of the posterior lobes, there is no approximation to the Lemurine, short hemisphered brain, in those monkeys which are commonly supposed to approach this family in other respects, viz. the lower members of the Platyrhine group."

So far as the structure of the adult brain is concerned, then, the very considerable additions to our knowledge, which have been made by the researches of so many investigators, during the past ten years, fully justify the statement which I made in 1863. But it has been said, that, admitting the similarity between the adult brains of man and apes, they are nevertheless, in reality, widely different, because they exhibit fundamental differences in the mode of their development. No one would be more ready than I to admit the force of this argument, if such fundamental differences of development really exist. But I deny that they do exist. On the contrary, there is a fundamental agreement in the development of the brain in men and apes.

Gratiolet originated the statement that there is a fundamental difference in the development of the brains of apes and that of man—consisting in this; that, in the apes, the sulci which first make their appearance are situated on the posterior region of the cerebral hemispheres, while, in the human fœtus, the sulci first become visible on the frontal lobes. [76]

[76] "Chez tous les singes, les plis postérieurs se développent les premiers; les plis antérieurs se développent plus tard, aussi la vertèbre occipitale et la pariétale sont-elles relativement très-grandes chez le fœtus. L'Homme présente une exception remarquable quant â l'époque de l'apparition des plis frontaux, qui sont les premiers indiqués; mais le développement général du lobe frontal, envisagé seulement par rapport à son volume, suit les mêmes lois que dans les singes:" Gratiolet, 'Mémoire sur les plis cérébres de l'Homme et des Primateaux,' p. 39, Tab. iv, fig. 3.

This general statement is based upon two observations, the one of a Gibbon almost ready to be born, in which the posterior gyri were "well developed," while those of the frontal lobes were "hardly indicated" [77] (l. c. p. 39) and the other of a human fœtus at the 22nd or 23rd week of uterogestation, in which Gratiolet notes that the insula was uncovered, but that nevertheless "des incisures sèment de lobe antérieur, une scissure peu profonde indique la séparation du lobe occipital, très-réduit, d'ailleurs dès cette époque. Le reste de la surface cérébrale est encore absolument lisse."

[77] Gratiolet's words are (l. c. p. 39): "Dans le fœtus dont il s'agit les plis cérébraux postérieurs sont bien développés, tandis que les plis du lobe frontal sont à peine indiqués." The figure, however (Pl. iv, fig. 3), shows the fissure of Rolando, and one of the frontal sulci plainly enough. Nevertheless, M. Alix, in his 'Notice sur les travaux anthropologiques de Gratiolet' ('Mém. de la Société d'Anthropologie de Paris,' 1868, page 32), writes thus: "Gratiolet a eu entre les mains le cerveau d'un fœtus de Gibbon, singe éminemment supérieur, et tellement rapproché de l'orang, que des naturalistes très-compétents l'ont rangé parmi les anthropoïdes. M. Huxley, par exemple, n'hésite pas sur ce point. Eh bien, c'est sur le cerveau d'un fœtus de Gibbon que Gratiolet a vu *les circonvolutions du lobe temporo-sphénoidal déjà*

développées lorsqu'il n'existent pas encore de plis sur le lobe frontal. Il etait donc bien autorise a dire que, chez l'homme les circonvolutions apparaissent d'a en w, tandis que chez les singes elles se développent d'ω en α."),

Three views of this brain are given in Plate II, figs. 1, 2, 3, of the work cited, showing the upper, lateral and inferior views of the hemispheres, but not the inner view. It is worthy of note that the figure by no means bears out Gratiolet's description, inasmuch as the fissure (antero-temporal) on the posterior half of the face of the hemisphere is more marked than any of those vaguely indicated in the anterior half. If the figure is correct, it in no way justifies Gratiolet's conclusion: "Il y a donc entre ces cerveaux [those of a Callithrix and of a Gibbon] et celui du fœtus humain une différence fondamental. Chez celui-ci, longtemps avant que les plis temporaux apparaissent, les plis frontaux, *essayent* d'exister."

Since Gratiolet's time, however, the development of the gyri and sulci of the brain has been made the subject of renewed investigation by Schmidt, Bischoff, Pansch [78], and more particularly by Ecker [79], whose work is not only the latest, but by far the most complete, memoir on the subject.

[78] 'Ueber die typische Anordnung der Furchen und Windungen auf den Grosshirn-Hemisphären des Menschen und der Affen,' 'Archiv für Anthropologie,' iii. 1868.
[79] 'Zur Entwicklungs Geschichte der Furchen und Windungen der Grosshirn-Hemisphären im Fœtus des Menschen.' 'Archiv für Anthropologie,' iii. 1868.

The final results of their inquiries may be summed up as follows:—

1. In the human fœtus, the sylvian fissure is formed in the course of the third month of uterogestation. In this, and in the fourth month, the cerebral hemispheres are smooth and rounded (with the exception of the sylvian depression), and they project backwards far beyond the cerebellum.

2. The sulci, properly so called, begin to appear in the interval between the end of the fourth and the beginning of the sixth month of fœtal life, but Ecker is careful to point out that, not only the time, but the order, of their appearance is subject to considerable individual variation. In no case, however, are either the frontal or the temporal sulci the earliest.

The first which appears, in fact, lies on the inner face of the hemisphere (whence doubtless Gratiolet, who does not seem to have examined that face in his fœtus, overlooked it), and is either the internal perpendicular (occipito-parietal), or the calcarine sulcus, these two being close together and eventually running into one another. As a rule the occipito-parietal is the earlier of the two.

3. At the latter part of this period, another sulcus, the "posterio-parietal," or "Fissure of Rolando" is developed, and it is followed, in the course of the sixth month, by the other principal sulci of the frontal, parietal, temporal and occipital lobes. There is, however, no clear evidence that one of these constantly appears before the other; and it is remarkable that, in the brain at the period described and figured by Ecker (loc. cit. pp. 212-213, Taf. II, figs. 1, 2, 3, 4), the antero-temporal sulcus (*scissure parallèle*) so characteristic of the ape's brain, is as well, if not better developed than the fissure of Rolando, and is much more marked than the proper frontal sulci.

Taking the facts as they now stand, it appears to me that the order of the appearance of the sulci and gyri in the fœtal human brain is in perfect harmony with the general doctrine of evolution, and with the view that man has been evolved from some ape-like form; though there can be no doubt that form was, in many respects, different from any member of the *Primates* now living.

Von Baer taught us, half a century ago, that, in the course of their development, allied animals put on at first, the characters of the greater groups to which they belong, and, by degrees, assume those which restrict them within the limits of their family, genus, and species; and he proved, at the same time, that no developmental stage of a higher animal is precisely similar to the adult condition of any lower animal. It is quite correct to say that a frog passes through the condition of a fish, inasmuch as at one period of its life the tadpole has all the characters of a fish, and if it went no further, would have to be grouped among fishes. But it is equally true that a tadpole is very different from any known fish.

In like manner, the brain of a human fœtus, at the fifth month, may correctly be said to be, not only the brain of an ape, but that of an Arctopithecine or marmoset-like ape; for its hemispheres, with their great posterior lobster, and with no sulci but the sylvian and the calcarine, present the characteristics found only in the group of the Arctopithecine *Primates*. But it is equally true, as Gratiolet remarks, that, in its widely open sylvian fissure, it differs from the brain of any actual marmoset. No doubt it would be much more similar to the brain of an advanced fœtus of a marmoset. But we know nothing whatever of the development of the brain in the marmosets. In the Platyrrhini proper, the only observation with which I am acquainted is due to Pansch, who found in the brain of a fœtal *Cebus Apella*, in addition to the sylvian fissure and the deep calcarine fissure, only a very shallow antero-temporal fissure (*scissure parallèle* of Gratiolet).

Now this fact, taken together with the circumstance that the antero-temporal sulcus is present in such Platyrrhini as the Saimiri, which present mere traces of sulci on the anterior half of the exterior of the cerebral hemispheres, or none at all, undoubtedly, so far as it goes, affords fair evidence in favour of Gratiolet's hypothesis, that the posterior sulci appear before the anterior, in the brains of the *Platyrrhini*. But, it by no means follows, that the rule which may hold good for the *Platyrrhini* extends to the *Catarrhini*. We have no information whatever respecting the development of the brain in the *Cynomorpha*; and, as regards the *Anthropomorpha*, nothing but the account of the brain of the Gibbon, near birth, already referred to. At the present moment there is not a shadow of evidence to show that the sulci of a chimpanzee's, or orang's, brain do not appear in the same order as a man's.

Gratiolet opens his preface with the aphorism: "Il est dangereux dans les sciences de conclure trop vite." I fear he must have forgotten this sound maxim by the time he had reached the discussion of the differences between men and apes, in the body of his work. No doubt, the excellent author of one of the most remarkable contributions to the just understanding of the mammalian brain which has ever been made, would have been the first to admit the insufficiency of his data had he lived to profit by the advance of inquiry. The misfortune is that his conclusions have been employed by persons incompetent to appreciate their foundation, as arguments in favour of obscurantism. [80]

[80] For example, M. l'Abbé Lecomte in his terrible pamphlet, 'Le Darwinisme et l'origine de l'Homme,' 1873.

But it is important to remark that, whether Gratiolet was right or wrong in his hypothesis respecting the relative order of appearance of the temporal and frontal sulci, the fact remains; that before either temporal or frontal sulci, appear, the fœtal brain of man presents characters which are found only in the lowest group of the Primates (leaving out the Lemurs); and that this is exactly what we should expect to be the case, if man has resulted from the gradual modification of the same form as that from which the other Primates have sprung.

PART II.

SEXUAL SELECTION.

CHAPTER VIII.

PRINCIPLES OF SEXUAL SELECTION.

Secondary sexual characters—Sexual selection—Manner of action—Excess of males—Polygamy—The male alone generally modified through sexual selection—Eagerness of the male—Variability of the male—Choice exerted by the female—Sexual compared with natural selection—Inheritance, at corresponding periods of life, at corresponding seasons of the year, and as limited by sex—Relations between the several forms of inheritance—Causes why one sex and the young are not modified through sexual selection—Supplement on the proportional numbers of the two sexes throughout the animal kingdom—The proportion of the sexes in relation to natural selection.

With animals which have their sexes separated, the males necessarily differ from the females in their organs of reproduction; and these are the primary sexual characters. But the sexes often differ in what Hunter has called secondary sexual characters, which are not directly connected with the act of reproduction; for instance, the male possesses certain organs of sense or locomotion, of which the female is quite destitute, or has them more highly-developed, in order that he may readily find or reach her; or again the male has special organs of prehension for holding her securely. These latter organs, of infinitely diversified kinds, graduate into those which are commonly ranked as primary, and in some cases can hardly be distinguished from them; we see instances of this in the complex appendages at the apex of the abdomen in male insects. Unless indeed we confine the term "primary" to the reproductive glands, it is scarcely possible to decide which ought to be called primary and which secondary.

The female often differs from the male in having organs for the nourishment or protection of her young, such as the mammary glands of mammals, and the abdominal sacks of the marsupials. In some few cases also the male possesses similar organs, which are wanting in the female, such as the receptacles for the ova in certain male fishes, and those temporarily developed in certain male frogs. The females of most bees are provided with a special apparatus for collecting and carrying pollen, and their ovipositor is modified into a sting for the defence of the larvæ and the community. Many similar cases could be given, but they do not here concern us. There are, however, other sexual differences quite unconnected with the primary reproductive organs, and it is with these that we are more especially concerned—such as the greater size, strength, and pugnacity of the male, his weapons of offence or means of defence against rivals, his gaudy coloring and various ornaments, his power of song, and other such characters.

Besides the primary and secondary sexual differences, such as the foregoing, the males and females of some animals differ in structures related to different habits of life, and not at all, or only indirectly, to the reproductive functions. Thus the females of certain flies (Culicidæ and Tabanidæ) are blood-suckers, whilst the males, living on flowers, have mouths destitute of mandibles. [1] The males of certain moths and of some crustaceans (*e.g.* Tanais) have imperfect, closed mouths, and cannot feed. The complemental males of certain Cirripedes live like epiphytic plants either on the female or the hermaphrodite form, and are destitute of a mouth and of prehensile limbs. In these cases it is the male which has been modified, and has lost certain important organs, which the females possess. In other cases it is the female which has lost such parts; for instance, the female glow-worm is destitute of wings, as also are many female moths, some of which never leave their cocoons. Many female parasitic crustaceans have lost their natatory legs. In some weevil-beetles (Curculionidæ) there is a great difference between the male and female in the length of the rostrum or snout [2]; but the meaning of this and of many analogous differences, is not at all understood. Differences of structure between the two sexes in relation to different habits of life are generally confined to the lower animals; but with some few birds the beak of the male differs from that of the female. In the Huia of New Zealand the difference is wonderfully great, and we hear from Dr. Buller [3] that the male uses his strong beak in chiselling the larvæ of insects out of decayed wood, whilst the female probes the softer parts with her far longer, much curved and pliant beak: and thus they mutually aid each other. In most cases, differences of structure between the sexes are more or less directly connected with the propagation of the species: thus a female, which has to nourish a multitude of ova, requires more food than the male, and consequently requires special means for procuring it. A male animal, which lives for a very short time, might lose its organs for procuring food through disuse, without detriment; but he would retain his locomotive organs in a perfect state, so that he might reach the female. The female, on the other hand, might safely lose her organs for flying, swimming, or walking, if she gradually acquired habits which rendered such powers useless.

[1] Westwood, 'Modern Classification of Insects,' vol. ii. 1840, p. 541. For the statement about Tanais, mentioned below, I am indebted to Fritz Müller.
[2] Kirby and Spence, 'Introduction to Entomology,' vol. iii. 1826, p. 309.
[3] 'Birds of New Zealand,' 1872, p. 66.

We are, however, here concerned only with sexual selection. This depends on the advantage which certain individuals have over others of the same sex and species solely in respect of reproduction. When, as in the cases above mentioned, the two sexes differ in structure in relation to different habits of life, they have no doubt been modified through natural selection, and by inheritance limited to one and the same sex. So again the primary sexual organs, and those for nourishing or protecting the young, come under the same influence; for those individuals which generated or nourished their offspring best, would leave, *ceteris paribus*, the greatest number to inherit their superiority; whilst those which generated or nourished their offspring badly, would leave but few to inherit their weaker powers. As the male has to find the female, he requires organs of sense and locomotion, but if these organs are necessary for the other purposes of life, as is generally the case, they will have been developed through natural selection. When the male has found the female, he sometimes absolutely requires prehensile organs to hold her; thus Dr. Wallace informs me that the males of certain moths cannot unite with the females if their tarsi or feet are broken. The males of many oceanic crustaceans, when adult, have their legs and antennæ modified in an extraordinary manner for the prehension of the female; hence we may suspect that it is because these animals are washed about by the waves of the open sea, that they require these organs in order to propagate their kind, and if so, their development has been the result of ordinary or natural selection. Some animals extremely low in the scale have been modified for this same purpose; thus the males of certain parasitic worms, when fully grown, have the lower surface of the terminal part of their bodies roughened like a rasp, and with this they coil round and permanently hold the females. [4]

[4] M. Perrier advances this case ('Revue Scientifique,' Feb. 1, 1873, p. 865) as one fatal to the belief in sexual election, inasmuch as he supposes that I attribute all the differences between the sexes to sexual selection. This distinguished naturalist, therefore, like so many other Frenchmen, has not taken the trouble to understand even the first principles of sexual selection. An English naturalist insists that the claspers of certain male animals could not have been developed through the choice of the female! Had I not met with this remark, I should not have thought it possible for any one to have read this chapter and to have imagined that I maintain that the choice of the female had anything to do with the development of the prehensile organs in the male.

When the two sexes follow exactly the same habits of life, and the male has the sensory or locomotive organs more highly developed than those of the female, it may be that the perfection of these is indispensable to the male for finding the female; but in the vast majority of cases, they serve only to give one male an advantage over another, for with sufficient time, the less well-endowed males would succeed in pairing with the females; and judging from the structure of the female, they would be in all other respects equally well adapted for their ordinary habits of life. Since in such cases the males have acquired their present structure, not from being better fitted to survive in the struggle for existence, but from having gained an advantage over other males, and from having transmitted this advantage to their male offspring alone, sexual selection must here have come into action. It was the importance of this distinction which led me to designate this form of selection as Sexual Selection. So again, if the chief service rendered to the male by his prehensile organs is to prevent the escape of the female before the arrival of other males, or when assaulted by them, these organs will have been perfected through sexual selection, that is by the advantage acquired by certain individuals over their rivals. But in most cases of this kind it is impossible to distinguish between the effects of natural and sexual selection. Whole chapters could be filled with details on the differences between the sexes in their sensory, locomotive, and prehensile organs. As, however, these structures are not more interesting than others adapted for the ordinary purposes of life I shall pass them over almost entirely, giving only a few instances under each class.

There are many other structures and instincts which must have been developed through sexual selection—such as the weapons of offence and the means of defence of the males for fighting with and driving away their rivals—their courage and pugnacity— their various ornaments—their contrivances for producing vocal or instrumental music— and their glands for emitting odors, most of these latter structures serving only to allure or excite the female. It is clear that these characters are the result of sexual and not of ordinary selection, since unarmed, unornamented, or unattractive males would succeed equally well in the battle for life and in leaving a numerous progeny, but for the presence of better endowed males. We may infer that this would be the case, because the females, which are unarmed and unornamented, are able to survive and procreate their kind. Secondary sexual characters of the kind just referred to, will be fully discussed in the following chapters, as being in many respects interesting, but especially as depending on the will, choice, and rivalry of the individuals of either sex. When we behold two males fighting for the possession of the female, or several male birds displaying their gorgeous plumage, and performing strange antics before an assembled body of females, we cannot doubt that, though led by instinct, they know what they are about, and consciously exert their mental and bodily powers.

Just as man can improve the breeds of his game-cocks by the selection of those birds which are victorious in the cockpit, so it appears that the strongest and most vigorous males, or those provided with the best weapons, have prevailed under nature, and have led to the improvement of the natural breed or species. A slight degree of variability leading to some advantage, however slight, in reiterated deadly contests would suffice for the work of sexual selection; and it is certain that secondary sexual characters are eminently variable. Just as man can give beauty, according to his standard of taste, to his male poultry, or more strictly can modify the beauty originally acquired by the parent species, can give to the Sebright bantam a new and elegant plumage, an erect and peculiar carriage—so it appears that female birds in a state of nature, have by a long selection of the more attractive males, added to their beauty or other attractive qualities. No doubt this implies powers of discrimination and taste on the part of the female which will at first appear extremely improbable; but by the facts to be adduced hereafter, I hope to be able to show that the females actually have these powers. When, however, it is said that the lower animals have a sense of beauty, it must not be supposed that such sense is comparable with that of a cultivated man, with his multiform and complex associated ideas. A more just comparison would be between the taste for the beautiful in animals, and that in the lowest savages, who admire and deck themselves with any brilliant, glittering, or curious object.

From our ignorance on several points, the precise manner in which sexual selection acts is somewhat uncertain. Nevertheless if those naturalists who already believe in the mutability of species, will read the following chapters, they will, I think, agree with me, that sexual selection has played an important part in the history of the organic world. It is certain that amongst almost all animals there is a struggle between the males for the possession of the female. This fact is so notorious that it would be superfluous to give instances. Hence the females have the opportunity of selecting one out of several males, on the supposition that their mental capacity suffices for the exertion of a choice. In many cases special circumstances tend to make the struggle between the males particularly severe. Thus the males of our migratory birds generally arrive at their places of breeding before the females, so that many males are ready to contend for each female. I am informed by Mr. Jenner Weir, that the bird-catchers assert that this is invariably the case with the nightingale and blackcap, and with respect to the latter he can himself confirm the statement.

Mr. Swaysland of Brighton has been in the habit, during the last forty years, of catching our migratory birds on their first arrival, and he has never known the females of any species to arrive before their males. During one spring he shot thirty-nine males of Ray's wagtail (*Budytes Raii*) before he saw a single female. Mr. Gould has ascertained by the dissection of those snipes which arrive the first in this country, that the males come before the females. And the like holds good with most of the migratory birds of the United States. [5] The majority of the male salmon in our rivers, on coming up from the sea, are ready to breed before the females. So it appears to be with frogs and toads. Throughout the great class of insects the males almost always are the first to emerge from the pupal state, so that they generally abound for a time before any females can be seen. [6] The cause of this difference between the males and females in their periods of arrival and maturity is sufficiently obvious. Those males which annually first migrated into any country, or which in the spring were first ready to breed, or were the most eager, would leave the largest number of offspring; and these would tend to inherit similar instincts and constitutions. It must be borne in mind that it would have been impossible to change very materially the time of sexual maturity in the females, without at the same time interfering with the period of the production of the young—a period which must be determined by the seasons of the year. On the whole there can be no doubt that with almost all animals, in which the sexes are separate, there is a constantly recurrent struggle between the males for the possession of the females.

[5] J.A. Allen, on the 'Mammals and Winter Birds of Florida,' Bulletin of Comparative Zoology, Harvard College, p. 268.

[6] Even with those plants in which the sexes are separate, the male flowers are generally mature before the female. As first shown by C.K. Sprengel, many hermaphrodite plants are dichogamous; that is, their male and female organs are not ready at the same time, so that they cannot be self-fertilized. Now in such flowers, the pollen is in general matured before the stigma, though there are exceptional cases in which the female organs are beforehand.

Our difficulty in regard to sexual selection lies in understanding how it is that the males which conquer other males, or those which prove the most attractive to the females, leave a greater number of offspring to inherit their superiority than their beaten and less attractive rivals. Unless this result does follow, the characters which give to certain males an advantage over others, could not be perfected and augmented through sexual selection. When the sexes exist in exactly equal numbers, the worst-endowed males will (except where polygamy prevails), ultimately find females, and leave as many offspring, as well fitted for their general habits of life, as the best-endowed males. From various facts and considerations, I formerly inferred that with most animals, in which secondary sexual characters are well developed, the males considerably exceeded the females in number; but this is not by any means always true. If the males were to the females as two to one, or as three to two, or even in a somewhat lower ratio, the whole affair would be simple; for the better-armed or more attractive males would leave the largest number of offspring. But after investigating, as far as possible, the numerical proportion of the sexes, I do not believe that any great inequality in number commonly exists. In most cases sexual selection appears to have been effective in the following manner.

Let us take any species, a bird for instance, and divide the females inhabiting a district into two equal bodies, the one consisting of the more vigorous and better-nourished individuals, and the other of the less vigorous and healthy. The former, there can be little doubt, would be ready to breed in the spring before the others; and this is the opinion of Mr. Jenner Weir, who has carefully attended to the habits of birds during many years. There can also be no doubt that the most vigorous, best-nourished and earliest breeders would on an average succeed in rearing the largest number of fine offspring. [7] The males, as we have seen, are generally ready to breed before the females; the strongest, and with some species the best armed of the males, drive away the weaker; and the former would then unite with the more vigorous and better-nourished females, because they are the first to breed. [8] Such vigorous pairs would surely rear a larger number of offspring than the retarded females, which would be compelled to unite with the conquered and less powerful males, supposing the sexes to be numerically equal; and this is all that is wanted to add, in the course of successive generations, to the size, strength and courage of the males, or to improve their weapons.

[7] Here is excellent evidence on the character of the offspring from an experienced ornithologist. Mr. J.A. Allen, in speaking ('Mammals and Winter Birds of E. Florida,' p. 229) of the later broods, after the accidental destruction of the first, says, that these "are found to be smaller and paler-colored than those hatched earlier in the season. In cases where several broods are reared each year, as a general rule the birds of the earlier broods seem in all respects the most perfect and vigorous."

[8] Hermann Müller has come to this same conclusion with respect to those female bees which are the first to emerge from the pupa each year. See his remarkable essay, 'Anwendung der Darwin'schen Lehre auf Bienen,' 'Verh. d. V. Jahrg.' xxix. p. 45.

But in very many cases the males which conquer their rivals, do not obtain possession of the females, independently of the choice of the latter. The courtship of animals is by no means so simple and short an affair as might be thought. The females are most excited by, or prefer pairing with, the more ornamented males, or those which are the best songsters, or play the best antics; but it is obviously probable that they would at the same time prefer the more vigorous and lively males, and this has in some cases been confirmed by actual observation. [9] Thus the more vigorous females, which are the first to breed, will have the choice of many males; and though they may not always select the strongest or best armed, they will select those which are vigorous and well armed, and in other respects the most attractive. Both sexes, therefore, of such early pairs would as above explained, have an advantage over others in rearing offspring; and this apparently has sufficed during a long course of generations to add not only to the strength and fighting powers of the males, but likewise to their various ornaments or other attractions.

[9] With respect to poultry, I have received information, hereafter to be given, to this effect. Even birds, such as pigeons, which pair for life, the female, as I hear from Mr. Jenner Weir, will desert her mate if he is injured or grows weak.

In the converse and much rarer case of the males selecting particular females, it is plain that those which were the most vigorous and had conquered others, would have the freest choice; and it is almost certain that they would select vigorous as well as attractive females. Such pairs would have an advantage in rearing offspring, more especially if the male had the power to defend the female during the pairing-season as occurs with some of the higher animals, or aided her in providing for the young. The same principles would apply if each sex preferred and selected certain individuals of the opposite sex; supposing that they selected not only the more attractive, but likewise the more vigorous individuals.

NUMERICAL PROPORTION OF THE TWO SEXES.

I have remarked that sexual selection would be a simple affair if the males were considerably more numerous than the females. Hence I was led to investigate, as far as I could, the proportions between the two sexes of as many animals as possible; but the materials are scanty. I will here give only a brief abstract of the results, retaining the details for a supplementary discussion, so as not to interfere with the course of my argument. Domesticated animals alone afford the means of ascertaining the proportional numbers at birth; but no records have been specially kept for this purpose. By indirect means, however, I have collected a considerable body of statistics, from which it appears that with most of our domestic animals the sexes are nearly equal at birth. Thus 25,560 births of race-horses have been recorded during twenty-one years, and the male births were to the female births as 99.7 to 100. In greyhounds the inequality is greater than with any other animal, for out of 6878 births during twelve years, the male births were to the female as 110.1 to 100. It is, however, in some degree doubtful whether it is safe to infer that the proportion would be the same under natural conditions as under domestication; for slight and unknown differences in the conditions affect the proportion of the sexes. Thus with mankind, the male births in England are as 104.5, in Russia as 108.9, and with the Jews of Livonia as 120, to 100 female births. But I shall recur to this curious point of the excess of male births in the supplement to this chapter. At the Cape of Good Hope, however, male children of European extraction have been born during several years in the proportion of between 90 and 99 to 100 female children.

For our present purpose we are concerned with the proportions of the sexes, not only at birth, but also at maturity, and this adds another element of doubt; for it is a well-ascertained fact that with man the number of males dying before or during birth, and during the first two years of infancy, is considerably larger than that of females. So it almost certainly is with male lambs, and probably with some other animals. The males of some species kill one another by fighting; or they drive one another about until they become greatly emaciated. They must also be often exposed to various dangers, whilst wandering about in eager search for the females. In many kinds of fish the males are much smaller than the females, and they are believed often to be devoured by the latter, or by other fishes. The females of some birds appear to die earlier than the males; they are also liable to be destroyed on their nests, or whilst in charge of their young. With insects the female larvæ are often larger than those of the males, and would consequently be more likely to be devoured. In some cases the mature females are less active and less rapid in their movements than the males, and could not escape so well from danger. Hence, with animals in a state of nature, we must rely on mere estimation, in order to judge of the proportions of the sexes at maturity; and this is but little trustworthy, except when the inequality is strongly marked. Nevertheless, as far as a judgment can be formed, we may conclude from the facts given in the supplement, that the males of some few mammals, of many birds, of some fish and insects, are considerably more numerous than the females.

The proportion between the sexes fluctuates slightly during successive years: thus with race-horses, for every 100 mares born the stallions varied from 107.1 in one year to 92.6 in another year, and with greyhounds from 116.3 to 95.3. But had larger numbers been tabulated throughout an area more extensive than England, these fluctuations would probably have disappeared; and such as they are, would hardly suffice to lead to effective sexual selection in a state of nature. Nevertheless, in the cases of some few wild animals, as shown in the supplement, the proportions seem to fluctuate either during different seasons or in different localities in a sufficient degree to lead to such selection. For it should be observed that any advantage, gained during certain years or in certain localities by those males which were able to conquer their rivals, or were the most attractive to the females, would probably be transmitted to the offspring, and would not subsequently be eliminated. During the succeeding seasons, when, from the equality of the sexes, every male was able to procure a female, the stronger or more attractive males previously produced would still have at least as good a chance of leaving offspring as the weaker or less attractive.

POLYGAMY.

The practice of polygamy leads to the same results as would follow from an actual inequality in the number of the sexes; for if each male secures two or more females, many males cannot pair; and the latter assuredly will be the weaker or less attractive individuals. Many mammals and some few birds are polygamous, but with animals belonging to the lower classes I have found no evidence of this habit. The intellectual powers of such animals are, perhaps, not sufficient to lead them to collect and guard a harem of females. That some relation exists between polygamy and the development of secondary sexual characters, appears nearly certain; and this supports the view that a numerical preponderance of males would be eminently favourable to the action of sexual selection. Nevertheless many animals, which are strictly monogamous, especially birds, display strongly-marked secondary sexual characters; whilst some few animals, which are polygamous, do not have such characters.

We will first briefly run through the mammals, and then turn to birds. The gorilla seems to be polygamous, and the male differs considerably from the female; so it is with some baboons, which live in herds containing twice as many adult females as males. In South America the Mycetes caraya presents well-marked sexual differences, in color, beard, and vocal organs; and the male generally lives with two or three wives: the male of the *Cebus capucinus* differs somewhat from the female, and appears to be polygamous. [10] Little is known on this head with respect to most other monkeys, but some species are strictly monogamous. The ruminants are eminently polygamous, and they present sexual differences more frequently than almost any other group of mammals; this holds good, especially in their weapons, but also in other characters. Most deer, cattle, and sheep are polygamous; as are most antelopes, though some are monogamous. Sir Andrew Smith, in speaking of the antelopes of South Africa, says that in herds of about a dozen there was rarely more than one mature male. The Asiatic *Antilope saiga* appears to be the most inordinate polygamist in the world; for Pallas [11] states that the male drives away all rivals, and collects a herd of about a hundred females and kids together; the female is hornless and has softer hair, but does not otherwise differ much from the male. The wild horse of the Falkland Islands and of the Western States of N. America is polygamous, but, except in his greater size and in the proportions of his body, differs but little from the mare. The wild boar presents well-marked sexual characters, in his great tusks and some other points. In Europe and in India he leads a solitary life, except during the breeding-season; but as is believed by Sir W. Elliot, who has had many opportunities in India of observing this animal, he consorts at this season with several females. Whether this holds good in Europe is doubtful, but it is supported by some evidence. The adult male Indian elephant, like the boar, passes much of his time in solitude; but as Dr. Campbell states, when with others, "It is rare to find more than one male with a whole herd of females"; the larger males expelling or killing the smaller and weaker ones. The male differs from the female in his immense tusks, greater size, strength, and endurance; so great is the difference in these respects that the males when caught are valued at one-fifth more than the females. [12] The sexes of other pachydermatous animals differ very little or not at all, and, as far as known, they are not polygamists. Nor have I heard of any species in the Orders of Cheiroptera, Edentata, Insectivora and Rodents being polygamous, excepting that amongst the Rodents, the common rat, according to some rat-catchers, lives with several females. Nevertheless the two sexes of some sloths (Edentata) differ in the character and color of certain patches of hair on their shoulders. [13] And many kinds of bats (Cheiroptera) present well-marked sexual differences, chiefly in the males possessing odoriferous glands and pouches, and by their being of a lighter color. [14] In the great order of Rodents, as far as I can learn, the sexes rarely differ, and when they do so, it is but slightly in the tint of the fur.

[10] On the Gorilla, Savage and Wyman, 'Boston Journal of Natural History,' vol. v. 1845-47, p. 423. On Cynocephalus, Brehm, 'Thierleben,' B. i. 1864, s. 77. On Mycetes, Rengger, 'Naturgesch.: Säugethiere von Paraguay,' 1830, ss. 14, 20. On Cebus, Brehm, ibid. s. 108.
[11] Pallas, 'Spicilegia Zoolog., Fasc.' xii. 1777, p. 29. Sir Andrew Smith, 'Illustrations of the Zoology of S. Africa,' 1849, pl. 29, on the Kobus. Owen, in his 'Anatomy of Vertebrates' (vol. iii. 1868, p. 633) gives a table showing incidentally which species of antelopes are gregarious.
[12] Dr. Campbell, in 'Proc. Zoolog. Soc.' 1869, p. 138. See also an interesting paper by Lieut. Johnstone, in 'Proceedings, Asiatic Society of Bengal,' May 1868.
[13] Dr. Gray, in 'Annals and Magazine of Natural History,' 1871, p. 302.
[14] See Dr. Dobson's excellent paper in 'Proceedings of the Zoological Society,' 1873, p. 241.

As I hear from Sir Andrew Smith, the lion in South Africa sometimes lives with a single female, but generally with more, and, in one case, was found with as many as five females; so that he is polygamous. As far as I can discover, he is the only polygamist amongst all the terrestrial Carnivora, and he alone presents well-marked sexual characters. If, however, we turn to the marine Carnivora, as we shall hereafter see, the case is widely different; for many species of seals offer extraordinary sexual differences, and they are eminently polygamous. Thus, according to Peron, the male sea-elephant of the Southern Ocean always possesses several females, and the sea-lion of Forster is said to be surrounded by from twenty to thirty females. In the North, the male sea-bear of Steller is accompanied by even a greater number of females. It is an interesting fact, as Dr. Gill remarks [15], that in the monogamous species, "or those living in small communities, there is little difference in size between the males and females; in the social species, or rather those of which the males have harems, the males are vastly larger than the females."

[15] 'The Eared Seals,' American Naturalist, vol. iv. Jan. 1871.

Amongst birds, many species, the sexes of which differ greatly from each other, are

quits the cell in which it is born, whilst the female has well-developed wings. Audouin believes that the females of this species are impregnated by the males which are born in the same cells with them; but it is much more probable that the females visit other cells, so that close inter-breeding is thus avoided. We shall hereafter meet in various classes, with a few exceptional cases, in which the female, instead of the male, is the seeker and wooer.

The female, on the other hand, with the rarest exceptions, is less eager than the male. As the illustrious Hunter [20] long ago observed, she generally "requires to be courted;" she is coy, and may often be seen endeavouring for a long time to escape from the male. Every observer of the habits of animals will be able to call to mind instances of this kind. It is shown by various facts, given hereafter, and by the results fairly attributable to sexual selection, that the female, though comparatively passive, generally exerts some choice and accepts one male in preference to others. Or she may accept, as appearances would sometimes lead us to believe, not the male which is the most attractive to her, but the one which is the least distasteful. The exertion of some choice on the part of the female seems a law almost as general as the eagerness of the male.

[20] 'Essays and Observations,' edited by Owen, vol. i. 1861, p. 194.

We are naturally led to enquire why the male, in so many and such distinct classes, has become more eager than the female, so that he searches for her, and plays the more active part in courtship. It would be no advantage and some loss of power if each sex searched for the other; but why should the male almost always be the seeker? The ovules of plants after fertilization have to be nourished for a time; hence the pollen is necessarily brought to the female organs—being placed on the stigma, by means of insects or the wind, or by the spontaneous movements of the stamens; and in the Algae, etc., by the locomotive power of the antherozooids. With lowly-organised aquatic animals, permanently affixed to the same spot and having their sexes separate, the male element is invariably brought to the female; and of this we can see the reason, for even if the ova were detached before fertilization, and did not require subsequent nourishment or protection, there would yet be greater difficulty in transporting them than the male element, because, being larger than the latter, they are produced in far smaller numbers. So that many of the lower animals are, in this respect, analogous with plants. [21] The males of affixed and aquatic animals having been led to emit their fertilising element in this way, it is natural that any of their descendants, which rose in the scale and became locomotive, should retain the same habit; and they would approach the female as closely as possible, in order not to risk the loss of the fertilising element in a long passage of it through the water. With some few of the lower animals, the females alone are fixed, and the males of these must be the seekers. But it is difficult to understand why the males of species, of which the progenitors were primordially free, should invariably have acquired the habit of approaching the females, instead of being approached by them. But in all cases, in order that the males should seek efficiently, it would be necessary that they should be endowed with strong passions; and the acquirement of such passions would naturally follow from the more eager leaving a larger number of offspring than the less eager.

[21] Prof. Sachs ('Lehrbuch der Botanik,' 1870, S. 633) in speaking of the male and

female reproductive cells, remarks, "verhät sich die eine bei der Vereinigung activ, ... die andere erscheint bei der Vereinigung passiv."

The great eagerness of the males has thus indirectly led to their much more frequently developing secondary sexual characters than the females. But the development of such characters would be much aided, if the males were more liable to vary than the females—as I concluded they were—after a long study of domesticated animals. Von Nathusius, who has had very wide experience, is strongly of the same opinion. [22] Good evidence also in favour of this conclusion can be produced by a comparison of the two sexes in mankind. During the Novara Expedition [23] a vast number of measurements was made of various parts of the body in different races, and the men were found in almost every case to present a greater range of variation than the women; but I shall have to recur to this subject in a future chapter. Mr. J. Wood [24], who has carefully attended to the variation of the muscles in man, puts in italics the conclusion that "the greatest number of abnormalities in each subject is found in the males." He had previously remarked that "altogether in 102 subjects, the varieties of redundancy were found to be half as many again as in females, contrasting widely with the greater frequency of deficiency in females before described." Professor Macalister likewise remarks [25] that variations in the muscles "are probably more common in males than females." Certain muscles which are not normally present in mankind are also more frequently developed in the male than in the female sex, although exceptions to this rule are said to occur. Dr. Burt Wilder [26] has tabulated the cases of 152 individuals with supernumerary digits, of which 86 were males, and 39, or less than half, females, the remaining 27 being of unknown sex. It should not, however, be overlooked that women would more frequently endeavour to conceal a deformity of this kind than men. Again, Dr. L. Meyer asserts that the ears of man are more variable in form than those of a woman. [27] Lastly the temperature is more variable in man than in woman. [28]

[22] 'Vortrage über Viehzucht,' 1872, p. 63.
[23] 'Reise der Novara: Anthropolog. Theil,' 1867, ss. 216-269. The results were calculated by Dr. Weisbach from measurements made by Drs. K. Scherzer and Schwarz. On the greater variability of the males of domesticated animals, see my 'Variation of Animals and Plants under Domestication,' vol. ii. 1868, p. 75.
[24] 'Proceedings of the Royal Society,' vol. xvi. July 1868, pp. 519 and 524.
[25] 'Proc. Royal Irish Academy,' vol. x. 1868, p. 123.
[26] 'Massachusetts Medical Society,' vol. ii. No. 3, 1868, p. 9.
[27] 'Archiv für Path. Anat. und Phys.' 1871, p. 488.
[28] The conclusions recently arrived at by Dr. J. Stockton Hough, on the temperature of man, are given in the 'Pop. Sci. Review,' Jan. 1st, 1874, p. 97.

The cause of the greater general variability in the male sex, than in the female is unknown, except in so far as secondary sexual characters are extraordinarily variable, and are usually confined to the males; and, as we shall presently see, this fact is, to a certain extent, intelligible. Through the action of sexual and natural selection male animals have been rendered in very many instances widely different from their females; but independently of selection the two sexes, from differing constitutionally, tend to vary in a somewhat different manner. The female has to expend much organic matter in the formation of her ova, whereas the male expends much force in fierce contests with his rivals, in wandering about in search of the female, in exerting his voice, pouring out odoriferous secretions, etc.: and this expenditure is generally concentrated within a short period. The great vigor of the male during the season of love seems often to intensify his colors, independently of any marked difference from the female. [29] In mankind, and even as low down in the organic scale as in the Lepidoptera, the temperature of the body is higher in the male than in the female, accompanied in the case of man by a slower pulse. [30] On the whole the expenditure of matter and force by the two sexes is probably nearly equal, though effected in very different ways and at different rates.

[29] Prof. Mantegazza is inclined to believe ('Lettera a Carlo Darwin,' 'Archivio per l'Anthropologia,' 1871, p. 306) that the bright colors, common in so many male animals, are due to the presence and retention by them of the spermatic fluid; but this can hardly be the case; for many male birds, for instance young pheasants, become brightly colored in the autumn of their first year.
[30] For mankind, see Dr. J. Stockton Hough, whose conclusions are given in the 'Popular Science Review,' 1874, p. 97. See Girard's observations on the Lepidoptera, as given in the 'Zoological Record,' 1869, p. 347.

From the causes just specified the two sexes can hardly fail to differ somewhat in constitution, at least during the breeding-season; and, although they may be subjected to exactly the same conditions, they will tend to vary in a different manner. If such variations are of no service to either sex, they will not be accumulated and increased by sexual or natural selection. Nevertheless, they may become permanent if the exciting cause acts permanently; and in accordance with a frequent form of inheritance they may be transmitted to that sex alone in which they first appeared. In this case the two sexes will come to present permanent, yet unimportant, differences of character. For instance, Mr. Allen shows that with a large number of birds inhabiting the northern and southern United States, the specimens from the south are darker-colored than those from the north; and this seems to be the direct result of the difference in temperature, light, etc., between the two regions. Now, in some few cases, the two sexes of the same species appear to have been differently affected; in the *Agelæus phœniceus* the males have had their colors greatly intensified in the south; whereas with *Cardinalis virginianus* it is the females which have been thus affected; with *Quiscalus major* the females have been rendered extremely variable in tint, whilst the males remain nearly uniform. [31]

[31] 'Mammals and Birds of E. Florida,' pp. 234, 280, 295.

A few exceptional cases occur in various classes of animals, in which the females instead of the males have acquired well pronounced secondary sexual characters, such as brighter colors, greater size, strength, or pugnacity. With birds there has sometimes been a complete transposition of the ordinary characters proper to each sex; the females having become the more eager in courtship, the males remaining comparatively passive, but apparently selecting the more attractive females, as we may infer from the results. Certain hen birds have thus been rendered more highly colored or otherwise ornamented, as well as more powerful and pugnacious than the cocks; these characters being transmitted to the female offspring alone.

It may be suggested that in some cases a double process of selection has been carried on; that the males have selected the more attractive females, and the latter the more attractive males. This process, however, though it might lead to the modification of both sexes, would not make the one sex different from the other, unless indeed their tastes for the beautiful differed; but this is a supposition too improbable to be worth considering in the case of any animal, excepting man. There are, however, many animals in which the sexes resemble each other, both being furnished with the same ornaments, which analogy would lead us to attribute to the agency of sexual selection. In such cases it may be suggested with more plausibility, that there has been a double or mutual process of sexual selection; the more vigorous and precocious females selecting the more attractive and vigorous males, the latter rejecting all except the more attractive females. But from what we know of the habits of animals, this view is hardly probable, for the male is generally eager to pair with any female. It is more probable that the ornaments common to both sexes were acquired by one sex, generally the male, and then transmitted to the offspring of both sexes. If, indeed, during a lengthened period the males of any species were greatly to exceed the females in number, and then during another lengthened period, but under different conditions, the reverse were to occur, a double, but not simultaneous, process of sexual selection might easily be carried on, by which the two sexes might be rendered widely different.

We shall hereafter see that many animals exist, of which neither sex is brilliantly colored or provided with special ornaments, and yet the members of both sexes or of one alone have probably acquired simple colors, such as white or black, through sexual selection. The absence of bright tints or other ornaments may be the result of variations of the right kind never having occurred, or of the animals themselves having preferred plain black or white. Obscure tints have often been developed through natural selection for the sake of protection, and the acquirement through sexual selection of conspicuous colors, appears to have been sometimes checked from the danger thus incurred. But in other cases the males during long ages may have struggled together for the possession of the females, and yet no effect will have been produced, unless a larger number of offspring were left by the more successful males to inherit their superiority, than by the less successful: and this, as previously shown, depends on many complex contingencies.

Sexual selection acts in a less rigorous manner than natural selection. The latter produces its effects by the life or death at all ages of the more or less successful individuals. Death, indeed, not rarely ensues from the conflicts of rival males. But generally the less successful male merely fails to obtain a female, or obtains a retarded and less vigorous female later in the season, or, if polygamous, obtains fewer females; so that they leave fewer, less vigorous, or no offspring. In regard to structures acquired through ordinary or natural selection, there is in most cases, as long as the conditions of life remain the same, a limit to the amount of advantageous modification in relation to certain special purposes; but in regard to structures adapted to make one male victorious over another, either in fighting or in charming the female, there is no definite limit to the amount of advantageous modification; so that as long as the proper variations arise the work of sexual selection will go on. This circumstance may partly account for the frequent and extraordinary amount of variability presented by secondary sexual characters. Nevertheless, natural selection will determine that such characters shall not be acquired by the victorious males, if they would be highly injurious, either by expending too much of their vital powers, or by exposing them to any great danger. The development, however, of certain structures—of the horns, for instance, in certain stags—has been carried to a wonderful extreme; and in some cases to an extreme which, as far as the general conditions of life are concerned, must be slightly injurious to the male. From this fact we learn that the advantages which favoured males derive from conquering other males in battle or courtship, and thus leaving a numerous progeny, are in the long run greater than those derived from rather more perfect adaptation to their conditions of life. We shall further see, and it could never have been anticipated, that the power to charm the female has sometimes been more important than the power to conquer other males in battle.

LAWS OF INHERITANCE.

In order to understand how sexual selection has acted on many animals of many classes, and in the course of ages has produced a conspicuous result, it is necessary to bear in mind the laws of inheritance, as far as they are known. Two distinct elements are included under the term "inheritance"—the transmission, and the development of characters; but as these generally go together, the distinction is often overlooked. We see this distinction in those characters which are transmitted through the early years of life, but are developed only at maturity or during old age. We see the same distinction more clearly with secondary sexual characters, for these are transmitted through both sexes, though developed in one alone. That they are present in both sexes, is manifest when two species, having strongly-marked sexual characters, are crossed, for each transmits the characters proper to its own male and female sex to the hybrid offspring of either sex. The same fact is likewise manifest, when characters proper to the male are occasionally developed in the female when she grows old or becomes diseased, as, for instance, when the common hen assumes the flowing tail-feathers, hackles, comb, spurs, voice, and even pugnacity of the cock. Conversely, the same thing is evident, more or less plainly, with castrated males. Again, independently of old age or disease, characters are occasionally transferred from the male to the female, as when, in certain breeds of the fowl, spurs regularly appear in the young and healthy females. But in truth they are simply developed in the female; for in every breed each detail in the structure of the spur is transmitted through the female to her male offspring. Many cases will hereafter be given, where the female exhibits, more or less perfectly, characters proper to the male, in whom they must have been first developed, and then transferred to the female. The converse case of the first development of characters in the female and of transference to the male, is less frequent; it will therefore be well to give one striking instance. With bees the pollen-collecting apparatus is used by the female alone for gathering pollen for the larvæ, yet in most of the species it is partially developed in the males to whom it is quite useless, and it is perfectly developed in the males of Bombus or the humble-bee. [32] As not a single other Hymenopterous insect, not even the wasp, which is closely allied to the bee, is provided with a pollen-collecting apparatus, we have no grounds for supposing that male bees primordially collected pollen as well as the females; although we have some reason to suspect that male mammals primordially suckled their young as well as the females. Lastly, in all cases of reversion, characters are transmitted through two, three, or many more generations, and are then developed under certain unknown favourable conditions. This important distinction between transmission and development will be best kept in mind by the aid of the hypothesis of pangenesis. According to this hypothesis, every unit or cell of the body throws off gemmules or undeveloped atoms, which are transmitted to the offspring of both sexes, and are multiplied by self-division. They may remain undeveloped during the early years of life or during successive generations; and their development into units or cells, like those from which they were derived, depends on their affinity for, and union with other units or cells previously developed in the due order of growth.

[32] H. Müller, 'Anwendung der Darwin'schen Lehre,' etc., Verh. d. n. V. Jahrg., xxix. p. 42.

INHERITANCE AT CORRESPONDING PERIODS OF LIFE.

This tendency is well established. A new character, appearing in a young animal, whether it lasts throughout life or is only transient, will, in general, reappear in the offspring at the same age and last for the same time. If, on the other hand, a new character appears at maturity, or even during old age, it tends to reappear in the offspring at the same advanced age. When deviations from this rule occur, the transmitted characters much oftener appear before, than after the corresponding age. As I have dwelt on this subject sufficiently in another work [33], I will here merely give two or three instances, for the sake of recalling the subject to the reader's mind. In several breeds of the Fowl, the down-covered chickens, the young birds in their first true plumage, and the adults differ greatly from one another, as well as from their common parent-form, the *Gallus bankiva*; and these characters are faithfully transmitted by each breed to their offspring at the corresponding periods of life. For instance, the chickens of spangled Hamburgs, whilst covered with down, have a few dark spots on the head and rump, but are not striped longitudinally, as in many other breeds; in their first true plumage, "they are beautifully pencilled," that is each feather is transversely marked by numerous dark bars; but in their second plumage the feathers all become spangled or tipped with a dark round spot. [34] Hence in this breed variations have occurred at, and been transmitted to, three distinct periods of life. The Pigeon offers a more remarkable case, because the aboriginal parent species does not undergo any change of plumage with advancing age, excepting that at maturity the breast becomes more iridescent; yet there are breeds which do not acquire their characteristic colors until they have moulted two, three, or four times; and these modifications of plumage are regularly transmitted.

[33] The 'Variation of Animals and Plants under Domestication,' vol. ii. 1868, p. 75. In the last chapter but one, the provisional hypothesis of pangenesis, above alluded to, is fully explained.
[34] These facts are given on the high authority of a great breeder, Mr. Teebay; see Tegetmeier's 'Poultry Book,' 1868, p. 158. On the characters of chickens of different breeds, and on the breeds of the pigeon, alluded to in the following paragraph, see 'Variation of Animals,' etc., vol. i. pp. 160, 249; vol. ii. p. 77.

INHERITANCE AT CORRESPONDING SEASONS OF THE YEAR.

With animals in a state of nature, innumerable instances occur of characters appearing periodically at different seasons. We see this in the horns of the stag, and in the fur of Artic animals which becomes thick and white during the winter. Many birds acquire bright colors and other decorations during the breeding-season alone. Pallas states [35], that in Siberia domestic cattle and horses become lighter-colored during the winter; and I have myself observed, and heard of similar strongly marked changes of color, that is, from brownish cream-color or reddish-brown to a perfect white, in several ponies in

England. Although I do not know that this tendency to change the color of the coat during different seasons is transmitted, yet it probably is so, as all shades of color are strongly inherited by the horse. Nor is this form of inheritance, as limited by the seasons, more remarkable than its limitation by age or sex.

[35] 'Novae species Quadrupedum e Glirium ordine,' 1778, p. 7. On the transmission of color by the horse, see 'Variation of Animals and Plants under Domestication,' vol. i. p. 51. Also vol. ii. p. 71, for a general discussion on 'Inheritance as limited by Sex.'

INHERITANCE AS LIMITED BY SEX.

The equal transmission of characters to both sexes is the commonest form of inheritance, at least with those animals which do not present strongly-marked sexual differences, and indeed with many of these. But characters are somewhat commonly transferred exclusively to that sex, in which they first appear. Ample evidence on this head has been advanced in my work on 'Variation under Domestication,' but a few instances may here be given. There are breeds of the sheep and goat, in which the horns of the male differ greatly in shape from those of the female; and these differences, acquired under domestication, are regularly transmitted to the same sex. As a rule, it is the females alone in cats which are tortoise-shell, the corresponding color in the males being rusty-red. With most breeds of the fowl, the characters proper to each sex are transmitted to the same sex alone. So general is this form of transmission that it is an anomaly when variations in certain breeds are transmitted equally to both sexes. There are also certain sub-breeds of the fowl in which the males can hardly be distinguished from one another, whilst the females differ considerably in color. The sexes of the pigeon in the parent-species do not differ in any external character; nevertheless, in certain domesticated breeds the male is colored differently from the female. [36] The wattle in the English Carrier pigeon, and the crop in the Pouter, are more highly developed in the male than in the female; and although these characters have been gained through long-continued selection by man, the slight differences between the sexes are wholly due to the form of inheritance which has prevailed; for they have arisen, not from, but rather in opposition to, the wish of the breeder.

[36] Dr. Chapuis, 'Le Pigeon Voyageur Belge,' 1865, p. 87. Boitard et Corbié, 'Les Pigeons de Volière,' etc., 1824, p. 173. See, also, on similar differences in certain breeds at Modena, 'Le variazioni dei Colombi domestici,' del Paolo Bonizzi, 1873.

Most of our domestic races have been formed by the accumulation of many slight variations; and as some of the successive steps have been transmitted to one sex alone, and some to both sexes, we find in the different breeds of the same species all gradations between great sexual dissimilarity and complete similarity. Instances have already been given with the breeds of the fowl and pigeon, and under nature analogous cases are common. With animals under domestication, but whether in nature I will not venture to say, one sex may lose characters proper to it, and may thus come somewhat to resemble the opposite sex; for instance, the males of some breeds of the fowl have lost their masculine tail-plumes and hackles. On the other hand, the differences between the sexes may be increased under domestication, as with merino sheep, in which the ewes have lost their horns. Again, characters proper to one sex may suddenly appear in the other sex; as in those sub-breeds of the fowl in which the hens acquire spurs whilst young; or, as in certain Polish sub-breeds, in which the females, as there is reason to believe, originally acquired a crest, and subsequently transferred it to the males. All these cases are intelligible on the hypothesis of pangenesis; for they depend on the gemmules of certain parts, although present in both sexes, becoming, through the influence of domestication, either dormant or developed in either sex.

There is one difficult question which it will be convenient to defer to a future chapter; namely, whether a character at first developed in both sexes, could through selection be limited in its development to one sex alone. If, for instance, a breeder observed that some of his pigeons (of which the characters are usually transferred in an equal degree to both sexes) varied into pale blue, could he by long-continued selection make a breed, in which the males alone should be of this tint, whilst the females remained unchanged? I will here only say, that this, though perhaps not impossible, would be extremely difficult; for the natural result of breeding from the pale-blue males would be to change the whole stock of both sexes to this tint. If, however, variations of the desired tint appeared, which were from the first limited in their development to the male sex, there would not be the least difficulty in making a breed with the two sexes of a different color, as indeed has been effected with a Belgian breed, in which the males alone are streaked with black. In a similar manner, if any variation appeared in a female pigeon, which was from the first sexually limited in its development to the females, it would be easy to make a breed with the females alone thus characterized; but if the variation was not thus originally limited, the process would be extremely difficult, perhaps impossible. [37]

[37] Since the publication of the first edition of this work, it has been highly satisfactory to me to find the following remarks (the 'Field,' Sept. 1872) from so experienced a breeder as Mr. Tegetmeier. After describing some curious cases in pigeons, of the transmission of color by one sex alone, and the formation of a sub-breed with this character, he says: "It is a singular circumstance that Mr. Darwin should have suggested the possibility of modifying the sexual colors of birds by a course of artificial selection. When he did so, he was in ignorance of these facts that I have related; but it is remarkable how very closely he suggested the right method of procedure."

ON THE RELATION BETWEEN THE PERIOD OF DEVELOPMENT OF A CHARACTER AND ITS TRANSMISSION TO ONE SEX OR TO BOTH SEXES.

Why certain characters should be inherited by both sexes, and other characters by one sex alone, namely by that sex in which the character first appeared, is in most cases quite unknown. We cannot even conjecture why with certain sub-breeds of the pigeon, black striæ, though transmitted through the female, should be developed in the male alone, whilst every other character is equally transferred to both sexes. Why, again, with cats, the tortoise-shell color should, with rare exceptions, be developed in the female alone. The very same character, such as deficient or supernumerary digits, color-blindness, etc., may with mankind be inherited by the males alone of one family, and in another family by the females alone, though in both cases transmitted through the opposite as well as through the same sex. [38] Although we are thus ignorant, the two following rules seem often to hold good—that variations which first appear in either sex at a late period of life, tend to be developed in the same sex alone; whilst variations which first appear early in life in either sex tend to be developed in both sexes. I am, however, far from supposing that this is the sole determining cause. As I have not elsewhere discussed this subject, and it has an important bearing on sexual selection, I must here enter into lengthy and somewhat intricate details.

[38] References are given in my 'Variation of Animals and Plants under Domestication,' vol. ii. p. 72.

It is in itself probable that any character appearing at an early age would tend to be inherited equally by both sexes, for the sexes do not differ much in constitution before the power of reproduction is gained. On the other hand, after this power has been gained and the sexes have come to differ in constitution, the gemmules (if I may again use the language of pangenesis) which are cast off from each varying part in the one sex would be much more likely to possess the proper affinities for uniting with the tissues of the same sex, and thus becoming developed, than with those of the opposite sex.

I was first led to infer that a relation of this kind exists, from the fact that whenever and in whatever manner the adult male differs from the adult female, he differs in the same manner from the young of both sexes. The generality of this fact is quite remarkable: it holds good with almost all mammals, birds, amphibians, and fishes; also with many crustaceans, spiders, and some few insects, such as certain orthoptera and libellulæ. In all these cases the variations, through the accumulation of which the male acquired his proper masculine characters, must have occurred at a somewhat late period of life; otherwise the young males would have been similarly characterized; and conformably with our rule, the variations are transmitted to and developed in the adult males alone. When, on the other hand, the adult male closely resembles the young of both sexes (these, with rare exceptions, being alike), he generally resembles the adult female; and in most of these cases the variations through which the young and old acquired their present characters, probably occurred, according to our rule, during youth. But there is here room for doubt, for characters are sometimes transferred to the offspring at an earlier age than that at which they first appeared in the parents, so that the parents may have varied when adult, and have transferred their characters to their offspring whilst young. There are, moreover, many animals, in which the two sexes closely resemble each other, and yet both differ from their young: and here the characters of the adults must have been acquired late in life; nevertheless, these characters, in apparent contradiction to our rule, are transferred to both sexes. We must not however, overlook the possibility or even probability of successive variations of the same nature occurring, under exposure to similar conditions, simultaneously in both sexes at a rather late period of life; and in this case the variations would be transferred to the offspring of both sexes at a corresponding late age; and there would then be no real contradiction to the rule that variations occurring late in life are transferred exclusively to the sex in which they first appeared. This latter rule seems to hold true more generally than the second one, namely, that variations which occur in either sex early in life tend to be transferred to both sexes. As it was obviously impossible even to estimate in how large a number of cases throughout the animal kingdom these two propositions held good, it occurred to me to investigate some striking or crucial instances, and to rely on the result.

An excellent case for investigation is afforded by the Deer family. In all the species, but one, the horns are developed only in the males, though certainly transmitted through the females, and capable of abnormal development in them. In the reindeer, on the other hand, the female is provided with horns; so that in this species, the horns ought, according to our rule, to appear early in life, long before the two sexes are mature and have come to differ much in constitution. In all the other species the horns ought to appear later in life, which would lead to their development in that sex alone, in which they first appeared in the progenitor of the whole Family. Now in seven species, belonging to distinct sections of the family and inhabiting different regions, in which the stags alone bear horns, I find that the horns first appear at periods, varying from nine months after birth in the roebuck, to ten, twelve or even more months in the stags of the six other and larger species. [39] But with the reindeer the case is widely different; for, as I hear from Prof. Nilsson, who kindly made special enquiries for me in Lapland, the horns appear in the young animals within four or five weeks after birth, and at the same time in both sexes. So that here we have a structure, developed at a most unusually early age in one species of the family, and likewise common to both sexes in this one species alone.

[39] I am much obliged to Mr. Cupples for having made enquiries for me in regard to the Roebuck and Red Deer of Scotland from Mr. Robertson, the experienced head-forester to the Marquis of Breadalbane. In regard to Fallow-deer, I have to thank Mr. Eyton and others for information. For the *Cervus alces* of N. America, see 'Land and Water,' 1868, pp. 221 and 254; and for the C. *Virginianus* and *strongyloceros* of the same continent, see J.D. Caton, in 'Ottawa Acad. of Nat. Sc.' 1868, p. 13. For *Cervus Eldi* of Pegu, see Lieut. Beaven, 'Proc. of the Zoolog. Soc.' 1867, p. 762.

In several kinds of antelopes, only the males are provided with horns, whilst in the greater number both sexes bear horns. With respect to the period of development, Mr. Blyth informs me that there was at one time in the Zoological Gardens a young koodoo (*Ant. strepsiceros*), of which the males alone are horned, and also the young of a closely-allied species, the eland (*Ant. oreas*), in which both sexes are horned. Now it is in strict conformity with our rule, that in the young male koodoo, although ten months old, the horns were remarkably small, considering the size ultimately attained by them; whilst in the young male eland, although only three months old, the horns were already very much larger than in the koodoo. It is also a noticeable fact that in the prong-horned antelope [40], only a few of the females, about one in five, have horns, and these are in a rudimentary state, though sometimes above four inches long: so that as far as concerns the possession of horns by the males alone, this species is in an intermediate condition, and the horns do not appear until about five or six months after birth. Therefore in comparison with what little we know of the development of the horns in other antelopes, and from what we do know with respect to the horns of deer, cattle, etc., those of the prong-horned antelope appear at an intermediate period of life,—that is, not very early, as in cattle and sheep, nor very late, as in the larger deer and antelopes. The horns of sheep, goats, and cattle, which are well developed in both sexes, though not quite equal in size, can be felt, or even seen, at birth or soon afterwards. [41] Our rule, however, seems to fail in some breeds of sheep, for instance merinos, in which the rams alone are horned; for I cannot find on enquiry [42], that the horns are developed later in life in this breed than in ordinary sheep in which both sexes are horned. But with domesticated sheep the presence or absence of horns is not a firmly fixed character; for a certain proportion of the merino ewes bear small horns, and some of the rams are hornless; and in most breeds hornless ewes are occasionally produced.

[40] *Antilocapra Americana.* I have to thank Dr. Canfield for information with respect to the horns of the female: see also his paper in 'Proceedings of the Zoological Society,' 1866, p. 109. Also Owen, 'Anatomy of Vertebrates,' vol. iii. p. 627

[41] I have been assured that the horns of the sheep in North Wales can always be felt, and are sometimes even an inch in length, at birth. Youatt says ('Cattle,' 1834, p. 277), that the prominence of the frontal bone in cattle penetrates the cutis at birth, and that the horny matter is soon formed over it.

[42] I am greatly indebted to Prof. Victor Carus for having made enquiries for me, from the highest authorities, with respect to the merino sheep of Saxony. On the Guinea coast of Africa there is, however, a breed of sheep in which, as with merinos, the rams alone bear horns; and Mr. Winwood Reade informs me that in one case observed by him, a young ram, born on Feb. 10th, first showed horns on March 6th, so that in this instance, in conformity with rule, the development of the horns occurred at a later period of life than in Welsh sheep, in which both sexes are horned.

Dr. W. Marshall has lately made a special study of the protuberances so common on the heads of birds [43], and he comes to the following conclusion:—that with those species in which they are confined to the males, they are developed late in life; whereas with those species in which they are common to the two sexes, they are developed at a very early period. This is certainly a striking confirmation of my two laws of inheritance.

[43] 'Ueber die knöchernen Schädelhöcker der Vogel,' in the 'Niederlandischen Archiv für Zoologie,' Band I. Heft 2, 1872.

In most of the species of the splendid family of the Pheasants, the males differ conspicuously from the females, and they acquire their ornaments at a rather late period of life. The eared pheasant (*Crossoptilon auritum*), however, offers a remarkable exception, for both sexes possess the fine caudal plumes, the large ear-tufts and the crimson velvet about the head; I find that all these characters appear very early in life in accordance with rule. The adult male can, however, be distinguished from the adult female by the presence of spurs; and conformably with our rule, these do not begin to be developed before the age of six months, as I am assured by Mr. Bartlett, and even at this age, the two sexes can hardly be distinguished. [44] The male and female Peacock differ conspicuously from each other in almost every part of their plumage, except in the elegant head-crest, which is common to both sexes; and this is developed very early in life, long before the other ornaments, which are confined to the male. The wild-duck offers an analogous case, for the beautiful green speculum on the wings is common to both sexes, though duller and somewhat smaller in the female, and it is developed early in life, whilst the curled tail-feathers and other ornaments of the male are developed later. [45] Between such extreme cases of close sexual resemblance and wide dissimilarity, as those of the Crossoptilon and peacock, many intermediate ones could be given, in which the characters follow our two rules in their order of development.

[44] In the common peacock (*Pavo cristatus*) the male alone possesses spurs, whilst both sexes of the Java Peacock (*P. muticus*) offer the unusual case of being furnished with spurs. Hence I fully expected that in the latter species they would have been

developed earlier in life than in the common peacock; but M. Hegt of Amsterdam informs me, that with young birds of the previous year, of both species, compared on April 23rd, 1869, there was no difference in the development of the spurs. The spurs, however, were as yet represented merely by slight knobs or elevations. I presume that I should have been informed if any difference in the rate of development had been observed subsequently.

[45] In some other species of the Duck family the speculum differs in a greater degree in the two sexes; but I have not been able to discover whether its full development occurs later in life in the males of such species, than in the male of the common duck, as ought to be the case according to our rule. With the allied *Mergus cucullatus* we have, however, a case of this kind: the two sexes differ conspicuously in general plumage, and to a considerable degree in the speculum, which is pure white in the male and grayish-white in the female. Now the young males at first entirely resemble the females, and have a grayish-white speculum, which becomes pure white at an earlier age than that at which the adult male acquires his other and more strongly-marked sexual differences: see Audubon, 'Ornithological Biography,' vol. iii. 1835, pp. 249-250.

As most insects emerge from the pupal state in a mature condition, it is doubtful whether the period of development can determine the transference of their characters to one or to both sexes. But we do not know that the colored scales, for instance, in two species of butterflies, in one of which the sexes differ in color, whilst in the other they are alike, are developed at the same relative age in the cocoon. Nor do we know whether all the scales are simultaneously developed on the wings of the same species of butterfly, in which certain colored marks are confined to one sex, whilst others are common to both sexes. A difference of this kind in the period of development is not so improbable as it may at first appear; for with the Orthoptera, which assume their adult state, not by a single metamorphosis, but by a succession of moults, the young males of some species at first resemble the females, and acquire their distinctive masculine characters only at a later moult. Strictly analogous cases occur at the successive moults of certain male crustaceans.

We have as yet considered the transference of characters, relatively to their period of development, only in species in a natural state; we will now turn to domesticated animals, and first touch on monstrosities and diseases. The presence of supernumerary digits, and the absence of certain phalanges, must be determined at an early embryonic period—the tendency to profuse bleeding is at least congenital, as is probably color-blindness—yet these peculiarities, and other similar ones, are often limited in their transmission to one sex; so that the rule that characters, developed at an early period, tend to be transmitted to both sexes, here wholly fails. But this rule, as before remarked, does not appear to be nearly so general as the converse one, namely, that characters which appear late in life in one sex are transmitted exclusively to the same sex. From the fact of the above abnormal peculiarities becoming attached to one sex, long before the sexual functions are active, we may infer that there must be some difference between the sexes at an extremely early age. With respect to sexually-limited diseases, we know too little of the period at which they originate, to draw any safe conclusion. Gout, however, seems to fall under our rule, for it is generally caused by intemperance during manhood, and is transmitted from the father to his sons in a much more marked manner than to his daughters.

In the various domestic breeds of sheep, goats, and cattle, the males differ from their respective females in the shape or development of their horns, forehead, mane, dewlap, tail, and hump on the shoulders; and these peculiarities, in accordance with our rule, are not fully developed until a rather late period of life. The sexes of dogs do not differ, except that in certain breeds, especially in the Scotch deer-hound, the male is much larger and heavier than the female; and, as we shall see in a future chapter, the male goes on increasing in size to an unusually late period of life, which, according to rule, will account for his increased size being transmitted to his male offspring alone. On the other hand, the tortoise-shell color, which is confined to female cats, is quite distinct at birth, and this case violates the rule. There is a breed of pigeons in which the males alone are streaked with black, and the streaks can be detected even in the nestlings; but they become more conspicuous at each successive moult, so that this case partly opposes and partly supports the rule. With the English Carrier and Pouter pigeons, the full development of the wattle and the crop occurs rather late in life, and conformably with the rule, these characters are transmitted in full perfection to the males alone. The following cases perhaps come within the class previously alluded to, in which both sexes have varied in the same manner at a rather late period of life, and have consequently transferred their new characters to both sexes at a corresponding late period; and if so, these cases are not opposed to our rule:—there exist sub-breeds of the pigeon, described by Neumeister [46], in which both sexes change their color during two or three moults (as is likewise the case with the Almond Tumbler); nevertheless, these changes, though occurring rather late in life, are common to both sexes. One variety of the Canary-bird, namely the London Prize, offers a nearly analogous case.

[46] 'Das Ganze der Taubenzucht,' 1837, ss. 21, 24. For the case of the streaked pigeons, see Dr. Chapuis, 'Le pigeon voyageur Belge,' 1865, p. 87.

With the breeds of the Fowl the inheritance of various characters by one or both sexes, seems generally determined by the period at which such characters are developed. Thus in all the many breeds in which the adult male differs greatly in color from the female, as well as from the wild parent-species, he differs also from the young male, so that the newly-acquired characters must have appeared at a rather late period of life. On the other hand, in most of the breeds in which the two sexes resemble each other, the young are colored in nearly the same manner as their parents, and this renders it probable that their colors first appeared early in life. We have instances of this fact in all black and white breeds, in which the young and old of both sexes are alike; nor can it be maintained that there is something peculiar in a black or white plumage, which leads to its transference to both sexes; for the males alone of many natural species are either black or white, the females being differently colored. With the so-called Cuckoo sub-breeds of the fowl, in which the feathers are transversely pencilled with dark stripes, both sexes and the chickens are colored in nearly the same manner. The laced plumage of the Sebright bantam is the same in both sexes, and in the young chickens the wing-feathers are distinctly, though imperfectly laced. Spangled Hamburgs, however, offer a partial exception; for the two sexes, though not quite alike, resemble each other more closely than do the sexes of the aboriginal parent-species; yet they acquire their characteristic plumage late in life, for the chickens are distinctly pencilled. With respect to other characters besides color, in the wild-parent species and in most of the domestic breeds, the males alone possess a well-developed comb; but in the young of the Spanish fowl it is largely developed at a very early age, and, in accordance with this early development in the male, it is of unusual size in the adult female. In the Game breeds pugnacity is developed at a wonderfully early age, of which curious proofs could be given; and this character is transmitted to both sexes, so that the hens, from their extreme pugnacity, are now generally exhibited in separate pens. With the Polish breeds the bony protuberance of the skull which supports the crest is partially developed even before the chickens are hatched, and the crest itself soon begins to grow, though at first feebly [47]; and in this breed the adults of both sexes are characterized by a great bony protuberance and an immense crest.

[47] For full particulars and references on all these points respecting the several breeds of the Fowl, see 'Variation of Animals and Plants under Domestication,' vol. i. pp. 250, 256. In regard to the higher animals, the sexual differences which have arisen under domestication are described in the same work under the head of each species.

Finally, from what we have now seen of the relation which exists in many natural species and domesticated races, between the period of the development of their characters and the manner of their transmission—for example, the striking fact of the early growth of the horns in the reindeer, in which both sexes bear horns, in comparison with their much later growth in the other species in which the male alone bears horns—we may conclude that one, though not the sole cause of characters being exclusively inherited by one sex, is their development at a late age. And secondly, that one, though apparently a less efficient cause of characters being inherited by both sexes, is their development at an early age, whilst the sexes differ but little in constitution. It appears, however, that some difference must exist between the sexes even during a very early embryonic period, for characters developed at this age not rarely become attached to one sex.

SUMMARY AND CONCLUDING REMARKS.

From the foregoing discussion on the various laws of inheritance, we learn that the characters of the parents often, or even generally, tend to become developed in the offspring of the same sex, at the same age, and periodically at the same season of the year, in which they first appeared in the parents. But these rules, owing to unknown causes, are far from being fixed. Hence during the modification of a species, the successive changes may readily be transmitted in different ways; some to one sex, and some to both; some to the offspring at one age, and some to the offspring at all ages. Not only are the laws of inheritance extremely complex, but so are the causes which induce and govern variability. The variations thus induced are preserved and accumulated by sexual selection, which is in itself an extremely complex affair, depending, as it does, on the ardour in love, the courage, and the rivalry of the males, as well as on the powers of perception, the taste, and will of the female. Sexual selection will also be largely dominated by natural selection tending towards the general welfare of the species. Hence the manner in which the individuals of either or both sexes have been affected through sexual selection cannot fail to be complex in the highest degree.

When variations occur late in life in one sex, and are transmitted to the same sex at the same age, the other sex and the young are left unmodified. When they occur late in life, but are transmitted to both sexes at the same age, the young alone are left unmodified. Variations, however, may occur at any period of life in one sex or in both, and be transmitted to both sexes at all ages, and then all the individuals of the species are similarly modified. In the following chapters it will be seen that all these cases frequently occur in nature.

Sexual selection can never act on any animal before the age for reproduction arrives. From the great eagerness of the male it has generally acted on this sex and not on the females. The males have thus become provided with weapons for fighting with their rivals, with organs for discovering and securely holding the female, and for exciting or charming her. When the sexes differ in these respects, it is also, as we have seen, an extremely general law that the adult male differs more or less from the young male; and we may conclude from this fact that the successive variations, by which the adult male became modified, did not generally occur much before the age for reproduction. Whenever some or many of the variations occurred early in life, the young males would partake more or less of the characters of the adult males; and differences of this kind between the old and young males may be observed in many species of animals.

It is probable that young male animals have often tended to vary in a manner which would not only have been of no use to them at an early age, but would have been actually injurious—as by acquiring bright colors, which would render them conspicuous to their enemies, or by acquiring structures, such as great horns, which would expend much vital force in their development. Variations of this kind occurring in the young males would almost certainly be eliminated through natural selection. With the adult and experienced males, on the other hand, the advantages derived from the acquisition of such characters, would more than counterbalance some exposure to danger, and some loss of vital force.

As variations which give to the male a better chance of conquering other males, or of finding, securing, or charming the opposite sex, would, if they happened to arise in the female, be of no service to her, they would not be preserved in her through sexual selection. We have also good evidence with domesticated animals, that variations of all kinds are, if not carefully selected, soon lost through intercrossing and accidental deaths. Consequently in a state of nature, if variations of the above kind chanced to arise in the female line, and to be transmitted exclusively in this line, they would be extremely liable to be lost. If, however, the females varied and transmitted their newly acquired characters to their offspring of both sexes, the characters which were advantageous to the males would be preserved by them through sexual selection, and the two sexes would in consequence be modified in the same manner, although such characters were of no use to the females: but I shall hereafter have to recur to these more intricate contingencies. Lastly, the females may acquire, and apparently have often acquired by transference, characters from the male sex.

As variations occurring later in life, and transmitted to one sex alone, have incessantly been taken advantage of and accumulated through sexual selection in relation to the reproduction of the species; therefore it appears, at first sight, an unaccountable fact that similar variations have not frequently been accumulated through natural selection, in relation to the ordinary habits of life. If this had occurred, the two sexes would often have been differently modified, for the sake, for instance, of capturing prey or of escaping from danger. Differences of this kind between the two sexes do occasionally occur, especially in the lower classes. But this implies that the two sexes follow different habits in their struggles for existence, which is a rare circumstance with the higher animals. The case, however, is widely different with the reproductive functions, in which respect the sexes necessarily differ. For variations in structure which are related to these functions, have often proved of value to one sex, and from having arisen at a late period of life, have been transmitted to one sex alone; and such variations, thus preserved and transmitted, have given rise to secondary sexual characters.

In the following chapters, I shall treat of the secondary sexual characters in animals of all classes, and shall endeavour in each case to apply the principles explained in the present chapter. The lowest classes will detain us for a very short time, but the higher animals, especially birds, must be treated at considerable length. It should be borne in mind that for reasons already assigned, I intend to give only a few illustrative instances of the innumerable structures by the aid of which the male finds the female, or, when found, holds her. On the other hand, all structures and instincts by the aid of which the male conquers other males, and by which he allures or excites the female, will be fully discussed, as these are in many ways the most interesting.

Supplement on the proportional numbers of the two sexes in animals belonging to various classes.

As no one, as far as I can discover, has paid attention to the relative numbers of the two sexes throughout the animal kingdom, I will here give such materials as I have been able to collect, although they are extremely imperfect. They consist in only a few instances of actual enumeration, and the numbers are not very large. As the proportions are known with certainty only in mankind, I will first give them as a standard of comparison.

MAN.

In England during ten years (from 1857 to 1866) the average number of children born alive yearly was 707,120, in the proportion of 104.5 males to 100 females. But in 1857 the male births throughout England were as 105.2, and in 1865 as 104.0 to 100. Looking to separate districts, in Buckinghamshire (where about 5000 children are annually born) the *mean* proportion of male to female births, during the whole period of the above ten years, was as 102.8 to 100; whilst in N. Wales (where the average annual births are 12,873) it was as high as 106.2 to 100. Taking a still smaller district, viz., Rutlandshire (where the annual births average only 739), in 1864 the male births were as 114.6, and in 1862 as only 97.0 to 100; but even in this small district the average of the 7385 births during the whole ten years, was as 104.5 to 100: that is in the same ratio as throughout England. [48] The proportions are sometimes slightly disturbed by unknown causes; thus Prof. Faye states "that in some districts of Norway there has been during a decennial period a steady deficiency of boys, whilst in others the opposite condition has existed." In France during forty-four years the male to the female births have been as 106.2 to 100; but during this period it has occurred five times in one department, and six times in another, that the female births have exceeded the males. In Russia the average proportion is as high as 108.9, and in Philadelphia in the United States as 110.5 to 100. [49] The average for Europe, deduced by Bickes from about seventy million births, is 106 males to 100 females. On the other hand, with white children born at the Cape of Good Hope, the proportion of males is so low as to fluctuate during successive years between 90 and 99 males for every 100 females. It is a singular fact that with Jews the proportion of male births is decidedly larger than with Christians: thus in Prussia the proportion is as 113, in Breslau as 114, and in Livonia as 120 to 100; the Christian births in these countries being the same as usual, for instance, in Livonia as 104 to 100. [50]

[48] 'Twenty-ninth Annual Report of the Registrar-General for 1866.' In this report (p. xii.) a special decennial table is given.
[49] For Norway and Russia, see abstract of Prof. Faye's researches, in 'British and Foreign Medico-Chirurg. Review,' April 1867, pp. 343, 345. For France, the 'Annuaire pour l'An 1867,' p. 213. For Philadelphia, Dr. Stockton Hough, 'Social Science Assoc.' 1874. For the Cape of Good Hope, Quetelet as quoted by Dr. H.H. Zouteveen, in the Dutch Translation of this work (vol. i. p. 417), where much information is given on the proportion of the sexes.
[50] In regard to the Jews, see M. Thury, 'La Loi de Production des Sexes,' 1863, p. 25.

Prof. Faye remarks that "a still greater preponderance of males would be met with, if death struck both sexes in equal proportion in the womb and during birth. But the fact is, that for every 100 still-born females, we have in several countries from 134.6 to 144.9 still-born males. During the first four or five years of life, also, more male children die than females, for example in England, during the first year, 126 boys die for every 100 girls—a proportion which in France is still more unfavorable." [51] Dr. Stockton Hough accounts for these facts in part by the more frequent defective development of males than of females. We have before seen that the male sex is more variable in structure than the female; and variations in important organs would generally be injurious. But the size of the body, and especially of the head, being greater in male than female infants is another cause: for the males are thus more liable to be injured during parturition. Consequently the still-born males are more numerous; and, as a highly competent judge, Dr. Crichton Browne [52], believes, male infants often suffer in health for some years after birth. Owing to this excess in the death-rate of male children, both at birth and for some time subsequently, and owing to the exposure of grown men to various dangers, and to their tendency to emigrate, the females in all old-settled countries, where statistical records have been kept, [53] are found to preponderate considerably over the males.

[51] 'British and Foreign Medico-Chirurg. Review,' April 1867, p. 343. Dr. Stark also remarks ('Tenth Annual Report of Births, Deaths, etc., in Scotland,' 1867, p. xxviii.) that "These examples may suffice to show that, at almost every stage of life, the males in Scotland have a greater liability to death and a higher death-rate than the females. The fact, however, of this peculiarity being most strongly developed at that infantile period of life when the dress, food, and general treatment of both sexes are alike, seems to prove that the higher male death-rate is an impressed, natural, and constitutional peculiarity due to sex alone."
[52] 'West Riding Lunatic Asylum Reports,' vol. i. 1871, p. 8. Sir J. Simpson has proved that the head of the male infant exceeds that of the female by ⅜ths of an inch in circumference, and by ⅛th in transverse diameter. Quetelet has shown that woman is born smaller than man; see Dr. Duncan, 'Fecundity, Fertility, and Sterility,' 1871, p. 382.
[53] With the savage Guaranys of Paraguay, according to the accurate Azara ('Voyages dans l'Amérique merid.' tom. ii. 1809, pp. 60, 179), the women are to the men in the proportion of 14 to 13.

It seems at first sight a mysterious fact that in different nations, under different conditions and climates, in Naples, Prussia, Westphalia, Holland, France, England and the United States, the excess of male over female births is less when they are illegitimate than when legitimate. [54] This has been explained by different writers in many different ways, as from the mothers being generally young, from the large proportion of first pregnancies, etc. But we have seen that male infants, from the large size of their heads, suffer more than female infants during parturition; and as the mothers of illegitimate children must be more liable than other women to undergo bad labors, from various causes, such as attempts at concealment by tight lacing, hard work, distress of mind, etc., their male infants would proportionably suffer. And this probably is the most efficient of all the causes of the proportion of males to females born alive being less amongst illegitimate children than amongst the legitimate. With most animals the greater size of the adult male than of the female, is due to the stronger males having conquered the weaker in their struggles for the possession of the females, and no doubt it is owing to this fact that the two sexes of at least some animals differ in size at birth. Thus we have the curious fact that we may attribute the more frequent deaths of male than female infants, especially amongst the illegitimate, at least in part to sexual selection.

[54] Babbage, 'Edinburgh Journal of Science,' 1829, vol. i. p. 88; also p. 90, on still-born children. On illegitimate children in England, see 'Report of Registrar-General for 1866,' p. xv.

It has often been supposed that the relative age of the two parents determine the sex of the offspring; and Prof. Leuckart [55] has advanced what he considers sufficient evidence, with respect to man and certain domesticated animals, that this is one important though not the sole factor in the result. So again the period of impregnation relatively to the state of the female has been thought by some to be the efficient cause; but recent observations discountenance this belief. According to Dr. Stockton Hough [56], the season of the year, the poverty or wealth of the parents, residence in the country or in cities, the crossing of foreign immigrants, etc., all influence the proportion of the sexes. With mankind, polygamy has also been supposed to lead to the birth of a greater proportion of female infants; but Dr. J. Campbell [57] carefully attended to this subject in the harems of Siam, and concludes that the proportion of male to female births is the same as from monogamous unions. Hardly any animal has been rendered so highly polygamous as the English race-horse, and we shall immediately see that his male and female offspring are almost exactly equal in number. I will now give the facts which I have collected with respect to the proportional numbers of the sexes of various animals; and will then briefly discuss how far selection has come into play in determining the result.

[55] Leuckart, in Wagner 'Handwörterbuch der Phys.' B. iv. 1853, s. 774.
[56] 'Social Science Association of Philadelphia,' 1874.
[57] 'Anthropological Review,' April 1870, p. cviii.

HORSES.

Mr. Tegetmeier has been so kind as to tabulate for me from the 'Racing Calendar' the births of race-horses during a period of twenty-one years, viz., from 1846 to 1867; 1849 being omitted, as no returns were that year published. The total births were 25,560, [58] consisting of 12,763 males and 12,797 females, or in the proportion of 99.7 males to 100 females. As these numbers are tolerably large, and as they are drawn from all parts of

England, during several years, we may with much confidence conclude that with the domestic horse, or at least with the race-horse, the two sexes are produced in almost equal numbers. The fluctuations in the proportions during successive years are closely like those which occur with mankind, when a small and thinly-populated area is considered; thus in 1856 the male horses were as 107.1, and in 1867 as only 92.6 to 100 females. In the tabulated returns the proportions vary in cycles, for the males exceeded the females during six successive years; and the females exceeded the males during two periods each of four years; this, however, may be accidental; at least I can detect nothing of the kind with man in the decennial table in the Registrar's Report for 1866.

[58] During eleven years a record was kept of the number of mares which proved barren or prematurely slipped their foals; and it deserves notice, as showing how infertile these highly-nurtured and rather closely-interbred animals have become, that not far from one-third of the mares failed to produce living foals. Thus during 1866, 809 male colts and 816 female colts were born, and 743 mares failed to produce offspring. During 1867, 836 males and 902 females were born, and 794 mares failed.

DOGS.

During a period of twelve years, from 1857 to 1868, the births of a large number of greyhounds, throughout England, were sent to the 'Field' newspaper; and I am again indebted to Mr. Tegetmeier for carefully tabulating the results. The recorded births were 6878, consisting of 3605 males and 3273 females, that is, in the proportion of 110.1 males to 100 females. The greatest fluctuations occurred in 1864, when the proportion was as 95.3 males, and in 1867, as 116.3 males to 100 females. The above average proportion of 110.1 to 100 is probably nearly correct in the case of the greyhound, but whether it would hold with other domesticated breeds is in some degree doubtful. Mr. Cupples has enquired from several great breeders of dogs, and finds that all without exception believe that females are produced in excess; but he suggests that this belief may have arisen from females being less valued, and from the consequent disappointment producing a stronger impression on the mind.

SHEEP.

The sexes of sheep are not ascertained by agriculturists until several months after birth, at the period when the males are castrated; so that the following returns do not give the proportions at birth. Moreover, I find that several great breeders in Scotland, who annually raise some thousand sheep, are firmly convinced that a larger proportion of males than of females die during the first year or two. Therefore the proportion of males would be somewhat larger at birth than at the age of castration. This is a remarkable coincidence with what, as we have seen, occurs with mankind, and both cases probably depend on the same cause. I have received returns from four gentlemen in England who have bred Lowland sheep, chiefly Leicesters, during the last ten to sixteen years; they amount altogether to 8965 births, consisting of 4407 males and 4558 females; that is in the proportion of 96.7 males to 100 females. With respect to Cheviot and black-faced sheep bred in Scotland, I have received returns from six breeders, two of them on a large scale, chiefly for the years 1867-1869, but some of the returns extend back to 1862. The total number recorded amounts to 50,685, consisting of 25,071 males and 25,614 females or in the proportion of 97.9 males to 100 females. If we take the English and Scotch returns together, the total number amounts to 59,650, consisting of 29,478 males and 30,172 females, or as 97.7 to 100. So that with sheep at the age of castration the females are certainly in excess of the males, but probably this would not hold good at birth. [59]

[59] I am much indebted to Mr. Cupples for having procured for me the above returns from Scotland, as well as some of the following returns on cattle. Mr. R. Elliot, of Laighwood, first called my attention to the premature deaths of the males, —a statement subsequently confirmed by Mr. Aitchison and others. To this latter gentleman, and to Mr. Payan, I owe my thanks for large returns as to sheep.

Of *Cattle* I have received returns from nine gentlemen of 982 births, too few to be trusted; these consisted of 477 bull-calves and 505 cow-calves; i.e., in the proportion of 94.4 males to 100 females. The Rev. W.D. Fox informs me that in 1867 out of 34 calves born on a farm in Derbyshire only one was a bull. Mr. Harrison Weir has enquired from several breeders of *Pigs*, and most of them estimate the male to the female births as about 7 to 6. This same gentleman has bred *rabbits* for many years, and has noticed that a far greater number of bucks are produced than does. But estimations are of little value.

Of mammalia in a state of nature I have been able to learn very little. In regard to the common rat, I have received conflicting statements. Mr. R. Elliot, of Laighwood, informs me that a rat-catcher assured him that he had always found the males in great excess, even with the young in the nest. In consequence of this, Mr. Elliot himself subsequently examined some hundred old ones, and found the statement true. Mr. F. Buckland has bred a large number of white rats, and he also believes that the males greatly exceed the females. In regard to Moles, it is said that "the males are much more numerous than the females" [60]: and as the catching of these animals is a special occupation, the statement may perhaps be trusted. Sir A. Smith, in describing an antelope of S. Africa [61] (*Kobus ellipsiprymnus*), remarks, that in the herds of this and other species, the males are few in number compared with the females: the natives believe that they are born in this proportion; others believe that the younger males are expelled from the herds, and Sir A. Smith says, that though he has himself never seen herds consisting of young males alone, others affirm that this does occur. It appears probable that the young when expelled from the herd, would often fall a prey to the many beasts of prey of the country.

[60] Bell, 'History of British Quadrupeds,' p. 100.
[61] 'Illustrations of the Zoology of S. Africa,' 1849, pl. 29.

BIRDS.

With respect to the *Fowl*, I have received only one account, namely, that out of 1001 chickens of a highly-bred stock of Cochins, reared during eight years by Mr. Stretch, 487 proved males and 514 females; i.e., as 94.7 to 100. In regard to domestic pigeons there is good evidence either that the males are produced in excess, or that they live longer; for these birds invariably pair, and single males, as Mr. Tegetmeier informs me, can always be purchased cheaper than females. Usually the two birds reared from the two eggs laid in the same nest are a male and a female; but Mr. Harrison Weir, who has been so large a breeder, says that he has often bred two cocks from the same nest, and seldom two hens; moreover, the hen is generally the weaker of the two, and more liable to perish.

With respect to birds in a state of nature, Mr. Gould and others [62] are convinced that the males are generally the more numerous; and as the young males of many species resemble the females, the latter would naturally appear to be the more numerous. Large numbers of pheasants are reared by Mr. Baker of Leadenhall from eggs laid by wild birds, and he informs Mr. Jenner Weir that four or five males to one female are generally produced. An experienced observer remarks [63], that in Scandinavia the broods of the capercailzie and black-cock contain more males than females; and that with the Dal-ripa (a kind of ptarmigan) more males than females attend the *leks* or places of courtship; but this latter circumstance is accounted for by some observers by a greater number of hen birds being killed by vermin. From various facts given by White of Selborne [64], it seems clear that the males of the partridge must be in considerable excess in the south of England; and I have been assured that this is the case in Scotland. Mr. Weir on enquiring from the dealers, who receive at certain seasons large numbers of ruffs (*Machetes pugnax*), was told that the males are much the more numerous. This same naturalist has also enquired for me from the birdcatchers, who annually catch an astonishing number of various small species alive for the London market, and he was unhesitatingly answered by an old and trustworthy man, that with the chaffinch the males are in large excess: he thought as high as 2 males to 1 female, or at least as high as 5 to 3. [65] The males of the blackbird, he likewise maintained, were by far the more numerous, whether caught by traps or by netting at night. These statements may apparently be trusted, because this same man said that the sexes are about equal with the lark, the twite (*Linaria montana*), and goldfinch. On the other hand, he is certain that with the common linnet, the females preponderate greatly, but unequally during different years; during some years he has found the females to the males as four to one. It should, however, be borne in mind, that the chief season for catching birds does not begin till September, so that with some species partial migrations may have begun, and the flocks at this period often consist of hens alone. Mr. Salvin paid particular attention to the sexes of the humming-birds in Central America, and is convinced that with most of the species the males are in excess; thus one year he procured 204 specimens belonging to ten species, and these consisted of 166 males and of only 38 females. With two other species the females were in excess: but the proportions apparently vary either during different seasons or in different localities; for on one occasion the males of *Campylopterus hemileucurus* were to the females as 5 to 2, and on another occasion [66] in exactly the reversed ratio. As bearing on this latter point, I may add, that Mr. Powys found in Corfu and Epirus the sexes of the chaffinch keeping apart, and "the females by far the most numerous"; whilst in Palestine Mr. Tristram found "the male flocks appearing greatly to exceed the female in number." [67] So again with the *Quiscalus major*, Mr. G. Taylor says, that in Florida there were "very few females in proportion to the males," [68] whilst in Honduras the proportion was the other way, the species there having the character of a polygamist.

[62] Brehm ('Thierleben,' B. iv. s. 990) comes to the same conclusion.

[63] On the authority of L. Lloyd, 'Game Birds of Sweden,' 1867, pp. 12, 132.

[64] 'Nat. Hist. of Selborne,' letter xxix. edit. of 1825, vol. i. p. 139.

[65] Mr. Jenner Weir received similar information, on making enquiries during the following year. To show the number of living chaffinches caught, I may mention that in 1869 there was a match between two experts, and one man caught in a day 62, and another 40, male chaffinches. The greatest number ever caught by one man in a single day was 70.

[66] 'Ibis,' vol. ii. p. 260, as quoted in Gould's 'Trochilidæ,' 1861, p. 52. For the foregoing proportions, I am indebted to Mr. Salvin for a table of his results.
[67] 'Ibis,' 1860, p. 137; and 1867, p. 369.
[68] 'Ibis,' 1862, p. 187.

FISH.

With fish the proportional numbers of the sexes can be ascertained only by catching them in the adult or nearly adult state; and there are many difficulties in arriving at any just conclusion. [69] Infertile females might readily be mistaken for males, as Dr. Günther has remarked to me in regard to trout. With some species the males are believed to die soon after fertilising the ova. With many species the males are of much smaller size than the females, so that a large number of males would escape from the same net by which the females were caught. M. Carbonnier [70], who has especially attended to the natural history of the pike (*Esox lucius*), states that many males, owing to their small size, are devoured by the larger females; and he believes that the males of almost all fish are exposed from this same cause to greater danger than the females. Nevertheless, in the few cases in which the proportional numbers have been actually observed, the males appear to be largely in excess. Thus Mr. R. Buist, the superintendent of the Stormontfield experiments, says that in 1865, out of 70 salmon first landed for the purpose of obtaining the ova, upwards of 60 were males. In 1867 he again "calls attention to the vast disproportion of the males to the females. We had at the outset at least ten males to one female." Afterwards females sufficient for obtaining ova were procured. He adds, "from the great proportion of the males, they are constantly fighting and tearing each other on the spawning-beds." [71] This disproportion, no doubt, can be accounted for in part, but whether wholly is doubtful, by the males ascending the rivers before the females. Mr. F. Buckland remarks in regard to trout, that "it is a curious fact that the males preponderate very largely in number over the females. It *invariably* happens that when the first rush of fish is made to the net, there will be at least seven or eight males to one female found captive. I cannot quite account for this; either the males are more numerous than the females, or the latter seek safety by concealment rather than flight." He then adds, that by carefully searching the banks sufficient females for obtaining ova can be found. [72] Mr. H. Lee informs me that out of 212 trout, taken for this purpose in Lord Portsmouth's park, 150 were males and 62 females.

[69] Leuckart quotes Bloch (Wagner, 'Handwörterbuch der Phys.' B. iv. 1853, s. 775), that with fish there are twice as many males as females.
[70] Quoted in the 'Farmer,' March 18, 1869, p. 369.
[71] 'The Stormontfield Piscicultural Experiments,' 1866, p. 23. The 'Field' newspaper, June 29, 1867.
[72] 'Land and Water,' 1868, p. 41.

The males of the Cyprinidæ likewise seem to be in excess; but several members of this Family, viz., the carp, tench, bream and minnow, appear regularly to follow the practice, rare in the animal kingdom, of polyandry; for the female whilst spawning is always attended by two males, one on each side, and in the case of the bream by three or four males. This fact is so well known, that it is always recommended to stock a pond with two male tenches to one female, or at least with three males to two females. With the minnow, an excellent observer states, that on the spawning-beds the males are ten times as numerous as the females; when a female comes amongst the males, "she is immediately pressed closely by a male on each side; and when they have been in that situation for a time, are superseded by other two males." [73]

[73] Yarrell, 'Hist. British Fishes,' vol. i. 1826, p. 307; on the *Cyprinus carpio*, p. 331; on the *Tinca vulgaris*, p. 331; on the *Abramis brama*, p. 336. See, for the minnow (*Leuciscus phoxinus*), 'Loudon's Magazine of Natural History,' vol. v. 1832, p. 682.

INSECTS.

In this great Class, the Lepidoptera almost alone afford means for judging of the proportional numbers of the sexes; for they have been collected with special care by many good observers, and have been largely bred from the egg or caterpillar state. I had hoped that some breeders of silk-moths might have kept an exact record, but after writing to France and Italy, and consulting various treatises, I cannot find that this has ever been done. The general opinion appears to be that the sexes are nearly equal, but in Italy, as I hear from Professor Canestrini, many breeders are convinced that the females are produced in excess. This same naturalist, however, informs me, that in the two yearly broods of the Ailanthus silk-moth (*Bombyx cynthia*), the males greatly preponderate in the first, whilst in the second the two sexes are nearly equal, or the females rather in excess.

In regard to Butterflies in a state of nature, several observers have been much struck by the apparently enormous preponderance of the males. [74] Thus Mr. Bates [75], in speaking of several species, about a hundred in number, which inhabit the upper Amazons, says that the males are much more numerous than the females, even in the proportion of a hundred to one. In North America, Edwards, who had great experience, estimates in the genus Papilio the males to the females as four to one; and Mr. Walsh, who informed me of this statement, says that with *P. turnus* this is certainly the case. In South Africa, Mr. R. Trimen found the males in excess in 19 species [76]; and in one of these, which swarms in open places, he estimated the number of males as fifty to one female. With another species, in which the males are numerous in certain localities, he collected only five females during seven years. In the island of Bourbon, M. Maillard states that the males of one species of Papilio are twenty times as numerous as the females. [77] Mr. Trimen informs me that as far as he has himself seen, or heard from others, it is rare for the females of any butterfly to exceed the males in number; but three South African species perhaps offer an exception. Mr. Wallace [78] states that the females of *Ornithoptera croesus*, in the Malay archipelago, are more common and more easily caught than the males; but this is a rare butterfly. I may here add, that in Hyperythra, a genus of moths, Guenée says, that from four to five females are sent in collections from India for one male.

[74] Leuckart quotes Meinecke (Wagner, 'Handwörterbuch der Phys.' B. iv. 1853, s. 775) that the males of Butterflies are three or four times as numerous as the females.

[75] 'The Naturalist on the Amazons,' vol. ii. 1863, pp. 228, 347.

[76] Four of these cases are given by Mr. Trimen in his 'Rhopalocera Africæ Australis.'

[77] Quoted by Trimen, 'Transactions of the Ent. Society,' vol. v. part iv. 1866, p. 330.

[78] 'Transactions, Linnean Society,' vol. xxv. p. 37.

When this subject of the proportional numbers of the sexes of insects was brought before the Entomological Society [79], it was generally admitted that the males of most Lepidoptera, in the adult or imago state, are caught in greater numbers than the females: but this fact was attributed by various observers to the more retiring habits of the females, and to the males emerging earlier from the cocoon. This latter circumstance is well known to occur with most Lepidoptera, as well as with other insects. So that, as M. Personnat remarks, the males of the domesticated *Bombyx Yamamai*, are useless at the beginning of the season, and the females at the end, from the want of mates. [80] I cannot, however, persuade myself that these causes suffice to explain the great excess of males, in the above cases of certain butterflies which are extremely common in their native countries. Mr. Stainton, who has paid very close attention during many years to the smaller moths, informs me that when he collected them in the imago state, he thought that the males were ten times as numerous as the females, but that since he has reared them on a large scale from the caterpillar state, he is convinced that the females are the more numerous. Several entomologists concur in this view. Mr. Doubleday, however, and some others, take an opposite view, and are convinced that they have reared from the eggs and caterpillars a larger proportion of males than of females.

[79] 'Proceedings, Entomological Society,' Feb. 17, 1868.
[80] Quoted by Dr. Wallace in 'Proceedings, Entomological Society,' 3rd series, vol. v. 1867, p. 487.

Besides the more active habits of the males, their earlier emergence from the cocoon, and in some cases their frequenting more open stations, other causes may be assigned for an apparent or real difference in the proportional numbers of the sexes of Lepidoptera, when captured in the imago state, and when reared from the egg or caterpillar state. I hear from Professor Canestrini, that it is believed by many breeders in Italy, that the female caterpillar of the silk-moth suffers more from the recent disease than the male; and Dr. Staudinger informs me that in rearing Lepidoptera more females die in the cocoon than males. With many species the female caterpillar is larger than the male, and a collector would naturally choose the finest specimens, and thus unintentionally collect a larger number of females. Three collectors have told me that this was their practice; but Dr. Wallace is sure that most collectors take all the specimens which they can find of the rarer kinds, which alone are worth the trouble of rearing. Birds when surrounded by caterpillars would probably devour the largest; and Professor Canestrini informs me that in Italy some breeders believe, though on insufficient evidence, that in the first broods of the Ailanthus silk-moth, the wasps destroy a larger number of the female than of the male caterpillars. Dr. Wallace further remarks that female caterpillars, from being larger than the males, require more time for their development, and consume more food and moisture: and thus they would be exposed during a longer time to danger from ichneumons, birds, etc., and in times of scarcity would perish in greater numbers. Hence it appears quite possible that in a state of nature, fewer female Lepidoptera may reach maturity than males; and for our special object we are concerned with their relative numbers at maturity, when the sexes are ready to propagate their kind.

The manner in which the males of certain moths congregate in extraordinary numbers round a single female, apparently indicates a great excess of males, though this fact may perhaps be accounted for by the earlier emergence of the males from their cocoons. Mr. Stainton informs me that from twelve to twenty males, may often be seen congregated round a female *Elachista rufocinerea*. It is well known that if a virgin *Lasiocampa quercus* or *Saturnia carpini* be exposed in a cage, vast numbers of males collect round her, and if confined in a room will even come down the chimney to her. Mr. Doubleday believes that he has seen from fifty to a hundred males of both these species attracted in the course of a single day by a female in confinement. In the Isle of Wight Mr. Trimen exposed a box in which a female of the Lasiocampa had been confined on the previous day, and five males soon endeavoured to gain admittance. In Australia, Mr. Verreaux, having placed the female of a small Bombyx in a box in his pocket, was followed by a crowd of males, so that about 200 entered the house with him. [81]

[81] Blanchard, 'Métamorphoses, Mœurs des Insectes,' 1868, pp. 225-226.

Mr. Doubleday has called my attention to M. Staudinger's [82] list of Lepidoptera, which gives the prices of the males and females of 300 species or well-marked varieties of butterflies (Rhopalocera). The prices for both sexes of the very common species are of course the same; but in 114 of the rarer species they differ; the males being in all cases, excepting one, the cheaper. On an average of the prices of the 113 species, the price of the male to that of the female is as 100 to 149; and this apparently indicates that inversely the males exceed the females in the same proportion. About 2000 species or varieties of moths (Heterocera) are catalogued, those with wingless females being here excluded on account of the difference in habits between the two sexes: of these 2000 species, 141 differ in price according to sex, the males of 130 being cheaper, and those of only 11 being dearer than the females. The average price of the males of the 130 species, to that of the females, is as 100 to 143. With respect to the butterflies in this priced list, Mr. Doubleday thinks (and no man in England has had more experience), that there is nothing in the habits of the species which can account for the difference in the prices of the two sexes, and that it can be accounted for only by an excess in the number of the males. But I am bound to add that Dr. Staudinger informs me, that he is himself of a different opinion. He thinks that the less active habits of the females and the earlier emergence of the males will account for his collectors securing a larger number of males than of females, and consequently for the lower prices of the former. With respect to specimens reared from the caterpillar-state, Dr. Staudinger believes, as previously stated, that a greater number of females than of males die whilst confined to the cocoons. He adds that with certain species one sex seems to preponderate over the other during certain years.

[82] 'Lepidopteren-Doubletten Liste,' Berlin, No. x. 1866.

Of direct observations on the sexes of Lepidoptera, reared either from eggs or caterpillars, I have received only the few following cases:

	Males.	Females.
The Rev. J. Hellins [83] of Exeter reared, during 1868, imagos of 73 species, which consisted of...............	153	137
Mr. Albert Jones of Elthan reared, during 1868, imagos of 9 species, which consisted of.................	159	126
During 1869 he reared imagos from 4 species, consisting of.........	114	112
Mr. Buckler of Emsworth, Hants, during 1869, reared imagos from 74 species, consisting of.........	180	169
Dr. Wallace of Colchester reared from one brood of Bombyx cynthia..............	52	48
Dr. Wallace raised, from cocoons of Bombyx Pernyi sent from China, during 1869...............	224	123
Dr. Wallace raised, during 1868 and 1869, from two lots of cocoons of Bombyx yama-mai..................	52	46
Total.........	934	761

So that in these eight lots of cocoons and eggs, males were produced in excess. Taken together the proportion of males is as 122.7 to 100 females. But the numbers are hardly large enough to be trustworthy.

On the whole, from these various sources of evidence, all pointing in the same direction, I infer that with most species of Lepidoptera, the mature males generally exceed the females in number, whatever the proportions may be at their first emergence from the egg.

[83] This naturalist has been so kind as to send me some results from former years, in which the females seemed to preponderate; but so many of the figures were estimates, that I found it impossible to tabulate them.

With reference to the other Orders of insects, I have been able to collect very little reliable information. With the stag-beetle (*Lucanus cervus*) "the males appear to be much more numerous than the females"; but when, as Cornelius remarked during 1867, an unusual number of these beetles appeared in one part of Germany, the females appeared to exceed the males as six to one. With one of the Elateridæ, the males are said to be much more numerous than the females, and "two or three are often found united with one female [84]; so that here polyandry seems to prevail." With Siagonium (Staphylinidæ), in which the males are furnished with horns, "the females are far more numerous than the opposite sex." Mr. Janson stated at the Entomological Society that the females of the bark feeding *Tomicus villosus* are so common as to be a plague, whilst the males are so rare as to be hardly known.

[84] Günther's 'Record of Zoological Literature,' 1867, p. 260. On the excess of female Lucanus, ibid, p. 250. On the males of Lucanus in England, Westwood,' 'Modern Classification of Insects,' vol. i. p. 187. On the Siagonium, ibid. p. 172.

It is hardly worth while saying anything about the proportion of the sexes in certain species and even groups of insects, for the males are unknown or very rare, and the females are parthenogenetic, that is, fertile without sexual union; examples of this are afforded by several of the Cynipidæ. [85] In all the gall-making Cynipidæ known to Mr. Walsh, the females are four or five times as numerous as the males; and so it is, as he informs me, with the gall-making Cecidomyiidæ (Diptera). With some common species of Saw-flies (Tenthredinæ) Mr. F. Smith has reared hundreds of specimens from larvæ of all sizes, but has never reared a single male; on the other hand, Curtis says [86], that with certain species (Athalia), bred by him, the males were to the females as six to one; whilst exactly the reverse occurred with the mature insects of the same species caught in the fields. In the family of bees, Hermann Müller [87], collected a large number of specimens of many species, and reared others from the cocoons, and counted the sexes. He found that the males of some species greatly exceeded the females in number; in others the reverse occurred; and in others the two sexes were nearly equal. But as in most cases the males emerge from the cocoons before the females, they are at the commencement of the breeding-season practically in excess. Müller also observed that the relative number of the two sexes in some species differed much in different localities. But as H. Müller has himself remarked to me, these remarks must be received with some caution, as one sex might more easily escape observation than the other. Thus his brother Fritz Müller has noticed in Brazil that the two sexes of the same species of bee sometimes frequent different kinds of flowers. With respect to the Orthoptera, I know hardly anything about the relative number of the sexes: Körte [88], however, says that out of 500 locusts which he examined, the males were to the females as five to six. With the Neuroptera, Mr. Walsh states that in many, but by no means in all the species of the Odonatous group, there is a great overplus of males: in the genus Hetærina, also, the males are generally at least four times as numerous as the females. In certain species in the genus Gomphus the males are equally in excess, whilst in two other species, the females are twice or thrice as numerous as the males. In some European species of Psocus thousands of females may be collected without a single male, whilst with other species of the same genus both sexes are common. [89] In England, Mr. MacLachlan has captured hundreds of the female *Apatania muliebris*, but has never seen the male; and of *Boreus hyemalis* only four or five males have been seen here. [90] With most of these species (excepting the Tenthredinæ) there is at present no evidence that the females are subject to parthenogenesis; and thus we see how ignorant we are of the causes of the apparent discrepancy in the proportion of the two sexes.

[85] Walsh in 'The American Entomologist,' vol. i. 1869, p. 103. F. Smith, 'Record of Zoological Lit.' 1867, p. 328.
[86] 'Farm Insects,' pp. 45-46.
[87] 'Anwendung der Darwinschen Lehre Verh. d. n. Jahrg., xxiv.'
[88] 'Die Strich, Zug oder Wanderheuschrecke,' 1828, p. 20.
[89] 'Observations on N. American Neuroptera,' by H. Hagen and B.D. Walsh, 'Proceedings, Ent. Soc. Philadelphia,' Oct. 1863, pp. 168, 223, 239.
[90] 'Proceedings, Ent. Soc. London,' Feb. 17, 1868.

In the other classes of the Articulata I have been able to collect still less information. With spiders, Mr. Blackwall, who has carefully attended to this class during many years, writes to me that the males from their more erratic habits are more commonly seen, and therefore appear more numerous. This is actually the case with a few species; but he mentions several species in six genera, in which the females appear to be much more numerous than the males. [91] The small size of the males in comparison with the females (a peculiarity which is sometimes carried to an extreme degree), and their widely different appearance, may account in some instances for their rarity in collections. [92]

[91] Another great authority with respect to this class, Prof. Thorell of Upsala ('On European Spiders,' 1869-70, part i. p. 205), speaks as if female spiders were generally commoner than the males.

[92] See, on this subject, Mr. O.P. Cambridge, as quoted in 'Quarterly Journal of Science,' 1868, page 429.

Some of the lower Crustaceans are able to propagate their kind sexually, and this will account for the extreme rarity of the males; thus von Siebold [93] carefully examined no less than 13,000 specimens of Apus from twenty-one localities, and amongst these he found only 319 males. With some other forms (as Tanais and Cypris), as Fritz Müller informs me, there is reason to believe that the males are much shorter-lived than the females; and this would explain their scarcity, supposing the two sexes to be at first equal in number. On the other hand, Müller has invariably taken far more males than females of the Diastylidæ and of Cypridina on the shores of Brazil: thus with a species in the latter genus, 63 specimens caught the same day included 57 males; but he suggests that this preponderance may be due to some unknown difference in the habits of the two sexes. With one of the higher Brazilian crabs, namely a Gelasimus, Fritz Müller found the males to be more numerous than the females. According to the large experience of Mr. C. Spence Bate, the reverse seems to be the case with six common British crabs, the names of which he has given me.

[93] 'Beit. zur Parthenogenesis,' p. 174.

THE PROPORTION OF THE SEXES IN RELATION TO NATURAL SELECTION.

There is reason to suspect that in some cases man has by selection indirectly influenced his own sex-producing powers. Certain women tend to produce during their whole lives more children of one sex than of the other: and the same holds good of many animals, for instance, cows and horses; thus Mr. Wright of Yeldersley House informs me that one of his Arab mares, though put seven times to different horses, produced seven fillies. Though I have very little evidence on this head, analogy would lead to the belief, that the tendency to produce either sex would be inherited like almost every other peculiarity, for instance, that of producing twins; and concerning the above tendency a good authority, Mr. J. Downing, has communicated to me facts which seem to prove that this does occur in certain families of short-horn cattle. Col. Marshall [94] has recently found on careful examination that the Todas, a hill-tribe of India, consist of 112 males and 84 females of all ages—that is in a ratio of 133.3 males to 100 females. The Todas, who are polyandrous in their marriages, during former times invariably practised female infanticide; but this practice has now been discontinued for a considerable period. Of the children born within late years, the males are more numerous than the females, in the proportion of 124 to 100. Colonel Marshall accounts for this fact in the following ingenious manner. "Let us for the purpose of illustration take three families as representing an average of the entire tribe; say that one mother gives birth to six daughters and no sons; a second mother has six sons only, whilst the third mother has three sons and three daughters. The first mother, following the tribal custom, destroys four daughters and preserves two. The second retains her six sons. The third kills two daughters and keeps one, as also her three sons. We have then from the three families, nine sons and three daughters, with which to continue the breed. But whilst the males belong to families in which the tendency to produce sons is great, the females are of those of a converse inclination. Thus the bias strengthens with each generation, until, as we find, families grow to have habitually more sons than daughters."

[94] 'The Todas,' 1873, pp. 100, 111, 194, 196.

That this result would follow from the above form of infanticide seems almost certain; that is if we assume that a sex-producing tendency is inherited. But as the above numbers are so extremely scanty, I have searched for additional evidence, but cannot decide whether what I have found is trustworthy; nevertheless the facts are, perhaps, worth giving. The Maories of New Zealand have long practised infanticide; and Mr. Fenton [95] states that he "has met with instances of women who have destroyed four, six, and even seven children, mostly females. However, the universal testimony of those best qualified to judge, is conclusive that this custom has for many years been almost extinct. Probably the year 1835 may be named as the period of its ceasing to exist." Now amongst the New Zealanders, as with the Todas, male births are considerably in excess. Mr. Fenton remarks (p. 30), "One fact is certain, although the exact period of the commencement of this singular condition of the disproportion of the sexes cannot be demonstratively fixed, it is quite clear that this course of decrease was in full operation during the years 1830 to 1844, when the non-adult population of 1844 was being produced, and has continued with great energy up to the present time." The following statements are taken from Mr. Fenton (p. 26), but as the numbers are not large, and as the census was not accurate, uniform results cannot be expected. It should be borne in mind in this and the following cases, that the normal state of every population is an excess of women, at least in all civilized countries, chiefly owing to the greater mortality of the male sex during youth, and partly to accidents of all kinds later in life. In 1858, the native population of New Zealand was estimated as consisting of 31,667 males and 24,303 females of all ages, that is in the ratio of 130.3 males to 100 females. But during this same year, and in certain limited districts, the numbers were ascertained with much care, and the males of all ages were here 753 and the females 616; that is in the ratio of 122.2 males to 100 females. It is more important for us that during this same year of 1858, the *non-adult* males within the same district were found to be 178, and the *non-adult* females 142, that is in the ratio of 125.3 to 100. It may be added that in 1844, at which period female infanticide had only lately ceased, the *non-adult* males in one district were 281, and the *non-adult* females only 194, that is in the ratio of 144.8 males to 100 females.

[95] 'Aboriginal Inhabitants of New Zealand: Government Report,' 1859, p. 36.

In the Sandwich Islands, the males exceed the females in number. Infanticide was formerly practised there to a frightful extent, but was by no means confined to female infants, as is shown by Mr. Ellis [96], and as I have been informed by Bishop Staley and the Rev. Mr. Coan. Nevertheless, another apparently trustworthy writer, Mr. Jarves [97], whose observations apply to the whole archipelago, remarks:—"Numbers of women are to be found, who confess to the murder of from three to six or eight children," and he adds, "females from being considered less useful than males were more often destroyed."

From what is known to occur in other parts of the world, this statement is probable; but must be received with much caution. The practice of infanticide ceased about the year 1819, when idolatry was abolished and missionaries settled in the Islands. A careful census in 1839 of the adult and taxable men and women in the island of Kauai and in one district of Oahu (Jarves, p. 404), gives 4723 males and 3776 females; that is in the ratio of 125.08 to 100. At the same time the number of males under fourteen years in Kauai and under eighteen in Oahu was 1797, and of females of the same ages 1429; and here we have the ratio of 125.75 males to 100 females.

[96] 'Narrative of a Tour through Hawaii,' 1826, p. 298.
[97] 'History of the Sandwich Islands,' 1843, p. 93.

In a census of all the islands in 1850 [98], the males of all ages amount to 36,272, and the females to 33,128, or as 109.49 to 100. The males under seventeen years amounted to 10,773, and the females under the same age to 9593, or as 112.3 to 100. From the census of 1872, the proportion of males of all ages (including half-castes) to females, is as 125.36 to 100. It must be borne in mind that all these returns for the Sandwich Islands give the proportion of living males to living females, and not of the births; and judging from all civilized countries the proportion of males would have been considerably higher if the numbers had referred to births. [99]

[98] This is given in the Rev. H.T. Cheever's 'Life in the Sandwich Islands,' 1851, p. 277.
[99] Dr. Coulter, in describing ('Journal R. Geograph. Soc.' vol. v. 1835, p. 67) the state of California about the year 1830, says that the natives, reclaimed by the Spanish missionaries, have nearly all perished, or are perishing, although well treated, not driven from their native land, and kept from the use of spirits. He attributes this, in great part, to the undoubted fact that the men greatly exceed the women in number; but he does not know whether this is due to a failure of female offspring, or to more females dying during early youth. The latter alternative, according to all analogy, is very improbable. He adds that "infanticide, properly so called, is not common, though very frequent recourse is had to abortion." If Dr. Coulter is correct about infanticide, this case cannot be advanced in support of Colonel Marshall's view. From the rapid decrease of the reclaimed natives, we may suspect that, as in the cases lately given, their fertility has been diminished from changed habits of life.
I had hoped to gain some light on this subject from the breeding of dogs; inasmuch as in most breeds, with the exception, perhaps, of greyhounds, many more female puppies are destroyed than males, just as with the Toda infants. Mr. Cupples assures me that this is usual with Scotch deer-hounds. Unfortunately, I know nothing of the proportion of the sexes in any breed, excepting greyhounds, and there the male births are to the females as 110.1 to 100. Now from enquiries made from many breeders, it seems that the females are in some respects more esteemed, though otherwise troublesome; and it does not appear that the female puppies of the best-bred dogs are systematically destroyed more than the males, though this does sometimes take place to a limited extent. Therefore I am unable to decide whether we can, on the above principles, account for the preponderance of male births in greyhounds. On the other hand, we have seen that with horses, cattle, and sheep, which are too valuable for the young of either sex to be destroyed, if there is any difference, the females are slightly in excess.

From the several foregoing cases we have some reason to believe that infanticide practised in the manner above explained, tends to make a male-producing race; but I am far from supposing that this practice in the case of man, or some analogous process with other species, has been the sole determining cause of an excess of males. There may be some unknown law leading to this result in decreasing races, which have already become somewhat infertile. Besides the several causes previously alluded to, the greater facility of parturition amongst savages, and the less consequent injury to their male infants, would tend to increase the proportion of live-born males to females. There does not, however, seem to be any necessary connection between savage life and a marked excess of males; that is if we may judge by the character of the scanty offspring of the lately existing Tasmanians and of the crossed offspring of the Tahitians now inhabiting Norfolk Island.

As the males and females of many animals differ somewhat in habits and are exposed in different degrees to danger, it is probable that in many cases, more of one sex than of the other are habitually destroyed. But as far as I can trace out the complication of causes, an indiscriminate though large destruction of either sex would not tend to modify the sex-producing power of the species. With strictly social animals, such as bees or ants, which produce a vast number of sterile and fertile females in comparison with the males, and to whom this preponderance is of paramount importance, we can see that those communities would flourish best which contained females having a strong inherited tendency to produce more and more females; and in such cases an unequal sex-producing tendency would be ultimately gained through natural selection. With animals living in herds or troops, in which the males come to the front and defend the herd, as with the bisons of North America and certain baboons, it is conceivable that a male-producing tendency might be gained by natural selection; for the individuals of the better defended herds would leave more numerous descendants. In the case of mankind the advantage arising from having a preponderance of men in the tribe is supposed to be one chief cause of the practice of female infanticide.

In no case, as far as we can see, would an inherited tendency to produce both sexes in equal numbers or to produce one sex in excess, be a direct advantage or disadvantage to certain individuals more than to others; for instance, an individual with a tendency to produce more males than females would not succeed better in the battle for life than an individual with an opposite tendency; and therefore a tendency of this kind could not be gained through natural selection. Nevertheless, there are certain animals (for instance, fishes and cirripedes) in which two or more males appear to be necessary for the fertilization of the female; and the males accordingly largely preponderate, but it is by no means obvious how this male-producing tendency could have been acquired. I formerly thought that when a tendency to produce the two sexes in equal numbers was advantageous to the species, it would follow from natural selection, but I now see that the whole problem is so intricate that it is safer to leave its solution for the future.

CHAPTER IX.

SECONDARY SEXUAL CHARACTERS IN THE LOWER CLASSES OF THE ANIMAL KINGDOM.

These characters absent in the lowest classes—Brilliant colors—Mollusca—Annelids—Crustacea, secondary sexual characters strongly developed; dimorphism; color; characters not acquired before maturity—Spiders, sexual colors of; stridulation by the males—Myriapoda.

With animals belonging to the lower classes, the two sexes are not rarely united in the same individual, and therefore secondary sexual characters cannot be developed. In many cases where the sexes are separate, both are permanently attached to some support, and the one cannot search or struggle for the other. Moreover it is almost certain that these animals have too imperfect senses and much too low mental powers to appreciate each other's beauty or other attractions, or to feel rivalry.

Hence in these classes or sub-kingdoms, such as the Protozoa, Cœlenterata, Echinodermata, Scolecida, secondary sexual characters, of the kind which we have to consider, do not occur: and this fact agrees with the belief that such characters in the higher classes have been acquired through sexual selection, which depends on the will, desire, and choice of either sex. Nevertheless some few apparent exceptions occur; thus, as I hear from Dr. Baird, the males of certain Entozoa, or internal parasitic worms, differ slightly in color from the females; but we have no reason to suppose that such differences have been augmented through sexual selection. Contrivances by which the male holds the female, and which are indispensable for the propagation of the species, are independent of sexual selection, and have been acquired through ordinary selection.

Many of the lower animals, whether hermaphrodites or with separate sexes, are ornamented with the most brilliant tints, or are shaded and striped in an elegant manner; for instance, many corals and sea-anemones (Actiniæ), some jelly-fish (Medusæ, Porpita, etc.), some Planariæ, many star-fishes, Echini, Ascidians, etc.; but we may conclude from the reasons already indicated, namely, the union of the two sexes in some of these animals, the permanently affixed condition of others, and the low mental powers of all, that such colors do not serve as a sexual attraction, and have not been acquired through sexual selection. It should be borne in mind that in no case have we sufficient evidence that colors have been thus acquired, except where one sex is much more brilliantly or conspicuously colored than the other, and where there is no difference in habits between the sexes sufficient to account for their different colors. But the evidence is rendered as complete as it can ever be, only when the more ornamented individuals, almost always the males, voluntarily display their attractions before the other sex; for we cannot believe that such display is useless, and if it be advantageous, sexual selection will almost inevitably follow. We may, however, extend this conclusion to both sexes, when colored alike, if their colors are plainly analogous to those of one sex alone in certain other species of the same group.

How, then, are we to account for the beautiful or even gorgeous colors of many animals in the lowest classes? It appears doubtful whether such colors often serve as a protection; but that we may easily err on this head, will be admitted by every one who reads Mr. Wallace's excellent essay on this subject. It would not, for instance, at first occur to any one that the transparency of the Medusæ, or jelly-fish, is of the highest service to them as a protection; but when we are reminded by Häckel that not only the Medusæ, but many floating Mollusca, crustaceans, and even small oceanic fishes partake of this same glass-like appearance, often accompanied by prismatic colors, we can hardly doubt that they thus escape the notice of pelagic birds and other enemies. M. Giard is also convinced [1] that the bright tints of certain sponges and ascidians serve as a protection. Conspicuous colors are likewise beneficial to many animals as a warning to their would-be devourers that they are distasteful, or that they possess some special means of defence; but this subject will be discussed more conveniently hereafter.

[1] 'Archives de Zoolog. Expér.' Oct. 1872, p. 563.

We can, in our ignorance of most of the lowest animals, only say that their bright tints result either from the chemical nature or the minute structure of their tissues, independently of any benefit thus derived. Hardly any color is finer than that of arterial blood; but there is no reason to suppose that the color of the blood is in itself any advantage; and though it adds to the beauty of the maiden's cheek, no one will pretend that it has been acquired for this purpose. So again with many animals, especially the lower ones, the bile is richly colored; thus, as I am informed by Mr. Hancock, the extreme beauty of the Eolidæ (naked sea-slugs) is chiefly due to the biliary glands being seen through the translucent integuments—this beauty being probably of no service to these animals. The tints of the decaying leaves in an American forest are described by every one as gorgeous; yet no one supposes that these tints are of the least advantage to the trees. Bearing in mind how many substances closely analogous to natural organic compounds have been recently formed by chemists, and which exhibit the most splendid colors, it would have been a strange fact if substances similarly colored had not often originated, independently of any useful end thus gained, in the complex laboratory of living organisms.

THE SUB-KINGDOM OF THE MOLLUSCA.

Throughout this great division of the animal kingdom, as far as I can discover, secondary sexual characters, such as we are here considering, never occur. Nor could they be expected in the three lowest classes, namely, in the Ascidians, Polyzoa, and Brachiopods (constituting the Molluscoida of some authors), for most of these animals are permanently affixed to a support or have their sexes united in the same individual. In the Lamellibranchiata, or bivalve shells, hermaphroditism is not rare. In the next higher class of the Gasteropoda, or univalve shells, the sexes are either united or separate. But in the latter case the males never possess special organs for finding, securing, or charming the females, or for fighting with other males. As I am informed by Mr. Gwyn Jeffreys, the sole external difference between the sexes consists in the shell sometimes differing a little in form; for instance, the shell of the male periwinkle (Littorina littorea) is narrower and has a more elongated spire than that of the female. But differences of this nature, it may be presumed, are directly connected with the act of reproduction, or with the development of the ova.

The Gasteropoda, though capable of locomotion and furnished with imperfect eyes, do not appear to be endowed with sufficient mental powers for the members of the same sex to struggle together in rivalry, and thus to acquire secondary sexual characters.

Nevertheless with the pulmoniferous gasteropods, or land-snails, the pairing is preceded by courtship; for these animals, though hermaphrodites, are compelled by their structure to pair together. Agassiz remarks, "Quiconque a eu l'occasion d'observer les amours des limaçons, ne saurait mettre en doute la séduction déployée dans les mouvements et les allures qui préparent et accomplissent le double embrassement de ces hermaphrodites." [2] These animals appear also susceptible of some degree of permanent attachment: an accurate observer, Mr. Lonsdale, informs me that he placed a pair of land-snails, (*Helix pomatia*), one of which was weakly, into a small and ill-provided garden. After a short time the strong and healthy individual disappeared, and was traced by its track of slime over a wall into an adjoining well-stocked garden. Mr. Lonsdale concluded that it had deserted its sickly mate; but after an absence of twenty-four hours it returned, and apparently communicated the result of its successful exploration, for both then started along the same track and disappeared over the wall.

[2] 'De l'Espèce et de la Class.' etc., 1869, p. 106.

Even in the highest class of the Mollusca, the Cephalopoda or cuttlefishes, in which the sexes are separate, secondary sexual characters of the present kind do not, as far as I can discover, occur. This is a surprising circumstance, as these animals possess highly-developed sense-organs and have considerable mental powers, as will be admitted by every one who has watched their artful endeavours to escape from an enemy. [3] Certain Cephalopoda, however, are characterized by one extraordinary sexual character, namely that the male element collects within one of the arms or tentacles, which is then cast off, and clinging by its sucking-discs to the female, lives for a time an independent life. So completely does the cast-off arm resemble a separate animal, that it was described by Cuvier as a parasitic worm under the name of Hectocotyle. But this marvellous structure may be classed as a primary rather than as a secondary sexual character.

[3] See, for instance, the account which I have given in my 'Journal of Researches,' 1845, p. 7.

Although with the Mollusca sexual selection does not seem to have come into play; yet many univalve and bivalve shells, such as volutes, cones, scallops, etc., are beautifully colored and shaped. The colors do not appear in most cases to be of any use as a protection; they are probably the direct result, as in the lowest classes, of the nature of the tissues; the patterns and the sculpture of the shell depending on its manner of growth. The amount of light seems to be influential to a certain extent; for although, as repeatedly stated by Mr. Gwyn Jeffreys, the shells of some species living at a profound depth are brightly colored, yet we generally see the lower surfaces, as well as the parts covered by the mantle, less highly-colored than the upper and exposed surfaces. [4] In some cases, as with shells living amongst corals or brightly-tinted seaweeds, the bright colors may serve as a protection. [5] But that many of the nudibranch Mollusca, or sea-slugs, are as beautifully colored as any shells, may be seen in Messrs. Alder and Hancock's magnificent work; and from information kindly given me by Mr. Hancock, it seems extremely doubtful whether these colors usually serve as a protection. With some species this may be the case, as with one kind which lives on the green leaves of algae, and is itself bright-green. But many brightly-colored, white, or otherwise conspicuous species, do not seek concealment; whilst again some equally conspicuous species, as well as other dull-colored kinds live under stones and in dark recesses. So that with these nudibranch molluscs, color apparently does not stand in any close relation to the nature of the places which they inhabit.

[4] I have given ('Geological Observations on Volcanic Islands,' 1844, p. 53) a curious instance of the influence of light on the colors of a frondescent incrustation, deposited by the surf on the coast-rocks of Ascension and formed by the solution of triturated sea-shells.

[5] Dr. Morse has lately discussed this subject in his paper on the 'Adaptive Coloration of Mollusca,' 'Proc. Boston Soc. of Nat. Hist.' vol. xiv. April 1871.

These naked sea-slugs are hermaphrodites, yet they pair together, as do land-snails, many of which have extremely pretty shells. It is conceivable that two hermaphrodites, attracted by each other's greater beauty, might unite and leave offspring which would inherit their parents' greater beauty. But with such lowly-organized creatures this is extremely improbable. Nor is it at all obvious how the offspring from the more beautiful pairs of hermaphrodites would have any advantage over the offspring of the less beautiful, so as to increase in number, unless indeed vigor and beauty generally coincided. We have not here the case of a number of males becoming mature before the females, with the more beautiful males selected by the more vigorous females. If, indeed, brilliant colors were beneficial to a hermaphrodite animal in relation to its general habits of life, the more brightly-tinted individuals would succeed best and would increase in number; but this would be a case of natural and not of sexual selection.

SUB-KINGDOM OF THE VERMES: CLASS, ANNELIDA (OR SEA-WORMS).

In this class, although the sexes, when separate, sometimes differ from each other in characters of such importance that they have been placed under distinct genera or even families, yet the differences do not seem of the kind which can be safely attributed to sexual selection. These animals are often beautifully colored, but as the sexes do not differ in this respect, we are but little concerned with them. Even the Nemertians, though so lowly organized, "vie in beauty and variety of coloring with any other group in the invertebrate series"; yet Dr. McIntosh [6] cannot discover that these colors are of any service. The sedentary annelids become duller-colored, according to M. Quatrefages [7], after the period of reproduction; and this I presume may be attributed to their less vigorous condition at that time. All these worm-like animals apparently stand too low in the scale for the individuals of either sex to exert any choice in selecting a partner, or for the individuals of the same sex to struggle together in rivalry.

[6] See his beautiful monograph on 'British Annelids,' part i. 1873, p. 3.
[7] See M. Perrier: 'l'Origine de l'Homme d'après Darwin,' 'Revue Scientifique', Feb. 1873, p. 866.

SUB-KINGDOM OF THE ARTHROPODA: CLASS, CRUSTACEA.

In this great class we first meet with undoubted secondary sexual characters, often developed in a remarkable manner. Unfortunately the habits of crustaceans are very imperfectly known, and we cannot explain the uses of many structures peculiar to one sex. With the lower parasitic species the males are of small size, and they alone are furnished with perfect swimming-legs, antennæ and sense-organs; the females being destitute of these organs, with their bodies often consisting of a mere distorted mass. But these extraordinary differences between the two sexes are no doubt related to their widely different habits of life, and consequently do not concern us. In various crustaceans, belonging to distinct families, the anterior antennæ are furnished with peculiar thread-like bodies, which are believed to act as smelling-organs, and these are much more numerous in the males than in the females. As the males, without any unusual development of their olfactory organs, would almost certainly be able sooner or later to find the females, the increased number of the smelling-threads has probably been acquired through sexual selection, by the better provided males having been the more successful in finding partners and in producing offspring. Fritz Müller has described a remarkable dimorphic species of Tanais, in which the male is represented by two distinct forms, which never graduate into each other. In the one form the male is furnished with more numerous smelling-threads, and in the other form with more powerful and more elongated chelae or pincers, which serve to hold the female. Fritz Müller suggests that these differences between the two male forms of the same species may have originated in certain individuals having varied in the number of the smelling-threads, whilst other individuals varied in the shape and size of their chelae; so that of the former, those which were best able to find the female, and of the latter, those which were best able to hold her, have left the greatest number of progeny to inherit their respective advantages. [8]

[8] 'Facts and Arguments for Darwin,' English translat., 1869, p. 20. See the previous discussion on the olfactory threads. Sars has described a somewhat analogous case (as quoted in 'Nature,' 1870, p. 455) in a Norwegian crustacean, the *Pontoporeia affinis*.

In some of the lower crustaceans, the right anterior antenna of the male differs greatly in structure from the left, the latter resembling in its simple tapering joints the antennæ of the female. In the male the modified antenna is either swollen in the middle or angularly bent, or converted (Fig. 4) into an elegant, and sometimes wonderfully complex, prehensile organ. [9] It serves, as I hear from Sir J. Lubbock, to hold the female, and for this same purpose one of the two posterior legs (b) on the same side of the body is converted into a forceps. In another family the inferior or posterior antennæ are "curiously zigzagged" in the males alone.

[9] See Sir J. Lubbock in 'Annals and Mag. of Nat. Hist.' vol. xi. 1853, pl. i. and x.; and vol. xii. (1853), pl. vii. See also Lubbock in 'Transactions, Entomological Society,' vol. iv. new series, 1856-1858, p. 8. With respect to the zigzagged antennæ mentioned below, see Fritz Müller, 'Facts and Arguments for Darwin,' 1869, p. 40, foot-note.

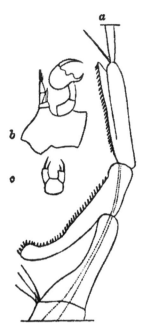

FIG. 4. — Labidocera Darwinii (from Lubbock).

a. Part of right anterior antenna of male, forming a prehensile organ.
b. Posterior pair of thoracic legs of male.
c. Ditto of female.

In the higher crustaceans the anterior legs are developed into chelae or pincers; and these are generally larger in the male than in the female,—so much so that the market value of the male edible crab (*Cancer pagurus*), according to Mr. C. Spence Bate, is five times as great as that of the female. In many species the chelae are of unequal size on the opposite side of the body, the right-hand one being, as I am informed by Mr. Bate, generally, though not invariably, the largest. This inequality is also often much greater in the male than in the female. The two chelae of the male often differ in structure (Figs. 5, 6, and 7), the smaller one resembling that of the female. What advantage is gained by their inequality in size on the opposite sides of the body, and by the inequality being much greater in the male than in the female; and why, when they are of equal size, both are often much larger in the male than in the female, is not known. As I hear from Mr. Bate, the chelae are sometimes of such length and size that they cannot possibly be used for carrying food to the mouth. In the males of certain fresh-water prawns (Palæmon) the right leg is actually longer than the whole body. [10] The great size of the one leg with its chelae may aid the male in fighting with his rivals; but this will not account for their inequality in the female on the opposite sides of the body. In Gelasimus, according to a statement quoted by Milne Edwards [11], the male and the female live in the same burrow, and this shows that they pair; the male closes the mouth of the burrow with one of its chelae, which is enormously developed; so that here it indirectly serves as a means of defence. Their main use, however, is probably to seize and to secure the female, and this in some instances, as with Gammarus, is known to be the case. The male of the hermit or soldier crab (*Pagurus*) for weeks together, carries about the shell inhabited by the female. [12] The sexes, however, of the common shore-crab (*Carcinus mœnas*), as Mr. Bate informs me, unite directly after the female has moulted her hard shell, when she is so soft that she would be injured if seized by the strong pincers of the male; but as she is caught and carried about by the male before moulting, she could then be seized with impunity.

[10] See a paper by Mr. C. Spence Bate, with figures, in 'Proceedings, Zoological Society,' 1868, p. 363; and on the nomenclature of the genus, ibid. p. 585. I am greatly indebted to Mr. Spence Bate for nearly all the above statements with respect to the chelae of the higher crustaceans.

[11] 'Hist. Nat. des Crust.' tom. ii. 1837, p. 50.

[12] Mr. C. Spence Bate, 'British Association, Fourth Report on the Fauna of S. Devon.'

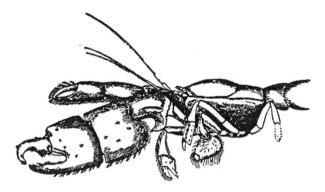

FIG. 5.—Anterior part of body of Callianassa (from Milne-Edwards), showing the unequal and differently-constructed right and left-hand chelæ of the male.
N. B.—The artist by mistake has reversed the drawing, and made the left-hand chela the largest.

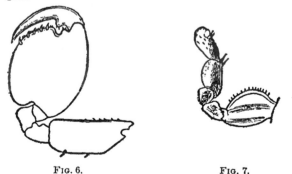

FIG. 6.   FIG. 7.

FIG. 6.—Second leg of male Orchestia Tucuratinga (from Fritz Müller).
FIG. 7.—Ditto of female.

Fritz Müller states that certain species of Melita are distinguished from all other amphipods by the females having "the coxal lamellae of the penultimate pair of feet produced into hook-like processes, of which the males lay hold with the hands of the first pair." The development of these hook-like processes has probably followed from those females which were the most securely held during the act of reproduction, having left the largest number of offspring. Another Brazilian amphipod (see *Orchestia darwinii*, Fig. 8) presents a case of dimorphism, like that of Tanais; for there are two male forms, which differ in the structure of their chelae. [13] As either chela would certainly suffice to hold the female,—for both are now used for this purpose,—the two male forms probably originated by some having varied in one manner and some in another; both forms having derived certain special, but nearly equal advantages, from their differently shaped organs.

[13] Fritz Müller, 'Facts and Arguments for Darwin,' 1869, pp. 25-28.

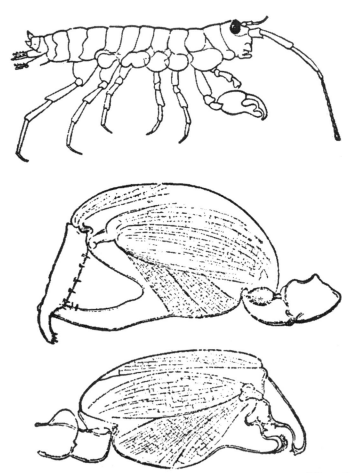

Fɪɢ. 8.—Orchestia Darwinii (from Fritz Müller), showing the differently constructed chelæ of the two male forms.

It is not known that male crustaceans fight together for the possession of the females, but it is probably the case; for with most animals when the male is larger than the female, he seems to owe his greater size to his ancestors having fought with other males during many generations. In most of the orders, especially in the highest or the Brachyura, the male is larger than the female; the parasitic genera, however, in which the sexes follow different habits of life, and most of the Entomostraca must be excepted. The chelae of many crustaceans are weapons well adapted for fighting. Thus when a Devil-crab (*Portunus puber*) was seen by a son of Mr. Bate fighting with a *Carcinus mœnas*, the latter was soon thrown on its back, and had every limb torn from its body. When several males of a Brazilian Gelasimus, a species furnished with immense pincers, were placed together in a glass vessel by Fritz Müller, they mutilated and killed one another. Mr. Bate put a large male *Carcinus mœnas* into a pan of water, inhabited by a female which was paired with a smaller male; but the latter was soon dispossessed. Mr. Bate adds, "if they fought, the victory was a bloodless one, for I saw no wounds." This same naturalist separated a male sand-skipper (so common on our sea-shores), *Gammarus marinus*, from its female, both of whom were imprisoned in the same vessel with many individuals of the same species. The female, when thus divorced, soon joined the others. After a time the male was put again into the same vessel; and he then, after swimming about for a time, dashed into the crowd, and without any fighting at once took away his wife. This fact shows that in the Amphopoda, an order low in the scale, the males and females recognize each other, and are mutually attached.

The mental powers of the Crustacea are probably higher than at first sight appears probable. Any one who tries to catch one of the shore-crabs, so common on tropical coasts, will perceive how wary and alert they are. There is a large crab (*Birgus latro*), found on coral islands, which makes a thick bed of the picked fibres of the cocoa-nut, at the bottom of a deep burrow. It feeds on the fallen fruit of this tree by tearing off the husk, fibre by fibre; and it always begins at that end where the three eye-like depressions are situated. It then breaks through one of these eyes by hammering with its heavy front pincers, and turning round, extracts the albuminous core with its narrow posterior pincers. But these actions are probably instinctive, so that they would be performed as well by a young animal as by an old one. The following case, however, can hardly be so considered: a trustworthy naturalist, Mr. Gardner [14], whilst watching a shore-crab (Gelasimus) making its burrow, threw some shells towards the hole. One rolled in, and three other shells remained within a few inches of the mouth. In about five minutes the crab brought out the shell which had fallen in, and carried it away to a distance of a foot; it then saw the three other shells lying near, and evidently thinking that they might likewise roll in, carried them to the spot where it had laid the first. It would, I think, be difficult to distinguish this act from one performed by man by the aid of reason.

[14] 'Travels in the Interior of Brazil,' 1846, p. 111. I have given, in my 'Journal of Researches,' p. 463, an account of the habits of the Birgus.

Mr. Bate does not know of any well-marked case of difference of color in the two sexes of our British crustaceans, in which respect the sexes of the higher animals so often differ. In some cases, however, the males and females differ slightly in tint, but Mr. Bate thinks not more than may be accounted for by their different habits of life, such as by the male wandering more about, and being thus more exposed to the light. Dr. Power tried to distinguish by color the sexes of the several species which inhabit the Mauritius, but failed, except with one species of Squilla, probably *S. stylifera*, the male of which is described as being "of a beautiful bluish-green," with some of the appendages cherry-red, whilst the female is clouded with brown and grey, "with the red about her much less vivid than in the male." [15] In this case, we may suspect the agency of sexual selection. From M. Bert's observations on Daphnia, when placed in a vessel illuminated by a prism, we have reason to believe that even the lowest crustaceans can distinguish colors. With Saphirina (an oceanic genus of Entomostraca), the males are furnished with minute shields or cell-like bodies, which exhibit beautiful changing colors; these are absent in the females, and in both sexes of one species. [16] It would, however, be extremely rash to conclude that these curious organs serve to attract the females. I am informed by Fritz Müller, that in the female of a Brazilian species of Gelasimus, the whole body is of a nearly uniform grayish-brown. In the male the posterior part of the cephalo-thorax is pure white, with the anterior part of a rich green, shading into dark brown; and it is remarkable that these colors are liable to change in the course of a few minutes—the white becoming dirty grey or even black, the green "losing much of its brilliancy." It deserves especial notice that the males do not acquire their bright colors until they become mature. They appear to be much more numerous than the females; they differ also in the larger size of their chelae. In some species of the genus, probably in all, the sexes pair and inhabit the same burrow. They are also, as we have seen, highly intelligent animals. From these various considerations it seems probable that the male in this species has become gaily ornamented in order to attract or excite the female.

[15] Mr. Ch. Fraser, in 'Proc. Zoolog. Soc.' 1869, p. 3. I am indebted to Mr. Bate for Dr. Power's statement.
[16] Claus, 'Die freilebenden Copepoden,' 1863, s. 35.

It has just been stated that the male Gelasimus does not acquire his conspicuous colors until mature and nearly ready to breed. This seems a general rule in the whole class in respect to the many remarkable structural differences between the sexes. We shall hereafter find the same law prevailing throughout the great sub-kingdom of the Vertebrata; and in all cases it is eminently distinctive of characters which have been acquired through sexual selection. Fritz Müller [17] gives some striking instances of this law; thus the male sand-hopper (Orchestia) does not, until nearly full grown, acquire his large claspers, which are very differently constructed from those of the female; whilst young, his claspers resemble those of the female.

[17] 'Facts and Arguments,' etc., p. 79.

CLASS, ARACHNIDA (SPIDERS).

The sexes do not generally differ much in color, but the males are often darker than the females, as may be seen in Mr. Blackwall's magnificent work. [18] In some species, however, the difference is conspicuous: thus the female of *Sparassus smaragdulus* is dullish green, whilst the adult male has the abdomen of a fine yellow, with three longitudinal stripes of rich red. In certain species of Thomisus the sexes closely resemble each other, in others they differ much; and analogous cases occur in many other genera. It is often difficult to say which of the two sexes departs most from the ordinary coloration of the genus to which the species belong; but Mr. Blackwall thinks that, as a general rule, it is the male; and Canestrini [19] remarks that in certain genera the males can be specifically distinguished with ease, but the females with great difficulty. I am informed by Mr. Blackwall that the sexes whilst young usually resemble each other; and both often undergo great changes in color during their successive moults, before arriving at maturity. In other cases the male alone appears to change color. Thus the male of the above bright-colored Sparassus at first resembles the female, and acquires his peculiar tints only when nearly adult. Spiders are possessed of acute senses, and exhibit much intelligence; as is well known, the females often show the strongest affection for their eggs, which they carry about enveloped in a silken web. The males search eagerly for the females, and have been seen by Canestrini and others to fight for possession of them. This same author says that the union of the two sexes has been observed in about twenty species; and he asserts positively that the female rejects some of the males who court her, threatens them with open mandibles, and at last after long hesitation accepts the chosen one. From these several considerations, we may admit with some confidence that the well-marked differences in color between the sexes of certain species are the results of sexual selection; though we have not here the best kind of evidence,—the display by the male of his ornaments. From the extreme variability of color in the male of some species, for instance of *Theridion lineatum*, it would appear that these sexual characters of the males have not as yet become well fixed. Canestrini draws the same conclusion from the fact that the males of certain species present two forms, differing from each other in the size and length of their jaws; and this reminds us of the above cases of dimorphic crustaceans.

[18] 'A History of the Spiders of Great Britain,' 1861-64. For the following facts, see pp. 77, 88, 102.
[19] This author has recently published a valuable essay on the 'Caratteri sessuali secondarii degli Arachnidi,' in the 'Atti della Soc. Veneto-Trentina di Sc. Nat. Padova,' vol. i. Fasc. 3, 1873.

The male is generally much smaller than the female, sometimes to an extraordinary degree [20], and he is forced to be extremely cautious in making his advances, as the female often carries her coyness to a dangerous pitch. De Geer saw a male that "in the midst of his preparatory caresses was seized by the object of his attentions, enveloped by her in a web and then devoured, a sight which, as he adds, filled him with horror and indignation." [21] The Rev. O.P. Cambridge [22] accounts in the following manner for the extreme smallness of the male in the genus Nephila. "M. Vinson gives a graphic account of the agile way in which the diminutive male escapes from the ferocity of the female, by gliding about and playing hide and seek over her body and along her gigantic limbs: in such a pursuit it is evident that the chances of escape would be in favour of the smallest males, while the larger ones would fall early victims; thus gradually a diminutive race of males would be selected, until at last they would dwindle to the smallest possible size compatible with the exercise of their generative functions,—in fact, probably to the size we now see them, i.e., so small as to be a sort of parasite upon the female, and either beneath her notice, or too agile and too small for her to catch without great difficulty."

[20] Aug. Vinson ('Aranéides des Iles de la Réunion,' pl. vi. figs. 1 and 2) gives a good instance of the small size of the male, in *Epeira nigra*. In this species, as I may add, the male is testaceous and the female black with legs banded with red. Other even more striking cases of inequality in size between the sexes have been recorded ('Quarterly Journal of Science,' July 1868, p. 429); but I have not seen the original accounts.
[21] Kirby and Spence, 'Introduction to Entomology,' vol. i. 1818, p. 280.
[22] 'Proceedings, Zoological Society,' 1871, p. 621.

Westring has made the interesting discovery that the males of several species of Theridion [23] have the power of making a stridulating sound, whilst the females are mute. The apparatus consists of a serrated ridge at the base of the abdomen, against which the hard hinder part of the thorax is rubbed; and of this structure not a trace can be detected in the females. It deserves notice that several writers, including the well-known arachnologist Walckenaer, have declared that spiders are attracted by music. [24] From the analogy of the Orthoptera and Homoptera, to be described in the next chapter, we may feel almost sure that the stridulation serves, as Westring also believes, to call or to excite the female; and this is the first case known to me in the ascending scale of the animal kingdom of sounds emitted for this purpose. [25]

[23] *Theridion* (*Asagena*, Sund.) *serratipes, 4-punctatum et guttatum*; see Westring, in Kroyer, 'Naturhist. Tidskrift,' vol. iv. 1842-1843, p. 349; and vol. ii. 1846-1849, p. 342. See, also, for other species, 'Araneæ Suecicæ,' p. 184.

[24] Dr. H.H. van Zouteveen, in his Dutch translation of this work (vol. i. p. 444), has collected several cases.

[25] Hilgendorf, however, has lately called attention to an analogous structure in some of the higher crustaceans, which seems adapted to produce sound; see 'Zoological Record,' 1869, p. 603.

CLASS, MYRIAPODA.

In neither of the two orders in this class, the millipedes and centipedes, can I find any well-marked instances of such sexual differences as more particularly concern us. In *Glomeris limbata*, however, and perhaps in some few other species, the males differ slightly in color from the females; but this Glomeris is a highly variable species. In the males of the Diplopoda, the legs belonging either to one of the anterior or of the posterior segments of the body are modified into prehensile hooks which serve to secure the female. In some species of Iulus the tarsi of the male are furnished with membranous suckers for the same purpose. As we shall see when we treat of Insects, it is a much more unusual circumstance, that it is the female in Lithobius, which is furnished with prehensile appendages at the extremity of her body for holding the male. [26]

[26] Walckenaer et P. Gervais, 'Hist. Nat. des Insectes: Apteres,' tom. iv. 1847, pp. 17, 19, 68.

CHAPTER X.

SECONDARY SEXUAL CHARACTERS OF INSECTS.

Diversified structures possessed by the males for seizing the females—Differences between the sexes, of which the meaning is not understood—Difference in size between the sexes—Thysanura—Diptera—Hemiptera—Homoptera, musical powers possessed by the males alone—Orthoptera, musical instruments of the males, much diversified in structure; pugnacity; colors—Neuroptera, sexual differences in color—Hymenoptera, pugnacity and odors—Coleoptera, colors; furnished with great horns, apparently as an ornament; battles, stridulating organs generally common to both sexes.

In the immense class of insects the sexes sometimes differ in their locomotive-organs, and often in their sense-organs, as in the pectinated and beautifully plumose antennæ of the males of many species. In Chloëon, one of the Ephemerae, the male has great pillared eyes, of which the female is entirely destitute. [1] The ocelli are absent in the females of certain insects, as in the Mutillidæ; and here the females are likewise wingless. But we are chiefly concerned with structures by which one male is enabled to conquer another, either in battle or courtship, through his strength, pugnacity, ornaments, or music. The innumerable contrivances, therefore, by which the male is able to seize the female, may be briefly passed over. Besides the complex structures at the apex of the abdomen, which ought perhaps to be ranked as primary organs [2] "it is astonishing," as Mr. B.D. Walsh [3] has remarked, "how many different organs are worked in by nature for the seemingly insignificant object of enabling the male to grasp the female firmly." The mandibles or jaws are sometimes used for this purpose; thus the male *Corydalis cornutus* (a neuropterous insect in some degree allied to the Dragon flies, etc.) has immense curved jaws, many times longer than those of the female; and they are smooth instead of being toothed, so that he is thus enabled to seize her without injury. [4] One of the stag-beetles of North America (*Lucanus elaphus*) uses his jaws, which are much larger than those of the female, for the same purpose, but probably likewise for fighting. In one of the sand-wasps (*Ammophila*) the jaws in the two sexes are closely alike, but are used for widely different purposes: the males, as Professor Westwood observes, "are exceedingly ardent, seizing their partners round the neck with their sickle-shaped jaws" [5]; whilst the females use these organs for burrowing in sand-banks and making their nests.

[1] Sir J. Lubbock, 'Transact. Linnean Soc.' vol. xxv, 1866, p. 484. With respect to the Mutillidæ see Westwood, 'Modern Class. of Insects,' vol. ii. p. 213.

[2] These organs in the male often differ in closely-allied species, and afford excellent specific characters. But their importance, from a functional point of view, as Mr. R. MacLachlan has remarked to me, has probably been overrated. It has been suggested, that slight differences in these organs would suffice to prevent the intercrossing of well-marked varieties or incipient species, and would thus aid in their development. That this can hardly be the case, we may infer from the many recorded cases (see, for instance, Bronn, 'Geschichte der Natur,' B. ii. 1843, s. 164; and Westwood, 'Transact. Ent. Soc.' vol. iii. 1842, p. 195) of distinct species having been observed in union. Mr. MacLachlan informs me (vide 'Stett. Ent. Zeitung,' 1867, s. 155) that when several species of Phryganidæ, which present strongly-pronounced differences of this kind, were confined together by Dr. Aug. Meyer, *they coupled*, and one pair produced fertile ova.),

[3] 'The Practical Entomologist,' Philadelphia, vol. ii. May 1867, p 88.

[4] Mr. Walsh, ibid. p. 107.

[5] 'Modern Classification of Insects,' vol. ii. 1840, pp. 205, 206. Mr. Walsh, who called my attention to the double use of the jaws, says that he has repeatedly observed this fact.

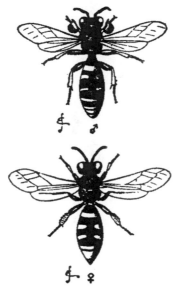

Fɪɢ. 9.--Crabro cribrarius. Upper
figure, male ; lower figure,
female.

The tarsi of the front-legs are dilated in many male beetles, or are furnished with broad cushions of hairs; and in many genera of water-beetles they are armed with a round flat sucker, so that the male may adhere to the slippery body of the female. It is a much more unusual circumstance that the females of some water-beetles (Dytiscus) have their elytra deeply grooved, and in *Acilius sulcatus* thickly set with hairs, as an aid to the male. The females of some other water-beetles (Hydroporus) have their elytra punctured for the same purpose. [6] In the male of *Crabro cribrarius* (Fig. 9), it is the tibia which is dilated into a broad horny plate, with minute membraneous dots, giving to it a singular appearance like that of a riddle. [7] In the male of Penthe (a genus of beetles) a few of the middle joints of the antennæ are dilated and furnished on the inferior surface with cushions of hair, exactly like those on the tarsi of the Carabidæ, "and obviously for the same end." In male dragon-flies, "the appendages at the tip of the tail are modified in an almost infinite variety of curious patterns to enable them to embrace the neck of the female." Lastly, in the males of many insects, the legs are furnished with peculiar spines, knobs or spurs; or the whole leg is bowed or thickened, but this is by no means invariably a sexual character; or one pair, or all three pairs are elongated, sometimes to an extravagant length. [8]

[6] We have here a curious and inexplicable case of dimorphism, for some of the females of four European species of Dytiscus, and of certain species of Hydroporus, have their elytra smooth; and no intermediate gradations between the sulcated or punctured, and the quite smooth elytra have been observed. See Dr. H. Schaum, as quoted in the 'Zoologist,' vols. v.-vi. 1847-48, p. 1896. Also Kirby and Spence, 'Introduction to Entomology,' vol. iii. 1826, p. 305.

[7] Westwood, 'Modern Class.' vol. ii. p. 193. The following statement about Penthe, and others in inverted commas, are taken from Mr. Walsh, 'Practical Entomologist,'

224

Philadelphia, vol. iii. p. 88.

[8] Kirby and Spence, 'Introduct.' etc., vol. iii. pp. 332-336.

FIG. 10. — Taphro-
deres distortus
(much enlarged).
Upper figure,
male; lower fig-
ure, female.

The sexes of many species in all the orders present differences, of which the meaning is not understood. One curious case is that of a beetle (Fig. 10), the male of which has left mandible much enlarged; so that the mouth is greatly distorted. In another Carabidous beetle, Eurygnathus, [9] we have the case, unique as far as known to Mr. Wollaston, of the head of the female being much broader and larger, though in a variable degree, than that of the male. Any number of such cases could be given. They abound in the Lepidoptera: one of the most extraordinary is that certain male butterflies have their fore-legs more or less atrophied, with the tibiae and tarsi reduced to mere rudimentary knobs. The wings, also, in the two sexes often differ in neuration, [10] and sometimes considerably in outline, as in the *Aricoris epitus*, which was shown to me in the British Museum by Mr. A. Butler. The males of certain South American butterflies have tufts of hair on the margins of the wings, and horny excrescences on the discs of the posterior pair. [11] In several British butterflies, as shown by Mr. Wonfor, the males alone are in parts clothed with peculiar scales.

[9] 'Insecta Maderensia,' 1854, page 20.
[10] E. Doubleday, 'Annals and Mag. of Nat. Hist.' vol. i. 1848, p. 379. I may add that the wings in certain Hymenoptera (see Shuckard, 'Fossorial Hymenoptera,' 1837, pp. 39-43) differ in neuration according to sex.
[11] H.W. Bates, in 'Journal of Proc. Linn. Soc.' vol. vi. 1862, p. 74. Mr. Wonfor's observations are quoted in 'Popular Science Review,' 1868, p. 343.

The use of the bright light of the female glow-worm has been subject to much discussion. The male is feebly luminous, as are the larvæ and even the eggs. It has been supposed by some authors that the light serves to frighten away enemies, and by others to guide the male to the female. At last, Mr. Belt [12] appears to have solved the difficulty: he finds that all the Lampyridæ which he has tried are highly distasteful to insectivorous mammals and birds. Hence it is in accordance with Mr. Bates' view, hereafter to be explained, that many insects mimic the Lampyridæ closely, in order to be mistaken for them, and thus to escape destruction. He further believes that the luminous species profit by being at once recognized as unpalatable. It is probable that the same explanation may be extended to the Elaters, both sexes of which are highly luminous. It is not known why the wings of the female glow-worm have not been developed; but in her present state she closely resembles a larva, and as larvæ are so largely preyed on by many animals, we can understand why she has been rendered so much more luminous and conspicuous than the male; and why the larvæ themselves are likewise luminous.

[12] 'The Naturalist in Nicaragua,' 1874, pp. 316-320. On the phosphorescence of the eggs, see 'Annals and Magazine of Natural History,' Nov. 1871, p. 372.

DIFFERENCE IN SIZE BETWEEN THE SEXES.

With insects of all kinds the males are commonly smaller than the females; and this difference can often be detected even in the larval state. So considerable is the difference between the male and female cocoons of the silk-moth (Bombyx mori), that in France they are separated by a particular mode of weighing. [13] In the lower classes of the animal kingdom, the greater size of the females seems generally to depend on their developing an enormous number of ova; and this may to a certain extent hold good with insects. But Dr. Wallace has suggested a much more probable explanation. He finds, after carefully attending to the development of the caterpillars of *Bombyx cynthia* and *yamamai*, and especially to that of some dwarfed caterpillars reared from a second brood on unnatural food, "that in proportion as the individual moth is finer, so is the time required for its metamorphosis longer; and for this reason the female, which is the larger and heavier insect, from having to carry her numerous eggs, will be preceded by the male, which is smaller and has less to mature." [14] Now as most insects are short-lived, and as they are exposed to many dangers, it would manifestly be advantageous to the female to be impregnated as soon as possible. This end would be gained by the males being first matured in large numbers ready for the advent of the females; and this again would naturally follow, as Mr. A.R. Wallace has remarked [15], through natural selection; for the smaller males would be first matured, and thus would procreate a large number of offspring which would inherit the reduced size of their male parents, whilst the larger males from being matured later would leave fewer offspring.

[13] Robinet, 'Vers à Soie,' 1848, p. 207.
[14] 'Transact. Ent. Soc.' 3rd series, vol. v. p. 486.
[15] 'Journal of Proc. Ent. Soc.' Feb. 4, 1867, p. lxxi.

There are, however, exceptions to the rule of male insects being smaller than the females: and some of these exceptions are intelligible. Size and strength would be an advantage to the males, which fight for the possession of the females; and in these cases, as with the stag-beetle (Lucanus), the males are larger than the females. There are, however, other beetles which are not known to fight together, of which the males exceed the females in size; and the meaning of this fact is not known; but in some of these cases, as with the huge Dynastes and Megasoma, we can at least see that there would be no necessity for the males to be smaller than the females, in order to be matured before them, for these beetles are not short-lived, and there would be ample time for the pairing of the sexes. So again, male dragon-flies (Libellulidæ) are sometimes sensibly larger, and never smaller, than the females; [16] and as Mr. MacLachlan believes, they do not generally pair with the females until a week or fortnight has elapsed, and until they have assumed their proper masculine colors. But the most curious case, showing on what complex and easily-overlooked relations, so trifling a character as difference in size between the sexes may depend, is that of the aculeate Hymenoptera; for Mr. F. Smith informs me that throughout nearly the whole of this large group, the males, in accordance with the general rule, are smaller than the females, and emerge about a week before them; but amongst the Bees, the males of *Apis mellifica, Anthidium manicatum,* and *Anthophora acervorum,* and amongst the Fossores, the males of the *Methoca ichneumonides,* are larger than the females. The explanation of this anomaly is that a marriage flight is absolutely necessary with these species, and the male requires great strength and size in order to carry the female through the air. Increased size has here been acquired in opposition to the usual relation between size and the period of development, for the males, though larger, emerge before the smaller females.

[16] For this and other statements on the size of the sexes, see Kirby and Spence, ibid. vol. iii. p. 300; on the duration of life in insects, see p. 344.

We will now review the several Orders, selecting such facts as more particularly concern us. The Lepidoptera (Butterflies and Moths) will be retained for a separate chapter.

ORDER, THYSANURA.

The members of this lowly organized order are wingless, dull-colored, minute insects, with ugly, almost misshapen heads and bodies. Their sexes do not differ, but they are interesting as showing us that the males pay sedulous court to the females even low down in the animal scale. Sir J. Lubbock [17] says: "it is very amusing to see these little creatures (*Smynthurus luteus*) coquetting together. The male, which is much smaller than the female, runs round her, and they butt one another, standing face to face and moving backward and forward like two playful lambs. Then the female pretends to run away and the male runs after her with a queer appearance of anger, gets in front and stands facing her again; then she turns coyly round, but he, quicker and more active, scuttles round too, and seems to whip her with his antennæ; then for a bit they stand face to face, play with their antennæ, and seem to be all in all to one another."

[17] 'Transact. Linnean Soc.' vol. xxvi. 1868, p. 296.

ORDER, DIPTERA (FLIES).

The sexes differ little in color. The greatest difference, known to Mr. F. Walker, is in the genus Bibio, in which the males are blackish or quite black, and the females obscure brownish-orange. The genus Elaphomyia, discovered by Mr. Wallace [18] in New Guinea, is highly remarkable, as the males are furnished with horns, of which the females are quite destitute. The horns spring from beneath the eyes, and curiously resemble those of a stag, being either branched or palmated. In one of the species, they equal the whole body in length. They might be thought to be adapted for fighting, but as in one species they are of a beautiful pink color, edged with black, with a pale central stripe, and as these insects have altogether a very elegant appearance, it is perhaps more probable that they serve as ornaments. That the males of some Diptera fight together is certain; Prof. Westwood [19] has several times seen this with the Tipulæ. The males of other Diptera apparently try to win the females by their music: H. Müller [20] watched for some time two males of an Eristalis courting a female; they hovered above her, and flew from side to side, making a high humming noise at the same time. Gnats and mosquitoes (Culicidæ) also seem to attract each other by humming; and Prof. Mayer has recently ascertained that the hairs on the antennæ of the male vibrate in unison with the notes of a tuning-fork, within the range of the sounds emitted by the female. The longer hairs vibrate sympathetically with the graver notes, and the shorter hairs with the higher ones. Landois also asserts that he has repeatedly drawn down a whole swarm of gnats by uttering a particular note. It may be added that the mental faculties of the Diptera are probably higher than in most other insects, in accordance with their highly-developed nervous system. [21]

[18] 'The Malay Archipelago,' vol. ii. 1869, p. 313.
[19] 'Modern Classification of Insects,' vol. ii. 1840, p. 526.
[20] 'Anwendung,' etc., 'Verh. d. n. V. Jahrg.' xxix. p. 80. Mayer, in 'American Naturalist,' 1874, p. 236.
[21] See Mr. B.T. Lowne's interesting work, 'On the Anatomy of the Blow-fly, Musca vomitoria,' 1870, p. 14. He remarks (p. 33) that, "the captured flies utter a peculiar plaintive note, and that this sound causes other flies to disappear."

ORDER, HEMIPTERA (FIELD-BUGS).

Mr. J.W. Douglas, who has particularly attended to the British species, has kindly given me an account of their sexual differences. The males of some species are furnished with wings, whilst the females are wingless; the sexes differ in the form of their bodies, elytra, antennæ and tarsi; but as the signification of these differences are unknown, they may be here passed over. The females are generally larger and more robust than the males. With British, and, as far as Mr. Douglas knows, with exotic species, the sexes do not commonly differ much in color; but in about six British species the male is considerably darker than the female, and in about four other species the female is darker than the male. Both sexes of some species are beautifully colored; and as these insects emit an extremely nauseous odor, their conspicuous colors may serve as a signal that they are unpalatable to insectivorous animals. In some few cases their colors appear to be directly protective: thus Prof. Hoffmann informs me that he could hardly distinguish a small pink and green species from the buds on the trunks of lime-trees, which this insect frequents.

Some species of Reduvidæ make a stridulating noise; and, in the case of *Pirates stridulus*, this is said [22] to be effected by the movement of the neck within the pro-thoracic cavity. According to Westring, *Reduvius personatus* also stridulates. But I have no reason to suppose that this is a sexual character, excepting that with non-social insects there seems to be no use for sound-producing organs, unless it be as a sexual call.

[22] Westwood, 'Modern Classification of Insects,' vol. ii. p. 473.

ORDER: HOMOPTERA.

Every one who has wandered in a tropical forest must have been astonished at the din made by the male Cicadæ. The females are mute; as the Grecian poet Xenarchus says, "Happy the Cicadas live, since they all have voiceless wives." The noise thus made could be plainly heard on board the "Beagle," when anchored at a quarter of a mile from the shore of Brazil; and Captain Hancock says it can be heard at the distance of a mile. The Greeks formerly kept, and the Chinese now keep these insects in cages for the sake of their song, so that it must be pleasing to the ears of some men. [23] The Cicadidæ usually sing during the day, whilst the Fulgoridæ appear to be night-songsters. The sound, according to Landois, [24] is produced by the vibration of the lips of the spiracles, which are set into motion by a current of air emitted from the tracheae; but this view has lately been disputed. Dr. Powell appears to have proved [25] that it is produced by the vibration of a membrane, set into action by a special muscle. In the living insect, whilst stridulating, this membrane can be seen to vibrate; and in the dead insect the proper sound is heard, if the muscle, when a little dried and hardened, is pulled with the point of a pin. In the female the whole complex musical apparatus is present, but is much less developed than in the male, and is never used for producing sound.

[23] These particulars are taken from Westwood's 'Modern Classification of Insects,' vol. ii. 1840, p. 422. See, also, on the Fulgoridæ, Kirby and Spence, 'Introduct.' vol. ii. p. 401.
[24] 'Zeitschrift für wissenschaft Zoolog.' B. xvii. 1867, ss. 152-158.
[25] 'Transactions of the New Zealand Institute,' vol. v. 1873, p. 286.

With respect to the object of the music, Dr. Hartman, in speaking of the *Cicada septemdecim* of the United States, says, [26] "the drums are now (June 6th and 7th, 1851) heard in all directions. This I believe to be the marital summons from the males. Standing in thick chestnut sprouts about as high as my head, where hundreds were around me, I observed the females coming around the drumming males." He adds, "this season (Aug. 1868) a dwarf pear-tree in my garden produced about fifty larvæ of *Cic. pruinosa*; and I several times noticed the females to alight near a male while he was uttering his clanging notes." Fritz Müller writes to me from S. Brazil that he has often listened to a musical contest between two or three males of a species with a particularly loud voice, seated at a considerable distance from each other: as soon as one had finished his song, another immediately began, and then another. As there is so much rivalry between the males, it is probable that the females not only find them by their sounds, but that, like female birds, they are excited or allured by the male with the most attractive voice.

[26] I am indebted to Mr. Walsh for having sent me this extract from 'A Journal of the Doings of Cicada septemdecim,' by Dr. Hartman.

I have not heard of any well-marked cases of ornamental differences between the sexes of the Homoptera. Mr. Douglas informs me that there are three British species, in which the male is black or marked with black bands, whilst the females are pale-colored or obscure.

ORDER, ORTHOPTERA (CRICKETS AND GRASSHOPPERS).

The males in the three saltatorial families in this Order are remarkable for their musical powers, namely the Achetidæ or crickets, the Locustidæ for which there is no equivalent English name, and the Acridiidæ or grasshoppers. The stridulation produced by some of the Locustidæ is so loud that it can be heard during the night at the distance of a mile; [27] and that made by certain species is not unmusical even to the human ear, so that the Indians on the Amazons keep them in wicker cages. All observers agree that the sounds serve either to call or excite the mute females. With respect to the migratory locusts of Russia, Körte has given [28] an interesting case of selection by the female of a male. The males of this species (*Pachytylus migratorius*) whilst coupled with the female stridulate from anger or jealousy, if approached by other males. The house-cricket when surprised at night uses its voice to warn its fellows. [29] In North America the Katy-did (*Platyphyllum concavum*, one of the Locustidæ) is described [30] as mounting on the upper branches of a tree, and in the evening beginning "his noisy babble, while rival notes issue from the neighbouring trees, and the groves resound with the call of *Katy-did-she-did* the live-long night." Mr. Bates, in speaking of the European field-cricket (one of the Achetidæ), says "the male has been observed to place himself in the evening at the entrance of his burrow, and stridulate until a female approaches, when the louder notes are succeeded by a more subdued tone, whilst the successful musician caresses with his antennæ the mate he has won." [31] Dr. Scudder was able to excite one of these insects to answer him, by rubbing on a file with a quill. [32] In both sexes a remarkable auditory apparatus has been discovered by Von Siebold, situated in the front legs. [33]

[27] L. Guilding, 'Transactions of the Linnean Society,' vol. xv. p. 154.
[28] I state this on the authority of Köppen, 'Ueber die Heuschrecken in Südrussland,' 1866, p. 32, for I have in vain endeavoured to procure Körte's work.
[29] Gilbert White, 'Natural History of Selborne,' vol. ii. 1825, p. 262.
[30] Harris, 'Insects of New England,' 1842, p. 128.
[31] 'The Naturalist on the Amazons,' vol. i. 1863, p. 252. Mr. Bates gives a very interesting discussion on the gradations in the musical apparatus of the three families. See also Westwood, 'Modern Classification of Insects,' vol. ii. pp. 445 and 453.

230

[32] 'Proceedings of the Boston Society of Natural History,' vol. xi. April 1868.
[33] 'Nouveau Manuel d'Anat. Comp.' (French translat.), tom. 1, 1850, p. 567.

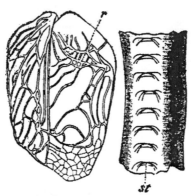

FIG. 11.—Gryllus campestris (from Landois).

Right-hand figure, under side of part of a wing-nervure, much magnified, showing the teeth (*st*). Left-hand figure, upper surface of wing-cover, with the projecting, smooth nervure, *r*, across which the teeth (*st*) are scraped.

In the three Families the sounds are differently produced. In the males of the Achetidæ both wing-covers have the same apparatus; and this in the field-cricket (see *Gryllus campestris*, Fig. 11) consists, as described by Landois, [34] of from 131 to 138 sharp, transverse ridges or teeth (*st*) on the under side of one of the nervures of the wing-cover. This toothed nervure is rapidly scraped across a projecting, smooth, hard nervure (*r*) on the upper surface of the opposite wing. First one wing is rubbed over the other, and then the movement is reversed. Both wings are raised a little at the same time, so as to increase the resonance. In some species the wing-covers of the males are furnished at the base with a talc-like plate. [35] I here give a drawing (Fig. 12) of the teeth on the under side of the nervure of another species of Gryllus, viz., *G. domesticus*. With respect to the formation of these teeth, Dr. Gruber has shown [36] that they have been developed by the aid of selection, from the minute scales and hairs with which the wings and body are covered, and I came to the same conclusion with respect to those of the Coleoptera. But Dr. Gruber further shows that their development is in part directly due to the stimulus from the friction of one wing over the other.

[34] 'Zeitschrift für wissenschaft. Zoolog.' B. xvii. 1867, s. 117.
[35] Westwood, 'Modern Classification of Insects,' vol. i. p. 440.
[36] 'Ueber der Tonapparat der Locustiden, ein Beitrag zum Darwinismus,' 'Zeitschrift fur wissenschaft. Zoolog.' B. xxii. 1872, p. 100.

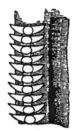

Fig. 12.—Teeth of Nervure of Gryllus domesticus (from Landois).

In the Locustidæ the opposite wing-covers differ from each other in structure (Fig. 13), and the action cannot, as in the last family, be reversed. The left wing, which acts as the bow, lies over the right wing which serves as the fiddle. One of the nervures (*a*) on the under surface of the former is finely serrated, and is scraped across the prominent nervures on the upper surface of the opposite or right wing. In our British *Phasgonura viridissima* it appeared to me that the serrated nervure is rubbed against the rounded hind-corner of the opposite wing, the edge of which is thickened, colored brown, and very sharp. In the right wing, but not in the left, there is a little plate, as transparent as talc, surrounded by nervures, and called the speculum. In *Ephippiger vitium*, a member of this same family, we have a curious subordinate modification; for the wing-covers are greatly reduced in size, but "the posterior part of the pro-thorax is elevated into a kind of dome over the wing-covers, and which has probably the effect of increasing the sound." [37]

[37] Westwood 'Modern Classification of Insects,' vol. i. p. 453.

We thus see that the musical apparatus is more differentiated or specialized in the Locustidæ (which include, I believe, the most powerful performers in the Order), than in the Achetidæ, in which both wing-covers have the same structure and the same function. [38] Landois, however, detected in one of the Locustidæ, namely in Decticus, a short and narrow row of small teeth, mere rudiments, on the inferior surface of the right wing-cover, which underlies the other and is never used as the bow. I observed the same rudimentary structure on the under side of the right wing-cover in *Phasgonura viridissima*. Hence we may infer with confidence that the Locustidæ are descended from a form, in which, as in the existing Achetidæ, both wing-covers had serrated nervures on the under surface, and could be indifferently used as the bow; but that in the Locustidæ the two wing-covers gradually became differentiated and perfected, on the principle of the division of labor, the one to act exclusively as the bow, and the other as the fiddle. Dr. Gruber takes the same view, and has shown that rudimentary teeth are commonly found on the inferior surface of the right wing. By what steps the more simple apparatus in the Achetidæ originated, we do not know, but it is probable that the basal portions of the wing-covers originally overlapped each other as they do at present; and that the friction of the nervures produced a grating sound, as is now the case with the wing-covers of the females. [39] A grating sound thus occasionally and accidentally made by the males, if it served them ever so little as a love-call to the females, might readily have been intensified through sexual selection, by variations in the roughness of the nervures having been continually preserved.

[38] Landois, 'Zeitschrift fur wissenschaft Zoolog.' B. xvii. 1867, ss. 121, 122.

[39] Mr. Walsh also informs me that he has noticed that the female of the *Platyphyllum concavum*, "when captured makes a feeble grating noise by shuffling her wing-covers together."

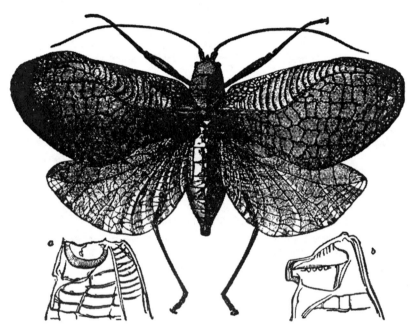

FIG. 13.—Chlorocœlus Tanana (from Bates). *a, b*, Lobes of opposite wing-covers.

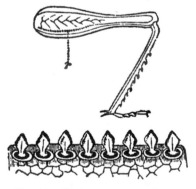

FIG. 14.—Hind-leg of Stenobothrus pratorum : *r*, the stridulating ridge ; lower figure, the teeth forming the ridge, much magnified (from Landois).

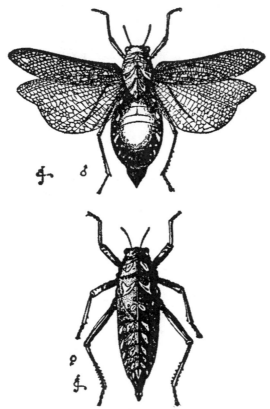

FIG. 15.—Pneumora (from specimens in the British Museum). Upper figure, male ; lower figure, female.

In the last and third family, namely the Acridiidæ or grasshoppers, the stridulation is produced in a very different manner, and according to Dr. Scudder, is not so shrill as in the preceding Families. The inner surface of the femur (Fig. 14, r) is furnished with a longitudinal row of minute, elegant, lancet-shaped, elastic teeth, from 85 to 93 in number; [40] and these are scraped across the sharp, projecting nervures on the wing-covers, which are thus made to vibrate and resound. Harris [41] says that when one of the males begins to play, he first "bends the shank of the hind-leg beneath the thigh, where it is lodged in a furrow designed to receive it, and then draws the leg briskly up and down. He does not play both fiddles together, but alternately, first upon one and then on the other." In many species, the base of the abdomen is hollowed out into a great cavity which is believed to act as a resounding board. In Pneumora (Fig. 15), a S. African genus belonging to the same family, we meet with a new and remarkable modification; in the males a small notched ridge projects obliquely from each side of the abdomen, against which the hind femora are rubbed. [42] As the male is furnished with wings (the female being wingless), it is remarkable that the thighs are not rubbed in the usual manner against the wing-covers; but this may perhaps be accounted for by the unusually small size of the hind-legs. I have not been able to examine the inner surface of the thighs, which, judging from analogy, would be finely serrated. The species of Pneumora have been more profoundly modified for the sake of stridulation than any other orthopterous insect; for in the male the whole body has been converted into a musical instrument, being distended with air, like a great pellucid bladder, so as to increase the resonance. Mr. Trimen informs me that at the Cape of Good Hope these insects make a wonderful noise during the night.

[40] Landois, ibid. s. 113.
[41] 'Insects of New England,' 1842, p. 133.
[42] Westwood, 'Modern Classification,' vol i. p. 462.

In the three foregoing families, the females are almost always destitute of an efficient musical apparatus. But there are a few exceptions to this rule, for Dr. Gruber has shown that both sexes of *Ephippiger vitium* are thus provided; though the organs differ in the male and female to a certain extent. Hence we cannot suppose that they have been transferred from the male to the female, as appears to have been the case with the secondary sexual characters of many other animals. They must have been independently developed in the two sexes, which no doubt mutually call to each other during the season of love. In most other Locustidæ (but not according to Landois in Decticus) the females have rudiments of the stridulatory organs proper to the male; from whom it is probable that these have been transferred. Landois also found such rudiments on the under surface of the wing-covers of the female Achetidæ, and on the femora of the female Acridiidæ. In the Homoptera, also, the females have the proper musical apparatus in a functionless state; and we shall hereafter meet in other divisions of the animal kingdom with many instances of structures proper to the male being present in a rudimentary condition of the female.

Landois has observed another important fact, namely, that in the females of the Acridiidæ, the stridulating teeth on the femora remain throughout life in the same condition in which they first appear during the larval state in both sexes. In the males, on the other hand, they become further developed, and acquire their perfect structure at the last moult, when the insect is mature and ready to breed.

From the facts now given, we see that the means by which the males of the Orthoptera produce their sounds are extremely diversified, and are altogether different from those employed by the Homoptera. [43] But throughout the animal kingdom we often find the same object gained by the most diversified means; this seems due to the whole organization having undergone multifarious changes in the course of ages, and as part after part varied different variations were taken advantage of for the same general purpose. The diversity of means for producing sound in the three families of the Orthoptera and in the Homoptera, impresses the mind with the high importance of these structures to the males, for the sake of calling or alluring the females. We need feel no surprise at the amount of modification which the Orthoptera have undergone in this respect, as we now know, from Dr. Scudder's remarkable discovery, [44] that there has been more than ample time. This naturalist has lately found a fossil insect in the Devonian formation of New Brunswick, which is furnished with "the well-known tympanum or stridulating apparatus of the male Locustidæ." The insect, though in most respects related to the Neuroptera, appears, as is so often the case with very ancient forms, to connect the two related Orders of the Neuroptera and Orthoptera.

[43] Landois has recently found in certain Orthoptera rudimentary structures closely

similar to the sound-producing organs in the Homoptera; and this is a surprising fact. See 'Zeitschrift fur wissenschaft Zoolog.' B. xxii. Heft 3, 1871, p. 348.

[44] 'Transactions, Entomological Society,' 3rd series, vol. ii. ('Journal of Proceedings,' p. 117).

I have but little more to say on the Orthoptera. Some of the species are very pugnacious: when two male field-crickets (*Gryllus campestris*) are confined together, they fight till one kills the other; and the species of Mantis are described as manœuvring with their sword-like front-limbs, like hussars with their sabres. The Chinese keep these insects in little bamboo cages, and match them like game-cocks. [45] With respect to color, some exotic locusts are beautifully ornamented; the posterior wings being marked with red, blue, and black; but as throughout the Order the sexes rarely differ much in color, it is not probable that they owe their bright tints to sexual selection. Conspicuous colors may be of use to these insects, by giving notice that they are unpalatable. Thus it has been observed [46] that a bright-colored Indian locust was invariably rejected when offered to birds and lizards. Some cases, however, are known of sexual differences in color in this Order. The male of an American cricket [47] is described as being as white as ivory, whilst the female varies from almost white to greenish-yellow or dusky. Mr. Walsh informs me that the adult male of *Spectrum femoratum* (one of the Phasmidæ) "is of a shining brownish-yellow color; the adult female being of a dull, opaque, cinereous brown; the young of both sexes being green." Lastly, I may mention that the male of one curious kind of cricket [48] is furnished with "a long membranous appendage, which falls over the face like a veil;" but what its use may be, is not known.

[45] Westwood, 'Modern Classification of Insects,' vol. i. p. 427; for crickets, p. 445.
[46] Mr. Ch. Horne, in 'Proceedings of the Entomological Society,' May 3, 1869, p. xii.
[47] The *Œcanthus nivalis*, Harris, 'Insects of New England,' 1842, p. 124. The two sexes of *Œ. pellucidus* of Europe differ, as I hear from Victor Carus, in nearly the same manner.
[48] Platyblemnus: Westwood, 'Modern Classification,' vol. i. p. 447.

ORDER, NEUROPTERA.

Little need here be said, except as to color. In the Ephemeridæ the sexes often differ slightly in their obscure tints; [49] but it is not probable that the males are thus rendered attractive to the females. The Libellulidæ, or dragon-flies, are ornamented with splendid green, blue, yellow, and vermilion metallic tints; and the sexes often differ. Thus, as Prof. Westwood remarks, [50] the males of some of the Agrionidæ, "are of a rich blue with black wings, whilst the females are fine green with colorless wings." But in *Agrion Ramburii* these colors are exactly reversed in the two sexes. [51] In the extensive N. American genus of Hetærina, the males alone have a beautiful carmine spot at the base of each wing. In *anax junius* the basal part of the abdomen in the male is a vivid ultramarine blue, and in the female grass-green. In the allied genus Gomphus, on the other hand, and in some other genera, the sexes differ but little in color. In closely-allied forms throughout the animal kingdom, similar cases of the sexes differing greatly, or very little, or not at all, are of frequent occurrence. Although there is so wide a difference in color between the sexes of many Libellulidæ, it is often difficult to say which is the more brilliant; and the ordinary coloration of the two sexes is reversed, as we have just seen, in one species of Agrion. It is not probable that their colors in any case have been gained as a protection. Mr. MacLachlan, who has closely attended to this family, writes to me that dragon-flies—the tyrants of the insect-world—are the least liable of any insect to be attacked by birds or other enemies, and he believes that their bright colors serve as a sexual attraction. Certain dragon-flies apparently are attracted by particular colors: Mr. Patterson observed [52] that the Agrionidæ, of which the males are blue, settled in numbers on the blue float of a fishing line; whilst two other species were attracted by shining white colors.

[49] B.D. Walsh, the 'Pseudo-neuroptera of Illinois,' in 'Proceedings of the Entomological Society of Philadelphia,' 1862, p. 361.
[50] 'Modern Classification,' vol. ii. p. 37.
[51] Walsh, ibid. p. 381. I am indebted to this naturalist for the following facts on Hetærina, Anax, and Gomphus.
[52] 'Transactions, Ent. Soc.' vol. i. 1836, p. lxxxi.

It is an interesting fact, first noticed by Schelver, that, in several genera belonging to two sub-families, the males on first emergence from the pupal state, are colored exactly like the females; but that their bodies in a short time assume a conspicuous milky-blue tint, owing to the exudation of a kind of oil, soluble in ether and alcohol. Mr. MacLachlan believes that in the male of *Libellula depressa* this change of color does not occur until nearly a fortnight after the metamorphosis, when the sexes are ready to pair.

Certain species of Neurothemis present, according to Brauer, [53] a curious case of dimorphism, some of the females having ordinary wings, whilst others have them "very richly netted, as in the males of the same species." Brauer "explains the phenomenon on Darwinian principles by the supposition that the close netting of the veins is a secondary sexual character in the males, which has been abruptly transferred to some of the females, instead of, as generally occurs, to all of them." Mr. MacLachlan informs me of another instance of dimorphism in several species of Agrion, in which some individuals are of an orange color, and these are invariably females. This is probably a case of reversion; for in the true Libellulæ, when the sexes differ in color, the females are orange or yellow; so that supposing Agrion to be descended from some primordial form which resembled the typical Libellulæ in its sexual characters, it would not be surprising that a tendency to vary in this manner should occur in the females alone.

[53] See abstract in the 'Zoological Record' for 1867, p. 450.

Although many dragon-flies are large, powerful, and fierce insects, the males have not been observed by Mr. MacLachlan to fight together, excepting, as he believes, in some of the smaller species of Agrion. In another group in this Order, namely, the Termites or white ants, both sexes at the time of swarming may be seen running about, "the male after the female, sometimes two chasing one female, and contending with great eagerness who shall win the prize." [54] The *Atropos pulsatorius* is said to make a noise with its jaws, which is answered by other individuals. [55]

[54] Kirby and Spence, 'Introduction to Entomology,' vol. ii. 1818, p. 35.
[55] Houzeau, 'Les Facultés Mentales,' etc. Tom. i. p. 104.

ORDER, HYMENOPTERA.

That inimitable observer, M. Fabre, [56] in describing the habits of Cerceris, a wasp-like insect, remarks that "fights frequently ensue between the males for the possession of some particular female, who sits an apparently unconcerned beholder of the struggle for supremacy, and when the victory is decided, quietly flies away in company with the conqueror." Westwood [57] says that the males of one of the saw-flies (Tenthredinæ) "have been found fighting together, with their mandibles locked." As M. Fabre speaks of the males of Cerceris striving to obtain a particular female, it may be well to bear in mind that insects belonging to this Order have the power of recognising each other after long intervals of time, and are deeply attached. For instance, Pierre Huber, whose accuracy no one doubts, separated some ants, and when, after an interval of four months, they met others which had formerly belonged to the same community, they recognized and caressed one another with their antennæ. Had they been strangers they would have fought together. Again, when two communities engage in a battle, the ants on the same side sometimes attack each other in the general confusion, but they soon perceive their mistake, and the one ant soothes the other. [58]

[56] See an interesting article, 'The Writings of Fabre,' in 'Nat. Hist. Review,' April 1862, p. 122.
[57] 'Journal of Proceedings of Entomological Society,' Sept. 7, 1863, p. 169.
[58] P. Huber, 'Recherches sur les Mœurs des Fourmis,' 1810, pp. 150, 165.

In this Order slight differences in color, according to sex, are common, but conspicuous differences are rare except in the family of Bees; yet both sexes of certain groups are so brilliantly colored—for instance in Chrysis, in which vermilion and metallic greens prevail—that we are tempted to attribute the result to sexual selection. In the Ichneumonidæ, according to Mr. Walsh, [59] the males are almost universally lighter-colored than the females. On the other hand, in the Tenthredinidæ the males are generally darker than the females. In the Siricidæ the sexes frequently differ; thus the male of *Sirex juvencus* is banded with orange, whilst the female is dark purple; but it is difficult to say which sex is the more ornamented. In *Tremex columbæ* the female is much brighter colored than the male. I am informed by Mr. F. Smith, that the male ants of several species are black, the females being testaceous.

[59] 'Proceedings of the Entomological Society of Philadelphia,' 1866, pp. 238, 239.

In the family of Bees, especially in the solitary species, as I hear from the same entomologist, the sexes often differ in color. The males are generally the brighter, and in Bombus as well as in Apathus, much more variable in color than the females. In *Anthophora retusa* the male is of a rich fulvous-brown, whilst the female is quite black: so are the females of several species of Xylocopa, the males being bright yellow. On the other hand the females of some species, as of *Andraena fulva*, are much brighter colored than the males. Such differences in color can hardly be accounted for by the males being defenceless and thus requiring protection, whilst the females are well defended by their stings. H. Müller, [60] who has particularly attended to the habits of bees, attributes these differences in color in chief part to sexual selection. That bees have a keen perception of color is certain. He says that the males search eagerly and fight for the possession of the females; and he accounts through such contests for the mandibles of the males being in certain species larger than those of the females. In some cases the males are far more numerous than the females, either early in the season, or at all times and places, or locally; whereas the females in other cases are apparently in excess. In some species the more beautiful males appear to have been selected by the females; and in others the more beautiful females by the males. Consequently in certain genera (Müller, p. 42), the males of the several species differ much in appearance, whilst the females are almost indistinguishable; in other genera the reverse occurs. H. Müller believes (p. 82) that the colors gained by one sex through sexual selection have often been transferred in a variable degree to the other sex, just as the pollen-collecting apparatus of the female has often been transferred to the male, to whom it is absolutely useless. [61]

[60] 'Anwendung der Darwinschen Lehre auf Bienen,' Verh. d. n. V. Jahrg. xxix.

[61] M. Perrier in his article 'la Sélection sexuelle d'après Darwin' ('Revue Scientifique,' Feb. 1873, p. 868), without apparently having reflected much on the subject, objects that as the males of social bees are known to be produced from unfertilized ova, they could not transmit new characters to their male offspring. This is an extraordinary objection. A female bee fertilized by a male, which presented some character facilitating the union of the sexes, or rendering him more attractive to the female, would lay eggs which would produce only females; but these young females would next year produce males; and will it be pretended that such males would not inherit the characters of their male grandfathers? To take a case with ordinary animals as nearly parallel as possible: if a female of any white quadruped or bird were crossed by a male of a black breed, and the male and female offspring were paired together, will it be pretended that the grandchildren would not inherit a tendency to blackness from their male grandfather? The acquirement of new characters by the sterile worker-bees is a much more difficult case, but I have endeavoured to show in my 'Origin of Species,' how these sterile beings are subjected to the power of natural selection.

Mutilla Europæa makes a stridulating noise; and according to Goureau [62] both sexes have this power. He attributes the sound to the friction of the third and preceding abdominal segments, and I find that these surfaces are marked with very fine concentric ridges; but so is the projecting thoracic collar into which the head articulates, and this collar, when scratched with the point of a needle, emits the proper sound. It is rather surprising that both sexes should have the power of stridulating, as the male is winged and the female wingless. It is notorious that Bees express certain emotions, as of anger, by the tone of their humming; and according to H. Müller (p. 80), the males of some species make a peculiar singing noise whilst pursuing the females.

[62] Quoted by Westwood, 'Modern Classification of Insects,' vol. ii. p. 214.

ORDER, COLEOPTERA (BEETLES).

Many beetles are colored so as to resemble the surfaces which they habitually frequent, and they thus escape detection by their enemies. Other species, for instance diamond-beetles, are ornamented with splendid colors, which are often arranged in stripes, spots, crosses, and other elegant patterns. Such colors can hardly serve directly as a protection, except in the case of certain flower-feeding species; but they may serve as a warning or means of recognition, on the same principle as the phosphorescence of the glow-worm. As with beetles the colors of the two sexes are generally alike, we have no evidence that they have been gained through sexual selection; but this is at least possible, for they have been developed in one sex and then transferred to the other; and this view is even in some degree probable in those groups which possess other well-marked secondary sexual characters. Blind beetles, which cannot of course behold each other's beauty, never, as I hear from Mr. Waterhouse, jun., exhibit bright colors, though they often have polished coats; but the explanation of their obscurity may be that they generally inhabit caves and other obscure stations.

Some Longicorns, especially certain Prionidæ, offer an exception to the rule that the sexes of beetles do not differ in color. Most of these insects are large and splendidly colored. The males in the genus Pyrodes, [63] which I saw in Mr. Bates's collection, are generally redder but rather duller than the females, the latter being colored of a more or less splendid golden-green. On the other hand, in one species the male is golden-green, the female being richly tinted with red and purple. In the genus Esmeralda the sexes differ so greatly in color that they have been ranked as distinct species; in one species both are of a beautiful shining green, but the male has a red thorax. On the whole, as far as I could judge, the females of those Prionidæ, in which the sexes differ, are colored more richly than the males, and this does not accord with the common rule in regard to color, when acquired through sexual selection.

[63] *Pyrodes pulcherrimus*, in which the sexes differ conspicuously, has been described by Mr. Bates in 'Transact. Ent. Soc.' 1869, p. 50. I will specify the few other cases in which I have heard of a difference in color between the sexes of beetles. Kirby and Spence ('Introduct. to Entomology,' vol. iii. p. 301) mention a Cantharis, Meloe, Rhagium, and the *Leptura testacea*; the male of the latter being testaceous, with a black thorax, and the female of a dull red all over. These two latter beetles belong to the family of Longicorns. Messrs. R. Trimen and Waterhouse, jun., inform me of two Lamellicorns, viz., a Peritrichia and Trichius, the male of the latter being more obscurely colored than the female. In *Tillus elongatus* the male is black, and the female always, as it is believed, of a dark blue color, with a red thorax. The male, also, of *Orsodacna atra*, as I hear from Mr. Walsh, is black, the female (the so-called *O. ruficollis*) having a rufous thorax.

240

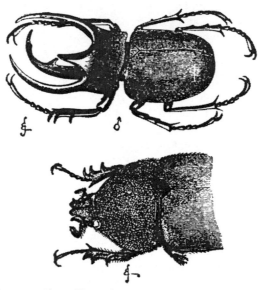

Fɪɢ. 16.—Chalcosoma atlas. Upper figure, male (reduced) ; lower figure, female
(nat. size).

A most remarkable distinction between the sexes of many beetles is presented by the great horns which rise from the head, thorax, and clypeus of the males; and in some few cases from the under surface of the body. These horns, in the great family of the Lamellicorns, resemble those of various quadrupeds, such as stags, rhinoceroses, etc., and are wonderful both from their size and diversified shapes. Instead of describing them, I have given figures of the males and females of some of the more remarkable forms. (Figs. 16 to 20.) The females generally exhibit rudiments of the horns in the form of small knobs or ridges; but some are destitute of even the slightest rudiment. On the other hand, the horns are nearly as well developed in the female as in the male *Phanæus lancifer*; and only a little less well developed in the females of some other species of this genus and of Copris. I am informed by Mr. Bates that the horns do not differ in any manner corresponding with the more important characteristic differences between the several subdivisions of the family: thus within the same section of the genus Onthophagus, there are species which have a single horn, and others which have two.

In almost all cases, the horns are remarkable from their excessive variability; so that a graduated series can be formed, from the most highly developed males to others so degenerate that they can barely be distinguished from the females. Mr. Walsh [64] found that in *Phanæus carnifex* the horns were thrice as long in some males as in others. Mr. Bates, after examining above a hundred males of *Onthophagus rangifer* (Fig. 20), thought that he had at last discovered a species in which the horns did not vary; but further research proved the contrary.

[64] 'Proceedings of the Entomological Society of Philadephia,' 1864, p. 228.

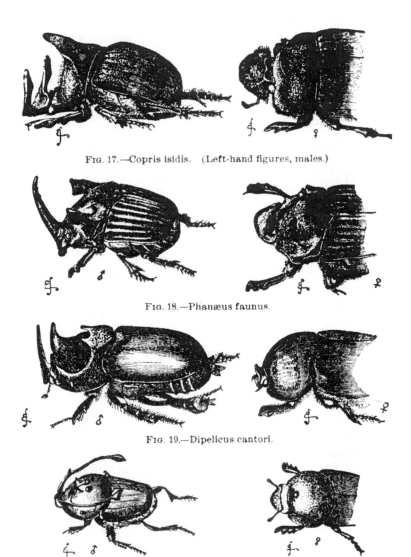

FIG. 17.—Copris isidis. (Left-hand figures, males.)

FIG. 18.—Phanæus faunus.

FIG. 19.—Dipelicus cantori.

FIG. 20.—Onthophagus rangifer, enlarged.

The extraordinary size of the horns, and their widely different structure in closely-allied forms, indicate that they have been formed for some purpose; but their excessive variability in the males of the same species leads to the inference that this purpose cannot be of a definite nature. The horns do not show marks of friction, as if used for any ordinary work. Some authors suppose [65] that as the males wander about much more than the females, they require horns as a defence against their enemies; but as the horns are often blunt, they do not seem well adapted for defence. The most obvious conjecture is that they are used by the males for fighting together; but the males have never been observed to fight; nor could Mr. Bates, after a careful examination of numerous species, find any sufficient evidence, in their mutilated or broken condition, of their having been

242

thus used. If the males had been habitual fighters, the size of their bodies would probably have been increased through sexual selection, so as to have exceeded that of the females; but Mr. Bates, after comparing the two sexes in above a hundred species of the Copridæ, did not find any marked difference in this respect amongst well-developed individuals. In Lethrus, moreover, a beetle belonging to the same great division of the Lamellicorns, the males are known to fight, but are not provided with horns, though their mandibles are much larger than those of the female.

[65] Kirby and Spence, 'Introduction to Entomology,' vol. iii. P. 300.

The conclusion that the horns have been acquired as ornaments is that which best agrees with the fact of their having been so immensely, yet not fixedly, developed,—as shown by their extreme variability in the same species, and by their extreme diversity in closely-allied species. This view will at first appear extremely improbable; but we shall hereafter find with many animals standing much higher in the scale, namely fishes, amphibians, reptiles and birds, that various kinds of crests, knobs, horns and combs have been developed apparently for this sole purpose.

Fig. 21.—Onitis furcifer, male, viewed from beneath.

Fig. 22—Left-hand figure, male of Onitis furcifer, viewed laterally. Right-hand figure, female. a. Rudiment of cephalic horn. b. Trace of thoracic horn or crest.

The males of *Onitis furcifer* (Fig. 21), and of some other species of the genus, are furnished with singular projections on their anterior femora, and with a great fork or pair of horns on the lower surface of the thorax. Judging from other insects, these may aid the male in clinging to the female. Although the males have not even a trace of a horn on the upper surface of the body, yet the females plainly exhibit a rudiment of a single horn on the head (Fig. 22, *a*), and of a crest (*b*) on the thorax. That the slight thoracic crest in the female is a rudiment of a projection proper to the male, though entirely absent in the male of this particular species, is clear: for the female of *Bubas bison* (a genus which comes next to Onitis) has a similar slight crest on the thorax, and the male bears a great projection in the same situation. So, again, there can hardly be a doubt that the little point (*a*) on the head of the female *Onitis furcifer*, as well as on the head of the females of two or three allied species, is a rudimentary representative of the cephalic horn, which is common to the males of so many Lamellicorn beetles, as in Phanæus (Fig. 18).

The old belief that rudiments have been created to complete the scheme of nature is here so far from holding good, that we have a complete inversion of the ordinary state of things in the family. We may reasonably suspect that the males originally bore horns and transferred them to the females in a rudimentary condition, as in so many other Lamellicorns. Why the males subsequently lost their horns, we know not; but this may have been caused through the principle of compensation, owing to the development of the large horns and projections on the lower surface; and as these are confined to the males, the rudiments of the upper horns on the females would not have been thus obliterated.

Fɪɢ. 23.—Bledius taurus, magnified.   Left-hand figure, male ; right-hand figure, female.

The cases hitherto given refer to the Lamellicorns, but the males of some few other beetles, belonging to two widely distinct groups, namely, the Curculionidæ and Staphylinidæ, are furnished with horns—in the former on the lower surface of the body, [66] in the latter on the upper surface of the head and thorax. In the Staphylinidæ, the horns of the males are extraordinarily variable in the same species, just as we have seen with the Lamellicorns. In Siagonium we have a case of dimorphism, for the males can be divided into two sets, differing greatly in the size of their bodies and in the development of their horns, without intermediate gradations. In a species of Bledius (Fig. 23), also belonging to the Staphylinidæ, Professor Westwood states that, "male specimens can be found in the same locality in which the central horn of the thorax is very large, but the horns of the head quite rudimental; and others, in which the thoracic horn is much shorter, whilst the protuberances on the head are long." [67] Here we apparently have a case of compensation, which throws light on that just given, of the supposed loss of the upper horns by the males of *Onitis*.

[66] Kirby and Spence, 'Introduction to Entomology,' vol. iii. p. 329.
[67] 'Modern Classification of Insects,' vol. i. p. 172: Siagonium, p. 172. In the British Museum I noticed one male specimen of Siagonium in an intermediate condition, so that the dimorphism is not strict.

LAW OF BATTLE.

Some male beetles, which seem ill-fitted for fighting, nevertheless engage in conflicts for the possession of the females. Mr. Wallace [68] saw two males of *Leptorhynchus angustatus*, a linear beetle with a much elongated rostrum, "fighting for a female, who stood close by busy at her boring. They pushed at each other with their rostra, and clawed and thumped, apparently in the greatest rage." The smaller male, however, "soon ran away, acknowledging himself vanquished." In some few cases male beetles are well adapted for fighting, by possessing great toothed mandibles, much larger than those of the females. This is the case with the common stag-beetle (*Lucanus cervus*), the males of which emerge from the pupal state about a week before the other sex, so that several may often be seen pursuing the same female. At this season they engage in fierce conflicts. When Mr. A.H. Davis [69] enclosed two males with one female in a box, the larger male severely pinched the smaller one, until he resigned his pretensions. A friend informs me that when a boy he often put the males together to see them fight, and he noticed that they were much bolder and fiercer than the females, as with the higher animals. The males would seize hold of his finger, if held in front of them, but not so the females, although they have stronger jaws. The males of many of the Lucanidæ, as well as of the above-mentioned Leptorhynchus, are larger and more powerful insects than the females. The two sexes of *Lethrus cephalotes* (one of the Lamellicorns) inhabit the same burrow; and the male has larger mandibles than the female. If, during the breeding-season, a strange male attempts to enter the burrow, he is attacked; the female does not remain passive, but closes the mouth of the burrow, and encourages her mate by continually pushing him on from behind; and the battle lasts until the aggressor is killed or runs away. [70] The two sexes of another Lamellicorn beetle, the *Ateuchus cicatricosus*, live in pairs, and seem much attached to each other; the male excites the females to roll the balls of dung in which the ova are deposited; and if she is removed, he becomes much agitated. If the male is removed the female ceases all work, and as M. Brulerie believes, would remain on the same spot until she died. [71]

[68] 'The Malay Archipelago,' vol. ii. 1869, p. 276. Riley, Sixth 'Report on Insects of Missouri,' 1874, p. 115.

[69] 'Entomological Magazine,' vol. i. 1833, p. 82. See also on the conflicts of this species, Kirby and Spence, ibid. vol. iii. p. 314; and Westwood, ibid. vol. i. p. 187.

[70] Quoted from Fischer, in 'Dict. Class. d'Hist. Nat.' tom. x. p. 324.

[71] 'Ann. Soc. Entomolog. France,' 1866, as quoted in 'Journal of Travel,' by A. Murray, 1868, p. 135.

The great mandibles of the male Lucanidæ are extremely variable both in size and structure, and in this respect resemble the horns on the head and thorax of many male Lamellicorns and Staphylinidæ. A perfect series can be formed from the best-provided to the worst-provided or degenerate males. Although the mandibles of the common stag-beetle, and probably of many other species, are used as efficient weapons for fighting, it is doubtful whether their great size can thus be accounted for. We have seen that they are used by the *Lucanus elaphus* of N. America for seizing the female. As they are so conspicuous and so elegantly branched, and as owing to their great length they are not well adapted for pinching, the suspicion has crossed my mind that they may in addition serve as an ornament, like the horns on the head and thorax of the various species above described. The male *Chiasognathus Grantii* of S. Chile—a splendid beetle belonging to the same family—has enormously developed mandibles (Fig. 24); he is bold and pugnacious; when threatened he faces round, opens his great jaws, and at the same time stridulates loudly. But the mandibles were not strong enough to pinch my finger so as to cause actual pain.

Sexual selection, which implies the possession of considerable perceptive powers and of strong passions, seems to have been more effective with the Lamellicorns than with any other family of beetles. With some species the males are provided with weapons for fighting; some live in pairs and show mutual affection; many have the power of stridulating when excited; many are furnished with the most extraordinary horns, apparently for the sake of ornament; and some, which are diurnal in their habits, are gorgeously colored. Lastly, several of the largest beetles in the world belong to this family, which was placed by Linnæus and Fabricius as the head of the Order. [72]

[72] Westwood, 'Modern Classification,' vol. i. p. 184.

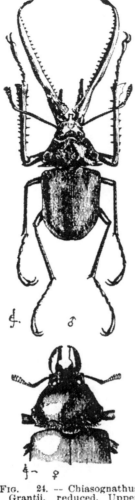

FIG. 24. -- Chiasognathus Grantii, reduced. Upper figure, male ; lower figure, female.

STRIDULATING ORGANS.

Beetles belonging to many and widely distinct families possess these organs. The sound thus produced can sometimes be heard at the distance of several feet or even yards, [73] but it is not comparable with that made by the Orthoptera. The rasp generally consists of a narrow, slightly-raised surface, crossed by very fine, parallel ribs, sometimes so fine as to cause iridescent colors, and having a very elegant appearance under the microscope. In some cases, as with Typhœus, minute, bristly or scale-like prominences, with which the whole surrounding surface is covered in approximately parallel lines, could be traced passing into the ribs of the rasp. The transition takes place by their becoming confluent and straight, and at the same time more prominent and smooth. A hard ridge on an adjoining part of the body serves as the scraper for the rasp, but this

246

scraper in some cases has been specially modified for the purpose. It is rapidly moved across the rasp, or conversely the rasp across the scraper.

[73] Wollaston, 'On Certain Musical Curculionidæ,' 'Annals and Mag. of Nat. Hist.' vol. vi. 1860, p. 14.

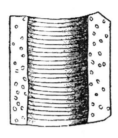

FIG. 25.—Necrophorus (from Landois). *r*. The two rasps. Left-hand figure, part of the rasp highly magnified.

These organs are situated in widely different positions. In the carrion-beetles (Necrophorus) two parallel rasps (*r*, Fig. 25) stand on the dorsal surface of the fifth abdominal segment, each rasp [74] consisting of 126 to 140 fine ribs. These ribs are scraped against the posterior margins of the elytra, a small portion of which projects beyond the general outline. In many Crioceridæ, and in *Clythra 4-punctata* (one of the Chrysomelidæ), and in some Tenebrionidæ, etc., [75] the rasp is seated on the dorsal apex of the abdomen, on the pygidium or pro-pygidium, and is scraped in the same manner by the elytra. In Heterocerus, which belongs to another family, the rasps are placed on the sides of the first abdominal segment, and are scraped by ridges on the femora. [76] In certain Curculionidæ and Carabidæ [77] the parts are completely reversed in position, for the rasps are seated on the inferior surface of the elytra, near their apices, or along their outer margins, and the edges of the abdominal segments serve as the scrapers. In *Pelobius Hermanni* (one of Dytiscidæ or water-beetles) a strong ridge runs parallel and near to the sutural margin of the elytra, and is crossed by ribs, coarse in the middle part, but becoming gradually finer at both ends, especially at the upper end; when this insect is held under water or in the air, a stridulating noise is produced by the extreme horny margin of the abdomen being scraped against the rasps. In a great number of long-horned beetles (Longicornia) the organs are situated quite otherwise, the rasp being on the meso-thorax, which is rubbed against the pro-thorax; Landois counted 238 very fine ribs on the rasp of *Cerambyx heros*.

[74] Landois, 'Zeitschrift fur wissenschaft Zoolog.' B. xvii. 1867, s. 127.
[75] I am greatly indebted to Mr. G.R. Crotch for having sent me many prepared specimens of various beetles belonging to these three families and to others, as well as for valuable information. He believes that the power of stridulation in the Clythra has not been previously observed. I am also much indebted to Mr. E.W. Janson, for information and specimens. I may add that my son, Mr. F. Darwin, finds that *Dermestes murinus* stridulates, but he searched in vain for the apparatus. Scolytus has lately been described by Dr. Chapman as a stridulator, in the 'Entomologist's Monthly Magazine,' vol. vi. p. 130.
[76] Schiödte, translated, in 'Annals and Magazine of Natural History,' vol. xx. 1867, p.

37.

[77] Westring has described (Kroyer, 'Naturhist. Tidskrift,' B. ii. 1848-49, p. 334) the stridulating organs in these two, as well as in other families. In the Carabidæ I have examined *Elaphrus uliginosus* and *Blethisa multipunctata*, sent to me by Mr. Crotch. In Blethisa the transverse ridges on the furrowed border of the abdominal segment do not, as far as I could judge, come into play in scraping the rasps on the elytra.),

8

FIG. 26.—Hind-leg of Geotrupes stercorarius (from Landois). *r*. Rasp. *c*. Coxa. *f*. Femur. *t*. Tibia. *tr*. Tarsi.

Many Lamellicorns have the power of stridulating, and the organs differ greatly in position. Some species stridulate very loudly, so that when Mr. F. Smith caught a *Trox sabulosus*, a gamekeeper, who stood by, thought he had caught a mouse; but I failed to discover the proper organs in this beetle. In Geotrupes and Typhœus, a narrow ridge runs obliquely across (*r*, Fig. 26) the coxa of each hind-leg (having in *G. stercorarius* 84 ribs), which is scraped by a specially projecting part of one of the abdominal segments. In the nearly allied *Copris lunaris*, an excessively narrow fine rasp runs along the sutural margin of the elytra, with another short rasp near the basal outer margin; but in some other Coprini the rasp is seated, according to Leconte, [78] on the dorsal surface of the abdomen. In Oryctes it is seated on the pro-pygidium; and, according to the same entomologist, in some other Dynastini, on the under surface of the elytra. Lastly, Westring states that in *Omaloplia brunnea* the rasp is placed on the pro-sternum, and the scraper on the meta-sternum, the parts thus occupying the under surface of the body, instead of the upper surface as in the Longicorns.

[78] I am indebted to Mr. Walsh, of Illinois, for having sent me extracts from Leconte's 'Introduction to Entomology,' pp. 101, 143.

We thus see that in the different coleopterous families the stridulating organs are wonderfully diversified in position, but not much in structure. Within the same family some species are provided with these organs, and others are destitute of them. This diversity is intelligible, if we suppose that originally various beetles made a shuffling or hissing noise by the rubbing together of any hard and rough parts of their bodies, which happened to be in contact; and that from the noise thus produced being in some way useful, the rough surfaces were gradually developed into regular stridulating organs. Some beetles as they move, now produce, either intentionally or unintentionally, a shuffling noise, without possessing any proper organs for the purpose. Mr. Wallace informs me that the *Euchirus longimanus* (a Lamellicorn, with the anterior legs wonderfully elongated in the male) "makes, whilst moving, a low hissing sound by the protrusion and contraction of the abdomen; and when seized it produces a grating sound by rubbing its hind-legs against the edges of the elytra." The hissing sound is clearly due to a narrow rasp running along the sutural margin of each elytron; and I could likewise make the grating sound by rubbing the shagreened surface of the femur against the granulated margin of the corresponding elytron; but I could not here detect any proper rasp; nor is it likely that I could have overlooked it in so large an insect. After examining Cychrus, and reading what Westring has written about this beetle, it seems very doubtful whether it possesses any true rasp, though it has the power of emitting a sound.

From the analogy of the Orthoptera and Homoptera, I expected to find the stridulating organs in the Coleoptera differing according to sex; but Landois, who has carefully examined several species, observed no such difference; nor did Westring; nor did Mr. G.R. Crotch in preparing the many specimens which he had the kindness to send me. Any difference in these organs, if slight, would, however, be difficult to detect, on account of their great variability. Thus, in the first pair of specimens of *Necrophorus humator* and of Pelobius which I examined, the rasp was considerably larger in the male than in the female; but not so with succeeding specimens. In *Geotrupes stercorarius* the rasp appeared to me thicker, opaquer, and more prominent in three males than in the same number of females; in order, therefore, to discover whether the sexes differed in their power of stridulating, my son, Mr. F. Darwin, collected fifty-seven living specimens, which he separated into two lots, according as they made a greater or lesser noise, when held in the same manner. He then examined all these specimens, and found that the males were very nearly in the same proportion to the females in both the lots. Mr. F. Smith has kept alive numerous specimens of *Monoynchus pseudacori* (Curculionidæ), and is convinced that both sexes stridulate, and apparently in an equal degree.

Nevertheless, the power of stridulating is certainly a sexual character in some few Coleoptera. Mr. Crotch discovered that the males alone of two species of Heliopathes (Tenebrionidæ) possess stridulating organs. I examined five males of *H. gibbus*, and in all these there was a well-developed rasp, partially divided into two, on the dorsal surface of the terminal abdominal segment; whilst in the same number of females there was not even a rudiment of the rasp, the membrane of this segment being transparent, and much thinner than in the male. In *H. cribratostriatus* the male has a similar rasp, excepting that it is not partially divided into two portions, and the female is completely destitute of this organ; the male in addition has on the apical margins of the elytra, on each side of the suture, three or four short longitudinal ridges, which are crossed by extremely fine ribs, parallel to and resembling those on the abdominal rasp; whether these ridges serve as an independent rasp, or as a scraper for the abdominal rasp, I could not decide: the female exhibits no trace of this latter structure.

Again, in three species of the Lamellicorn genus Oryctes, we have a nearly parallel case. In the females of *O. gryphus* and *nasicornis* the ribs on the rasp of the pro-pygidium are less continuous and less distinct than in the males; but the chief difference is that the whole upper surface of this segment, when held in the proper light, is seen to be clothed with hairs, which are absent or are represented by excessively fine down in the males. It should be noticed that in all Coleoptera the effective part of the rasp is destitute of hairs. In *O. senegalensis* the difference between the sexes is more strongly marked, and this is best seen when the proper abdominal segment is cleaned and viewed as a transparent object. In the female the whole surface is covered with little separate crests, bearing spines; whilst in the male these crests in proceeding towards the apex, become more and more confluent, regular, and naked; so that three-fourths of the segment is covered with extremely fine parallel ribs, which are quite absent in the female. In the females, however, of all three species of Oryctes, a slight grating or stridulating sound is produced, when the abdomen of a softened specimen is pushed backwards and forwards.

In the case of the Heliopathes and Oryctes there can hardly be a doubt that the males stridulate in order to call or to excite the females; but with most beetles the stridulation apparently serves both sexes as a mutual call. Beetles stridulate under various emotions, in the same manner as birds use their voices for many purposes besides singing to their mates. The great Chiasognathus stridulates in anger or defiance; many species do the same from distress or fear, if held so that they cannot escape; by striking the hollow stems of trees in the Canary Islands, Messrs. Wollaston and Crotch were able to discover the presence of beetles belonging to the genus Acalles by their stridulation. Lastly, the male Ateuchus stridulates to encourage the female in her work, and from distress when she is removed. [79] Some naturalists believe that beetles make this noise to frighten away their enemies; but I cannot think that a quadruped or bird, able to devour a large beetle, would be frightened by so slight a sound. The belief that the stridulation serves as a sexual call is supported by the fact that death-ticks (*Anobium tessellatum*) are well known to answer each other's ticking, and, as I have myself observed, a tapping noise artificially made. Mr. Doubleday also informs me that he has sometimes observed a female ticking, [80] and in an hour or two afterwards has found her united with a male, and on one occasion surrounded by several males. Finally, it is probable that the two sexes of many kinds of beetles were at first enabled to find each other by the slight shuffling noise produced by the rubbing together of the adjoining hard parts of their bodies; and that as those males or females which made the greatest noise succeeded best in finding partners, rugosities on various parts of their bodies were gradually developed by means of sexual selection into true stridulating organs.

[79] M. P. de la Brulerie, as quoted in 'Journal of Travel,' A. Murray, vol. i. 1868, p. 135.

[80] According to Mr. Doubleday, "the noise is produced by the insect raising itself on its legs as high as it can, and then striking its thorax five or six times, in rapid succession, against the substance upon which it is sitting." For references on this subject see Landois, 'Zeitschrift fur wissen. Zoolog.' B. xvii. s. 131. Olivier says (as quoted by Kirby and Spence, 'Introduction to Entomology,' vol. ii. p. 395) that the female of *Pimelia striata* produces a rather loud sound by striking her abdomen against any hard substance, "and that the male, obedient to this call, soon attends her, and they pair."

CHAPTER XI.

INSECTS, continued.—ORDER LEPIDOPTERA.

(BUTTERFLIES AND MOTHS.)

Courtship of butterflies—Battles—Ticking noise—Colors common to both sexes, or more brilliant in the males—Examples—Not due to the direct action of the conditions of life—Colors adapted for protection—Colors of moths—Display—Perceptive powers of the Lepidoptera—Variability—Causes of the difference in color between the males and females—Mimicry, female butterflies more brilliantly colored than the males—Bright colors of caterpillars—Summary and concluding remarks on the secondary sexual characters of insects—Birds and insects compared.

In this great Order the most interesting points for us are the differences in color between the sexes of the same species, and between the distinct species of the same genus. Nearly the whole of the following chapter will be devoted to this subject; but I will first make a few remarks on one or two other points. Several males may often be seen pursuing and crowding round the same female. Their courtship appears to be a prolonged affair, for I have frequently watched one or more males pirouetting round a female until I was tired, without seeing the end of the courtship. Mr. A.G. Butler also informs me that he has several times watched a male courting a female for a full quarter of an hour; but she pertinaciously refused him, and at last settled on the ground and closed her wings, so as to escape from his addresses.

Although butterflies are weak and fragile creatures, they are pugnacious, and an emperor butterfly [1] has been captured with the tips of its wings broken from a conflict with another male. Mr. Collingwood, in speaking of the frequent battles between the butterflies of Borneo, says, "They whirl round each other with the greatest rapidity, and appear to be incited by the greatest ferocity."

[1] *Apatura Iris*: 'The Entomologist's Weekly Intelligence,' 1859, p. 139. For the Bornean Butterflies, see C. Collingwood, 'Rambles of a Naturalist,' 1868, p. 183.

The *Ageronia feronia* makes a noise like that produced by a toothed wheel passing under a spring catch, and which can be heard at the distance of several yards: I noticed this sound at Rio de Janeiro, only when two of these butterflies were chasing each other in an irregular course, so that it is probably made during the courtship of the sexes. [2]

[2] See my 'Journal of Researches,' 1845, p. 33. Mr. Doubleday has detected ('Proc. Ent. Soc.' March 3, 1845, p. 123) a peculiar membranous sac at the base of the front wings, which is probably connected with the production of the sound. For the case of Thecophora, see 'Zoological Record,' 1869, p. 401. For Mr. Buchanan White's observations, the Scottish Naturalist, July 1872, p. 214.

Some moths also produce sounds; for instance, the males *Theocophora fovea*. On two occasions Mr. F. Buchanan White [3] heard a sharp quick noise made by the male of *Hylophila prasinana*, and which he believes to be produced, as in Cicada, by an elastic membrane, furnished with a muscle. He quotes, also, Guenée, that Setina produces a sound like the ticking of a watch, apparently by the aid of "two large tympaniform vesicles, situated in the pectoral region"; and these "are much more developed in the male than in the female." Hence the sound-producing organs in the Lepidoptera appear to stand in some relation with the sexual functions. I have not alluded to the well-known noise made by the Death's Head Sphinx, for it is generally heard soon after the moth has emerged from its cocoon.

[3] 'The Scottish Naturalist,' July 1872, p. 213.

Giard has always observed that the musky odor, which is emitted by two species of Sphinx moths, is peculiar to the males; [4] and in the higher classes we shall meet with many instances of the males alone being odoriferous.

Every one must have admired the extreme beauty of many butterflies and of some moths; and it may be asked, are their colors and diversified patterns the result of the direct action of the physical conditions to which these insects have been exposed, without any benefit being thus derived? Or have successive variations been accumulated and determined as a protection, or for some unknown purpose, or that one sex may be attractive to the other? And, again, what is the meaning of the colors being widely different in the males and females of certain species, and alike in the two sexes of other species of the same genus? Before attempting to answer these questions a body of facts must be given.

[4] 'Zoological Record,' 1869, p. 347.

With our beautiful English butterflies, the admiral, peacock, and painted lady (Vanessæ), as well as many others, the sexes are alike. This is also the case with the magnificent Heliconidæ, and most of the Danaidæ in the tropics. But in certain other tropical groups, and in some of our English butterflies, as the purple emperor, orange-tip, etc. (*Apatura Iris* and *Anthocharis cardamines*), the sexes differ either greatly or slightly in color. No language suffices to describe the splendor of the males of some tropical species. Even within the same genus we often find species presenting extraordinary differences between the sexes, whilst others have their sexes closely alike. Thus in the South American genus Epicalia, Mr. Bates, to whom I am indebted for most of the following facts, and for looking over this whole discussion, informs me that he knows twelve species, the two sexes of which haunt the same stations (and this is not always the case with butterflies), and which, therefore, cannot have been differently affected by external conditions. [5] In nine of these twelve species the males rank amongst the most brilliant of all butterflies, and differ so greatly from the comparatively plain females that they were formerly placed in distinct genera. The females of these nine species resemble each other in their general type of coloration; and they likewise resemble both sexes of the species in several allied genera found in various parts of the world. Hence we may infer that these nine species, and probably all the others of the genus, are descended from an ancestral form which was colored in nearly the same manner. In the tenth species the female still retains the same general coloring, but the male resembles her, so that he is colored in a much less gaudy and contrasted manner than the males of the previous species. In the eleventh and twelfth species, the females depart from the usual type, for they are gaily decorated almost like the males, but in a somewhat less degree. Hence in these two latter species the bright colors of the males seem to have been transferred to the females; whilst in the tenth species the male has either retained or recovered the plain colors of the female, as well as of the parent-form of the genus. The sexes in these three cases have thus been rendered nearly alike, though in an opposite manner. In the allied genus Eubagis, both sexes of some of the species are plain-colored and nearly alike; whilst with the greater number the males are decorated with beautiful metallic tints in a diversified manner, and differ much from their females. The females throughout the genus retain the same general style of coloring, so that they resemble one another much more closely than they resemble their own males.

[5] See also Mr. Bates's paper in 'Proc. Ent. Soc. of Philadelphia,' 1865, p. 206. Also Mr. Wallace on the same subject, in regard to Diadema, in 'Transactions, Entomological Society of London,' 1869, p. 278.

In the genus Papilio, all the species of the Aeneas group are remarkable for their conspicuous and strongly contrasted colors, and they illustrate the frequent tendency to gradation in the amount of difference between the sexes. In a few species, for instance in P. ascanius, the males and females are alike; in others the males are either a little brighter, or very much more superb than the females. The genus Junonia, allied to our Vanessæ, offers a nearly parallel case, for although the sexes of most of the species resemble each other, and are destitute of rich colors, yet in certain species, as in J. œnone, the male is rather more bright-colored than the female, and in a few (for instance J. andremiaja) the male is so different from the female that he might be mistaken for an entirely distinct species.

Another striking case was pointed out to me in the British Museum by Mr. A. Butler, namely, one of the tropical American Theclæ, in which both sexes are nearly alike and wonderfully splendid; in another species the male is colored in a similarly gorgeous manner, whilst the whole upper surface of the female is of a dull uniform brown. Our common little English blue butterflies of the genus Lycæna, illustrate the various differences in color between the sexes, almost as well, though not in so striking a manner, as the above exotic genera. In Lycæna agestis both sexes have wings of a brown color, bordered with small ocellated orange spots, and are thus alike. In L. ægon the wings of the males are of a fine blue, bordered with black, whilst those of the female are brown, with a similar border, closely resembling the wings of L. agestis. Lastly, in L. arion both sexes are of a blue color and are very like, though in the female the edges of the wings are rather duskier, with the black spots plainer; and in a bright blue Indian species both sexes are still more alike.

I have given the foregoing details in order to show, in the first place, that when the sexes of butterflies differ, the male as a general rule is the more beautiful, and departs more from the usual type of coloring of the group to which the species belongs. Hence in most groups the females of the several species resemble each other much more closely than do the males. In some cases, however, to which I shall hereafter allude, the females are colored more splendidly than the males. In the second place, these details have been given to bring clearly before the mind that within the same genus, the two sexes frequently present every gradation from no difference in color, to so great a difference that it was long before the two were placed by entomologists in the same genus. In the third place, we have seen that when the sexes nearly resemble each other, this appears due either to the male having transferred his colors to the female, or to the male having retained, or perhaps recovered, the primordial colors of the group. It also deserves notice that in those groups in which the sexes differ, the females usually somewhat resemble the males, so that when the males are beautiful to an extraordinary degree, the females almost invariably exhibit some degree of beauty. From the many cases of gradation in the amount of difference between the sexes, and from the prevalence of the same general type of coloration throughout the whole of the same group, we may conclude that the causes have generally been the same which have determined the brilliant coloring of the males alone of some species, and of both sexes of other species.

As so many gorgeous butterflies inhabit the tropics, it has often been supposed that they owe their colors to the great heat and moisture of these zones; but Mr. Bates [6] has shown by the comparison of various closely-allied groups of insects from the temperate and tropical regions, that this view cannot be maintained; and the evidence becomes conclusive when brilliantly-colored males and plain-colored females of the same species inhabit the same district, feed on the same food, and follow exactly the same habits of life. Even when the sexes resemble each other, we can hardly believe that their brilliant and beautifully-arranged colors are the purposeless result of the nature of the tissues and of the action of the surrounding conditions.

[6] 'The Naturalist on the Amazons,' vol. i. 1863, p. 19.

With animals of all kinds, whenever color has been modified for some special purpose, this has been, as far as we can judge, either for direct or indirect protection, or as an attraction between the sexes. With many species of butterflies the upper surfaces of the wings are obscure; and this in all probability leads to their escaping observation and danger. But butterflies would be particularly liable to be attacked by their enemies when at rest; and most kinds whilst resting raise their wings vertically over their backs, so that the lower surface alone is exposed to view. Hence it is this side which is often colored so as to imitate the objects on which these insects commonly rest. Dr. Rössler, I believe, first noticed the similarity of the closed wings of certain Vanessæ and other butterflies to the bark of trees. Many analogous and striking facts could be given. The most interesting one is that recorded by Mr. Wallace [7] of a common Indian and Sumatran butterfly (Kallima) which disappears like magic when it settles on a bush; for it hides its head and antennæ between its closed wings, which, in form, color and veining, cannot be distinguished from a withered leaf with its footstalk. In some other cases the lower surfaces of the wings are brilliantly colored, and yet are protective; thus in *Thecla rubi* the wings when closed are of an emerald green, and resemble the young leaves of the bramble, on which in spring this butterfly may often be seen seated. It is also remarkable that in very many species in which the sexes differ greatly in color on their upper surface, the lower surface is closely similar or identical in both sexes, and serves as a protection. [8]

[7] See the interesting article in the 'Westminster Review,' July 1867, p. 10. A woodcut of the Kallima is given by Mr. Wallace in 'Hardwicke's Science Gossip,' September 1867, p. 196.
[8] Mr. G. Fraser, in 'Nature,' April 1871, p. 489.

Although the obscure tints both of the upper and under sides of many butterflies no doubt serve to conceal them, yet we cannot extend this view to the brilliant and conspicuous colors on the upper surface of such species as our admiral and peacock Vanessæ, our white cabbage-butterflies (Pieris), or the great swallow-tail Papilio which haunts the open fens—for these butterflies are thus rendered visible to every living creature. In these species both sexes are alike; but in the common brimstone butterfly (*Gonepteryx rhamni*), the male is of an intense yellow, whilst the female is much paler; and in the orange-tip (*Anthocharis cardamines*) the males alone have their wings tipped with bright orange. Both the males and females in these cases are conspicuous, and it is not credible that their difference in color should stand in any relation to ordinary protection. Prof. Weismann remarks, [9] that the female of one of the Lycænæ expands her brown wings when she settles on the ground, and is then almost invisible; the male, on the other hand, as if aware of the danger incurred from the bright blue of the upper surface of his wings, rests with them closed; and this shows that the blue color cannot be in any way protective. Nevertheless, it is probable that conspicuous colors are indirectly beneficial to many species, as a warning that they are unpalatable. For in certain other cases, beauty has been gained through the imitation of other beautiful species, which inhabit the same district and enjoy an immunity from attack by being in some way offensive to their enemies; but then we have to account for the beauty of the imitated species.

[9] 'Einfluss der Isolirung auf die Artbildung,' 1872, p. 58.

As Mr. Walsh has remarked to me, the females of our orange-tip butterfly, above referred to, and of an American species (*Anth. genutia*) probably show us the primordial colors of the parent-species of the genus; for both sexes of four or five widely-distributed species are colored in nearly the same manner. As in several previous cases, we may here infer that it is the males of *Anth. cardamines* and *genutia* which have departed from the usual type of the genus. In the *Anth. sara* from California, the orange-tips to the wings have been partially developed in the female; but they are paler than in the male, and slightly different in some other respects. In an allied Indian form, the *Iphias glaucippe*, the orange-tips are fully developed in both sexes. In this Iphias, as pointed out to me by Mr. A. Butler, the under surface of the wings marvellously resembles a pale-colored leaf; and in our English orange-tip, the under surface resembles the flower-head of the wild parsley, on which the butterfly often rests at night. [10] The same reason which compels us to believe that the lower surfaces have here been colored for the sake of protection, leads us to deny that the wings have been tipped with bright orange for the same purpose, especially when this character is confined to the males.

[10] See the interesting observations by T.W. Wood, 'The Student,' Sept. 1868, p. 81.

Most Moths rest motionless during the whole or greater part of the day with their wings depressed; and the whole upper surface is often shaded and colored in an admirable manner, as Mr. Wallace has remarked, for escaping detection. The front-wings of the Bombycidæ and Noctuidæ, [11] when at rest, generally overlap and conceal the hind-wings; so that the latter might be brightly colored without much risk; and they are in fact often thus colored. During flight, moths would often be able to escape from their enemies; nevertheless, as the hind-wings are then fully exposed to view, their bright colors must generally have been acquired at some little risk. But the following fact shows how cautious we ought to be in drawing conclusions on this head. The common Yellow

Under-wings (Triphæna) often fly about during the day or early evening, and are then conspicuous from the color of their hind-wings. It would naturally be thought that this would be a source of danger; but Mr. J. Jenner Weir believes that it actually serves them as a means of escape, for birds strike at these brightly colored and fragile surfaces, instead of at the body. For instance, Mr. Weir turned into his aviary a vigorous specimen of *Triphæna pronuba*, which was instantly pursued by a robin; but the bird's attention being caught by the colored wings, the moth was not captured until after about fifty attempts, and small portions of the wings were repeatedly broken off. He tried the same experiment, in the open air, with a swallow and *T. fimbria*; but the large size of this moth probably interfered with its capture. [12] We are thus reminded of a statement made by Mr. Wallace, [13] namely, that in the Brazilian forests and Malayan islands, many common and highly-decorated butterflies are weak flyers, though furnished with a broad expanse of wing; and they "are often captured with pierced and broken wings, as if they had been seized by birds, from which they had escaped: if the wings had been much smaller in proportion to the body, it seems probable that the insect would more frequently have been struck or pierced in a vital part, and thus the increased expanse of the wings may have been indirectly beneficial."

[11] Mr. Wallace in 'Hardwicke's Science Gossip,' September 1867, p. 193.
[12] See also, on this subject, Mr. Weir's paper in 'Transactions, Entomological Society,' 1869, p. 23.
[13] 'Westminster Review,' July 1867, p. 16.

DISPLAY.

The bright colors of many butterflies and of some moths are specially arranged for display, so that they may be readily seen. During the night colors are not visible, and there can be no doubt that the nocturnal moths, taken as a body, are much less gaily decorated than butterflies, all of which are diurnal in their habits. But the moths of certain families, such as the Zygænidæ, several Sphingidæ, Uraniidæ, some Arctiidæ and Saturniidæ, fly about during the day or early evening, and many of these are extremely beautiful, being far brighter colored than the strictly nocturnal kinds. A few exceptional cases, however, of bright-colored nocturnal species have been recorded. [14]

[14] For instance, Lithosia; but Prof. Westwood ('Modern Class. of Insects,' vol. ii. p. 390) seems surprised at this case. On the relative colors of diurnal and nocturnal Lepidoptera, see ibid. pp. 333 and 392; also Harris, 'Treatise on the Insects of New England,' 1842, p. 315.

There is evidence of another kind in regard to display. Butterflies, as before remarked, elevate their wings when at rest, but whilst basking in the sunshine often alternately raise and depress them, thus exposing both surfaces to full view; and although the lower surface is often colored in an obscure manner as a protection, yet in many species it is as highly decorated as the upper surface, and sometimes in a very different manner. In some tropical species the lower surface is even more brilliantly colored than the upper. [15] In the English fritillaries (*Argynnis*) the lower surface alone is ornamented with shining silver. Nevertheless, as a general rule, the upper surface, which is probably more fully exposed, is colored more brightly and diversely than the lower. Hence the lower surface generally affords to entomologists the more useful character for detecting the affinities of the various species. Fritz Müller informs me that three species of Castnia are found near his house in S. Brazil: of two of them the hind-wings are obscure, and are always covered by the front-wings when these butterflies are at rest; but the third species has black hind-wings, beautifully spotted with red and white, and these are fully expanded and displayed whenever the butterfly rests. Other such cases could be added.

[15] Such differences between the upper and lower surfaces of the wings of several species of Papilio may be seen in the beautiful plates to Mr. Wallace's 'Memoir on the Papilionidæ of the Malayan Region,' in 'Transactions of the Linnean Society,' vol. xxv. part i. 1865.

If we now turn to the enormous group of moths, which, as I hear from Mr. Stainton, do not habitually expose the under surface of their wings to full view, we find this side very rarely colored with a brightness greater than, or even equal to, that of the upper side. Some exceptions to the rule, either real or apparent, must be noticed, as the case of Hypopyra. [16] Mr. Trimen informs me that in Guenée's great work, three moths are figured, in which the under surface is much the more brilliant. For instance, in the Australian Gastrophora the upper surface of the fore-wing is pale grayish-ochreous, while the lower surface is magnificently ornamented by an ocellus of cobalt-blue, placed in the midst of a black mark, surrounded by orange-yellow, and this by bluish-white. But the habits of these three moths are unknown; so that no explanation can be given of their unusual style of coloring. Mr. Trimen also informs me that the lower surface of the wings in certain other Geometræ [17] and quadrifid Noctuæ are either more variegated or more brightly-colored than the upper surface; but some of these species have the habit of "holding their wings quite erect over their backs, retaining them in this position for a considerable time," and thus exposing the under surface to view. Other species, when settled on the ground or herbage, now and then suddenly and slightly lift up their wings. Hence the lower surface of the wings being brighter than the upper surface in certain moths is not so anomalous as it at first appears. The Saturniidæ include some of the most beautiful of all moths, their wings being decorated, as in our British Emperor moth, with fine ocelli; and Mr. T.W. Wood [18] observes that they resemble butterflies in some of their movements; "for instance, in the gentle waving up and down of the wings as if for display, which is more characteristic of diurnal than of nocturnal Lepidoptera."

[16] See Mr. Wormald on this moth: 'Proceedings of the Entomological Society,' March 2, 1868.
[17] See also an account of the S. American genus Erateina (one of the Geometræ) in 'Transactions, Ent. Soc.' new series, vol. v. pl. xv. and xvi.
[18] 'Proc Ent. Soc. of London,' July 6, 1868, p. xxvii.

It is a singular fact that no British moths which are brilliantly colored, and, as far as I can discover, hardly any foreign species, differ much in color according to sex; though this is the case with many brilliant butterflies. The male, however, of one American moth, the *Saturnia Io*, is described as having its fore-wings deep yellow, curiously marked with purplish-red spots; whilst the wings of the female are purple-brown, marked with grey lines. [19] The British moths which differ sexually in color are all brown, or of various dull yellow tints, or nearly white. In several species the males are much darker than the females, [20] and these belong to groups which generally fly about during the afternoon. On the other hand, in many genera, as Mr. Stainton informs me, the males have the hind-wings whiter than those of the female—of which fact *Agrotis exclamationis* offers a good instance. In the Ghost Moth (*Hepialus humuli*) the difference is more strongly marked; the males being white, and the females yellow with darker markings. [21]. Mr. G. Fraser suggests ('Nature,' April 1871, p. 489) that at the season of the year when the ghost-moth appears in these northern islands, the whiteness of the males would not be needed to render them visible to the females in the twilight night.) It is probable that in these cases the males are thus rendered more conspicuous, and more easily seen by the females whilst flying about in the dusk.

[19] Harris, 'Treatise,' etc., edited by Flint, 1862, p. 395.

[20] For instance, I observe in my son's cabinet that the males are darker than the females in the *Lasiocampa quercus, Odonestis potatoria, Hypogymna dispar, Dasychira pudibunda,* and *Cycnia mendica.* In this latter species the difference in color between the two sexes is strongly marked; and Mr. Wallace informs me that we here have, as he believes, an instance of protective mimicry confined to one sex, as will hereafter be more fully explained. The white female of the Cycnia resembles the very common *Spilosoma menthrasti,* both sexes of which are white; and Mr. Stainton observed that this latter moth was rejected with utter disgust by a whole brood of young turkeys, which were fond of eating other moths; so that if the Cycnia was commonly mistaken by British birds for the Spilosoma, it would escape being devoured, and its white deceptive color would thus be highly beneficial.

[21] It is remarkable, that in the Shetland Islands the male of this moth, instead of differing widely from the female, frequently resembles her closely in color (see Mr. MacLachlan, 'Transactions, Entomological Society,' vol. ii. 1866, p. 459

From the several foregoing facts it is impossible to admit that the brilliant colors of butterflies, and of some few moths, have commonly been acquired for the sake of protection. We have seen that their colors and elegant patterns are arranged and exhibited as if for display. Hence I am led to believe that the females prefer or are most excited by the more brilliant males; for on any other supposition the males would, as far as we can see, be ornamented to no purpose. We know that ants and certain Lamellicorn beetles are capable of feeling an attachment for each other, and that ants recognize their fellows after an interval of several months. Hence there is no abstract improbability in the Lepidoptera, which probably stand nearly or quite as high in the scale as these insects, having sufficient mental capacity to admire bright colors. They certainly discover flowers by color. The Humming-bird Sphinx may often be seen to swoop down from a distance on a bunch of flowers in the midst of green foliage; and I have been assured by two persons abroad, that these moths repeatedly visit flowers painted on the walls of a room, and vainly endeavour to insert their proboscis into them. Fritz Müller informs me that several kinds of butterflies in S. Brazil show an unmistakable preference for certain colors over others: he observed that they very often visited the brilliant red flowers of five or six genera of plants, but never the white or yellow flowering species of the same and other genera, growing in the same garden; and I have received other accounts to the same effect. As I hear from Mr. Doubleday, the common white butterfly often flies down to a bit of paper on the ground, no doubt mistaking it for one of its own species. Mr. Collingwood [22] in speaking of the difficulty in collecting certain butterflies in the

Malay Archipelago, states that "a dead specimen pinned upon a conspicuous twig will often arrest an insect of the same species in its headlong flight, and bring it down within easy reach of the net, especially if it be of the opposite sex."

[22] 'Rambles of a Naturalist in the Chinese Seas,' 1868, p. 182.

The courtship of butterflies is, as before remarked, a prolonged affair. The males sometimes fight together in rivalry; and many may be seen pursuing or crowding round the same female. Unless, then, the females prefer one male to another, the pairing must be left to mere chance, and this does not appear probable. If, on the other band, the females habitually, or even occasionally, prefer the more beautiful males, the colors of the latter will have been rendered brighter by degrees, and will have been transmitted to both sexes or to one sex, according to the law of inheritance which has prevailed. The process of sexual selection will have been much facilitated, if the conclusion can be trusted, arrived at from various kinds of evidence in the supplement to the ninth chapter; namely, that the males of many Lepidoptera, at least in the imago state, greatly exceed the females in number.

Some facts, however, are opposed to the belief that female butterflies prefer the more beautiful males; thus, as I have been assured by several collectors, fresh females may frequently be seen paired with battered, faded, or dingy males; but this is a circumstance which could hardly fail often to follow from the males emerging from their cocoons earlier than the females. With moths of the family of the Bombycidæ, the sexes pair immediately after assuming the imago state; for they cannot feed, owing to the rudimentary condition of their mouths. The females, as several entomologists have remarked to me, lie in an almost torpid state, and appear not to evince the least choice in regard to their partners. This is the case with the common silk-moth (*B. mori*), as I have been told by some continental and English breeders. Dr. Wallace, who has had great experience in breeding *Bombyx cynthia*, is convinced that the females evince no choice or preference. He has kept above 300 of these moths together, and has often found the most vigorous females mated with stunted males. The reverse appears to occur seldom; for, as he believes, the more vigorous males pass over the weakly females, and are attracted by those endowed with most vitality. Nevertheless, the Bombycidæ, though obscurely-colored, are often beautiful to our eyes from their elegant and mottled shades.

I have as yet only referred to the species in which the males are brighter colored than the females, and I have attributed their beauty to the females for many generations having chosen and paired with the more attractive males. But converse cases occur, though rarely, in which the females are more brilliant than the males; and here, as I believe, the males have selected the more beautiful females, and have thus slowly added to their beauty. We do not know why in various classes of animals the males of some few species have selected the more beautiful females instead of having gladly accepted any female, as seems to be the general rule in the animal kingdom: but if, contrary to what generally occurs with the Lepidoptera, the females were much more numerous than the males, the latter would be likely to pick out the more beautiful females. Mr. Butler showed me several species of Callidryas in the British Museum, in some of which the females equalled, and in others greatly surpassed the males in beauty; for the females alone have the borders of their wings suffused with crimson and orange, and spotted with black. The plainer males of these species closely resemble each other, showing that here the females have been modified; whereas in those cases, where the males are the more ornate, it is these which have been modified, the females remaining closely alike.

In England we have some analogous cases, though not so marked. The females alone of two species of Thecla have a bright-purple or orange patch on their fore-wings. In Hipparchia the sexes do not differ much; but it is the female of *H. janira* which has a conspicuous light-brown patch on her wings; and the females of some of the other species are brighter colored than their males. Again, the females of *Colias edusa* and *hyale* have "orange or yellow spots on the black marginal border, represented in the males only by thin streaks"; and in Pieris it is the females which "are ornamented with black spots on the fore-wings, and these are only partially present in the males." Now the males of many butterflies are known to support the females during their marriage flight; but in the species just named it is the females which support the males; so that the part which the two sexes play is reversed, as is their relative beauty. Throughout the animal kingdom the males commonly take the more active share in wooing, and their beauty seems to have been increased by the females having accepted the more attractive individuals; but with these butterflies, the females take the more active part in the final marriage ceremony, so that we may suppose that they likewise do so in the wooing; and in this case we can understand how it is that they have been rendered the more beautiful. Mr. Meldola, from whom the foregoing statements have been taken, says in conclusion: "Though I am not convinced of the action of sexual selection in producing the colors of insects, it cannot be denied that these facts are strikingly corroborative of Mr. Darwin's views." [23]

[23] 'Nature,' April 27, 1871, p. 508. Mr. Meldola quotes Donzel, in 'Soc. Ent. de France,' 1837, p. 77, on the flight of butterflies whilst pairing. See also Mr. G. Fraser, in 'Nature,' April 20, 1871, p. 489, on the sexual differences of several British butterflies.

As sexual selection primarily depends on variability, a few words must be added on this subject. In respect to color there is no difficulty, for any number of highly variable Lepidoptera could be named. One good instance will suffice. Mr. Bates showed me a whole series of specimens of *Papilio sesostris* and *P. childrenæ*; in the latter the males varied much in the extent of the beautifully enamelled green patch on the fore-wings, and in the size of the white mark, and of the splendid crimson stripe on the hind-wings; so that there was a great contrast amongst the males between the most and the least gaudy. The male of *Papilio sesostris* is much less beautiful than of *P. childrenæ*; and it likewise varies a little in the size of the green patch on the fore-wings, and in the occasional appearance of the small crimson stripe on the hind-wings, borrowed, as it would seem, from its own female; for the females of this and of many other species in the Æneas group possess this crimson stripe. Hence between the brightest specimens of *P. sesostris* and the dullest of *P. childrenæ*, there was but a small interval; and it was evident that as far as mere variability is concerned, there would be no difficulty in permanently increasing the beauty of either species by means of selection. The variability is here almost confined to the male sex; but Mr. Wallace and Mr. Bates have shown [24] that the females of some species are extremely variable, the males being nearly constant. In a future chapter I shall have occasion to show that the beautiful eye-like spots, or ocelli, found on the wings of many Lepidoptera, are eminently variable. I may here add that these ocelli offer a difficulty on the theory of sexual selection; for though appearing to us so ornamental, they are never present in one sex and absent in the other, nor do they ever differ much in the two sexes. [25] This fact is at present inexplicable; but if it should hereafter be found that the formation of an ocellus is due to some change in the tissues of the wings, for instance, occurring at a very early period of development, we might expect, from what we know of the laws of inheritance, that it would be transmitted to both sexes, though arising and perfected in one sex alone.

[24] Wallace on the Papilionidæ of the Malayan Region, in 'Transact. Linn. Soc.' vol. xxv. 1865, pp. 8, 36. A striking case of a rare variety, strictly intermediate between two other well-marked female varieties, is given by Mr. Wallace. See also Mr. Bates, in 'Proc. Entomolog. Soc.' Nov. 19, 1866, p. xl.

[25] Mr. Bates was so kind as to lay this subject before the Entomological Society, and I have received answers to this effect from several entomologists.

On the whole, although many serious objections may be urged, it seems probable that most of the brilliantly-colored species of Lepidoptera owe their colors to sexual selection, excepting in certain cases, presently to be mentioned, in which conspicuous colors have been gained through mimicry as a protection. From the ardour of the male throughout the animal kingdom, he is generally willing to accept any female; and it is the female which usually exerts a choice. Hence, if sexual selection has been efficient with the Lepidoptera, the male, when the sexes differ, ought to be the more brilliantly colored, and this undoubtedly is the case. When both sexes are brilliantly colored and resemble each other, the characters acquired by the males appear to have been transmitted to both. We are led to this conclusion by cases, even within the same genus, of gradation from an extraordinary amount of difference to identity in color between the two sexes.

But it may be asked whether the difference in color between the sexes may not be accounted for by other means besides sexual selection. Thus the males and females of the same species of butterfly are in several cases known [26] to inhabit different stations, the former commonly basking in the sunshine, the latter haunting gloomy forests. It is therefore possible that different conditions of life may have acted directly on the two sexes; but this is not probable [27] as in the adult state they are exposed to different conditions during a very short period; and the larvæ of both are exposed to the same conditions. Mr. Wallace believes that the difference between the sexes is due not so much to the males having been modified, as to the females having in all or almost all cases acquired dull colors for the sake of protection. It seems to me, on the contrary, far more probable that it is the males which have been chiefly modified through sexual selection, the females having been comparatively little changed. We can thus understand how it is that the females of allied species generally resemble one another so much more closely than do the males. They thus show us approximately the primordial coloring of the parent-species of the group to which they belong. They have, however, almost always been somewhat modified by the transfer to them of some of the successive variations, through the accumulation of which the males were rendered beautiful. But I do not wish to deny that the females alone of some species may have been specially modified for protection. In most cases the males and females of distinct species will have been exposed during their prolonged larval state to different conditions, and may have been thus affected; though with the males any slight change of color thus caused will generally have been masked by the brilliant tints gained through sexual selection. When we treat of Birds, I shall have to discuss the whole question, as to how far the differences in color between the sexes are due to the males having been modified through sexual selection for ornamental purposes, or to the females having been modified through natural selection for the sake of protection, so that I will here say but little on the subject.

[26] H.W. Bates, 'The Naturalist on the Amazons,' vol. ii. 1863, p. 228. A.R. Wallace, in 'Transactions, Linnean Society,' vol. xxv. 1865, p. 10.
[27] On this whole subject see 'The Variation of Animals and Plants under Domestication,' 1868, vol. ii. chap. xxiii.

In all the cases in which the more common form of equal inheritance by both sexes has prevailed, the selection of bright-colored males would tend to make the females bright-colored; and the selection of dull-colored females would tend to make the males dull. If both processes were carried on simultaneously, they would tend to counteract each other; and the final result would depend on whether a greater number of females from being well protected by obscure colors, or a greater number of males by being brightly-colored and thus finding partners, succeeded in leaving more numerous offspring.

In order to account for the frequent transmission of characters to one sex alone, Mr. Wallace expresses his belief that the more common form of equal inheritance by both sexes can be changed through natural selection into inheritance by one sex alone, but in favour of this view I can discover no evidence. We know from what occurs under domestication that new characters often appear, which from the first are transmitted to one sex alone; and by the selection of such variations there would not be the slightest difficulty in giving bright colors to the males alone, and at the same time or subsequently, dull colors to the females alone. In this manner the females of some butterflies and moths have, it is probable, been rendered inconspicuous for the sake of protection, and widely different from their males.

I am, however, unwilling without distinct evidence to admit that two complex processes of selection, each requiring the transference of new characters to one sex alone, have been carried on with a multitude of species,—that the males have been rendered more brilliant by beating their rivals, and the females more dull-colored by having escaped from their enemies. The male, for instance, of the common brimstone butterfly (Gonepteryx), is of a far more intense yellow than the female, though she is equally conspicuous; and it does not seem probable that she specially acquired her pale tints as a protection, though it is probable that the male acquired his bright colors as a sexual attraction. The female of *Anthocharis cardamines* does not possess the beautiful orange wing-tips of the male; consequently she closely resembles the white butterflies (Pieris) so common in our gardens; but we have no evidence that this resemblance is beneficial to her. As, on the other hand, she resembles both sexes of several other species of the genus inhabiting various quarters of the world, it is probable that she has simply retained to a large extent her primordial colors.

Finally, as we have seen, various considerations lead to the conclusion that with the greater number of brilliantly-colored Lepidoptera it is the male which has been chiefly modified through sexual selection; the amount of difference between the sexes mostly depending on the form of inheritance which has prevailed. Inheritance is governed by so many unknown laws or conditions, that it seems to us to act in a capricious manner; [28] and we can thus, to a certain extent, understand how it is that with closely allied species the sexes either differ to an astonishing degree, or are identical in color. As all the successive steps in the process of variation are necessarily transmitted through the female, a greater or less number of such steps might readily become developed in her;

262

and thus we can understand the frequent gradations from an extreme difference to none at all between the sexes of allied species. These cases of gradation, it may be added, are much too common to favour the supposition that we here see females actually undergoing the process of transition and losing their brightness for the sake of protection; for we have every reason to conclude that at any one time the greater number of species are in a fixed condition.

[28] The 'Variation of Animals and Plants under Domestication,' vol. ii. chap. xii. p. 17.

MIMICRY.

This principle was first made clear in an admirable paper by Mr. Bates, [29] who thus threw a flood of light on many obscure problems. It had previously been observed that certain butterflies in S. America belonging to quite distinct families, resembled the Heliconidæ so closely in every stripe and shade of color, that they could not be distinguished save by an experienced entomologist. As the Heliconidæ are colored in their usual manner, whilst the others depart from the usual coloring of the groups to which they belong, it is clear that the latter are the imitators, and the Heliconidæ the imitated. Mr. Bates further observed that the imitating species are comparatively rare, whilst the imitated abound, and that the two sets live mingled together. From the fact of the Heliconidæ being conspicuous and beautiful insects, yet so numerous in individuals and species, he concluded that they must be protected from the attacks of enemies by some secretion or odor; and this conclusion has now been amply confirmed, [30] especially by Mr. Belt. Hence Mr. Bates inferred that the butterflies which imitate the protected species have acquired their present marvellously deceptive appearance through variation and natural selection, in order to be mistaken for the protected kinds, and thus to escape being devoured. No explanation is here attempted of the brilliant colors of the imitated, but only of the imitating butterflies. We must account for the colors of the former in the same general manner, as in the cases previously discussed in this chapter. Since the publication of Mr. Bates' paper, similar and equally striking facts have been observed by Mr. Wallace in the Malayan region, by Mr. Trimen in South Africa, and by Mr. Riley in the United States. [31]

[29] 'Transact. Linn. Soc.' vol. xxiii. 1862, p. 495.
[30] 'Proc. Entomological Soc.' Dec. 3, 1866, p. xlv.
[31] Wallace, 'Transact. Linn. Soc.' vol. xxv. 1865 p. i.; also, 'Transact. Ent. Soc.' vol. iv. (3rd series), 1867, p. 301. Trimen, 'Linn. Transact.' vol. xxvi. 1869, p. 497. Riley, 'Third Annual Report on the Noxious Insects of Missouri,' 1871, pp. 163-168. This latter essay is valuable, as Mr. Riley here discusses all the objections which have been raised against Mr. Bates's theory.

As some writers have felt much difficulty in understanding how the first steps in the process of mimicry could have been effected through natural selection, it may be well to remark that the process probably commenced long ago between forms not widely dissimilar in color. In this case even a slight variation would be beneficial, if it rendered the one species more like the other; and afterwards the imitated species might be modified to an extreme degree through sexual selection or other means, and if the changes were gradual, the imitators might easily be led along the same track, until they differed to an equally extreme degree from their original condition; and they would thus ultimately assume an appearance or coloring wholly unlike that of the other members of the group to which they belonged. It should also be remembered that many species of Lepidoptera are liable to considerable and abrupt variations in color. A few instances have been given in this chapter; and many more may be found in the papers of Mr. Bates and Mr. Wallace.

With several species the sexes are alike, and imitate the two sexes of another species. But Mr. Trimen gives, in the paper already referred to, three cases in which the sexes of the imitated form differ from each other in color, and the sexes of the imitating form differ in a like manner. Several cases have also been recorded where the females alone imitate brilliantly-colored and protected species, the males retaining "the normal aspect of their immediate congeners." It is here obvious that the successive variations by which the female has been modified have been transmitted to her alone. It is, however, probable that some of the many successive variations would have been transmitted to, and developed in, the males had not such males been eliminated by being thus rendered less attractive to the females; so that only those variations were preserved which were from the first strictly limited in their transmission to the female sex. We have a partial illustration of these remarks in a statement by Mr. Belt; [32] that the males of some of the Leptalides, which imitate protected species, still retain in a concealed manner some of their original characters. Thus in the males "the upper half of the lower wing is of a pure white, whilst all the rest of the wings is barred and spotted with black, red and yellow, like the species they mimic. The females have not this white patch, and the males usually conceal it by covering it with the upper wing, so that I cannot imagine its being of any other use to them than as an attraction in courtship, when they exhibit it to the females, and thus gratify their deep-seated preference for the normal color of the Order to which the Leptalides belong."

[32] 'The Naturalist in Nicaragua,' 1874, p. 385.

BRIGHT COLORS OF CATERPILLARS.

Whilst reflecting on the beauty of many butterflies, it occurred to me that some caterpillars were splendidly colored; and as sexual selection could not possibly have here acted, it appeared rash to attribute the beauty of the mature insect to this agency, unless the bright colors of their larvæ could be somehow explained. In the first place, it may be observed that the colors of caterpillars do not stand in any close correlation with those of the mature insect. Secondly, their bright colors do not serve in any ordinary manner as a protection. Mr. Bates informs me, as an instance of this, that the most conspicuous caterpillar which he ever beheld (that of a Sphinx) lived on the large green leaves of a tree on the open llanos of South America; it was about four inches in length, transversely banded with black and yellow, and with its head, legs, and tail of a bright red. Hence it caught the eye of any one who passed by, even at the distance of many yards, and no doubt that of every passing bird.

I then applied to Mr. Wallace, who has an innate genius for solving difficulties. After some consideration he replied: "Most caterpillars require protection, as may be inferred from some kinds being furnished with spines or irritating hairs, and from many being colored green like the leaves on which they feed, or being curiously like the twigs of the trees on which they live." Another instance of protection, furnished me by Mr. J. Mansel

Weale, may be added, namely, that there is a caterpillar of a moth which lives on the mimosas in South Africa, and fabricates for itself a case quite indistinguishable from the surrounding thorns. From such considerations Mr. Wallace thought it probable that conspicuously colored caterpillars were protected by having a nauseous taste; but as their skin is extremely tender, and as their intestines readily protrude from a wound, a slight peck from the beak of a bird would be as fatal to them as if they had been devoured. Hence, as Mr. Wallace remarks, "distastefulness alone would be insufficient to protect a caterpillar unless some outward sign indicated to its would-be destroyer that its prey was a disgusting morsel." Under these circumstances it would be highly advantageous to a caterpillar to be instantaneously and certainly recognized as unpalatable by all birds and other animals. Thus the most gaudy colors would be serviceable, and might have been gained by variation and the survival of the most easily-recognized individuals.

This hypothesis appears at first sight very bold, but when it was brought before the Entomological Society [33] it was supported by various statements; and Mr. J. Jenner Weir, who keeps a large number of birds in an aviary, informs me that he has made many trials, and finds no exception to the rule, that all caterpillars of nocturnal and retiring habits with smooth skins, all of a green color, and all which imitate twigs, are greedily devoured by his birds. The hairy and spinose kinds are invariably rejected, as were four conspicuously-colored species. When the birds rejected a caterpillar, they plainly showed, by shaking their heads, and cleansing their beaks, that they were disgusted by the taste. [34] Three conspicuous kinds of caterpillars and moths were also given to some lizards and frogs, by Mr. A. Butler, and were rejected, though other kinds were eagerly eaten. Thus the probability of Mr. Wallace's view is confirmed, namely, that certain caterpillars have been made conspicuous for their own good, so as to be easily recognized by their enemies, on nearly the same principle that poisons are sold in colored bottles by druggists for the good of man. We cannot, however, at present thus explain the elegant diversity in the colors of many caterpillars; but any species which had at some former period acquired a dull, mottled, or striped appearance, either in imitation of surrounding objects, or from the direct action of climate, etc., almost certainly would not become uniform in color, when its tints were rendered intense and bright; for in order to make a caterpillar merely conspicuous, there would be no selection in any definite direction.

[33] 'Proceedings, Entomological Society,' Dec. 3, 1866, p. xlv. and March 4, 1867, p. lxxx.

[34] See Mr. J. Jenner Weir's paper on Insects and Insectivorous Birds, in 'Transact. Ent. Soc.' 1869, p. 21; also Mr. Butler's paper, ibid. p. 27. Mr. Riley has given analogous facts in the 'Third Annual Report on the Noxious Insects of Missouri,' 1871, p. 148. Some opposed cases are, however, given by Dr. Wallace and M. H. d'Orville; see 'Zoological Record,' 1869, p. 349.

SUMMARY AND CONCLUDING REMARKS ON INSECTS.

Looking back to the several Orders, we see that the sexes often differ in various characters, the meaning of which is not in the least understood. The sexes, also, often differ in their organs of sense and means of locomotion, so that the males may quickly discover and reach the females. They differ still oftener in the males possessing diversified contrivances for retaining the females when found. We are, however, here concerned only in a secondary degree with sexual differences of these kinds.

In almost all the Orders, the males of some species, even of weak and delicate kinds, are known to be highly pugnacious; and some few are furnished with special weapons for fighting with their rivals. But the law of battle does not prevail nearly so widely with insects as with the higher animals. Hence it probably arises, that it is in only a few cases that the males have been rendered larger and stronger than the females. On the contrary, they are usually smaller, so that they may be developed within a shorter time, to be ready in large numbers for the emergence of the females.

In two families of the Homoptera and in three of the Orthoptera, the males alone possess sound-producing organs in an efficient state. These are used incessantly during the breeding-season, not only for calling the females, but apparently for charming or exciting them in rivalry with other males. No one who admits the agency of selection of any kind, will, after reading the above discussion, dispute that these musical instruments have been acquired through sexual selection. In four other Orders the members of one sex, or more commonly of both sexes, are provided with organs for producing various sounds, which apparently serve merely as call-notes. When both sexes are thus provided, the individuals which were able to make the loudest or most continuous noise would gain partners before those which were less noisy, so that their organs have probably been gained through sexual selection. It is instructive to reflect on the wonderful diversity of the means for producing sound, possessed by the males alone, or by both sexes, in no less than six Orders. We thus learn how effectual sexual selection has been in leading to modifications which sometimes, as with the Homoptera, relate to important parts of the organization.

From the reasons assigned in the last chapter, it is probable that the great horns possessed by the males of many Lamellicorn, and some other beetles, have been acquired as ornaments. From the small size of insects, we are apt to undervalue their appearance. If we could imagine a male Chalcosoma (Fig. 16), with its polished bronzed coat of mail, and its vast complex horns, magnified to the size of a horse, or even of a dog, it would be one of the most imposing animals in the world.

The coloring of insects is a complex and obscure subject. When the male differs slightly from the female, and neither are brilliantly-colored, it is probable that the sexes have varied in a slightly different manner, and that the variations have been transmitted by each sex to the same without any benefit or evil thus accruing. When the male is brilliantly-colored and differs conspicuously from the female, as with some dragon-flies and many butterflies, it is probable that he owes his colors to sexual selection; whilst the female has retained a primordial or very ancient type of coloring, slightly modified by the agencies before explained. But in some cases the female has apparently been made obscure by variations transmitted to her alone, as a means of direct protection; and it is almost certain that she has sometimes been made brilliant, so as to imitate other protected species inhabiting the same district. When the sexes resemble each other and both are obscurely colored, there is no doubt that they have been in a multitude of cases so colored for the sake of protection. So it is in some instances when both are brightly-colored, for they thus imitate protected species, or resemble surrounding objects such as flowers; or they give notice to their enemies that they are unpalatable. In other cases in which the sexes resemble each other and are both brilliant, especially when the colors are arranged for display, we may conclude that they have been gained by the male sex as an attraction, and have been transferred to the female. We are more especially led to this conclusion whenever the same type of coloration prevails throughout a whole group, and we find that the males of some species differ widely in color from the females, whilst others differ slightly or not at all with intermediate gradations connecting these extreme states.

In the same manner as bright colors have often been partially transferred from the males to the females, so it has been with the extraordinary horns of many Lamellicorn and some other beetles. So again, the sound-producing organs proper to the males of the Homoptera and Orthoptera have generally been transferred in a rudimentary, or even in a nearly perfect condition, to the females; yet not sufficiently perfect to be of any use. It is also an interesting fact, as bearing on sexual selection, that the stridulating organs of certain male Orthoptera are not fully developed until the last moult; and that the colors of certain male dragon-flies are not fully developed until some little time after their emergence from the pupal state, and when they are ready to breed.

Sexual selection implies that the more attractive individuals are preferred by the opposite sex; and as with insects, when the sexes differ, it is the male which, with some rare exceptions, is the more ornamented, and departs more from the type to which the species belongs;—and as it is the male which searches eagerly for the female, we must suppose that the females habitually or occasionally prefer the more beautiful males, and that these have thus acquired their beauty. That the females in most or all the Orders would have the power of rejecting any particular male, is probable from the many singular contrivances possessed by the males, such as great jaws, adhesive cushions, spines, elongated legs, etc., for seizing the female; for these contrivances show that there is some difficulty in the act, so that her concurrence would seem necessary. Judging from what we know of the perceptive powers and affections of various insects, there is no antecedent improbability in sexual selection having come largely into play; but we have as yet no direct evidence on this head, and some facts are opposed to the belief. Nevertheless, when we see many males pursuing the same female, we can hardly believe that the pairing is left to blind chance—that the female exerts no choice, and is not influenced by the gorgeous colors or other ornaments with which the male is decorated.

If we admit that the females of the Homoptera and Orthoptera appreciate the musical tones of their male partners, and that the various instruments have been perfected through sexual selection, there is little improbability in the females of other insects appreciating beauty in form or color, and consequently in such characters having been thus gained by the males. But from the circumstance of color being so variable, and from its having been so often modified for the sake of protection, it is difficult to decide in how large a proportion of cases sexual selection has played a part. This is more especially difficult in those Orders, such as Orthoptera, Hymenoptera, and Coleoptera, in which the two sexes rarely differ much in color; for we are then left to mere analogy. With the Coleoptera, however, as before remarked, it is in the great Lamellicorn group, placed by some authors at the head of the Order, and in which we sometimes see a mutual attachment between the sexes, that we find the males of some species possessing weapons for sexual strife, others furnished with wonderful horns, many with stridulating organs, and others ornamented with splendid metallic tints. Hence it seems probable that all these characters have been gained through the same means, namely sexual selection. With butterflies we have the best evidence, as the males sometimes take pains to display their beautiful colors; and we cannot believe that they would act thus, unless the display was of use to them in their courtship.

When we treat of Birds, we shall see that they present in their secondary sexual characters the closest analogy with insects. Thus, many male birds are highly pugnacious, and some are furnished with special weapons for fighting with their rivals. They possess organs which are used during the breeding-season for producing vocal and instrumental music. They are frequently ornamented with combs, horns, wattles and plumes of the most diversified kinds, and are decorated with beautiful colors, all evidently for the sake of display. We shall find that, as with insects, both sexes in certain groups are equally beautiful, and are equally provided with ornaments which are usually confined to the male sex. In other groups both sexes are equally plain-colored and unornamented. Lastly, in some few anomalous cases, the females are more beautiful than the males. We shall often find, in the same group of birds, every gradation from no difference between the sexes, to an extreme difference. We shall see that female birds, like female insects, often possess more or less plain traces or rudiments of characters which properly belong to the males and are of use only to them. The analogy, indeed, in all these respects between birds and insects is curiously close. Whatever explanation applies to the one class probably applies to the other; and this explanation, as we shall hereafter attempt to show in further detail, is sexual selection.

CHAPTER XII.

SECONDARY SEXUAL CHARACTERS OF
FISHES, AMPHIBIANS, AND REPTILES.

FISHES: Courtship and battles of the males—Larger size of the females—Males, bright colors and ornamental appendages; other strange characters—Colors and appendages acquired by the males during the breeding-season alone—Fishes with both sexes brilliantly colored—Protective colors—The less conspicuous colors of the female cannot be accounted for on the principle of protection—Male fishes building nests, and taking charge of the ova and young. AMPHIBIANS: Differences in structure and color between the sexes—Vocal organs. REPTILES: Chelonians—Crocodiles—Snakes, colors in some cases protective—Lizards, battles of—Ornamental appendages—Strange differences in structure between the sexes—Colors—Sexual differences almost as great as with birds.

We have now arrived at the great sub-kingdom of the Vertebrata, and will commence with the lowest class, that of fishes. The males of Plagiostomous fishes (sharks, rays) and of Chimæroid fishes are provided with claspers which serve to retain the female, like the various structures possessed by many of the lower animals. Besides the claspers, the males of many rays have clusters of strong sharp spines on their heads, and several rows along "the upper outer surface of their pectoral fins." These are present in the males of some species, which have other parts of their bodies smooth. They are only temporarily developed during the breeding-season; and Dr. Günther suspects that they are brought into action as prehensile organs by the doubling inwards and downwards of the two sides of the body. It is a remarkable fact that the females and not the males of some species, as of *Raia clavata*, have their backs studded with large hook-formed spines. [1]

[1] Yarrell's 'Hist. of British Fishes,' vol. ii. 1836, pp 417, 425, 436. Dr. Günther informs me that the spines in *R. clavata* are peculiar to the female.

The males alone of the capelin (*Mallotus villosus*, one of Salmonidæ), are provided with a ridge of closely-set, brush-like scales, by the aid of which two males, one on each side, hold the female, whilst she runs with great swiftness on the sandy beach, and there

268

deposits her spawn. [2] The widely distinct *Monacanthus scopas* presents a somewhat analogous structure. The male, as Dr. Günther informs me, has a cluster of stiff, straight spines, like those of a comb, on the sides of the tail; and these in a specimen six inches long were nearly one and a half inches in length; the female has in the same place a cluster of bristles, which may be compared with those of a tooth-brush. In another species, *M. peronii*, the male has a brush like that possessed by the female of the last species, whilst the sides of the tail in the female are smooth. In some other species of the same genus the tail can be perceived to be a little roughened in the male and perfectly smooth in the female; and lastly in others, both sexes have smooth sides.

[2] The 'American Naturalist,' April 1871, p. 119.

The males of many fish fight for the possession of the females. Thus the male stickleback (*Gasterosteus leiurus*) has been described as "mad with delight," when the female comes out of her hiding-place and surveys the nest which he has made for her. "He darts round her in every direction, then to his accumulated materials for the nest, then back again in an instant; and as she does not advance he endeavours to push her with his snout, and then tries to pull her by the tail and side-spine to the nest." [3] The males are said to be polygamists; [4] they are extraordinarily bold and pugnacious, whilst "the females are quite pacific." Their battles are at times desperate; "for these puny combatants fasten tight on each other for several seconds, tumbling over and over again until their strength appears completely exhausted." With the rough-tailed stickleback (*G. trachurus*) the males whilst fighting swim round and round each other, biting and endeavouring to pierce each other with their raised lateral spines. The same writer adds, [5] "the bite of these little furies is very severe. They also use their lateral spines with such fatal effect, that I have seen one during a battle absolutely rip his opponent quite open, so that he sank to the bottom and died." When a fish is conquered, "his gallant bearing forsakes him; his gay colors fade away; and he hides his disgrace among his peaceable companions, but is for some time the constant object of his conqueror's persecution."

[3] See Mr. R. Warington's interesting articles in 'Annals and Magazine of Natural History,' October 1852, and November 1855.
[4] Noel Humphreys, 'River Gardens,' 1857.
[5] Loudon's 'Magazine of Natural History,' vol. iii. 1830, p. 331.

The male salmon is as pugnacious as the little stickleback; and so is the male trout, as I hear from Dr. Günther. Mr. Shaw saw a violent contest between two male salmon which lasted the whole day; and Mr. R. Buist, Superintendent of Fisheries, informs me that he has often watched from the bridge at Perth the males driving away their rivals, whilst the females were spawning. The males "are constantly fighting and tearing each other on the spawning-beds, and many so injure each other as to cause the death of numbers, many being seen swimming near the banks of the river in a state of exhaustion, and apparently in a dying state." [6] Mr. Buist informs me, that in June 1868, the keeper of the Stormontfield breeding-ponds visited the northern Tyne and found about 300 dead salmon, all of which with one exception were males; and he was convinced that they had lost their lives by fighting.

[6] The 'Field,' June 29, 1867. For Mr. Shaw's Statement, see 'Edinburgh Review,' 1843. Another experienced observer (Scrope's 'Days of Salmon Fishing,' p. 60) remarks that like the stag, the male would, if he could, keep all other males away.

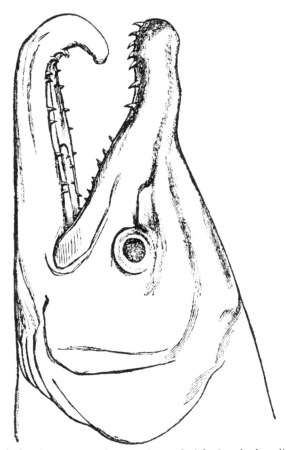

Fɪɢ. 27.—Head of male common salmon (*salmo salar*) during the breeding-season.

[This drawing, as well as all the others in the present chapter, have been executed by the well-known artist, Mr. G. Ford, from specimens in the British Museum, under the kind superintendence of Dr. Günther.]

The most curious point about the male salmon is that during the breeding-season, besides a slight change in color, "the lower jaw elongates, and a cartilaginous projection turns upwards from the point, which, when the jaws are closed, occupies a deep cavity between the intermaxillary bones of the upper jaw." [7] (Figs. 27 and 28.) In our salmon this change of structure lasts only during the breeding-season; but in the *Salmo lycaodon* of N.W. America the change, as Mr. J.K. Lord [8] believes, is permanent, and best marked in the older males which have previously ascended the rivers. In these old males the jaw becomes developed into an immense hook-like projection, and the teeth grow into regular fangs, often more than half an inch in length. With the European salmon, according to Mr. Lloyd, [9] the temporary hook-like structure serves to strengthen and protect the jaws, when one male charges another with wonderful violence; but the greatly developed teeth of the male American salmon may be compared with the tusks of many male mammals, and they indicate an offensive rather than a protective purpose.

[7] Yarrell, 'History of British Fishes,' vol. ii. 1836, p. 10.
[8] 'The Naturalist in Vancouver's Island,' vol. i. 1866, p. 54.
[9] 'Scandinavian Adventures,' vol. i. 1854, pp. 100, 104.

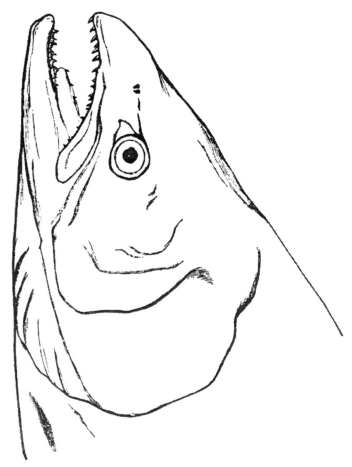

FIG. 28.—Head of female salmon.

The salmon is not the only fish in which the teeth differ in the two sexes; as this is the case with many rays. In the thornback (*Raia clavata*) the adult male has sharp, pointed teeth, directed backwards, whilst those of the female are broad and flat, and form a pavement; so that these teeth differ in the two sexes of the same species more than is usual in distinct genera of the same family. The teeth of the male become sharp only when he is adult: whilst young they are broad and flat like those of the female. As so frequently occurs with secondary sexual characters, both sexes of some species of rays (for instance *R. batis*), when adult, possess sharp pointed teeth; and here a character, proper to and primarily gained by the male, appears to have been transmitted to the offspring of both sexes. The teeth are likewise pointed in both sexes of *R. maculata*, but only when quite adult; the males acquiring them at an earlier age than the females. We shall hereafter meet with analogous cases in certain birds, in which the male acquires the plumage common to both sexes when adult, at a somewhat earlier age than does the female. With other species of rays the males even when old never possess sharp teeth, and consequently the adults of both sexes are provided with broad, flat teeth like those of the young, and like those of the mature females of the above-mentioned species. [10] As the rays are bold, strong and voracious fish, we may suspect that the males require their sharp teeth for fighting with their rivals; but as they possess many parts modified and adapted for the prehension of the female, it is possible that their teeth may be used for this purpose.

[10] See Yarrell's account of the rays in his 'History of British Fishes,' vol. ii. 1836, p. 416, with an excellent figure, and pp. 422, 432.

In regard to size, M. Carbonnier [11] maintains that the female of almost all fishes is larger than the male; and Dr. Günther does not know of a single instance in which the male is actually larger than the female. With some Cyprinodonts the male is not even half as large. As in many kinds of fishes the males habitually fight together, it is surprising that they have not generally become larger and stronger than the females through the effects of sexual selection. The males suffer from their small size, for according to M. Carbonnier, they are liable to be devoured by the females of their own species when carnivorous, and no doubt by other species. Increased size must be in some manner of more importance to the females, than strength and size are to the males for fighting with other males; and this perhaps is to allow of the production of a vast number of ova.

[11] As quoted in 'The Farmer,' 1868, p. 369.

In many species the male alone is ornamented with bright colors; or these are much brighter in the male than the female. The male, also, is sometimes provided with appendages which appear to be of no more use to him for the ordinary purposes of life, than are the tail feathers to the peacock. I am indebted for most of the following facts to the kindness of Dr. Günther. There is reason to suspect that many tropical fishes differ sexually in color and structure; and there are some striking cases with our British fishes. The male *Callionymus lyra* has been called the *gemmeous dragonet* "from its brilliant gem-like colors." When fresh caught from the sea the body is yellow of various shades, striped and spotted with vivid blue on the head; the dorsal fins are pale brown with dark longitudinal bands; the ventral, caudal, and anal fins being bluish-black. The female, or sordid dragonet, was considered by Linnæus, and by many subsequent naturalists, as a distinct species; it is of a dingy reddish-brown, with the dorsal fin brown and the other fins white. The sexes differ also in the proportional size of the head and mouth, and in the position of the eyes; [12] but the most striking difference is the extraordinary elongation in the male (Fig. 29) of the dorsal fin. Mr. W. Saville Kent remarks that this "singular appendage appears from my observations of the species in confinement, to be subservient to the same end as the wattles, crests, and other abnormal adjuncts of the male in gallinaceous birds, for the purpose of fascinating their mates." [13] The young males resemble the adult females in structure and color. Throughout the genus Callionymus, [14] the male is generally much more brightly spotted than the female, and in several species, not only the dorsal, but the anal fin is much elongated in the males.

[12] I have drawn up this description from Yarrell's 'British Fishes,' vol. i. 1836, pp. 261 and 266.

[13] 'Nature,' July 1873, p. 264.

[14] 'Catalogue of Acanth. Fishes in the British Museum,' by Dr. Günther, 1861, pp. 138-151.

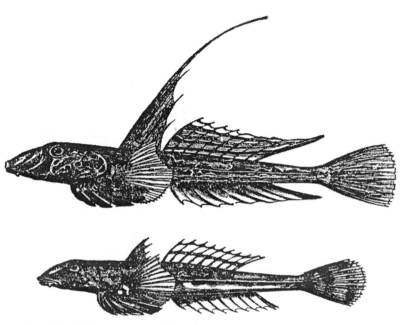

FIG. 29.—Callionymus lyra. Upper figure, male ; lower figure, female. N. B. The lower figure is more reduced than the upper.

The male of the *Cottus scorpius*, or sea-scorpion, is slenderer and smaller than the female. There is also a great difference in color between them. It is difficult, as Mr. Lloyd [15] remarks, "for any one, who has not seen this fish during the spawning-season, when its hues are brightest, to conceive the admixture of brilliant colors with which it, in other respects so ill-favoured, is at that time adorned." Both sexes of the *Labrus mixtus*, although very different in color, are beautiful; the male being orange with bright blue stripes, and the female bright red with some black spots on the back.

[15] 'Game Birds of Sweden,' etc., 1867, p. 466.

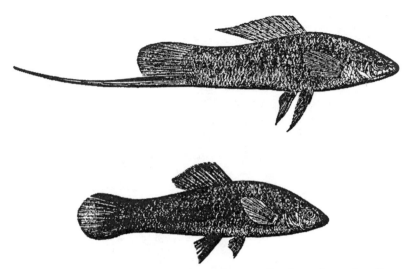

Fɪɢ. 30.—Xiphophorus Hellerii.   Upper figure, male ; lower figure, female.

In the very distinct family of the Cyprinodontidæ—inhabitants of the fresh waters of foreign lands—the sexes sometimes differ much in various characters. In the male of the *Mollienesia petenensis*, [16] the dorsal fin is greatly developed and is marked with a row of large, round, ocellated, bright-colored spots; whilst the same fin in the female is smaller, of a different shape, and marked only with irregularly curved brown spots. In the male the basal margin of the anal fin is also a little produced and dark colored. In the male of an allied form, the *Xiphophorus Hellerii* (Fig. 30), the inferior margin of the caudal fin is developed into a long filament, which, as I hear from Dr. Günther, is striped with bright colors. This filament does not contain any muscles, and apparently cannot be of any direct use to the fish. As in the case of the Callionymus, the males whilst young resemble the adult females in color and structure. Sexual differences such as these may be strictly compared with those which are so frequent with gallinaceous birds. [17]

[16] With respect to this and the following species I am indebted to Dr. Günther for information: see also his paper on the 'Fishes of Central America,' in 'Transact. Zoological Soc.' vol. vi. 1868, p. 485.

[17] Dr. Günther makes this remark; 'Catalogue of Fishes in the British Museum,' vol. iii. 1861, p. 141.

274

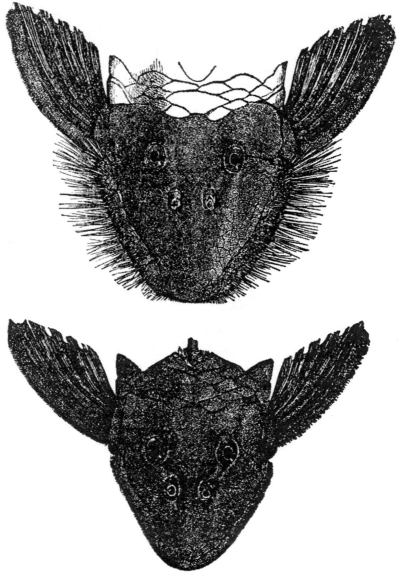

FIG. 31.—Plecostomus barbatus.
Upper figure, head of male ; lower figure, female.

In a siluroid fish, inhabiting the fresh waters of South America, the *Plecostomus barbatus* [18] (Fig. 31), the male has its mouth and inter-operculum fringed with a beard of stiff hairs, of which the female shows hardly a trace. These hairs are of the nature of scales. In another species of the same genus, soft flexible tentacles project from the front part of the head of the male, which are absent in the female. These tentacles are prolongations of the true skin, and therefore are not homologous with the stiff hairs of the former species; but it can hardly be doubted that both serve the same purpose. What this purpose may be, it is difficult to conjecture; ornament does not here seem probable, but we can hardly suppose that stiff hairs and flexible filaments can be useful in any ordinary way to the males alone. In that strange monster, the *Chimæra monstrosa*, the male has a hook-shaped bone on the top of the head, directed forwards, with its end rounded and covered with sharp spines; in the female "this crown is altogether absent," but what its use may be to the male is utterly unknown. [19]

[18] See Dr. Günther on this genus, in 'Proceedings of the Zoological Society,' 1868, p. 232.
[19] F. Buckland, in 'Land and Water,' July 1868, p. 377, with a figure. Many other cases could be added of structures peculiar to the male, of which the uses are not known.

The structures as yet referred to are permanent in the male after he has arrived at maturity; but with some Blennies, and in another allied genus, [20] a crest is developed on the head of the male only during the breeding-season, and the body at the same time becomes more brightly-colored. There can be little doubt that this crest serves as a temporary sexual ornament, for the female does not exhibit a trace of it. In other species of the same genus both sexes possess a crest, and in at least one species neither sex is thus provided. In many of the Chromidæ, for instance in Geophagus and especially in Cichla, the males, as I hear from Professor Agassiz, [21] have a conspicuous protuberance on the forehead, which is wholly wanting in the females and in the young males. Professor Agassiz adds, "I have often observed these fishes at the time of spawning when the protuberance is largest, and at other seasons when it is totally wanting, and the two sexes show no difference whatever in the outline of the profile of the head. I never could ascertain that it subserves any special function, and the Indians on the Amazon know nothing about its use." These protuberances resemble, in their periodical appearance, the fleshy carbuncles on the heads of certain birds; but whether they serve as ornaments must remain at present doubtful.

[20] Dr. Günther, 'Catalogue of Fishes,' vol. iii. pp. 221 and 240.
[21] See also 'A Journey in Brazil,' by Prof. and Mrs. Agassiz, 1868, p. 220.

I hear from Professor Agassiz and Dr. Günther, that the males of those fishes, which differ permanently in color from the females, often become more brilliant during the breeding-season. This is likewise the case with a multitude of fishes, the sexes of which are identical in color at all other seasons of the year. The tench, roach, and perch may be given as instances. The male salmon at this season is "marked on the cheeks with orange-colored stripes, which give it the appearance of a Labrus, and the body partakes of a golden orange tinge. The females are dark in color, and are commonly called black-fish." [22] An analogous and even greater change takes place with the *Salmo eriox* or bull trout; the males of the char (*S. umbla*) are likewise at this season rather brighter in color than the females. [23] The colors of the pike (*Esox reticulatus*) of the United States, especially of the male, become, during the breeding-season, exceedingly intense, brilliant, and iridescent. [24] Another striking instance out of many is afforded by the male stickleback (*Gasterosteus leiurus*), which is described by Mr. Warington, [25] as being then "beautiful beyond description." The back and eyes of the female are simply brown, and the belly white. The eyes of the male, on the other hand, are "of the most splendid green, having a metallic lustre like the green feathers of some humming-birds. The throat and belly are of a bright crimson, the back of an ashy-green, and the whole fish appears as though it were somewhat translucent and glowed with an internal incandescence." After the breeding season these colors all change, the throat and belly become of a paler red, the back more green, and the glowing tints subside.

[22] Yarrell, 'History of British Fishes,' vol. ii. 1836, pp. 10, 12, 35.
[23] W. Thompson, in 'Annals and Magazine of Natural History,' vol. vi. 1841, p. 440.
[24] 'The American Agriculturalist,' 1868, p. 100.
[25] 'Annals and Mag. of Nat. Hist.' Oct. 1852.

With respect to the courtship of fishes, other cases have been observed since the first edition of this book appeared, besides that already given of the stickleback. Mr. W.S. Kent says that the male of the *Labrus mixtus*, which, as we have seen, differs in color from the female, makes "a deep hollow in the sand of the tank, and then endeavours in the most persuasive manner to induce a female of the same species to share it with him, swimming backwards and forwards between her and the completed nest, and plainly exhibiting the greatest anxiety for her to follow." The males of *Cantharus lineatus* become, during the breeding-season, of deep leaden-black; they then retire from the shoal, and excavate a hollow as a nest. "Each male now mounts vigilant guard over his respective hollow, and vigorously attacks and drives away any other fish of the same sex. Towards his companions of the opposite sex his conduct is far different; many of the latter are now distended with spawn, and these he endeavours by all the means in his power to lure singly to his prepared hollow, and there to deposit the myriad ova with which they are laden, which he then protects and guards with the greatest care." [26]

[26] 'Nature,' May 1873, p. 25.

A more striking case of courtship, as well as of display, by the males of a Chinese Macropus has been given by M. Carbonnier, who carefully observed these fishes under confinement. [27] The males are most beautifully colored, more so than the females. During the breeding-season they contend for the possession of the females; and, in the act of courtship, expand their fins, which are spotted and ornamented with brightly colored rays, in the same manner, according to M. Carbonnier, as the peacock. They then also bound about the females with much vivacity, and appear by "l'étalage de leurs vives couleurs chercher a attirer l'attention des femelles, lesquelles ne paraissaient indifférentes à ce manége, elles nageaient avec une molle lenteur vers les mâles et semblaient se complaire dans leur voisinage." After the male has won his bride, he makes a little disc of froth by blowing air and mucus out of his mouth. He then collects the fertilized ova, dropped by the female, in his mouth; and this caused M. Carbonnier much alarm, as he thought that they were going to be devoured. But the male soon deposits them in the disc of froth, afterwards guarding them, repairing the froth, and taking care of the young when hatched. I mention these particulars because, as we shall presently see, there are fishes, the males of which hatch their eggs in their mouths; and those who do not believe in the principle of gradual evolution might ask how could such a habit have originated; but the difficulty is much diminished when we know that there are fishes which thus collect and carry the eggs; for if delayed by any cause in depositing them, the habit of hatching them in their mouths might have been acquired.

[27] 'Bulletin de la Société d'Acclimat.' Paris, July 1869, and Jan. 1870.

To return to our more immediate subject. The case stands thus: female fishes, as far as I can learn, never willingly spawn except in the presence of the males; and the males never fertilize the ova except in the presence of the females. The males fight for the possession of the females. In many species, the males whilst young resemble the females in color; but when adult become much more brilliant, and retain their colors throughout life. In other species the males become brighter than the females and otherwise more highly ornamented, only during the season of love. The males sedulously court the females, and in one case, as we have seen, take pains in displaying their beauty before them. Can it be believed that they would thus act to no purpose during their courtship? And this would be the case, unless the females exert some choice and select those males which please or excite them most. If the female exerts such choice, all the above facts on the ornamentation of the males become at once intelligible by the aid of sexual selection.

We have next to inquire whether this view of the bright colors of certain male fishes having been acquired through sexual selection can, through the law of the equal transmission of characters to both sexes, be extended to those groups in which the males and females are brilliant in the same, or nearly the same degree and manner. In such a genus as Labrus, which includes some of the most splendid fishes in the world—for instance, the Peacock Labrus (L. pavo), described, [28] with pardonable exaggeration, as formed of polished scales of gold, encrusting lapis-lazuli, rubies, sapphires, emeralds, and amethysts—we may, with much probability, accept this belief; for we have seen that the sexes in at least one species of the genus differ greatly in color. With some fishes, as with many of the lowest animals, splendid colors may be the direct result of the nature of their tissues and of the surrounding conditions, without the aid of selection of any kind. The gold-fish (Cyprinus auratus), judging from the analogy of the golden variety of the common carp, is perhaps a case in point, as it may owe its splendid colors to a single abrupt variation, due to the conditions to which this fish has been subjected under confinement. It is, however, more probable that these colors have been intensified through artificial selection, as this species has been carefully bred in China from a remote period. [29] Under natural conditions it does not seem probable that beings so highly organized as fishes, and which live under such complex relations, should become brilliantly colored without suffering some evil or receiving some benefit from so great a change, and consequently without the intervention of natural selection.

[28] Bory Saint Vincent, in 'Dict. Class. d'Hist. Nat.' tom. ix. 1826, p. 151.
[29] Owing to some remarks on this subject, made in my work 'On the Variation of Animals under Domestication,' Mr. W.F. Mayers ('Chinese Notes and Queries,' Aug. 1868, p. 123) has searched the ancient Chinese encyclopedias. He finds that gold-fish were first reared in confinement during the Sung Dynasty, which commenced A.D. 960. In the year 1129 these fishes abounded. In another place it is said that since the year 1548 there has been "produced at Hangchow a variety called the fire-fish, from its intensely red color. It is universally admired, and there is not a household where it is not cultivated, in rivalry as to its color, and as a source of profit."

What, then, are we to conclude in regard to the many fishes, both sexes of which are splendidly colored? Mr. Wallace [30] believes that the species which frequent reefs, where corals and other brightly-colored organisms abound, are brightly colored in order to escape detection by their enemies; but according to my recollection they were thus rendered highly conspicuous. In the fresh-waters of the tropics there are no brilliantly-colored corals or other organisms for the fishes to resemble; yet many species in the Amazons are beautifully colored, and many of the carnivorous Cyprinidæ in India are ornamented with "bright longitudinal lines of various tints." [31] Mr. M'Clelland, in describing these fishes, goes so far as to suppose that "the peculiar brilliancy of their colors" serves as "a better mark for king-fishers, terns, and other birds which are destined to keep the number of these fishes in check"; but at the present day few naturalists will admit that any animal has been made conspicuous as an aid to its own destruction. It is possible that certain fishes may have been rendered conspicuous in order to warn birds and beasts of prey that they were unpalatable, as explained when treating of caterpillars; but it is not, I believe, known that any fish, at least any fresh-water fish, is rejected from being distasteful to fish-devouring animals. On the whole, the most probable view in regard to the fishes, of which both sexes are brilliantly colored, is that their colors were acquired by the males as a sexual ornament, and were transferred equally, or nearly so, to the other sex.

[30] 'Westminster Review,' July 1867, p. 7.
[31] 'Indian Cyprinidæ,' by Mr. M'Clelland, 'Asiatic Researches,' vol. xix. part ii. 1839, p. 230.

We have now to consider whether, when the male differs in a marked manner from the female in color or in other ornaments, he alone has been modified, the variations being inherited by his male offspring alone; or whether the female has been specially modified and rendered inconspicuous for the sake of protection, such modifications being inherited only by the females. It is impossible to doubt that color has been gained by many fishes as a protection: no one can examine the speckled upper surface of a flounder, and overlook its resemblance to the sandy bed of the sea on which it lives. Certain fishes, moreover, can through the action of the nervous system change their colors in adaptation to surrounding objects, and that within a short time. [32] One of the most striking instances ever recorded of an animal being protected by its color (as far as it can be judged of in preserved specimens), as well as by its form, is that given by Dr. Günther [33] of a pipe-fish, which, with its reddish streaming filaments, is hardly distinguishable from the sea-weed to which it clings with its prehensile tail. But the question now under consideration is whether the females alone have been modified for this object. We can see that one sex will not be modified through natural selection for the sake of protection more than the other, supposing both to vary, unless one sex is exposed for a longer period to danger, or has less power of escaping from such danger than the other; and it does not appear that with fishes the sexes differ in these respects. As far as there is any difference, the males, from being generally smaller and from wandering more about, are exposed to greater danger than the females; and yet, when the sexes differ, the males are almost always the more conspicuously colored. The ova are fertilized immediately after being deposited; and when this process lasts for several days, as in the case of the salmon, [34] the female, during the whole time, is attended by the male. After the ova are fertilized they are, in most cases, left unprotected by both parents, so that the males and females, as far as oviposition is concerned, are equally exposed to danger, and both are equally important for the production of fertile ova; consequently the more or less brightly-colored individuals of either sex would be equally liable to be destroyed or preserved, and both would have an equal influence on the colors of their offspring.

[32] G. Pouchet, 'L'Institut.' Nov. 1, 1871, p. 134.
[33] 'Proc. Zoolog. Soc.' 1865, p. 327, pl. xiv. and xv.
[34] Yarrell, 'British Fishes,' vol. ii. p. 11.

Certain fishes, belonging to several families, make nests, and some of them take care of their young when hatched. Both sexes of the bright colored *Crenilabrus massa* and *Melops* work together in building their nests with sea-weed, shells, etc. [35] But the males of certain fishes do all the work, and afterwards take exclusive charge of the young. This is the case with the dull-colored gobies, [36] in which the sexes are not known to differ in color, and likewise with the sticklebacks (Gasterosteus), in which the males become brilliantly colored during the spawning season. The male of the smooth-tailed stickleback (*G. leiurus*) performs the duties of a nurse with exemplary care and vigilance during a long time, and is continually employed in gently leading back the young to the nest, when they stray too far. He courageously drives away all enemies including the females of his own species. It would indeed be no small relief to the male, if the female, after depositing her eggs, were immediately devoured by some enemy, for he is forced incessantly to drive her from the nest. [37]

[35] According to the observations of M. Gerbe; see Günther's 'Record of Zoolog. Literature,' 1865, p. 194.
[36] Cuvier, 'Règne Animal,' vol. ii. 1829, p. 242.
[37] See Mr. Warington's most interesting description of the habits of the *Gasterosteus leiurus* in 'Annals and Magazine of Nat. History,' November 1855.

The males of certain other fishes inhabiting South America and Ceylon, belonging to two distinct Orders, have the extraordinary habit of hatching within their mouths, or branchial cavities, the eggs laid by the females. [38] I am informed by Professor Agassiz that the males of the Amazonian species which follow this habit, "not only are generally brighter than the females, but the difference is greater at the spawning-season than at any other time." The species of Geophagus act in the same manner; and in this genus, a conspicuous protuberance becomes developed on the forehead of the males during the breeding-season. With the various species of Chromids, as Professor Agassiz likewise informs me, sexual differences in color may be observed, "whether they lay their eggs in the water among aquatic plants, or deposit them in holes, leaving them to come out without further care, or build shallow nests in the river mud, over which they sit, as our Pomotis does. It ought also to be observed that these sitters are among the brightest species in their respective families; for instance, Hygrogonus is bright green, with large black ocelli, encircled with the most brilliant red." Whether with all the species of Chromids it is the male alone which sits on the eggs is not known. It is, however, manifest that the fact of the eggs being protected or unprotected by the parents, has had little or no influence on the differences in color between the sexes. It is further manifest, in all the cases in which the males take exclusive charge of the nests and young, that the destruction of the brighter-colored males would be far more influential on the character of the race, than the destruction of the brighter-colored females; for the death of the male during the period of incubation or nursing would entail the death of the young, so that they could not inherit his peculiarities; yet, in many of these very cases the males are more conspicuously colored than the females.

[38] Prof. Wyman, in 'Proc. Boston Soc. of Nat. Hist.' Sept. 15, 1857. Also Prof. Turner, in 'Journal of Anatomy and Physiology,' Nov. 1, 1866, p. 78. Dr. Günther has likewise described other cases.

In most of the Lophobranchii (Pipe-fish, Hippocampi, etc.) the males have either marsupial sacks or hemispherical depressions on the abdomen, in which the ova laid by the female are hatched. The males also show great attachment to their young. [39] The sexes do not commonly differ much in color; but Dr. Günther believes that the male Hippocampi are rather brighter than the females. The genus Solenostoma, however, offers a curious exceptional case, [40] for the female is much more vividly-colored and spotted than the male, and she alone has a marsupial sack and hatches the eggs; so that the female of Solenostoma differs from all the other Lophobranchii in this latter respect, and from almost all other fishes, in being more brightly-colored than the male. It is improbable that this remarkable double inversion of character in the female should be an accidental coincidence. As the males of several fishes, which take exclusive charge of the eggs and young, are more brightly colored than the females, and as here the female Solenostoma takes the same charge and is brighter than the male, it might be argued that the conspicuous colors of that sex which is the more important of the two for the welfare of the offspring, must be in some manner protective. But from the large number of fishes, of which the males are either permanently or periodically brighter than the females, but whose life is not at all more important for the welfare of the species than that of the female, this view can hardly be maintained. When we treat of birds we shall meet with analogous cases, where there has been a complete inversion of the usual attributes of the two sexes, and we shall then give what appears to be the probable explanation, namely, that the males have selected the more attractive females, instead of the latter having selected, in accordance with the usual rule throughout the animal kingdom, the more attractive males.

[39] Yarrell, 'History of British Fishes,' vol. ii. 1836, pp. 329, 338.
[40] Dr. Günther, since publishing an account of this species in 'The Fishes of Zanzibar,' by Col. Playfair, 1866, p. 137, has re-examined the specimens, and has given me the above information.

On the whole we may conclude, that with most fishes, in which the sexes differ in color or in other ornamental characters, the males originally varied, with their variations transmitted to the same sex, and accumulated through sexual selection by attracting or exciting the females. In many cases, however, such characters have been transferred, either partially or completely, to the females. In other cases, again, both sexes have been colored alike for the sake of protection; but in no instance does it appear that the female alone has had her colors or other characters specially modified for this latter purpose.

The last point which need be noticed is that fishes are known to make various noises, some of which are described as being musical. Dr. Dufossé, who has especially attended to this subject, says that the sounds are voluntarily produced in several ways by different fishes: by the friction of the pharyngeal bones—by the vibration of certain muscles attached to the swim bladder, which serves as a resounding board—and by the vibration of the intrinsic muscles of the swim bladder. By this latter means the Trigla produces pure and long-drawn sounds which range over nearly an octave. But the most interesting case for us is that of two species of Ophidium, in which the males alone are provided with a sound-producing apparatus, consisting of small movable bones, with proper muscles, in connection with the swim bladder. [41] The drumming of the Umbrinas in the European seas is said to be audible from a depth of twenty fathoms; and the fishermen of Rochelle assert "that the males alone make the noise during the spawning-time; and that it is possible by imitating it, to take them without bait." [42] From this statement, and more especially from the case of Ophidium, it is almost certain that in this, the lowest class of the Vertebrata, as with so many insects and spiders, sound-producing instruments have, at least in some cases, been developed through sexual selection, as a means for bringing the sexes together.

[41] 'Comptes-Rendus,' tom. xlvi. 1858, p. 353; tom. xlvii. 1858, p. 916; tom. liv. 1862, p. 393. The noise made by the Umbrinas (*Sciæna aquila*), is said by some authors to be more like that of a flute or organ, than drumming: Dr. Zouteveen, in the Dutch translation of this work (vol. ii. p. 36), gives some further particulars on the sounds made by fishes.

[42] The Rev. C. Kingsley, in 'Nature,' May 1870, p. 40.

AMPHIBIANS.

URODELA.

I will begin with the tailed amphibians. The sexes of salamanders or newts often differ much both in color and structure. In some species prehensile claws are developed on the fore-legs of the males during the breeding-season: and at this season in the male *Triton palmipes* the hind-feet are provided with a swimming-web, which is almost completely absorbed during the winter; so that their feet then resemble those of the female. [43] This structure no doubt aids the male in his eager search and pursuit of the female. Whilst courting her he rapidly vibrates the end of his tail. With our common newts (*Triton punctatus and cristatus*) a deep, much indented crest is developed along the back and tail of the male during the breeding-season, which disappears during the winter. Mr. St. George Mivart informs me that it is not furnished with muscles, and therefore cannot be used for locomotion. As during the season of courtship it becomes edged with bright colors, there can hardly be a doubt that it is a masculine ornament. In many species the body presents strongly contrasted, though lurid tints, and these become more vivid during the breeding-season. The male, for instance, of our common little newt (*Triton punctatus*) is "brownish-grey above, passing into yellow beneath, which in the spring becomes a rich bright orange, marked everywhere with round dark spots." The edge of the crest also is then tipped with bright red or violet. The female is usually of a yellowish-brown color with scattered brown dots, and the lower surface is often quite plain. [44] The young are obscurely tinted. The ova are fertilized during the act of deposition, and are not subsequently tended by either parent. We may therefore conclude that the males have acquired their strongly-marked colors and ornamental appendages through sexual selection; these being transmitted either to the male offspring alone, or to both sexes.

[43] Bell, 'History of British Reptiles,' 2nd ed., 1849, pp. 156-159.
[44] Bell, 'History of British Reptiles,' 2nd ed., 1849, pp. 146, 151.

FIG. 32.--Triton cristatus (half natural size, from Bell's 'British Reptiles').
Upper figure, male during the breeding season ; lower figure, female.

ANURA OR BATRACHIA.

With many frogs and toads the colors evidently serve as a protection, such as the bright green tints of tree frogs and the obscure mottled shades of many terrestrial species. The most conspicuously-colored toad which I ever saw, the *Phryniscus nigricans*, [45] had the whole upper surface of the body as black as ink, with the soles of the feet and parts of the abdomen spotted with the brightest vermilion. It crawled about the bare sandy or open grassy plains of La Plata under a scorching sun, and could not fail to catch the eye of every passing creature. These colors are probably beneficial by making this animal known to all birds of prey as a nauseous mouthful.

[45] 'Zoology of the Voyage of the "Beagle,"' 1843. Bell, ibid. p. 49.

In Nicaragua there is a little frog "dressed in a bright livery of red and blue" which does not conceal itself like most other species, but hops about during the daytime, and Mr. Belt says [46] that as soon as he saw its happy sense of security, he felt sure that it was uneatable. After several trials he succeeded in tempting a young duck to snatch up a young one, but it was instantly rejected; and the duck "went about jerking its head, as if trying to throw off some unpleasant taste."

[46] 'The Naturalist in Nicaragua,' 1874, p. 321.

With respect to sexual differences of color, Dr. Günther does not know of any striking instance either with frogs or toads; yet he can often distinguish the male from the female by the tints of the former being a little more intense. Nor does he know of any striking difference in external structure between the sexes, excepting the prominences which become developed during the breeding-season on the front legs of the male, by which he is enabled to hold the female. [47] It is surprising that these animals have not acquired more strongly-marked sexual characters; for though cold-blooded their passions are strong. Dr. Günther informs me that he has several times found an unfortunate female toad dead and smothered from having been so closely embraced by three or four males. Frogs have been observed by Professor Hoffman in Giessen fighting all day long during the breeding-season, and with so much violence that one had its body ripped open.

[47] The male alone of the *Bufo sikimmensis* (Dr. Anderson, 'Proc. Zoolog. Soc.' 1871, p. 204) has two plate-like callosities on the thorax and certain rugosities on the fingers, which perhaps subserve the same end as the above-mentioned prominences.

Frogs and toads offer one interesting sexual difference, namely, in the musical powers possessed by the males; but to speak of music, when applied to the discordant and overwhelming sounds emitted by male bull-frogs and some other species, seems, according to our taste, a singularly inappropriate expression. Nevertheless, certain frogs sing in a decidedly pleasing manner. Near Rio Janeiro I used often to sit in the evening to listen to a number of little Hylæ, perched on blades of grass close to the water, which sent forth sweet chirping notes in harmony. The various sounds are emitted chiefly by the males during the breeding-season, as in the case of the croaking of our common frog. [48] In accordance with this fact the vocal organs of the males are more highly-developed than those of the females. In some genera the males alone are provided with sacs which open into the larynx. [49] For instance, in the edible frog (*Rana esculenta*) "the sacs are peculiar to the males, and become, when filled with air in the act of croaking, large globular bladders, standing out one on each side of the head, near the corners of the mouth." The croak of the male is thus rendered exceedingly powerful; whilst that of the female is only a slight groaning noise. [50] In the several genera of the family the vocal organs differ considerably in structure, and their development in all cases may be attributed to sexual selection.

[48] Bell, 'History British Reptiles,' 1849, p. 93.
[49] J. Bishop, in 'Todd's Cyclop. of Anatomy and Physiology,' vol. iv. p. 1503.
[50] Bell, ibid. pp. 112-114.

REPTILES.

CHELONIA.

Tortoises and turtles do not offer well-marked sexual differences. In some species, the tail of the male is longer than that of the female. In some, the plastron or lower surface of the shell of the male is slightly concave in relation to the back of the female. The male of the mud-turtle of the United States (*Chrysemys picta*) has claws on its front feet twice as long as those of the female; and these are used when the sexes unite. [51] With the huge tortoise of the Galapagos Islands (*Testudo nigra*) the males are said to grow to a larger size than the females: during the pairing-season, and at no other time, the male utters a hoarse bellowing noise, which can be heard at the distance of more than a hundred yards; the female, on the other hand, never uses her voice. [52]

[51] Mr. C.J. Maynard, 'The American Naturalist,' Dec. 1869, p. 555.
[52] See my 'Journal of Researches during the Voyage of the "Beagle,"' 1845, p. 384.

With the *Testudo elegans* of India, it is said "that the combats of the males may be heard at some distance, from the noise they produce in butting against each other." [53]

[53] Dr. Günther, 'Reptiles of British India,' 1864, p. 7.

CROCODILIA.

The sexes apparently do not differ in color; nor do I know that the males fight together, though this is probable, for some kinds make a prodigious display before the females. Bartram [54] describes the male alligator as striving to win the female by splashing and roaring in the midst of a lagoon, "swollen to an extent ready to burst, with its head and tail lifted up, he springs or twirls round on the surface of the water, like an Indian chief rehearsing his feats of war." During the season of love, a musky odor is emitted by the submaxillary glands of the crocodile, and pervades their haunts. [55]

[54] 'Travels through Carolina,' etc., 1791, p. 128.
[55] Owen, 'Anatomy of Vertebrates,' vol. i. 1866, p. 615.

OPHIDIA.

Dr. Günther informs me that the males are always smaller than the females, and generally have longer and slenderer tails; but he knows of no other difference in external structure. In regard to color, be can almost always distinguish the male from the female, by his more strongly-pronounced tints; thus the black zigzag band on the back of the male English viper is more distinctly defined than in the female. The difference is much plainer in the rattle-snakes of N. America, the male of which, as the keeper in the Zoological Gardens showed me, can at once be distinguished from the female by having more lurid yellow about its whole body. In S. Africa the *Bucephalus capensis* presents an analogous difference, for the female "is never so fully variegated with yellow on the sides as the male." [56] The male of the Indian *Dipsas cynodon*, on the other hand, is blackish-brown, with the belly partly black, whilst the female is reddish or yellowish-olive, with the belly either uniform yellowish or marbled with black. In the *Tragops dispar* of the same country the male is bright green, and the female bronze-colored. [57] No doubt the colors of some snakes are protective, as shown by the green tints of tree-snakes, and the various mottled shades of the species which live in sandy places; but it is doubtful whether the colors of many kinds, for instance of the common English snake and viper, serve to conceal them; and this is still more doubtful with the many foreign species which are colored with extreme elegance. The colors of certain species are very different in the adult and young states. [58]

[56] Sir Andrew Smith, 'Zoology of S. Africa: Reptilia,' 1849, pl. x.
[57] Dr. A. Günther, 'Reptiles of British India,' Ray Soc., 1864, pp. 304, 308.
[58] Dr. Stoliczka, 'Journal of Asiatic Society of Bengal,' vol. xxxix, 1870, pp. 205, 211.

During the breeding-season the anal scent-glands of snakes are in active function; [59] and so it is with the same glands in lizards, and as we have seen with the submaxillary glands of crocodiles. As the males of most animals search for the females, these odoriferous glands probably serve to excite or charm the female, rather than to guide her to the spot where the male may be found. Male snakes, though appearing so sluggish, are amorous; for many have been observed crowding round the same female, and even round her dead body. They are not known to fight together from rivalry. Their intellectual powers are higher than might have been anticipated. In the Zoological Gardens they soon learn not to strike at the iron bar with which their cages are cleaned; and Dr. Keen of Philadelphia informs me that some snakes which he kept learned after four or five times to avoid a noose, with which they were at first easily caught. An excellent observer in Ceylon, Mr. E. Layard, saw [60] a cobra thrust its head through a narrow hole and swallow a toad. "With this encumbrance he could not withdraw himself; finding this, he reluctantly disgorged the precious morsel, which began to move off; this was too much for snake philosophy to bear, and the toad was again seized, and again was the snake, after violent efforts to escape, compelled to part with its prey. This time, however, a lesson had been learnt, and the toad was seized by one leg, withdrawn, and then swallowed in triumph."

[59] Owen, 'Anatomy of Vertebrates,' vol. i. 1866, p. 615.
[60] 'Rambles in Ceylon,' in 'Annals and Magazine of Natural History,' 2nd series, vol. ix. 1852, p. 333.

The keeper in the Zoological Gardens is positive that certain snakes, for instance Crotalus and Python, distinguish him from all other persons. Cobras kept together in the same cage apparently feel some attachment towards each other. [61]

[61] Dr. Günther, 'Reptiles of British India,' 1864, p. 340.

It does not, however, follow because snakes have some reasoning power, strong passions and mutual affection, that they should likewise be endowed with sufficient taste to admire brilliant colors in their partners, so as to lead to the adornment of the species through sexual selection. Nevertheless, it is difficult to account in any other manner for the extreme beauty of certain species; for instance, of the coral-snakes of S. America, which are of a rich red with black and yellow transverse bands. I well remember how much surprise I felt at the beauty of the first coral-snake which I saw gliding across a path in Brazil. Snakes colored in this peculiar manner, as Mr. Wallace states on the authority of Dr. Günther, [62] are found nowhere else in the world except in S. America, and here no less than four genera occur. One of these, Elaps, is venomous; a second and widely-distinct genus is doubtfully venomous, and the two others are quite harmless. The species belonging to these distinct genera inhabit the same districts, and are so like each other that no one "but a naturalist would distinguish the harmless from the poisonous kinds." Hence, as Mr. Wallace believes, the innocuous kinds have probably acquired their colors as a protection, on the principle of imitation; for they would naturally be thought dangerous by their enemies. The cause, however, of the bright colors of the venomous Elaps remains to be explained, and this may perhaps be sexual selection.

[62] 'Westminster Review,' July 1st, 1867, p. 32.

Snakes produce other sounds besides hissing. The deadly *Echis carinata* has on its sides some oblique rows of scales of a peculiar structure with serrated edges; and when this snake is excited these scales are rubbed against each other, which produces "a curious prolonged, almost hissing sound." [63] With respect to the rattling of the rattle-snake, we have at last some definite information: for Professor Aughey states, [64] that on two occasions, being himself unseen, he watched from a little distance a rattle-snake coiled up with head erect, which continued to rattle at short intervals for half an hour: and at last he saw another snake approach, and when they met they paired. Hence he is satisfied that one of the uses of the rattle is to bring the sexes together. Unfortunately he did not ascertain whether it was the male or the female which remained stationary and called for the other. But it by no means follows from the above fact that the rattle may not be of use to these snakes in other ways, as a warning to animals which would otherwise attack them. Nor can I quite disbelieve the several accounts which have appeared of their thus paralyzing their prey with fear. Some other snakes also make a distinct noise by rapidly vibrating their tails against the surrounding stalks of plants; and I have myself heard this in the case of a Trigonocephalus in S. America.

[63] Dr. Anderson, 'Proc. Zoolog. Soc.' 1871, p. 196.
[64] The 'American Naturalist,' 1873, p. 85.

LACERTILIA.

The males of some, probably of many kinds of lizards, fight together from rivalry. Thus the arboreal *Anolis cristatellus* of S. America is extremely pugnacious: "During the spring and early part of the summer, two adult males rarely meet without a contest. On first seeing one another, they nod their heads up and down three or four times, and at the same time expanding the frill or pouch beneath the throat; their eyes glisten with rage, and after waving their tails from side to side for a few seconds, as if to gather energy, they dart at each other furiously, rolling over and over, and holding firmly with their teeth. The conflict generally ends in one of the combatants losing his tail, which is often devoured by the victor." The male of this species is considerably larger than the female; [65] and this, as far as Dr. Günther has been able to ascertain, is the general rule with lizards of all kinds. The male alone of the *Cyrtodactylus rubidus* of the Andaman Islands possesses pre-anal pores; and these pores, judging from analogy, probably serve to emit an odor. [66]

[65] Mr. N.L. Austen kept these animals alive for a considerable time; see 'Land and Water,' July 1867, p. 9.
[66] Stoliczka, 'Journal of the Asiatic Society of Bengal,' vol. xxxiv. 1870, p. 166.

Fig. 33.—*Sitana minor*. Male with the gular pouch expanded (from Günther's ' Reptiles of India ').

The sexes often differ greatly in various external characters. The male of the above-mentioned Anolis is furnished with a crest which runs along the back and tail, and can be erected at pleasure; but of this crest the female does not exhibit a trace. In the Indian *Cophotis ceylanica*, the female has a dorsal crest, though much less developed than in the male; and so it is, as Dr. Günther informs me, with the females of many Iguanas, Chameleons, and other lizards. In some species, however, the crest is equally developed in both sexes, as in the Iguana tuberculata. In the genus Sitana, the males alone are furnished with a large throat pouch (Fig. 33), which can be folded up like a fan, and is colored blue, black, and red; but these splendid colors are exhibited only during the pairing-season. The female does not possess even a rudiment of this appendage. In the *Anolis cristatellus*, according to Mr. Austen, the throat pouch, which is bright red marbled with yellow, is present in the female, though in a rudimental condition. Again, in certain other lizards, both sexes are equally well provided with throat pouches. Here we see with species belonging to the same group, as in so many previous cases, the same character either confined to the males, or more largely developed in them than in the females, or again equally developed in both sexes. The little lizards of the genus Draco, which glide through the air on their rib-supported parachutes, and which in the beauty of their colors baffle description, are furnished with skinny appendages to the throat "like the wattles of gallinaceous birds." These become erected when the animal is excited. They occur in both sexes, but are best developed when the male arrives at maturity, at which age the middle appendage is sometimes twice as long as the head. Most of the species likewise have a low crest running along the neck; and this is much more developed in the full-grown males than in the females or young males. [67]

[67] All the foregoing statements and quotations, in regard to Cophotis, Sitana and Draco, as well as the following facts in regard to Ceratophora and Chamæleon, are from Dr. Günther himself, or from his magnificent work on the 'Reptiles of British India,' Ray Soc., 1864, pp. 122, 130, 135.

A Chinese species is said to live in pairs during the spring; "and if one is caught, the other falls from the tree to the ground, and allows itself to be captured with impunity"—I presume from despair. [68]

[68] Mr. Swinhoe, 'Proc. Zoolog. Soc.' 1870, p. 240.

There are other and much more remarkable differences between the sexes of certain lizards. The male of Ceratophora aspera bears on the extremity of his snout an appendage half as long as the head. It is cylindrical, covered with scales, flexible, and apparently capable of erection: in the female it is quite rudimental. In a second species of the same genus a terminal scale forms a minute horn on the summit of the flexible appendage; and in a third species (*C. Stoddartii*, fig. 34) the whole appendage is converted into a horn, which is usually of a white color, but assumes a purplish tint when the animal is excited. In the adult male of this latter species the horn is half an inch in length, but it is of quite minute size in the female and in the young. These appendages, as Dr. Günther has remarked to me, may be compared with the combs of gallinaceous birds, and apparently serve as ornaments.

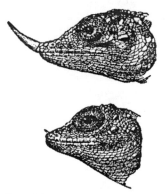

FIG. 34. — Ceratophora Stoddartii. Upper figure, male ; Lower figure, female.

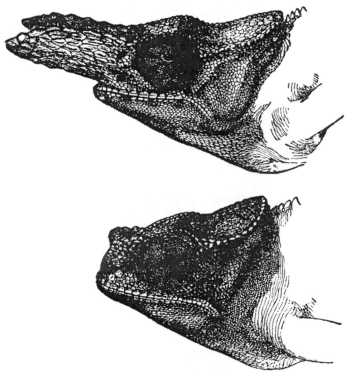

FIG. 35.—Chamæleo bifurcus. Upper figure, male ; lower figure, female.

In the genus Chamæleon we come to the acme of difference between the sexes. The upper part of the skull of the male *C. bifurcus* (Fig. 35), an inhabitant of Madagascar, is produced into two great, solid, bony projections, covered with scales like the rest of the head; and of this wonderful modification of structure the female exhibits only a rudiment.

Again, in *Chamæleo Owenii* (Fig. 36), from the West Coast of Africa, the male bears on his snout and forehead three curious horns, of which the female has not a trace. These horns consist of an excrescence of bone covered with a smooth sheath, forming part of the general integuments of the body, so that they are identical in structure with those of a bull, goat, or other sheath-horned ruminant. Although the three horns differ so much in appearance from the two great prolongations of the skull in *C. bifurcus*, we can hardly doubt that they serve the same general purpose in the economy of these two animals. The first conjecture, which will occur to every one, is that they are used by the males for fighting together; and as these animals are very quarrelsome, [69] this is probably a correct view. Mr. T.W. Wood also informs me that he once watched two individuals of *C. pumilus* fighting violently on the branch of a tree; they flung their heads about and tried to bite each other; they then rested for a time and afterwards continued their battle.

[69] Dr. Buchholz, 'Monatsbericht K. Preuss. Akad.' Jan. 1874, p. 78.

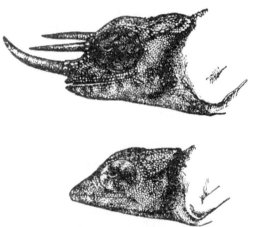

FIG. 36.—Chamælo Owenii. Upper figure, male;
lower figure, female.

With many lizards the sexes differ slightly in color, the tints and stripes of the males being brighter and more distinctly defined than in the females. This, for instance, is the case with the above Cophotis and with the *Acanthodactylus capensis* of S. Africa. In a Cordylus of the latter country, the male is either much redder or greener than the female. In the Indian *Calotes nigrilabris* there is a still greater difference; the lips also of the male are black, whilst those of the female are green. In our common little viviparous lizard (*Zootoca vivipara*) "the under side of the body and base of the tail in the male are bright orange, spotted with black; in the female these parts are pale-grayish-green without spots." [70] We have seen that the males alone of Sitana possess a throat-pouch; and this is splendidly tinted with blue, black, and red. In the *Proctotretus tenuis* of Chile the male alone is marked with spots of blue, green, and coppery-red. [71] In many cases the males retain the same colors throughout the year, but in others they become much brighter during the breeding-season; I may give as an additional instance the *Calotes maria*, which at this season has a bright red head, the rest of the body being green. [72]

[70] Bell, 'History of British Reptiles,' 2nd ed., 1849, p. 40.

[71] For Proctotretus, see 'Zoology of the Voyage of the "Beagle"; Reptiles,' by Mr. Bell, p. 8. For the Lizards of S. Africa, see 'Zoology of S. Africa: Reptiles,' by Sir Andrew Smith, pl. 25 and 39. For the Indian Calotes, see 'Reptiles of British India,' by Dr. Günther, p. 143.

[72] Günther in 'Proceedings, Zoological Society,' 1870, p. 778, with a colored figure.

Both sexes of many species are beautifully colored exactly alike; and there is no reason to suppose that such colors are protective. No doubt with the bright green kinds which live in the midst of vegetation, this color serves to conceal them; and in N. Patagonia I saw a lizard (*Proctotretus multimaculatus*) which, when frightened, flattened its body, closed its eyes, and then from its mottled tints was hardly distinguishable from the surrounding sand. But the bright colors with which so many lizards are ornamented, as well as their various curious appendages, were probably acquired by the males as an attraction, and then transmitted either to their male offspring alone, or to both sexes. Sexual selection, indeed, seems to have played almost as important a part with reptiles as with birds; and the less conspicuous colors of the females in comparison with the males cannot be accounted for, as Mr. Wallace believes to be the case with birds, by the greater exposure of the females to danger during incubation.

CHAPTER XIII.

SECONDARY SEXUAL CHARACTERS OF BIRDS.

Sexual differences—Law of battle—Special weapons—Vocal organs—Instrumental music—Love-antics and dances—Decorations, permanent and seasonal—Double and single annual moults—Display of ornaments by the males.

Secondary sexual characters are more diversified and conspicuous in birds, though not perhaps entailing more important changes of structure, than in any other class of animals. I shall, therefore, treat the subject at considerable length. Male birds sometimes, though rarely, possess special weapons for fighting with each other. They charm the female by vocal or instrumental music of the most varied kinds. They are ornamented by all sorts of combs, wattles, protuberances, horns, air-distended sacks, top-knots, naked shafts, plumes and lengthened feathers gracefully springing from all parts of the body. The beak and naked skin about the head, and the feathers, are often gorgeously colored. The males sometimes pay their court by dancing, or by fantastic antics performed either on the ground or in the air. In one instance, at least, the male emits a musky odor, which we may suppose serves to charm or excite the female; for that excellent observer, Mr. Ramsay, [1] says of the Australian musk-duck (*Biziura lobata*) that "the smell which the male emits during the summer months is confined to that sex, and in some individuals is retained throughout the year; I have never, even in the breeding-season, shot a female which had any smell of musk." So powerful is this odor during the pairing-season, that it can be detected long before the bird can be seen. [2] On the whole, birds appear to be the most aesthetic of all animals, excepting of course man, and they have nearly the same taste for the beautiful as we have. This is shown by our enjoyment of the singing of birds, and by our women, both civilized and savage, decking their heads with borrowed plumes, and using gems which are hardly more brilliantly colored than the naked skin and wattles of certain birds. In man, however, when cultivated, the sense of beauty is manifestly a far more complex feeling, and is associated with various intellectual ideas.

[1] 'Ibis,' vol. iii. (new series), 1867, p. 414.
[2] Gould, 'Handbook of the Birds of Australia,' 1865, vol. ii. p. 383.

Before treating of the sexual characters with which we are here more particularly concerned, I may just allude to certain differences between the sexes which apparently depend on differences in their habits of life; for such cases, though common in the lower, are rare in the higher classes. Two humming-birds belonging to the genus Eustephanus, which inhabit the island of Juan Fernandez, were long thought to be specifically distinct, but are now known, as Mr. Gould informs me, to be the male and female of the same species, and they differ slightly in the form of the beak. In another genus of humming-birds (*Grypus*), the beak of the male is serrated along the margin and hooked at the extremity, thus differing much from that of the female. In the Neomorpha of New Zealand, there is, as we have seen, a still wider difference in the form of the beak in relation to the manner of feeding of the two sexes. Something of the same kind has been observed with the goldfinch (*Carduelis elegans*), for I am assured by Mr. J. Jenner Weir that the bird-catchers can distinguish the males by their slightly longer beaks. The flocks of males are often found feeding on the seeds of the teazle (Dipsacus), which they can reach with their elongated beaks, whilst the females more commonly feed on the seeds of the betony or Scrophularia. With a slight difference of this kind as a foundation, we can see how the beaks of the two sexes might be made to differ greatly through natural selection. In some of the above cases, however, it is possible that the beaks of the males may have been first modified in relation to their contests with other males; and that this afterwards led to slightly changed habits of life.

LAW OF BATTLE.

Almost all male birds are extremely pugnacious, using their beaks, wings, and legs for fighting together. We see this every spring with our robins and sparrows. The smallest of all birds, namely the humming-bird, is one of the most quarrelsome. Mr. Gosse [3] describes a battle in which a pair seized hold of each other's beaks, and whirled round and round, till they almost fell to the ground; and M. Montes de Oca, in speaking or another genus of humming-bird, says that two males rarely meet without a fierce aerial encounter: when kept in cages "their fighting has mostly ended in the splitting of the tongue of one of the two, which then surely dies from being unable to feed." [4] With waders, the males of the common water-hen (*Gallinula chloropus*) "when pairing, fight violently for the females: they stand nearly upright in the water and strike with their feet." Two were seen to be thus engaged for half an hour, until one got hold of the head of the other, which would have been killed had not the observer interfered; the female all the time looking on as a quiet spectator. [5] Mr. Blyth informs me that the males of an allied bird (*Gallicrex cristatus*) are a third larger than the females, and are so pugnacious during the breeding-season that they are kept by the natives of Eastern Bengal for the sake of fighting. Various other birds are kept in India for the same purpose, for instance, the bulbuls (*Pycnonotus hœmorrhous*) which "fight with great spirit." [6]

[3] Quoted by Mr. Gould, 'Introduction to the Trochilidæ,' 1861, page 29.

[4] Gould, ibid. p. 52.

[5] W. Thompson, 'Natural History of Ireland: Birds,' vol. ii. 1850, p. 327.

[6] Jerdon, 'Birds of India,' 1863, vol. ii. p. 96.

FIG. 37.—The Ruff or Machetes pugnax (from Brehm's 'Thierleden').

The polygamous ruff (*Machetes pugnax*, Fig. 37) is notorious for his extreme pugnacity; and in the spring, the males, which are considerably larger than the females, congregate day after day at a particular spot, where the females propose to lay their eggs. The fowlers discover these spots by the turf being trampled somewhat bare. Here they fight very much like game-cocks, seizing each other with their beaks and striking with their wings. The great ruff of feathers round the neck is then erected, and according to Col. Montagu "sweeps the ground as a shield to defend the more tender parts"; and this is the only instance known to me in the case of birds of any structure serving as a shield. The ruff of feathers, however, from its varied and rich colors probably serves in chief part as an ornament. Like most pugnacious birds, they seem always ready to fight, and when closely confined, often kill each other; but Montagu observed that their pugnacity becomes greater during the spring, when the long feathers on their necks are fully developed; and at this period the least movement by any one bird provokes a general battle. [7] Of the pugnacity of web-footed birds, two instances will suffice: in Guiana "bloody fights occur during the breeding-season between the males of the wild musk-duck (*Cairina moschata*); and where these fights have occurred the river is covered for some distance with feathers." [8] Birds which seem ill-adapted for fighting engage in fierce conflicts; thus the stronger males of the pelican drive away the weaker ones, snapping with their huge beaks and giving heavy blows with their wings. Male snipe fight together, "tugging and pushing each other with their bills in the most curious manner imaginable." Some few birds are believed never to fight; this is the case, according to Audubon, with one of the woodpeckers of the United States (*Picu sauratus*), although "the hens are followed by even half a dozen of their gay suitors." [9]

[7] Macgillivray, 'History of British Birds,' vol. iv. 1852, pp. 177-181.
[8] Sir R. Schomburgk, in 'Journal of Royal Geographic Society,' vol. xiii. 1843, p. 31.
[9] 'Ornithological Biography,' vol. i. p. 191. For pelicans and snipes, see vol. iii. pp. 138, 477.

The males of many birds are larger than the females, and this no doubt is the result of the advantage gained by the larger and stronger males over their rivals during many generations. The difference in size between the two sexes is carried to an extreme point in several Australian species; thus the male musk-duck (Biziura), and the male *Cincloramphus cruralis* (allied to our pipits) are by measurement actually twice as large as their respective females. [10] With many other birds the females are larger than the males; and, as formerly remarked, the explanation often given, namely, that the females have most of the work in feeding their young, will not suffice. In some few cases, as we shall hereafter see, the females apparently have acquired their greater size and strength for the sake of conquering other females and obtaining possession of the males.

[10] Gould, 'Handbook of Birds of Australia,' vol. i. p. 395; vol. ii. p. 383.

The males of many gallinaceous birds, especially of the polygamous kinds, are furnished with special weapons for fighting with their rivals, namely spurs, which can be used with fearful effect. It has been recorded by a trustworthy writer [11] that in Derbyshire a kite struck at a game-hen accompanied by her chickens, when the cock rushed to the rescue, and drove his spur right through the eye and skull of the aggressor. The spur was with difficulty drawn from the skull, and as the kite, though dead, retained his grasp, the two birds were firmly locked together; but the cock when disentangled was very little injured. The invincible courage of the game-cock is notorious: a gentleman who long ago witnessed the brutal scene, told me that a bird had both its legs broken by some accident in the cockpit, and the owner laid a wager that if the legs could be spliced so that the bird could stand upright, he would continue fighting. This was effected on the spot, and the bird fought with undaunted courage until he received his death-stroke. In Ceylon a closely allied, wild species, the *Gallus Stanleyi*, is known to fight desperately "in defence of his seraglio," so that one of the combatants is frequently found dead. [12] An Indian partridge (*Ortygornis gularis*), the male of which is furnished with strong and sharp spurs, is so quarrelsome "that the scars of former fights disfigure the breast of almost every bird you kill." [13]

[11] Mr. Hewitt, in the 'Poultry Book' by Tegetmeier, 1866, p. 137.
[12] Layard, 'Annals and Magazine of Natural History,' vol. xiv. 1854, p. 63.
[13] Jerdon, 'Birds of India,' vol. iii. p. 574.

The males of almost all gallinaceous birds, even those which are not furnished with spurs, engage during the breeding-season in fierce conflicts. The Capercailzie and Black-cock (*Tetrao urogallus* and *T. tetrix*), which are both polygamists, have regular appointed places, where during many weeks they congregate in numbers to fight together and to display their charms before the females. Dr. W. Kovalevsky informs me that in Russia he has seen the snow all bloody on the arenas where the capercailzie have fought; and the black-cocks "make the feathers fly in every direction," when several "engage in a battle royal." The elder Brehm gives a curious account of the Balz, as the love-dances and love-songs of the Black-cock are called in Germany. The bird utters almost continuously the strangest noises: "he holds his tail up and spreads it out like a fan, he lifts up his head and neck with all the feathers erect, and stretches his wings from the body. Then he takes a few jumps in different directions, sometimes in a circle, and presses the under part of his beak so hard against the ground that the chin feathers are rubbed off. During these movements he beats his wings and turns round and round. The more ardent he grows the more lively he becomes, until at last the bird appears like a frantic creature." At such times the black-cocks are so absorbed that they become almost blind and deaf, but less so than the capercailzie: hence bird after bird may be shot on the same spot, or even caught by the hand. After performing these antics the males begin to fight: and the same black-cock, in order to prove his strength over several antagonists, will visit in the course of one morning several Balz-places, which remain the same during successive years. [14]

[14] Brehm, 'Thierleben,' 1867, B. iv. s. 351. Some of the foregoing statements are taken from L. Lloyd, 'The Game Birds of Sweden,' etc., 1867, p. 79.

The peacock with his long train appears more like a dandy than a warrior, but he sometimes engages in fierce contests: the Rev. W. Darwin Fox informs me that at some little distance from Chester two peacocks became so excited whilst fighting, that they flew over the whole city, still engaged, until they alighted on the top of St. John's tower.

The spur, in those gallinaceous birds which are thus provided, is generally single; but Polyplectron (Fig. 51) has two or more on each leg; and one of the Blood-pheasants (*Ithaginis cruentus*) has been seen with five spurs. The spurs are generally confined to the male, being represented by mere knobs or rudiments in the female; but the females of the Java peacock (*Pavo muticus*) and, as I am informed by Mr. Blyth, of the small fire-backed pheasant (*Euplocamus erythrophthalmus*) possess spurs. In Galloperdix it is usual for the males to have two spurs, and for the females to have only one on each leg. [15] Hence spurs may be considered as a masculine structure, which has been occasionally more or less transferred to the females. Like most other secondary sexual characters, the spurs are highly variable, both in number and development, in the same species.

[15] Jerdon, 'Birds of India': on Ithaginis, vol. iii. p. 523; on Galloperdix, p. 541.

Fig. 38.—Palamedea cornuta (from Brehm), showing the double wing-spurs, and the filament on the head.

Various birds have spurs on their wings. But the Egyptian goose (*Chenalopex ægyptiacus*) has only "bare obtuse knobs," and these probably show us the first steps by which true spurs have been developed in other species. In the spur-winged goose, *Plectropterus gambensis*, the males have much larger spurs than the females; and they use them, as I am informed by Mr. Bartlett, in fighting together, so that, in this case, the wing-spurs serve as sexual weapons; but according to Livingstone, they are chiefly used in the defence of the young. The Palamedea (Fig. 38) is armed with a pair of spurs on each wing; and these are such formidable weapons that a single blow has been known to drive a dog howling away. But it does not appear that the spurs in this case, or in that of some of the spur-winged rails, are larger in the male than in the female. [16] In certain plovers, however, the wing-spurs must be considered as a sexual character. Thus in the male of our common peewit (*Vanellus cristatus*) the tubercle on the shoulder of the wing becomes more prominent during the breeding-season, and the males fight together. In some species of Lobivanellus a similar tubercle becomes developed during the breeding-season "into a short horny spur." In the Australian *L. lobatus* both sexes have spurs, but these are much larger in the males than in the females. In an allied bird, the *Hoplopterus armatus*, the spurs do not increase in size during the breeding-season; but these birds have been seen in Egypt to fight together, in the same manner as our peewits, by turning suddenly in the air and striking sideways at each other, sometimes with fatal results. Thus also they drive away other enemies. [17]

[16] For the Egyptian goose, see Macgillivray, 'British Birds,' vol. iv. p. 639. For Plectropterus, Livingstone's 'Travels,' p. 254. For Palamedea, Brehm's 'Thierleben,' B. iv. s. 740. See also on this bird Azara, 'Voyages dans l'Amérique mérid.' tom. iv. 1809, pp. 179, 253.

[17] See, on our peewit, Mr. R. Carr in 'Land and Water,' Aug. 8th, 1868, p. 46. In regard to Lobivanellus, see Jerdon's 'Birds of India,' vol. iii. p. 647, and Gould's 'Handbook of Birds of Australia,' vol. ii. p. 220. For the Hoplopterus, see Mr. Allen in the 'Ibis,' vol. v. 1863, p. 156.

The season of love is that of battle; but the males of some birds, as of the game-fowl and ruff, and even the young males of the wild turkey and grouse, [18] are ready to fight whenever they meet. The presence of the female is the *teterrima belli causa*. The Bengali baboos make the pretty little males of the amadavat (*Estrelda amandava*) fight together by placing three small cages in a row, with a female in the middle; after a little time the two males are turned loose, and immediately a desperate battle ensues. [19] When many males congregate at the same appointed spot and fight together, as in the case of grouse and various other birds, they are generally attended by the females, [20] which afterwards pair with the victorious combatants. But in some cases the pairing precedes instead of succeeding the combat: thus according to Audubon, [21] several males of the Virginian goat-sucker (*Caprimulgus virgianus*) "court, in a highly entertaining manner the female, and no sooner has she made her choice, than her approved gives chase to all intruders, and drives them beyond his dominions." Generally the males try to drive away or kill their rivals before they pair. It does not, however, appear that the females invariably prefer the victorious males. I have indeed been assured by Dr. W. Kovalevsky that the female capercailzie sometimes steals away with a young male who has not dared to enter the arena with the older cocks, in the same manner as occasionally happens with the does of the red-deer in Scotland. When two males contend in presence of a single female, the victor, no doubt, commonly gains his desire; but some of these battles are caused by wandering males trying to distract the peace of an already mated pair. [22]

[18] Audubon, 'Ornithological Biography,' vol. ii. p. 492; vol. i. pp. 4-13.

[19] Mr. Blyth, 'Land and Water,' 1867, p. 212.

[20] Richardson on *Tetrao umbellus*, 'Fauna Bor. Amer.: Birds,' 1831, p. 343. L. Lloyd, 'Game Birds of Sweden,' 1867, pp. 22, 79, on the capercailzie and black-cock. Brehm, however, asserts ('Thierleben,' B. iv. s. 352) that in Germany the grey-hens do not generally attend the Balzen of the black-cocks, but this is an exception to the common rule; possibly the hens may lie hidden in the surrounding bushes, as is known to be the case with the gray-hens in Scandinavia, and with other species in N. America.

[21] 'Ornithological Biography,' vol. ii. p. 275.

[22] Brehm, 'Thierleben,' etc., B. iv. 1867, p. 990. Audubon, 'Ornithological Biography,' vol. ii. p. 492.

Even with the most pugnacious species it is probable that the pairing does not depend exclusively on the mere strength and courage of the male; for such males are generally decorated with various ornaments, which often become more brilliant during the breeding-season, and which are sedulously displayed before the females. The males also endeavour to charm or excite their mates by love-notes, songs, and antics; and the courtship is, in many instances, a prolonged affair. Hence it is not probable that the females are indifferent to the charms of the opposite sex, or that they are invariably compelled to yield to the victorious males. It is more probable that the females are excited, either before or after the conflict, by certain males, and thus unconsciously prefer them. In the case of Tetrao umbellus, a good observer [23] goes so far as to believe that the battles of the male "are all a sham, performed to show themselves to the greatest advantage before the admiring females who assemble around; for I have never been able to find a maimed hero, and seldom more than a broken feather." I shall have to recur to this subject, but I may here add that with the Tetrao cupido of the United States, about a score of males assemble at a particular spot, and, strutting about, make the whole air resound with their extraordinary noises. At the first answer from a female the males begin to fight furiously, and the weaker give way; but then, according to Audubon, both the victors and vanquished search for the female, so that the females must either then exert a choice, or the battle must be renewed. So, again, with one of the field-starlings of the United States (Sturnella ludoviciana) the males engage in fierce conflicts, "but at the sight of a female they all fly after her as if mad." [24]

[23] 'Land and Water,' July 25, 1868, p. 14.
[24] Audubon's 'Ornithological Biography;' on Tetrao cupido, vol. ii. p. 492; on the Sturnus, vol. ii. p. 219.

VOCAL AND INSTRUMENTAL MUSIC.

With birds the voice serves to express various emotions, such as distress, fear, anger, triumph, or mere happiness. It is apparently sometimes used to excite terror, as in the case of the hissing noise made by some nestling-birds. Audubon, [25] relates that a night-heron (Ardea nycticorax, Linn.), which he kept tame, used to hide itself when a cat approached, and then "suddenly start up uttering one of the most frightful cries, apparently enjoying the cat's alarm and flight." The common domestic cock clucks to the hen, and the hen to her chickens, when a dainty morsel is found. The hen, when she has laid an egg, "repeats the same note very often, and concludes with the sixth above, which she holds for a longer time" [26]; and thus she expresses her joy. Some social birds apparently call to each other for aid; and as they flit from tree to tree, the flock is kept together by chirp answering chirp. During the nocturnal migrations of geese and other water-fowl, sonorous clangs from the van may be heard in the darkness overhead, answered by clangs in the rear. Certain cries serve as danger signals, which, as the sportsman knows to his cost, are understood by the same species and by others. The domestic cock crows, and the humming-bird chirps, in triumph over a defeated rival. The true song, however, of most birds and various strange cries are chiefly uttered during the breeding-season, and serve as a charm, or merely as a call-note, to the other sex.

[25] 'Ornithological Biography,' vol. v. p. 601.
[26] The Hon. Daines Barrington, 'Philosophical Transactions,' 1773, p. 252.

Naturalists are much divided with respect to the object of the singing of birds. Few more careful observers ever lived than Montagu, and he maintained that the "males of song-birds and of many others do not in general search for the female, but, on the contrary, their business in the spring is to perch on some conspicuous spot, breathing out their full and armorous notes, which, by instinct, the female knows, and repairs to the spot to choose her mate." [27] Mr. Jenner Weir informs me that this is certainly the case with the nightingale. Bechstein, who kept birds during his whole life, asserts, "that the female canary always chooses the best singer, and that in a state of nature the female finch selects that male out of a hundred whose notes please her most." [28] There can be no doubt that birds closely attend to each other's song. Mr. Weir has told me of the case of a bullfinch which had been taught to pipe a German waltz, and who was so good a performer that he cost ten guineas; when this bird was first introduced into a room where other birds were kept and he began to sing, all the others, consisting of about twenty linnets and canaries, ranged themselves on the nearest side of their cages, and listened with the greatest interest to the new performer. Many naturalists believe that the singing of birds is almost exclusively "the effect of rivalry and emulation," and not for the sake of charming their mates. This was the opinion of Daines Barrington and White of Selborne, who both especially attended to this subject. [29] Barrington, however, admits that "superiority in song gives to birds an amazing ascendancy over others, as is well known to bird-catchers."

[27] 'Ornithological Dictionary,' 1833, p. 475.
[28] 'Naturgeschichte der Stubenvögel,' 1840, s. 4. Mr. Harrison Weir likewise writes to me:—"I am informed that the best singing males generally get a mate first, when they are bred in the same room."
[29] 'Philosophical Transactions,' 1773, p. 263. White's 'Natural History of Selborne,' 1825, vol. i. p. 246.

It is certain that there is an intense degree of rivalry between the males in their singing. Bird-fanciers match their birds to see which will sing longest; and I was told by Mr. Yarrell that a first-rate bird will sometimes sing till he drops down almost dead, or according to Bechstein, [30] quite dead from rupturing a vessel in the lungs. Whatever the cause may be, male birds, as I hear from Mr. Weir, often die suddenly during the season of song. That the habit of singing is sometimes quite independent of love is clear, for a sterile, hybrid canary-bird has been described [31] as singing whilst viewing itself in a mirror, and then dashing at its own image; it likewise attacked with fury a female canary, when put into the same cage. The jealousy excited by the act of singing is constantly taken advantage of by bird-catchers; a male, in good song, is hidden and protected, whilst a stuffed bird, surrounded by limed twigs, is exposed to view. In this manner, as Mr. Weir informs me, a man has in the course of a single day caught fifty, and in one instance, seventy, male chaffinches. The power and inclination to sing differ so greatly with birds that although the price of an ordinary male chaffinch is only sixpence, Mr. Weir saw one bird for which the bird-catcher asked three pounds; the test of a really good singer being that it will continue to sing whilst the cage is swung round the owner's head.

[30] 'Naturgesch. der Stubenvögel,' 1840, s. 252.
[31] Mr. Bold, 'Zoologist,' 1843-44, p. 659.

That male birds should sing from emulation as well as for charming the female, is not at all incompatible; and it might have been expected that these two habits would have concurred, like those of display and pugnacity. Some authors, however, argue that the song of the male cannot serve to charm the female, because the females of some few species, such as of the canary, robin, lark, and bullfinch, especially when in a state of widowhood, as Bechstein remarks, pour forth fairly melodious strains. In some of these cases the habit of singing may be in part attributed to the females having been highly fed and confined, [32] for this disturbs all the functions connected with the reproduction of the species. Many instances have already been given of the partial transference of secondary masculine characters to the female, so that it is not at all surprising that the females of some species should possess the power of song. It has also been argued, that the song of the male cannot serve as a charm, because the males of certain species, for instance of the robin, sing during the autumn. [33] But nothing is more common than for animals to take pleasure in practising whatever instinct they follow at other times for some real good. How often do we see birds which fly easily, gliding and sailing through the air obviously for pleasure? The cat plays with the captured mouse, and the cormorant with the captured fish. The weaver-bird (Ploceus), when confined in a cage, amuses itself by neatly weaving blades of grass between the wires of its cage. Birds which habitually fight during the breeding-season are generally ready to fight at all times; and the males of the capercailzie sometimes hold their *Balzen* or *leks* at the usual place of assemblage during the autumn. [34] Hence it is not at all surprising that male birds should continue singing for their own amusement after the season for courtship is over.

[32] D. Barrington, 'Philosophical Transactions,' 1773, p. 262. Bechstein, 'Stubenvögel,' 1840, s. 4.
[33] This is likewise the case with the water-ouzel; see Mr. Hepburn in the 'Zoologist,' 1845-46, p. 1068.
[34] L. Lloyd, 'Game Birds of Sweden,' 1867, p. 25.

As shown in a previous chapter, singing is to a certain extent an art, and is much improved by practice. Birds can be taught various tunes, and even the unmelodious sparrow has learnt to sing like a linnet. They acquire the song of their foster parents, [35] and sometimes that of their neighbours. [36] All the common songsters belong to the Order of Insessores, and their vocal organs are much more complex than those of most other birds; yet it is a singular fact that some of the Insessores, such as ravens, crows, and magpies, possess the proper apparatus, [37] though they never sing, and do not naturally modulate their voices to any great extent. Hunter asserts [38] that with the true songsters the muscles of the larynx are stronger in the males than in the females; but with this slight exception there is no difference in the vocal organs of the two sexes, although the males of most species sing so much better and more continuously than the females.

[35] Barrington, ibid. p. 264, Bechstein, ibid. s. 5.
[36] Dureau de la Malle gives a curious instance ('Annales des Sc. Nat.' 3rd series,

Zoolog., tom. x. p. 118) of some wild blackbirds in his garden in Paris, which naturally learnt a republican air from a caged bird.

[37] Bishop, in 'Todd's Cyclopædia of Anatomy and Physiology,' vol. iv. p. 1496.

[38] As stated by Barrington in 'Philosophical Transactions,' 1773, p. 262.

It is remarkable that only small birds properly sing. The Australian genus Menura, however, must be excepted; for the *Menura Alberti*, which is about the size of a half-grown turkey, not only mocks other birds, but "its own whistle is exceedingly beautiful and varied." The males congregate and form "*corroborying* places," where they sing, raising and spreading their tails like peacocks, and drooping their wings. [39] It is also remarkable that birds which sing well are rarely decorated with brilliant colors or other ornaments. Of our British birds, excepting the bullfinch and goldfinch, the best songsters are plain-colored. The kingfisher, bee-eater, roller, hoopoe, woodpeckers, etc., utter harsh cries; and the brilliant birds of the tropics are hardly ever songsters. [40] Hence bright colors and the power of song seem to replace each other. We can perceive that if the plumage did not vary in brightness, or if bright colors were dangerous to the species, other means would be employed to charm the females; and melody of voice offers one such means.

[39] Gould, 'Handbook to the Birds of Australia,' vol. i. 1865, pp. 308-310. See also Mr. T.W. Wood in the 'Student,' April 1870, p. 125.

[40] See remarks to this effect in Gould's 'Introduction to the Trochilidæ,' 1861, p. 22.

In some birds the vocal organs differ greatly in the two sexes. In the *Tetrao cupido* (Fig. 39) the male has two bare, orange-colored sacks, one on each side of the neck; and these are largely inflated when the male, during the breeding-season, makes his curious hollow sound, audible at a great distance. Audubon proved that the sound was intimately connected with this apparatus (which reminds us of the air-sacks on each side of the mouth of certain male frogs), for he found that the sound was much diminished when one of the sacks of a tame bird was pricked, and when both were pricked it was altogether stopped. The female has "a somewhat similar, though smaller naked space of skin on the neck; but this is not capable of inflation." [41] The male of another kind of grouse (*Tetrao urophasianus*), whilst courting the female, has his "bare yellow œsophagus inflated to a prodigious size, fully half as large as the body"; and he then utters various grating, deep, hollow tones. With his neck-feathers erect, his wings lowered, and buzzing on the ground, and his long pointed tail spread out like a fan, he displays a variety of grotesque attitudes. The œsophagus of the female is not in any way remarkable. [42]

[41] 'The Sportsman and Naturalist in Canada,' by Major W. Ross King, 1866, pp. 144-146. Mr. T.W. Wood gives in the 'Student' (April 1870, p. 116) an excellent account of the attitude and habits of this bird during its courtship. He states that the ear-tufts or neck-plumes are erected, so that they meet over the crown of the head. See his drawing, Fig. 39.

[42] Richardson, 'Fauna Bor. American: Birds,' 1831, p. 359. Audubon, ibid. vol. iv. p. 507.

FIG. 39.—Tetrao cupido: male. (T. W. Wood.)

It seems now well made out that the great throat pouch of the European male bustard (*Otis tarda*), and of at least four other species, does not, as was formerly supposed, serve to hold water, but is connected with the utterance during the breeding-season of a peculiar sound resembling "oak." [43] A crow-like bird inhabiting South America (*Cephalopterus ornatus*, Fig. 40) is called the umbrella-bird, from its immense top knot, formed of bare white quills surmounted by dark-blue plumes, which it can elevate into a great dome no less than five inches in diameter, covering the whole head. This bird has on its neck a long, thin, cylindrical fleshy appendage, which is thickly clothed with scale-like blue feathers. It probably serves in part as an ornament, but likewise as a resounding apparatus; for Mr. Bates found that it is connected "with an unusual development of the trachea and vocal organs." It is dilated when the bird utters its singularly deep, loud and long sustained fluty note. The head-crest and neck-appendage are rudimentary in the female. [44]

[43] The following papers have been lately written on this subject: Prof. A. Newton, in the 'Ibis,' 1862, p. 107; Dr. Cullen, ibid. 1865, p. 145; Mr. Flower, in 'Proc. Zool. Soc.' 1865, p. 747; and Dr. Murie, in 'Proc. Zool. Soc.' 1868, p. 471. In this latter paper an excellent figure is given of the male Australian Bustard in full display with the sack distended. It is a singular fact that the sack is not developed in all the males of the same species.
[44] Bates, 'The Naturalist on the Amazons,' 1863, vol. ii. p. 284; Wallace, in 'Proceedings, Zoological Society,' 1850, p. 206. A new species, with a still larger neck-appendage (*C. penduliger*), has lately been discovered, see 'Ibis,' vol. i. p. 457.

FIG. 40.—The Umbrella-bird or Cephalopterus ornatus, male (from Brehm.)

The vocal organs of various web-footed and wading birds are extraordinarily complex, and differ to a certain extent in the two sexes. In some cases the trachea is convoluted, like a French horn, and is deeply embedded in the sternum. In the wild swan (*Cygnus ferus*) it is more deeply embedded in the adult male than in the adult female or young male. In the male Merganser the enlarged portion of the trachea is furnished with an additional pair of muscles. [45] In one of the ducks, however, namely *Anas punctata*, the bony enlargement is only a little more developed in the male than in the female. [46] But the meaning of these differences in the trachea of the two sexes of the Anatidæ is not understood; for the male is not always the more vociferous; thus with the common duck, the male hisses, whilst the female utters a loud quack. [47] is mute; but Mr. Blyth informs me that the convolutions are not constantly present, so that perhaps they are now tending towards abortion.) In both sexes of one of the cranes (*Grus virgo*) the trachea penetrates the sternum, but presents "certain sexual modifications." In the male of the black stork there is also a well-marked sexual difference in the length and curvature of the bronchi. [48] Highly important structures have, therefore, in these cases been modified according to sex.

[45] Bishop, in Todd's 'Cyclopædia of Anatomy and Physiology,' vol. iv. p. 1499.
[46] Prof. Newton, 'Proc. Zoolog. Soc.' 1871, p. 651.
[47] The spoonbill (Platalea) has its trachea convoluted into a figure of eight, and yet this bird (Jerdon, 'Birds of India,' vol. iii. p. 763
[48] 'Elements of Comparative Anatomy,' by R. Wagner, Eng. translat. 1845, p. 111.

With respect to the swan, as given above, Yarrell's 'History of British Birds,' 2nd edition, 1845, vol. iii. p. 193.

It is often difficult to conjecture whether the many strange cries and notes uttered by male birds during the breeding-season serve as a charm or merely as a call to the female. The soft cooing of the turtle-dove and of many pigeons, it may be presumed, pleases the female. When the female of the wild turkey utters her call in the morning, the male answers by a note which differs from the gobbling noise made, when with erected feathers, rustling wings and distended wattles, he puffs and struts before her. [49] The *spel* of the black-cock certainly serves as a call to the female, for it has been known to bring four or five females from a distance to a male under confinement; but as the black-cock continues his *spel* for hours during successive days, and in the case of the capercailzie "with an agony of passion," we are led to suppose that the females which are present are thus charmed. [50] The voice of the common rook is known to alter during the breeding-season, and is therefore in some way sexual. [51] But what shall we say about the harsh screams of, for instance, some kinds of macaws; have these birds as bad taste for musical sounds as they apparently have for color, judging by the inharmonious contrast of their bright yellow and blue plumage? It is indeed possible that without any advantage being thus gained, the loud voices of many male birds may be the result of the inherited effects of the continued use of their vocal organs when excited by the strong passions of love, jealousy and rage; but to this point we shall recur when we treat of quadrupeds.

[49] C.L. Bonaparte, quoted in the 'Naturalist Library: Birds,' vol. xiv. p. 126.
[50] L. Lloyd, 'The Game Birds of Sweden,' etc., 1867, pp. 22, 81.
[51] Jenner, 'Philosophical Transactions,' 1824, p. 20.

We have as yet spoken only of the voice, but the males of various birds practise, during their courtship, what may be called instrumental music. Peacocks and Birds of Paradise rattle their quills together. Turkey-cocks scrape their wings against the ground, and some kinds of grouse thus produce a buzzing sound. Another North American grouse, the *Tetrao umbellus*, when with his tail erect, his ruffs displayed, "he shows off his finery to the females, who lie hid in the neighbourhood," drums by rapidly striking his wings together above his back, according to Mr. R. Haymond, and not, as Audubon thought, by striking them against his sides. The sound thus produced is compared by some to distant thunder, and by others to the quick roll of a drum. The female never drums, "but flies directly to the place where the male is thus engaged." The male of the Kalij-pheasant, in the Himalayas, often makes a singular drumming noise with his wings, not unlike the sound produced by shaking a stiff piece of cloth." On the west coast of Africa the little black-weavers (Ploceus?) congregate in a small party on the bushes round a small open space, and sing and glide through the air with quivering wings, "which make a rapid whirring sound like a child's rattle." One bird after another thus performs for hours together, but only during the courting-season. At this season, and at no other time, the males of certain night-jars (Caprimulgus) make a strange booming noise with their wings. The various species of woodpeckers strike a sonorous branch with their beaks, with so rapid a vibratory movement that "the head appears to be in two places at once." The sound thus produced is audible at a considerable distance but cannot be described; and I feel sure that its source would never be conjectured by any one hearing it for the first time. As this jarring sound is made chiefly during the breeding-season, it has been considered as a love-song; but it is perhaps more strictly a love-call. The female, when driven from her nest, has been observed thus to call her mate, who answered in the same manner and soon appeared. Lastly, the male hoopoe (*Upupa epops*) combines vocal and instrumental music; for during the breeding-season this bird, as Mr. Swinhoe observed, first draws in air, and then taps the end of its beak perpendicularly down against a stone or the trunk of a tree, "when the breath being forced down the tubular bill produces the correct sound." If the beak is not thus struck against some object, the sound is quite different. Air is at the same time swallowed, and the œsophagus thus becomes much swollen; and this probably acts as a resonator, not only with the hoopoe, but with pigeons and other birds. [52]

[52] For the foregoing facts see, on Birds of Paradise, Brehm, 'Thierleben,' Band iii. s. 325. On Grouse, Richardson, 'Fauna Bor. Americ.: Birds,' pp. 343 and 359; Major W. Ross King, 'The Sportsman in Canada,' 1866, p. 156; Mr. Haymond, in Prof. Cox's 'Geol. Survey of Indiana,' p. 227; Audubon, 'American Ornitholog. Biograph.' vol. i. p. 216. On the Kalij-pheasant, Jerdon, 'Birds of India,' vol. iii. p. 533. On the Weavers, Livingstone's 'Expedition to the Zambesi,' 1865, p. 425. On Woodpeckers, Macgillivray, 'Hist. of British Birds,' vol. iii. 1840, pp. 84, 88, 89, and 95. On the Hoopoe, Mr. Swinhoe, in 'Proc. Zoolog. Soc.' June 23, 1863 and 1871, p. 348. On the Night-jar, Audubon, ibid. vol. ii. p. 255, and 'American Naturalist,' 1873, p. 672. The English Night-jar likewise makes in the spring a curious noise during its rapid flight.

FIG. 41.—Outer tail-feather of Scolopax gallinago (from ' Proc. Zool. Soc.' 1858).

In the foregoing cases sounds are made by the aid of structures already present and otherwise necessary; but in the following cases certain feathers have been specially modified for the express purpose of producing sounds. The drumming, bleating, neighing, or thundering noise (as expressed by different observers) made by the common snipe (*Scolopax gallinago*) must have surprised every one who has ever heard it. This bird, during the pairing-season, flies to "perhaps a thousand feet in height," and after zig-zagging about for a time descends to the earth in a curved line, with outspread tail and quivering pinions, and surprising velocity. The sound is emitted only during this rapid descent. No one was able to explain the cause until M. Meves observed that on each side of the tail the outer feathers are peculiarly formed (Fig. 41), having a stiff sabre-shaped shaft with the oblique barbs of unusual length, the outer webs being strongly bound together. He found that by blowing on these feathers, or by fastening them to a long thin stick and waving them rapidly through the air, he could reproduce the drumming noise made by the living bird. Both sexes are furnished with these feathers, but they are generally larger in the male than in the female, and emit a deeper note. In some species, as in *S. frenata* (Fig. 42), four feathers, and in *S. javensis* (Fig. 43), no less than eight on each side of the tail are greatly modified. Different tones are emitted by the feathers of the different species when waved through the air; and the *Scolopax Wilsonii* of the United

States makes a switching noise whilst descending rapidly to the earth. [53]

[53] See M. Meves' interesting paper in 'Proc. Zool. Soc.' 1858, p. 199. For the habits of the snipe, Macgillivray, 'History of British Birds,' vol. iv. p. 371. For the American snipe, Capt. Blakiston, 'Ibis,' vol. v. 1863, p. 131.

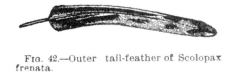

FIG. 42.—Outer tail-feather of Scolopax frenata.

FIG. 43.—Outer tail-feather of Scolopax javensis.

In the male of the *Chamæpetes unicolor* (a large gallinaceous bird of America), the first primary wing-feather is arched towards the tip and is much more attenuated than in the female. In an allied bird, the Penelope nigra, Mr. Salvin observed a male, which, whilst it flew downwards "with outstretched wings, gave forth a kind of crashing rushing noise," like the falling of a tree. [54] The male alone of one of the Indian bustards (*Sypheotides auritus*) has its primary wing-feathers greatly acuminated; and the male of an allied species is known to make a humming noise whilst courting the female. [55] In a widely different group of birds, namely Humming-birds, the males alone of certain kinds have either the shafts of their primary wing-feathers broadly dilated, or the webs abruptly excised towards the extremity. The male, for instance, of *Selasphorus platycercus*, when adult, has the first primary wing-feather (Fig. 44), thus excised. Whilst flying from flower to flower he makes "a shrill, almost whistling noise" [56]; but it did not appear to Mr. Salvin that the noise was intentionally made.

[54] Mr. Salvin, in 'Proceedings, Zoological Society,' 1867, p. 160. I am much indebted to this distinguished ornithologist for sketches of the feathers of the Chamæpetes, and for other information.
[55] Jerdon, 'Birds of India,' vol. iii. pp. 618, 621.
[56] Gould, 'Introduction to the Trochilidæ,' 1861, p. 49. Salvin, 'Proceedings, Zoological Society,' 1867, p. 160.

FIG. 44.—Primary wing-feather of a Humming-bird, the Selasphorus platycercus (from a sketch by Mr. Salvin). Upper figure, that of male; lower figure, corresponding feather of female.

Lastly, in several species of a sub-genus of Pipra or Manakin, the males, as described by Mr. Sclater, have their *secondary* wing-feathers modified in a still more remarkable manner. In the brilliantly-colored *P. deliciosa* the first three secondaries are thick-stemmed and curved towards the body; in the fourth and fifth (Fig. 45, *a*) the change is greater; and in the sixth and seventh (*b, c*) the shaft "is thickened to an extraordinary degree, forming a solid horny lump." The barbs also are greatly changed in shape, in comparison with the corresponding feathers (*d, e, f*) in the female. Even the bones of the wing, which support these singular feathers in the male, are said by Mr. Fraser to be much thickened. These little birds make an extraordinary noise, the first "sharp note being not unlike the crack of a whip." [57]

[57] Sclater, in 'Proceedings, Zoological Society,' 1860, p. 90, and in 'Ibis,' vol. iv. 1862, p. 175. Also Salvin, in 'Ibis,' 1860, p. 37.

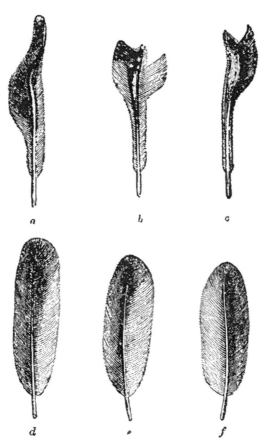

FIG. 45.—Secondary wing-feathers of *Pipra deliciosa* (from Mr. Sclater, in 'Proc. Zool. Soc.' 1860). The three upper feathers, *a, b, c,* from the male; the three lower corresponding feathers, *d, e, f,* from the female.

a and *d*, fifth secondary wing-feather of male and female, upper surface.
b and *e*, sixth secondary, upper surface. *c* and *f*, seventh secondary, lower surface.

The diversity of the sounds, both vocal and instrumental, made by the males of many birds during the breeding-season, and the diversity of the means for producing such sounds, are highly remarkable. We thus gain a high idea of their importance for sexual purposes, and are reminded of the conclusion arrived at as to insects. It is not difficult to imagine the steps by which the notes of a bird, primarily used as a mere call or for some other purpose, might have been improved into a melodious love song. In the case of the modified feathers, by which the drumming, whistling, or roaring noises are produced, we know that some birds during their courtship flutter, shake, or rattle their unmodified feathers together; and if the females were led to select the best performers, the males which possessed the strongest or thickest, or most attenuated feathers, situated on any part of the body, would be the most successful; and thus by slow degrees the feathers might be modified to almost any extent. The females, of course, would not notice each slight successive alteration in shape, but only the sounds thus produced. It is a curious fact that in the same class of animals, sounds so different as the drumming of the snipe's tail, the tapping of the woodpecker's beak, the harsh trumpet-like cry of certain water-fowl, the cooing of the turtle-dove, and the song of the nightingale, should all be pleasing to the females of the several species. But we must not judge of the tastes of distinct species by a uniform standard; nor must we judge by the standard of man's taste. Even with man, we should remember what discordant noises, the beating of tom-toms and the shrill notes of reeds, please the ears of savages. Sir S. Baker remarks, [58] that "as the stomach of the Arab prefers the raw meat and reeking liver taken hot from the animal, so does his ear prefer his equally coarse and discordant music to all other."

[58] 'The Nile Tributaries of Abyssinia,' 1867, p. 203.

LOVE ANTICS AND DANCES.

The curious love gestures of some birds have already been incidentally noticed; so that little need here be added. In Northern America large numbers of a grouse, the *Tetrao phasianellus*, meet every morning during the breeding-season on a selected level spot, and here they run round and round in a circle of about fifteen or twenty feet in diameter, so that the ground is worn quite bare, like a fairy-ring. In these Partridge-dances, as they are called by the hunters, the birds assume the strangest attitudes, and run round, some to the left and some to the right. Audubon describes the males of a heron (*Ardea herodias*) as walking about on their long legs with great dignity before the females, bidding defiance to their rivals. With one of the disgusting carrion-vultures (*Cathartes jota*) the same naturalist states that "the gesticulations and parade of the males at the beginning of the love-season are extremely ludicrous." Certain birds perform their love-antics on the wing, as we have seen with the black African weaver, instead of on the ground. During the spring our little white-throat (*Sylvia cinerea*) often rises a few feet or yards in the air above some bush, and "flutters with a fitful and fantastic motion, singing all the while, and then drops to its perch." The great English bustard throws himself into indescribably odd attitudes whilst courting the female, as has been figured by Wolf. An allied Indian bustard (*Otis bengalensis*) at such times "rises perpendicularly into the air with a hurried flapping of his wings, raising his crest and puffing out the feathers of his neck and breast, and then drops to the ground;" he repeats this manœuvre several times, at the same time humming in a peculiar tone. Such females as happen to be near "obey this saltatory summons," and when they approach he trails his wings and spreads his tail like a turkey-cock. [59]

[59] For *Tetrao phasianellus*, see Richardson, 'Fauna, Bor. America,' p. 361, and for further particulars Capt. Blakiston, 'Ibis,' 1863, p. 125. For the Cathartes and Ardea, Audubon, 'Ornithological Biography,' vol. ii. p. 51, and vol. iii. p. 89. On the White-throat, Macgillivray, 'History of British Birds,' vol. ii. p. 354. On the Indian Bustard, Jerdon, 'Birds of India,' vol. iii. p. 618.

FIG. 46.—Bower-bird, Chlamydera maculata, with bower (from Brehm).

But the most curious case is afforded by three allied genera of Australian birds, the famous Bower-birds,—no doubt the co-descendants of some ancient species which first acquired the strange instinct of constructing bowers for performing their love-antics. The bowers (Fig. 46), which, as we shall hereafter see, are decorated with feathers, shells, bones, and leaves, are built on the ground for the sole purpose of courtship, for their nests are formed in trees. Both sexes assist in the erection of the bowers, but the male is the principal workman. So strong is this instinct that it is practised under confinement, and Mr. Strange has described [60] the habits of some Satin Bower-birds which he kept in an aviary in New South Wales. "At times the male will chase the female all over the aviary, then go to the bower, pick up a gay feather or a large leaf, utter a curious kind of note, set all his feathers erect, run round the bower and become so excited that his eyes appear ready to start from his head; he continues opening first one wing then the other, uttering a low, whistling note, and, like the domestic cock, seems to be picking up something from the ground, until at last the female goes gently towards him." Captain Stokes has described the habits and "play-houses" of another species, the Great Bower-bird, which was seen "amusing itself by flying backwards and forwards, taking a shell alternately from each side, and carrying it through the archway in its mouth." These curious structures, formed solely as halls of assemblage, where both sexes amuse themselves and pay their court, must cost the birds much labor. The bower, for instance, of the Fawn-breasted species, is nearly four feet in length, eighteen inches in height, and is raised on a thick platform of sticks.

[60] Gould, 'Handbook to the Birds of Australia,' vol. i. pp. 444, 449, 455. The bower of the Satin Bower-bird may be seen in the Zoological Society's Gardens, Regent's Park.

DECORATION.

I will first discuss the cases in which the males are ornamented either exclusively or in a much higher degree than the females, and in a succeeding chapter those in which both sexes are equally ornamented, and finally the rare cases in which the female is somewhat more brightly-colored than the male. As with the artificial ornaments used by savage and civilized men, so with the natural ornaments of birds, the head is the chief seat of decoration. [61] The ornaments, as mentioned at the commencement of this chapter, are wonderfully diversified. The plumes on the front or back of the head consist of variously-shaped feathers, sometimes capable of erection or expansion, by which their beautiful colors are fully displayed. Elegant ear-tufts (Fig. 39) are occasionally present. The head is sometimes covered with velvety down, as with the pheasant; or is naked and vividly colored. The throat, also, is sometimes ornamented with a beard, wattles, or caruncles. Such appendages are generally brightly-colored, and no doubt serve as ornaments, though not always ornamental in our eyes; for whilst the male is in the act of courting the female, they often swell and assume vivid tints, as in the male turkey. At such times the fleshy appendages about the head of the male Tragopan pheasant (*Ceriornis Temminckii*) swell into a large lappet on the throat and into two horns, one on each side of the splendid top-knot; and these are then colored of the most intense blue which I have ever beheld. [62] The African hornbill (*Bucorax abyssinicus*) inflates the scarlet bladder-like wattle on its neck, and with its wings drooping and tail expanded "makes quite a grand appearance." [63] Even the iris of the eye is sometimes more brightly-colored in the male than in the female; and this is frequently the case with the beak, for instance, in our common blackbird. In *Buceros corrugatus*, the whole beak and immense casque are colored more conspicuously in the male than in the female; and "the oblique grooves upon the sides of the lower mandible are peculiar to the male sex." [64]

[61] See remarks to this effect, on the 'Feeling of Beauty among Animals,' by Mr. J. Shaw, in the 'Athenaeum,' Nov. 24th, 1866, p. 681.
[62] See Dr. Murie's account with colored figures in 'Proceedings, Zoological Society,' 1872, p. 730.
[63] Mr. Monteiro, 'Ibis,' vol. iv. 1862, p. 339.
[64] 'Land and Water,' 1868, p. 217.

The head, again, often supports fleshy appendages, filaments, and solid protuberances. These, if not common to both sexes, are always confined to the males. The solid protuberances have been described in detail by Dr. W. Marshall, [65] who shows that they are formed either of cancellated bone coated with skin, or of dermal and other tissues. With mammals true horns are always supported on the frontal bones, but with birds various bones have been modified for this purpose; and in species of the same group the protuberances may have cores of bone, or be quite destitute of them, with intermediate gradations connecting these two extremes. Hence, as Dr. Marshall justly remarks, variations of the most different kinds have served for the development through sexual selection of these ornamental appendages. Elongated feathers or plumes spring from almost every part of the body. The feathers on the throat and breast are sometimes developed into beautiful ruffs and collars. The tail-feathers are frequently increased in length; as we see in the tail-coverts of the peacock, and in the tail itself of the Argus pheasant. With the peacock even the bones of the tail have been modified to support the heavy tail-coverts. [66] The body of the Argus is not larger than that of a fowl; yet the length from the end of the beak to the extremity of the tail is no less than five feet three inches, [67] and that of the beautifully ocellated secondary wing-feathers nearly three feet. In a small African night-jar (*Cosmetornis vexillarius*) one of the primary wing-feathers, during the breeding-season, attains a length of twenty-six inches, whilst the bird itself is only ten inches in length. In another closely-allied genus of night-jars, the shafts of the elongated wing-feathers are naked, except at the extremity, where there is a disc. [68] Again, in another genus of night-jars, the tail-feathers are even still more prodigiously developed. In general the feathers of the tail are more often elongated than those of the wings, as any great elongation of the latter impedes flight. We thus see that in closely-allied birds ornaments of the same kind have been gained by the males through the development of widely different feathers.

[65] 'Ueber die Schädelhocker,' etc., 'Niederländischen Archiv. für Zoologie,' B. I. Heft 2, 1872.
[66] Dr. W. Marshall, 'Ueber den Vogelschwanz,' ibid. B. I. Heft 2, 1872.
[67] Jardine's 'Naturalist Library: Birds,' vol. xiv. p. 166.
[68] Sclater, in the 'Ibis,' vol. vi. 1864, p. 114; Livingstone, 'Expedition to the Zambesi,' 1865, p. 66.

It is a curious fact that the feathers of species belonging to very distinct groups have been modified in almost exactly the same peculiar manner. Thus the wing-feathers in one of the above-mentioned night-jars are bare along the shaft, and terminate in a disc; or are, as they are sometimes called, spoon or racket-shaped. Feathers of this kind occur in the tail of a motmot (*Eumomota superciliaris*), of a king-fisher, finch, humming-bird, parrot, several Indian drongos (Dicrurus and Edolius, in one of which the disc stands vertically), and in the tail of certain birds of paradise. In these latter birds, similar feathers, beautifully ocellated, ornament the head, as is likewise the case with some gallinaceous birds. In an Indian bustard (*Sypheotides auritus*) the feathers forming the ear-tufts, which are about four inches in length, also terminate in discs. [69] It is a most singular fact that the motmots, as Mr. Salvin has clearly shown, [70] give to their tail feathers the racket-shape by biting off the barbs, and, further, that this continued mutilation has produced a certain amount of inherited effect.

[69] Jerdon, 'Birds of India,' vol. iii. p. 620.
[70] 'Proceedings, Zoological Society,' 1873, p. 429.

Fig. 47.—Paradisea Papuana (T. W. Wood).

Again, the barbs of the feathers in various widely-distinct birds are filamentous or plumose, as with some herons, ibises, birds of paradise, and Gallinaceæ. In other cases the barbs disappear, leaving the shafts bare from end to end; and these in the tail of the *Paradisea apoda* attain a length of thirty-four inches [71]: in *P. Papuana* (Fig. 47) they are much shorter and thin. Smaller feathers when thus denuded appear like bristles, as on the breast of the turkey-cock. As any fleeting fashion in dress comes to be admired by man, so with birds a change of almost any kind in the structure or coloring of the feathers in the male appears to have been admired by the female. The fact of the feathers in widely distinct groups having been modified in an analogous manner no doubt depends primarily on all the feathers having nearly the same structure and manner of development, and consequently tending to vary in the same manner. We often see a tendency to analogous variability in the plumage of our domestic breeds belonging to distinct species. Thus top-knots have appeared in several species. In an extinct variety of the turkey, the top-knot consisted of bare quills surmounted with plumes of down, so that they somewhat resembled the racket-shaped feathers above described. In certain breeds of the pigeon and fowl the feathers are plumose, with some tendency in the shafts to be naked. In the Sebastopol goose the scapular feathers are greatly elongated, curled, or even spirally twisted, with the margins plumose. [72]

[71] Wallace, in 'Annals and Magazine of Natural History,' vol. xx. 1857, p. 416, and in his 'Malay Archipelago,' vol. ii. 1869, p. 390.
[72] See my work on 'The Variation of Animals and Plants under Domestication,' vol. i. pp. 289, 293.

FIG. 48.—Lophornis ornatus, male and female (from Brehm).

Fɪɢ. 49.—Spathura underwoodi, male and female (from Brehm).

In regard to color, hardly anything need here be said, for every one knows how splendid are the tints of many birds, and how harmoniously they are combined. The colors are often metallic and iridescent. Circular spots are sometimes surrounded by one or more differently shaded zones, and are thus converted into ocelli. Nor need much be said on the wonderful difference between the sexes of many birds. The common peacock offers a striking instance. Female birds of paradise are obscurely colored and destitute of all ornaments, whilst the males are probably the most highly decorated of all birds, and in so many different ways that they must be seen to be appreciated. The elongated and golden-orange plumes which spring from beneath the wings of the *Paradisea apoda*, when vertically erected and made to vibrate, are described as forming a sort of halo, in the centre of which the head "looks like a little emerald sun with its rays formed by the two plumes." [73] In another most beautiful species the head is bald, "and of a rich cobalt blue, crossed by several lines of black velvety feathers." [74]

[73] Quoted from M. de Lafresnaye in 'Annals and Mag. of Natural History,' vol. xiii. 1854, p. 157: see also Mr. Wallace's much fuller account in vol. xx. 1857, p. 412, and in his 'Malay Archipelago.'
[74] Wallace, 'The Malay Archipelago,' vol. ii. 1869, p. 405.

Male humming-birds (Figs. 48 and 49) almost vie with birds of paradise in their beauty, as every one will admit who has seen Mr. Gould's splendid volumes, or his rich collection. It is very remarkable in how many different ways these birds are ornamented. Almost every part of their plumage has been taken advantage of, and modified; and the modifications have been carried, as Mr. Gould showed me, to a wonderful extreme in some species belonging to nearly every sub-group. Such cases are curiously like those which we see in our fancy breeds, reared by man for the sake of ornament; certain individuals originally varied in one character, and other individuals of the same species in other characters; and these have been seized on by man and much augmented—as shown by the tail of the fantail-pigeon, the hood of the jacobin, the beak and wattle of the carrier, and so forth. The sole difference between these cases is that in the one, the result is due to man's selection, whilst in the other, as with humming-birds, birds of paradise, etc., it is due to the selection by the females of the more beautiful males.

I will mention only one other bird, remarkable from the extreme contrast in color between the sexes, namely the famous bell-bird (*Chasmorhynchus niveus*) of S. America, the note of which can be distinguished at the distance of nearly three miles, and astonishes every one when first hearing it. The male is pure white, whilst the female is dusky-green; and white is a very rare color in terrestrial species of moderate size and inoffensive habits. The male, also, as described by Waterton, has a spiral tube, nearly three inches in length, which rises from the base of the beak. It is jet-black, dotted over with minute downy feathers. This tube can be inflated with air, through a communication with the palate; and when not inflated hangs down on one side. The genus consists of four species, the males of which are very distinct, whilst the females, as described by Mr. Sclater in a very interesting paper, closely resemble each other, thus offering an excellent instance of the common rule that within the same group the males differ much more from each other than do the females. In a second species (*C. nudicollis*) the male is likewise snow-white, with the exception of a large space of naked skin on the throat and round the eyes, which during the breeding-season is of a fine green color. In a third species (*C. tricarunculatus*) the head and neck alone of the male are white, the rest of the body being chestnut-brown, and the male of this species is provided with three filamentous projections half as long as the body—one rising from the base of the beak, and the two others from the corners of the mouth. [75]

[75] Mr. Sclater, 'Intellectual Observer,' Jan. 1867. Waterton's 'Wanderings,' p. 118. See also Mr. Salvin's interesting paper, with a plate, in the 'Ibis,' 1865, p. 90.

The colored plumage and certain other ornaments of the adult males are either retained for life, or are periodically renewed during the summer and breeding-season. At this same season the beak and naked skin about the head frequently change color, as with some herons, ibises, gulls, one of the bell-birds just noticed, etc. In the white ibis, the cheeks, the inflatable skin of the throat, and the basal portion of the beak then become crimson. [76] In one of the rails, *Gallicrex cristatus*, a large red caruncle is developed during this period on the head of the male. So it is with a thin horny crest on the beak of one of the pelicans, *P. erythrorhynchus*; for, after the breeding-season, these horny crests are shed, like horns from the heads of stags, and the shore of an island in a lake in Nevada was found covered with these curious exuviæ. [77]

[76] 'Land and Water,' 1867, p. 394.
[77] Mr. D.G. Elliot, in 'Proc. Zool. Soc.' 1869, p. 589.

Changes of color in the plumage according to the season depend, firstly on a double annual moult, secondly on an actual change of color in the feathers themselves, and thirdly on their dull-colored margins being periodically shed, or on these three processes more or less combined. The shedding of the deciduary margins may be compared with the shedding of their down by very young birds; for the down in most cases arises from the summits of the first true feathers. [78]

[78] Nitzsch's 'Pterylography,' edited by P.L. Sclater, Ray Society, 1867, p. 14.

With respect to the birds which annually undergo a double moult, there are, firstly, some kinds, for instance snipes, swallow-plovers (Glareolæ), and curlews, in which the two sexes resemble each other, and do not change color at any season. I do not know whether the winter plumage is thicker and warmer than the summer plumage, but warmth seems the most probable end attained of a double moult, where there is no change of color. Secondly, there are birds, for instance, certain species of Totanus and other Grallatores, the sexes of which resemble each other, but in which the summer and winter plumage differ slightly in color. The difference, however, in these cases is so small that it can hardly be an advantage to them; and it may, perhaps, be attributed to the direct action of the different conditions to which the birds are exposed during the two seasons. Thirdly, there are many other birds the sexes of which are alike, but which are widely different in their summer and winter plumage. Fourthly, there are birds the sexes of which differ from each other in color; but the females, though moulting twice, retain the same colors throughout the year, whilst the males undergo a change of color, sometimes a great one, as with certain bustards. Fifthly and lastly, there are birds the sexes of which differ from each other in both their summer and winter plumage; but the male undergoes a greater amount of change at each recurrent season than the female—of which the ruff (*Machetes pugnax*) offers a good instance.

With respect to the cause or purpose of the differences in color between the summer and winter plumage, this may in some instances, as with the ptarmigan, [79] serve during both seasons as a protection. When the difference between the two plumages is slight it may perhaps be attributed, as already remarked, to the direct action of the conditions of life. But with many birds there can hardly be a doubt that the summer plumage is ornamental, even when both sexes are alike. We may conclude that this is the case with many herons, egrets, etc., for they acquire their beautiful plumes only during the breeding-season. Moreover, such plumes, top-knots, etc., though possessed by both sexes, are occasionally a little more developed in the male than in the female; and they resemble the plumes and ornaments possessed by the males alone of other birds. It is also known that confinement, by affecting the reproductive system of male birds, frequently checks the development of their secondary sexual characters, but has no immediate influence on any other characters; and I am informed by Mr. Bartlett that eight or nine specimens of the Knot (*Tringa canutus*) retained their unadorned winter plumage in the Zoological Gardens throughout the year, from which fact we may infer that the summer plumage, though common to both sexes, partakes of the nature of the exclusively masculine plumage of many other birds. [80]

[79] The brown mottled summer plumage of the ptarmigan is of as much importance to it, as a protection, as the white winter plumage; for in Scandinavia during the spring, when the snow has disappeared, this bird is known to suffer greatly from birds of prey, before it has acquired its summer dress: see Wilhelm von Wright, in Lloyd, 'Game Birds of Sweden,' 1867, p. 125.

[80] In regard to the previous statements on moulting, see, on snipes, etc., Macgillivray, 'Hist. Brit. Birds,' vol. iv. p. 371; on Glareolæ, curlews, and bustards, Jerdon, 'Birds of India,' vol. iii. pp. 615, 630, 683; on Totanus, ibid. p. 700; on the plumes of herons, ibid. p. 738, and Macgillivray, vol. iv. pp. 435 and 444, and Mr. Stafford Allen, in the 'Ibis,' vol. v. 1863, p. 33.

From the foregoing facts, more especially from neither sex of certain birds changing color during either annual moult, or changing so slightly that the change can hardly be of any service to them, and from the females of other species moulting twice yet retaining the same colors throughout the year, we may conclude that the habit of annually moulting twice has not been acquired in order that the male should assume an ornamental character during the breeding-season; but that the double moult, having been originally acquired for some distinct purpose, has subsequently been taken advantage of in certain cases for gaining a nuptial plumage.

It appears at first sight a surprising circumstance that some closely-allied species should regularly undergo a double annual moult, and others only a single one. The ptarmigan, for instance, moults twice or even thrice in the year, and the blackcock only once: some of the splendidly colored honey-suckers (Nectariniæ) of India and some sub-genera of obscurely colored pipits (Anthus) have a double, whilst others have only a single annual moult. [81] But the gradations in the manner of moulting, which are known to occur with various birds, show us how species, or whole groups, might have originally acquired their double annual moult, or having once gained the habit, have again lost it. With certain bustards and plovers the vernal moult is far from complete, some feathers being renewed, and some changed in color. There is also reason to believe that with certain bustards and rail-like birds, which properly undergo a double moult, some of the older males retain their nuptial plumage throughout the year. A few highly modified feathers may merely be added during the spring to the plumage, as occurs with the disc-formed tail-feathers of certain drongos (*Bhringa*) in India, and with the elongated feathers on the back, neck, and crest of certain herons. By such steps as these, the vernal moult might be rendered more and more complete, until a perfect double moult was acquired. Some of the birds of paradise retain their nuptial feathers throughout the year, and thus have only a single moult; others cast them directly after the breeding-season, and thus have a double moult; and others again cast them at this season during the first year, but not afterwards; so that these latter species are intermediate in their manner of moulting. There is also a great difference with many birds in the length of time during which the two annual plumages are retained; so that the one might come to be retained for the whole year, and the other completely lost. Thus in the spring Machetes pugnax retains his ruff for barely two months. In Natal the male widow-bird (*Chera progne*) acquires his fine plumage and long tail-feathers in December or January, and loses them in March; so that they are retained only for about three months. Most species, which undergo a double moult, keep their ornamental feathers for about six months. The male, however, of the wild *Gallus bankiva* retains his neck-hackles for nine or ten months; and when these are cast off, the underlying black feathers on the neck are fully exposed to view. But with the domesticated descendant of this species, the neck-hackles of the male are immediately replaced by new ones; so that we here see, as to part of the plumage, a double moult changed under domestication into a single moult. [82]

[81] On the moulting of the ptarmigan, see Gould's 'Birds of Great Britain.' On the honey-suckers, Jerdon, 'Birds of India,' vol. i. pp. 359, 365, 369. On the moulting of Anthus, see Blyth, in 'Ibis,' 1867, p. 32.

[82] For the foregoing statements in regard to partial moults, and on old males retaining their nuptial plumage, see Jerdon, on bustards and plovers, in 'Birds of India,' vol. iii. pp. 617, 637, 709, 711. Also Blyth in 'Land and Water,' 1867, p. 84. On the moulting of Paradisea, see an interesting article by Dr. W. Marshall, 'Archives Neerlandaises,' tom. vi. 1871. On the Vidua, 'Ibis,' vol. iii. 1861, p. 133. On the Drongo-shrikes, Jerdon, ibid. vol. i. p. 435. On the vernal moult of the *Herodias bubulcus*, Mr. S.S. Allen, in 'Ibis,' 1863, p. 33. On *Gallus bankiva*, Blyth, in 'Annals and Mag. of Natural History,' vol. i. 1848, p. 455; see, also, on this subject, my 'Variation of Animals under Domestication,' vol. i. p. 236.

The common drake (*Anas boschas*), after the breeding-season, is well known to lose his male plumage for a period of three months, during which time he assumes that of the female. The male pin-tail duck (*Anas acuta*) loses his plumage for the shorter period of six weeks or two months; and Montagu remarks that "this double moult within so short a time is a most extraordinary circumstance, that seems to bid defiance to all human reasoning." But the believer in the gradual modification of species will be far from feeling surprise at finding gradations of all kinds. If the male pin-tail were to acquire his new plumage within a still shorter period, the new male feathers would almost necessarily be mingled with the old, and both with some proper to the female; and this apparently is the case with the male of a not distantly-allied bird, namely the *Merganser serrator*, for the males are said to "undergo a change of plumage, which assimilates them in some measure to the female." By a little further acceleration in the process, the double moult would be completely lost. [83]

[83] See Macgillivray, 'Hist. British Birds' (vol. v. pp. 34, 70, and 223), on the moulting of the Anatidæ, with quotations from Waterton and Montagu. Also Yarrell, 'History of British Birds,' vol. iii. p. 243.

Some male birds, as before stated, become more brightly colored in the spring, not by a vernal moult, but either by an actual change of color in the feathers, or by their obscurely-colored deciduary margins being shed. Changes of color thus caused may last for a longer or shorter time. In the *Pelecanus onocrotalus* a beautiful rosy tint, with lemon-colored marks on the breast, overspreads the whole plumage in the spring; but these tints, as Mr. Sclater states, "do not last long, disappearing generally in about six weeks or two months after they have been attained." Certain finches shed the margins of their feathers in the spring, and then become brighter colored, while other finches undergo no such change. Thus the *Fringilla tristis* of the United States (as well as many other American species) exhibits its bright colors only when the winter is past, whilst our goldfinch, which exactly represents this bird in habits, and our siskin, which represents it still more closely in structure, undergo no such annual change. But a difference of this kind in the plumage of allied species is not surprising, for with the common linnet, which belongs to the same family, the crimson forehead and breast are displayed only during the summer in England, whilst in Madeira these colors are retained throughout the year. [84]

[84] On the pelican, see Sclater, in 'Proc. Zool. Soc.' 1868, p. 265. On the American finches, see Audubon, 'Ornithological Biography,' vol. i. pp. 174, 221, and Jerdon, 'Birds of India,' vol. ii. p. 383. On the *Fringilla cannabina* of Madeira, Mr. E. Vernon Harcourt, 'Ibis,' vol. v. 1863, p. 230.

DISPLAY BY MALE BIRDS OF THEIR PLUMAGE.

Ornaments of all kinds, whether permanently or temporarily gained, are sedulously displayed by the males, and apparently serve to excite, attract, or fascinate the females. But the males will sometimes display their ornaments, when not in the presence of the females, as occasionally occurs with grouse at their balz-places, and as may be noticed with the peacock; this latter bird, however, evidently wishes for a spectator of some kind, and, as I have often seen, will show off his finery before poultry, or even pigs. [85] All naturalists who have closely attended to the habits of birds, whether in a state of nature or under confinement, are unanimously of opinion that the males take delight in displaying their beauty. Audubon frequently speaks of the male as endeavouring in various ways to charm the female. Mr. Gould, after describing some peculiarities in a male humming-bird, says he has no doubt that it has the power of displaying them to the greatest advantage before the female. Dr. Jerdon [86] insists that the beautiful plumage of the male serves "to fascinate and attract the female." Mr. Bartlett, at the Zoological Gardens, expressed himself to me in the strongest terms to the same effect.

[85] See also 'Ornamental Poultry,' by Rev. E.S. Dixon, 1848, p. 8.
[86] 'Birds of India,' introduct., vol. i. p. xxiv.; on the peacock, vol. iii. p. 507. See Gould's 'Introduction to Trochilidæ,' 1861, pp. 15 and 111.

It must be a grand sight in the forests of India "to come suddenly on twenty or thirty pea-fowl, the males displaying their gorgeous trains, and strutting about in all the pomp of pride before the gratified females." The wild turkey-cock erects his glittering plumage, expands his finely-zoned tail and barred wing-feathers, and altogether, with his crimson and blue wattles, makes a superb, though, to our eyes, grotesque appearance. Similar facts have already been given with respect to grouse of various kinds. Turning to another Order: The male *Rupicola crocea* (Fig. 50) is one of the most beautiful birds in the world, being of a splendid orange, with some of the feathers curiously truncated and plumose. The female is brownish-green, shaded with red, and has a much smaller crest. Sir R. Schomburgk has described their courtship; he found one of their meeting-places where ten males and two females were present. The space was from four to five feet in diameter, and appeared to have been cleared of every blade of grass and smoothed as if by human hands. A male "was capering, to the apparent delight of several others. Now spreading its wings, throwing up its head, or opening its tail like a fan; now strutting about with a hopping gait until tired, when it gabbled some kind of note, and was relieved by another. Thus three of them successively took the field, and then, with self-approbation, withdrew to rest." The Indians, in order to obtain their skins, wait at one of the meeting-places till the birds are eagerly engaged in dancing, and then are able to kill with their poisoned arrows four or five males, one after the other. [87] With birds of paradise a dozen or more full-plumaged males congregate in a tree to hold a dancing-party, as it is called by the natives: and here they fly about, raise their wings, elevate their exquisite plumes, and make them vibrate, and the whole tree seems, as Mr. Wallace remarks, to be filled with waving plumes. When thus engaged, they become so absorbed that a skilful archer may shoot nearly the whole party. These birds, when kept in confinement in the Malay Archipelago, are said to take much care in keeping their feathers clean; often spreading them out, examining them, and removing every speck of dirt. One observer, who kept several pairs alive, did not doubt that the display of the male was intended to please the female. [88]

[87] 'Journal of R. Geograph. Soc.' vol. x. 1840, p. 236.
[88] 'Annals and Mag. of Nat. Hist.' vol. xiii. 1854, p. 157; also Wallace, ibid. vol. xx. 1857, p. 412, and 'The Malay Archipelago,' vol. ii. 1869, p. 252. Also Dr. Bennett, as quoted by Brehm, 'Thierleben,' B. iii. s. 326.

FIG. 50.—Rupicola crocea, male (T. W. Wood).

FIG. 51.—Polyplectron chinquis, male (T. W. Wood).

The Gold and Amherst pheasants during their courtship not only expand and raise their splendid frills, but twist them, as I have myself seen, obliquely towards the female on whichever side she may be standing, obviously in order that a large surface may be displayed before her. [89] They likewise turn their beautiful tails and tail-coverts a little towards the same side. Mr. Bartlett has observed a male Polyplectron (Fig. 51) in the act of courtship, and has shown me a specimen stuffed in the attitude then assumed. The tail and wing-feathers of this bird are ornamented with beautiful ocelli, like those on the peacock's train. Now when the peacock displays himself, he expands and erects his tail transversely to his body, for he stands in front of the female, and has to show off, at the same time, his rich blue throat and breast. But the breast of the Polyplectron is obscurely colored, and the ocelli are not confined to the tail-feathers. Consequently the Polyplectron does not stand in front of the female; but he erects and expands his tail-feathers a little obliquely, lowering the expanded wing on the same side, and raising that on the opposite side. In this attitude the ocelli over the whole body are exposed at the same time before the eyes of the admiring female in one grand bespangled expanse. To whichever side she may turn, the expanded wings and the obliquely-held tail are turned towards her. The male Tragopan pheasant acts in nearly the same manner, for he raises the feathers of the body, though not the wing itself, on the side which is opposite to the female, and which would otherwise be concealed, so that nearly all the beautifully spotted feathers are exhibited at the same time.

[89] Mr. T.W. Wood has given ('The Student,' April 1870, p. 115) a full account of this manner of display, by the Gold pheasant and by the Japanese pheasant, *Ph. versicolor*; and he calls it the lateral or one-sided display.

The Argus pheasant affords a much more remarkable case. The immensely developed secondary wing-feathers are confined to the male; and each is ornamented with a row of from twenty to twenty-three ocelli, above an inch in diameter. These feathers are also elegantly marked with oblique stripes and rows of spots of a dark color, like those on the skin of a tiger and leopard combined. These beautiful ornaments are hidden until the male shows himself off before the female. He then erects his tail, and expands his wing-feathers into a great, almost upright, circular fan or shield, which is carried in front of the body. The neck and head are held on one side, so that they are concealed by the fan; but the bird in order to see the female, before whom he is displaying himself, sometimes pushes his head between two of the long wing-feathers (as Mr. Bartlett has seen), and then presents a grotesque appearance. This must be a frequent habit with the bird in a state of nature, for Mr. Bartlett and his son on examining some perfect skins sent from the East, found a place between two of the feathers which was much frayed, as if the head had here frequently been pushed through. Mr. Wood thinks that the male can also peep at the female on one side, beyond the margin of the fan.

The ocelli on the wing-feathers are wonderful objects; for they are so shaded that, as the Duke of Argyll remarks, [90] they stand out like balls lying loosely within sockets. When I looked at the specimen in the British Museum, which is mounted with the wings expanded and trailing downwards, I was however greatly disappointed, for the ocelli appeared flat, or even concave. But Mr. Gould soon made the case clear to me, for he held the feathers erect, in the position in which they would naturally be displayed, and now, from the light shining on them from above, each ocellus at once resembled the ornament called a ball and socket. These feathers have been shown to several artists, and all have expressed their admiration at the perfect shading. It may well be asked, could such artistically shaded ornaments have been formed by means of sexual selection? But it will be convenient to defer giving an answer to this question until we treat in the next chapter of the principle of gradation.

[90] 'The Reign of Law,' 1867, p. 203.

Fɪɢ. 52.--Side view of male Argus pheasant, whilst displaying before the female
Observed and sketched from nature by Mr. T. W. Wood.

The foregoing remarks relate to the secondary wing-feathers, but the primary wing-feathers, which in most gallinaceous birds are uniformly colored, are in the Argus pheasant equally wonderful. They are of a soft brown tint with numerous dark spots, each of which consists of two or three black dots with a surrounding dark zone. But the chief ornament is a space parallel to the dark-blue shaft, which in outline forms a perfect second feather lying within the true feather. This inner part is colored of a lighter chestnut, and is thickly dotted with minute white points. I have shown this feather to several persons, and many have admired it even more than the ball and socket feathers, and have declared that it was more like a work of art than of nature. Now these feathers are quite hidden on all ordinary occasions, but are fully displayed, together with the long secondary feathers, when they are all expanded together so as to form the great fan or shield.

The case of the male Argus pheasant is eminently interesting, because it affords good evidence that the most refined beauty may serve as a sexual charm, and for no other purpose. We must conclude that this is the case, as the secondary and primary wing-feathers are not at all displayed, and the ball and socket ornaments are not exhibited in full perfection until the male assumes the attitude of courtship. The Argus pheasant does not possess brilliant colors, so that his success in love appears to depend on the great size of his plumes, and on the elaboration of the most elegant patterns. Many will declare that it is utterly incredible that a female bird should be able to appreciate fine shading and exquisite patterns. It is undoubtedly a marvellous fact that she should possess this almost human degree of taste. He who thinks that he can safely gauge the discrimination and taste of the lower animals may deny that the female Argus pheasant can appreciate such refined beauty; but he will then be compelled to admit that the extraordinary attitudes assumed by the male during the act of courtship, by which the wonderful beauty of his plumage is fully displayed, are purposeless; and this is a conclusion which I for one will never admit.

Although so many pheasants and allied gallinaceous birds carefully display their plumage before the females, it is remarkable, as Mr. Bartlett informs me, that this is not the case with the dull-colored Eared and Cheer pheasants (*Crossoptilon auritum* and *Phasianus wallichii*); so that these birds seem conscious that they have little beauty to display. Mr. Bartlett has never seen the males of either of these species fighting together, though he has not had such good opportunities for observing the Cheer as the Eared pheasant. Mr. Jenner Weir, also, finds that all male birds with rich or strongly-characterized plumage are more quarrelsome than the dull-colored species belonging to the same groups. The goldfinch, for instance, is far more pugnacious than the linnet, and the blackbird than the thrush. Those birds which undergo a seasonal change of plumage likewise become much more pugnacious at the period when they are most gaily ornamented. No doubt the males of some obscurely-colored birds fight desperately together, but it appears that when sexual selection has been highly influential, and has given bright colors to the males of any species, it has also very often given a strong tendency to pugnacity. We shall meet with nearly analogous cases when we treat of mammals. On the other hand, with birds the power of song and brilliant colors have rarely been both acquired by the males of the same species; but in this case the advantage gained would have been the same, namely success in charming the female. Nevertheless it must be owned that the males of several brilliantly colored birds have had their feathers specially modified for the sake of producing instrumental music, though the beauty of this cannot be compared, at least according to our taste, with that of the vocal music of many songsters.

We will now turn to male birds which are not ornamented in any high degree, but which nevertheless display during their courtship whatever attractions they may possess. These cases are in some respects more curious than the foregoing, and have been but little noticed. I owe the following facts to Mr. Weir, who has long kept confined birds of many kinds, including all the British Fringillidæ and Emberizidæ. The facts have been selected from a large body of valuable notes kindly sent me by him. The bullfinch makes his advances in front of the female, and then puffs out his breast, so that many more of the crimson feathers are seen at once than otherwise would be the case. At the same time he twists and bows his black tail from side to side in a ludicrous manner. The male chaffinch also stands in front of the female, thus showing his red breast and "blue bell," as the fanciers call his head; the wings at the same time being slightly expanded, with the pure white bands on the shoulders thus rendered conspicuous. The common linnet distends his rosy breast, slightly expands his brown wings and tail, so as to make the best of them by exhibiting their white edgings. We must, however, be cautious in concluding that the wings are spread out solely for display, as some birds do so whose wings are not beautiful. This is the case with the domestic cock, but it is always the wing on the side opposite to the female which is expanded, and at the same time scraped on the ground. The male goldfinch behaves differently from all other finches: his wings are beautiful, the shoulders being black, with the dark-tipped wing-feathers spotted with white and edged with golden yellow. When he courts the female, he sways his body from side to side, and quickly turns his slightly expanded wings first to one side, then to the other, with a golden flashing effect. Mr. Weir informs me that no other British finch turns thus from side to side during his courtship, not even the closely-allied male siskin, for he would not thus add to his beauty.

Most of the British Buntings are plain colored birds; but in the spring the feathers on the head of the male reed-bunting (*Emberiza schœniculus*) acquire a fine black color by the abrasion of the dusky tips; and these are erected during the act of courtship. Mr. Weir has kept two species of Amadina from Australia: the *A. castanotis* is a very small and chastely colored finch, with a dark tail, white rump, and jet-black upper tail-coverts, each of the latter being marked with three large conspicuous oval spots of white. [91] This species, when courting the female, slightly spreads out and vibrates these parti-colored tail-coverts in a very peculiar manner. The male *Amadina Lathami* behaves very differently, exhibiting before the female his brilliantly spotted breast, scarlet rump, and scarlet upper tail-coverts. I may here add from Dr. Jerdon that the Indian bulbul (*Pycnonotus hoemorrhous*) has its under tail-coverts of a crimson color, and these, it might be thought, could never be well exhibited; but the bird "when excited often spreads them out laterally, so that they can be seen even from above." [92] The crimson under tail-coverts of some other birds, as with one of the woodpeckers, *Picus major*, can be seen without any such display. The common pigeon has iridescent feathers on the breast, and every one must have seen how the male inflates his breast whilst courting the female, thus showing them off to the best advantage. One of the beautiful bronze-winged pigeons of Australia (*Ocyphaps lophotes*) behaves, as described to me by Mr. Weir, very differently: the male, whilst standing before the female, lowers his head almost to the ground, spreads out and raises his tail, and half expands his wings. He then alternately and slowly raises and depresses his body, so that the iridescent metallic feathers are all seen at once, and glitter in the sun.

[91] For the description of these birds, see Gould's 'Handbook to the Birds of Australia,' vol. i. 1865, p. 417.
[92] 'Birds of India,' vol. ii. p. 96.

Sufficient facts have now been given to show with what care male birds display their various charms, and this they do with the utmost skill. Whilst preening their feathers, they have frequent opportunities for admiring themselves, and of studying how best to exhibit their beauty. But as all the males of the same species display themselves in exactly the same manner, it appears that actions, at first perhaps intentional, have become instinctive. If so, we ought not to accuse birds of conscious vanity; yet when we see a peacock strutting about, with expanded and quivering tail-feathers, he seems the very emblem of pride and vanity.

The various ornaments possessed by the males are certainly of the highest importance to them, for in some cases they have been acquired at the expense of greatly impeded powers of flight or of running. The African night-jar (Cosmetornis), which during the pairing-season has one of its primary wing-feathers developed into a streamer of very great length, is thereby much retarded in its flight, although at other times remarkable for its swiftness. The "unwieldy size" of the secondary wing-feathers of the male Argus pheasant is said "almost entirely to deprive the bird of flight." The fine plumes of male birds of paradise trouble them during a high wind. The extremely long tail-feathers of the male widow-birds (Vidua) of Southern Africa render "their flight heavy;" but as soon as these are cast off they fly as well as the females. As birds always breed when food is abundant, the males probably do not suffer much inconvenience in searching for food from their impeded powers of movement; but there can hardly be a doubt that they must be much more liable to be struck down by birds of prey. Nor can we doubt that the long train of the peacock and the long tail and wing-feathers of the Argus pheasant must render them an easier prey to any prowling tiger-cat than would otherwise be the case. Even the bright colors of many male birds cannot fail to make them conspicuous to their enemies of all kinds. Hence, as Mr. Gould has remarked, it probably is that such birds are generally of a shy disposition, as if conscious that their beauty was a source of danger, and are much more difficult to discover or approach, than the sombre colored and comparatively tame females or than the young and as yet unadorned males. [93]

[93] On the Cosmetornis, see Livingstone's 'Expedition to the Zambesi,' 1865, p. 66. On the Argus pheasant, Jardine's 'Nat. Hist. Lib.: Birds,' vol. xiv. p. 167. On Birds of Paradise, Lesson, quoted by Brehm, 'Thierleben,' B. iii. s. 325. On the widow-bird, Barrow's 'Travels in Africa,' vol. i. p. 243, and 'Ibis,' vol. iii. 1861 p. 133. Mr. Gould, on the shyness of male birds, 'Handbook to Birds of Australia,' vol. i. 1865, pp. 210, 457.

It is a more curious fact that the males of some birds which are provided with special weapons for battle, and which in a state of nature are so pugnacious that they often kill each other, suffer from possessing certain ornaments. Cock-fighters trim the hackles and cut off the combs and gills of their cocks; and the birds are then said to be dubbed. An undubbed bird, as Mr. Tegetmeier insists, "is at a fearful disadvantage; the comb and gills offer an easy hold to his adversary's beak, and as a cock always strikes where he holds, when once he has seized his foe, he has him entirely in his power. Even supposing that the bird is not killed, the loss of blood suffered by an undubbed cock is much greater than that sustained by one that has been trimmed." [94] Young turkey-cocks in fighting always seize hold of each other's wattles; and I presume that the old birds fight in the same manner. It may perhaps be objected that the comb and wattles are not ornamental, and cannot be of service to the birds in this way; but even to our eyes, the beauty of the glossy black Spanish cock is much enhanced by his white face and crimson comb; and no one who has ever seen the splendid blue wattles of the male Tragopan pheasant distended in courtship can for a moment doubt that beauty is the object gained. From the foregoing facts we clearly see that the plumes and other ornaments of the males must be of the highest importance to them; and we further see that beauty is even sometimes more important than success in battle.

[94] Tegetmeier, 'The Poultry Book,' 1866, p. 139.

CHAPTER XIV.

BIRDS—continued.

Choice exerted by the female—Length of courtship—Unpaired birds—Mental qualities and taste for the beautiful—Preference or antipathy shown by the female for particular males—Variability of birds—Variations sometimes abrupt—Laws of variation—Formation of ocelli—Gradations of character—Case of Peacock, Argus pheasant, and Urosticte.

When the sexes differ in beauty or in the power of singing, or in producing what I have called instrumental music, it is almost invariably the male who surpasses the female. These qualities, as we have just seen, are evidently of high importance to the male. When they are gained for only a part of the year it is always before the breeding-season. It is the male alone who elaborately displays his varied attractions, and often performs strange antics on the ground or in the air, in the presence of the female. Each male drives away, or if he can, kills his rivals. Hence we may conclude that it is the object of the male to induce the female to pair with him, and for this purpose he tries to excite or charm her in various ways; and this is the opinion of all those who have carefully studied the habits of living birds. But there remains a question which has an all important bearing on sexual selection, namely, does every male of the same species excite and attract the female equally? Or does she exert a choice, and prefer certain males? This latter question can be answered in the affirmative by much direct and indirect evidence. It is far more difficult to decide what qualities determine the choice of the females; but here again we have some direct and indirect evidence that it is to a large extent the external attractions of the male; though no doubt his vigor, courage, and other mental qualities come into play. We will begin with the indirect evidence.

LENGTH OF COURTSHIP.

The lengthened period during which both sexes of certain birds meet day after day at an appointed place probably depends partly on the courtship being a prolonged affair, and partly on reiteration in the act of pairing. Thus in Germany and Scandinavia the balzen or leks of the black-cocks last from the middle of March, all through April into May. As many as forty or fifty, or even more birds congregate at the leks; and the same place is often frequented during successive years. The lek of the capercailzie lasts from the end of March to the middle or even end of May. In North America "the partridge dances" of the *Tetrao phasianellus* "last for a month or more." Other kinds of grouse, both in North America and Eastern Siberia, [1] follow nearly the same habits. The fowlers discover the hillocks where the ruffs congregate by the grass being trampled bare, and this shows that the same spot is long frequented. The Indians of Guiana are well acquainted with the cleared arenas, where they expect to find the beautiful cocks of the Rock; and the natives of New Guinea know the trees where from ten to twenty male birds of paradise in full plumage congregate. In this latter case it is not expressly stated that the females meet on the same trees, but the hunters, if not specially asked, would probably not mention their presence, as their skins are valueless. Small parties of an African weaver (*Ploceus*) congregate, during the breeding-season, and perform for hours their graceful evolutions. Large numbers of the Solitary snipe (*Scolopax major*) assemble during dusk in a morass; and the same place is frequented for the same purpose during successive years; here they may be seen running about "like so many large rats," puffing out their feathers, flapping their wings, and uttering the strangest cries. [2]

[1] Nordman describes ('Bull. Soc. Imp. des Nat. Moscou,' 1861, tom. xxxiv. p. 264) the balzen of *Tetrao urogalloides* in Amur Land. He estimated the number of birds assembled at above a hundred, not counting the females, which lie hid in the surrounding bushes. The noises uttered differ from those of *T. urogallus*.

[2] With respect to the assemblages of the above named grouse, see Brehm, 'Thierleben,' B. iv. s. 350; also L. Lloyd, 'Game Birds of Sweden,' 1867, pp. 19, 78. Richardson, 'Fauna Bor. Americana: Birds,' p. 362. References in regard to the assemblages of other birds have already been given. On Paradisea, see Wallace, in 'Annals and Mag. of Nat. Hist.' vol. xx. 1857, p. 412. On the snipe, Lloyd, ibid. p. 221.

Some of the above birds,—the black-cock, capercailzie, pheasant-grouse, ruff, solitary snipe, and perhaps others,—are, as is believed, polygamists. With such birds it might have been thought that the stronger males would simply have driven away the weaker, and then at once have taken possession of as many females as possible; but if it be indispensable for the male to excite or please the female, we can understand the length of the courtship and the congregation of so many individuals of both sexes at the same spot. Certain strictly monogamous species likewise hold nuptial assemblages; this seems to be the case in Scandinavia with one of the ptarmigans, and their leks last from the middle of March to the middle of May. In Australia the lyre-bird (*Menura superba*) forms "small round hillocks," and the *M. Alberti* scratches for itself shallow holes, or, as they are called by the natives, "*corroborying places*," where it is believed both sexes assemble. The meetings of the *M. superba* are sometimes very large; and an account has lately been published [3] by a traveller, who heard in a valley beneath him, thickly covered with scrub, "a din which completely astonished" him; on crawling onwards he beheld, to his amazement, about one hundred and fifty of the magnificent lyre-cocks, "ranged in order of battle, and fighting with indescribable fury." The bowers of the Bower-birds are the resort of both sexes during the breeding-season; and "here the males meet and contend with each other for the favours of the female, and here the latter assemble and coquet with the males." With two of the genera, the same bower is resorted to during many years. [4]

[3] Quoted by Mr. T.W. Wood, in the 'Student,' April 1870, p. 125.

[4] Gould, 'Handbook to the Birds of Australia,' vol. i. pp. 300, 308, 448, 451. On the ptarmigan, above alluded to, see Lloyd, ibid. p. 129.

The common magpie (*Corvus pica*, Linn.), as I have been informed by the Rev. W. Darwin Fox, used to assemble from all parts of Delamere Forest, in order to celebrate the "great magpie marriage." Some years ago these birds abounded in extraordinary numbers, so that a gamekeeper killed in one morning nineteen males, and another killed by a single shot seven birds at roost together. They then had the habit of assembling very early in the spring at particular spots, where they could be seen in flocks, chattering, sometimes fighting, bustling and flying about the trees. The whole affair was evidently considered by the birds as one of the highest importance. Shortly after the meeting they all separated, and were then observed by Mr. Fox and others to be paired for the season. In any district in which a species does not exist in large numbers, great assemblages cannot, of course, be held, and the same species may have different habits in different countries. For instance, I have heard of only one instance, from Mr. Wedderburn, of a regular assemblage of black game in Scotland, yet these assemblages are so well known in Germany and Scandinavia that they have received special names.

UNPAIRED BIRDS.

From the facts now given, we may conclude that the courtship of birds belonging to widely different groups, is often a prolonged, delicate, and troublesome affair. There is even reason to suspect, improbable as this will at first appear, that some males and females of the same species, inhabiting the same district, do not always please each other, and consequently do not pair. Many accounts have been published of either the male or female of a pair having been shot, and quickly replaced by another. This has been observed more frequently with the magpie than with any other bird, owing perhaps to its conspicuous appearance and nest. The illustrious Jenner states that in Wiltshire one of a pair was daily shot no less than seven times successively, "but all to no purpose, for the remaining magpie soon found another mate"; and the last pair reared their young. A new partner is generally found on the succeeding day; but Mr. Thompson gives the case of one being replaced on the evening of the same day. Even after the eggs are hatched, if one of the old birds is destroyed a mate will often be found; this occurred after an interval of two days, in a case recently observed by one of Sir J. Lubbock's keepers. [5] The first and most obvious conjecture is that male magpies must be much more numerous than females; and that in the above cases, as well as in many others which could be given, the males alone had been killed. This apparently holds good in some instances, for the gamekeepers in Delamere Forest assured Mr. Fox that the magpies and carrion-crows which they formerly killed in succession in large numbers near their nests, were all males; and they accounted for this fact by the males being easily killed whilst bringing food to the sitting females. Macgillivray, however, gives, on the authority of an excellent observer, an instance of three magpies successively killed on the same nest, which were all females; and another case of six magpies successively killed whilst sitting on the same eggs, which renders it probable that most of them were females; though, as I hear from Mr. Fox, the male will sit on the eggs when the female is killed.

[5] On magpies, Jenner, in 'Philosophical Transactions,' 1824, p. 21. Macgillivray, 'Hist. British Birds,' vol. i. p. 570. Thompson, in 'Annals and Magazine of Natural History,' vol. viii. 1842, p. 494.

Sir J. Lubbock's gamekeeper has repeatedly shot, but how often he could not say, one of a pair of jays (*Garrulus glandarius*), and has never failed shortly afterwards to find the survivor re-matched. Mr. Fox, Mr. F. Bond, and others have shot one of a pair of carrion-crows (*Corvus corone*), but the nest was soon again tenanted by a pair. These birds are rather common; but the peregrine-falcon (*Falco peregrinus*) is rare, yet Mr. Thompson states that in Ireland "if either an old male or female be killed in the breeding-season (not an uncommon circumstance), another mate is found within a very few days, so that the eyries, notwithstanding such casualties, are sure to turn out their complement of young." Mr. Jenner Weir has known the same thing with the peregrine-falcons at Beachy Head. The same observer informs me that three kestrels (*Falco tinnunculus*), all males, were killed one after the other whilst attending the same nest; two of these were in mature plumage, but the third was in the plumage of the previous year. Even with the rare golden eagle (*Aquila chrysaëtos*), Mr. Birkbeck was assured by a trustworthy gamekeeper in Scotland, that if one is killed, another is soon found. So with the white owl (*Strix flammea*), "the survivor readily found a mate, and the mischief went on."

White of Selborne, who gives the case of the owl, adds that he knew a man, who from believing that partridges when paired were disturbed by the males fighting, used to shoot them; and though he had widowed the same female several times, she always soon found a fresh partner. This same naturalist ordered the sparrows, which deprived the house-martins of their nests, to be shot; but the one which was left, "be it cock or hen, presently procured a mate, and so for several times following." I could add analogous cases relating to the chaffinch, nightingale, and redstart. With respect to the latter bird (*Phœnicura ruticilla*), a writer expresses much surprise how the sitting female could so soon have given effectual notice that she was a widow, for the species was not common in the neighbourhood. Mr. Jenner Weir has mentioned to me a nearly similar case; at Blackheath he never sees or hears the note of the wild bullfinch, yet when one of his caged males has died, a wild one in the course of a few days has generally come and perched near the widowed female, whose call-note is not loud. I will give only one other fact, on the authority of this same observer; one of a pair of starlings (*Sturnus vulgaris*) was shot in the morning; by noon a new mate was found; this was again shot, but before night the pair was complete; so that the disconsolate widow or widower was thrice consoled during the same day. Mr. Engleheart also informs me that he used during several years to shoot one of a pair of starlings which built in a hole in a house at Blackheath; but the loss was always immediately repaired. During one season he kept an account, and found that he had shot thirty-five birds from the same nest; these consisted of both males and females, but in what proportion he could not say: nevertheless, after all this destruction, a brood was reared. [6]

[6] On the peregrine falcon, see Thompson, 'Nat. Hist. of Ireland: Birds,' vol. i. 1849, p. 39. On owls, sparrows, and partridges, see White, 'Nat. Hist. of Selborne,' edit. of 1825, vol. i. p. 139. On the Phœnicura, see Loudon's 'Mag. of Nat. Hist.' vol. vii. 1834, p. 245. Brehm ('Thierleben,' B. iv. s. 991) also alludes to cases of birds thrice mated during the same day.

These facts well deserve attention. How is it that there are birds enough ready to replace immediately a lost mate of either sex? Magpies, jays, carrion-crows, partridges, and some other birds, are always seen during the spring in pairs, and never by themselves; and these offer at first sight the most perplexing cases. But birds of the same sex, although of course not truly paired, sometimes live in pairs or in small parties, as is known to be the case with pigeons and partridges. Birds also sometimes live in triplets, as has been observed with starlings, carrion-crows, parrots, and partridges. With partridges two females have been known to live with one male, and two males with one female. In all such cases it is probable that the union would be easily broken; and one of the three would readily pair with a widow or widower. The males of certain birds may occasionally be heard pouring forth their love-song long after the proper time, showing that they have either lost or never gained a mate. Death from accident or disease of one of a pair would leave the other free and single; and there is reason to believe that female birds during the breeding-season are especially liable to premature death. Again, birds which have had their nests destroyed, or barren pairs, or retarded individuals, would easily be induced to desert their mates, and would probably be glad to take what share they could of the pleasures and duties of rearing offspring although not their own. [7] Such contingencies as these probably explain most of the foregoing cases. [8] Nevertheless, it is a strange fact that within the same district, during the height of the breeding-season, there should be so many males and females always ready to repair the loss of a mated bird. Why do not such spare birds immediately pair together? Have we not some reason to suspect, and the suspicion has occurred to Mr. Jenner Weir, that as the courtship of birds appears to be in many cases prolonged and tedious, so it occasionally happens that certain males and females do not succeed, during the proper season, in exciting each other's love, and consequently do not pair? This suspicion will appear somewhat less improbable after we have seen what strong antipathies and preferences female birds occasionally evince towards particular males.

[7] See White ('Nat. Hist. of Selborne,' 1825, vol. i. p. 140) on the existence, early in the season, of small coveys of male partridges, of which fact I have heard other instances. See Jenner, on the retarded state of the generative organs in certain birds, in 'Phil. Transact.' 1824. In regard to birds living in triplets, I owe to Mr. Jenner Weir the cases of the starlings and parrots, and to Mr. Fox, of partridges; on carrion-crows, see the 'Field,' 1868, p. 415. On various male birds singing after the proper period, see Rev. L. Jenyns, 'Observations in Natural History,' 1846, p. 87.

[8] The following case has been given ('The Times,' Aug. 6, 1868) by the Rev. F.O. Morris, on the authority of the Hon. and Rev. O.W. Forester. "The gamekeeper here found a hawk's nest this year, with five young ones on it. He took four and killed them, but left one with its wings clipped as a decoy to destroy the old ones by. They were both shot next day, in the act of feeding the young one, and the keeper thought it was done with. The next day he came again and found two other charitable hawks, who had come with an adopted feeling to succour the orphan. These two he killed, and then left the nest. On returning afterwards he found two more charitable individuals on the same errand of mercy. One of these he killed; the other he also shot, but could not find. No more came on the like fruitless errand."

MENTAL QUALITIES OF BIRDS, AND THEIR TASTE FOR THE BEAUTIFUL.

Before we further discuss the question whether the females select the more attractive males or accept the first whom they may encounter, it will be advisable briefly to consider the mental powers of birds. Their reason is generally, and perhaps justly, ranked as low; yet some facts could be given leading to an opposite conclusion. [9] Low powers of reasoning, however, are compatible, as we see with mankind, with strong affections, acute perception, and a taste for the beautiful; and it is with these latter qualities that we are here concerned. It has often been said that parrots become so deeply attached to each other that when one dies the other pines for a long time; but Mr. Jenner Weir thinks that with most birds the strength of their affection has been much exaggerated. Nevertheless when one of a pair in a state of nature has been shot, the survivor has been heard for days afterwards uttering a plaintive call; and Mr. St. John gives various facts proving the attachment of mated birds. [10] Mr. Bennett relates [11] that in China after a drake of the beautiful mandarin Teal had been stolen, the duck remained disconsolate, though sedulously courted by another mandarin drake, who displayed before her all his charms. After an interval of three weeks the stolen drake was recovered, and instantly the pair recognized each other with extreme joy. On the other hand, starlings, as we have seen, may be consoled thrice in the same day for the loss of their mates. Pigeons have such excellent local memories, that they have been known to return to their former homes after an interval of nine months, yet, as I hear from Mr. Harrison Weir, if a pair which naturally would remain mated for life be separated for a few weeks during the winter, and afterwards matched with other birds, the two when brought together again, rarely, if ever, recognize each other.

[9] I am indebted to Prof. Newton for the following passage from Mr. Adam's 'Travels of a Naturalist,' 1870, p. 278. Speaking of Japanese nut-hatches in confinement, he says: "Instead of the more yielding fruit of the yew, which is the usual food of the nut-hatch of Japan, at one time I substituted hard hazel-nuts. As the bird was unable to crack them, he placed them one by one in his water-glass, evidently with the notion that they would in time become softer—an interesting proof of intelligence on the part of these birds."

[10] 'A Tour in Sutherlandshire,' vol. i. 1849, p. 185. Dr. Buller says ('Birds of New Zealand,' 1872, p. 56) that a male King Lory was killed; and the female "fretted and moped, refused her food, and died of a broken heart."

[11] 'Wanderings in New South Wales,' vol. ii. 1834, p. 62.

Birds sometimes exhibit benevolent feelings; they will feed the deserted young ones even of distinct species, but this perhaps ought to be considered as a mistaken instinct. They will feed, as shown in an earlier part of this work, adult birds of their own species which have become blind. Mr. Buxton gives a curious account of a parrot which took care of a frost-bitten and crippled bird of a distinct species, cleansed her feathers, and defended her from the attacks of the other parrots which roamed freely about his garden. It is a still more curious fact that these birds apparently evince some sympathy for the pleasures of their fellows. When a pair of cockatoos made a nest in an acacia tree, "it was ridiculous to see the extravagant interest taken in the matter by the others of the same species." These parrots, also, evinced unbounded curiosity, and clearly had "the idea of property and possession." [12] They have good memories, for in the Zoological Gardens they have plainly recognized their former masters after an interval of some months.

[12] 'Acclimatization of Parrots,' by C. Buxton, M.P., 'Annals and Mag. of Nat. Hist.' Nov. 1868, p. 381.

Birds possess acute powers of observation. Every mated bird, of course, recognizes its fellow. Audubon states that a certain number of mocking-thrushes (*Mimus polyglottus*) remain all the year round in Louisiana, whilst others migrate to the Eastern States; these latter, on their return, are instantly recognized, and always attacked, by their southern brethren. Birds under confinement distinguish different persons, as is proved by the strong and permanent antipathy or affection which they show, without any apparent cause, towards certain individuals. I have heard of numerous instances with jays, partridges, canaries, and especially bullfinches. Mr. Hussey has described in how extraordinary a manner a tamed partridge recognized everybody: and its likes and dislikes were very strong. This bird seemed "fond of gay colors, and no new gown or cap could be put on without catching his attention." [13] Mr. Hewitt has described the habits of some ducks (recently descended from wild birds), which, at the approach of a strange dog or cat, would rush headlong into the water, and exhaust themselves in their attempts to escape; but they knew Mr. Hewitt's own dogs and cats so well that they would lie down and bask in the sun close to them. They always moved away from a strange man, and so they would from the lady who attended them if she made any great change in her dress. Audubon relates that he reared and tamed a wild turkey which always ran away from any strange dog; this bird escaped into the woods, and some days afterwards Audubon saw, as he thought, a wild turkey, and made his dog chase it; but, to his astonishment, the bird did not run away, and the dog, when he came up, did not attack the bird, for they mutually recognized each other as old friends. [14]

[13] The 'Zoologist,' 1847-48, p. 1602.
[14] Hewitt on wild ducks, 'Journal of Horticulture,' Jan. 13, 1863, p. 39. Audubon on the wild turkey, 'Ornithological Biography,' vol. i. p. 14. On the mocking-thrush, ibid. vol. i. p. 110.

Mr. Jenner Weir is convinced that birds pay particular attention to the colors of other birds, sometimes out of jealousy, and sometimes as a sign of kinship. Thus he turned a reed-bunting (*Emberiza schoeniculus*), which had acquired its black head-dress, into his aviary, and the new-comer was not noticed by any bird, except by a bullfinch, which is likewise black-headed. This bullfinch was a very quiet bird, and had never before quarrelled with any of its comrades, including another reed-bunting, which had not as yet become black-headed: but the reed-bunting with a black head was so unmercifully treated that it had to be removed. *Spiza cyanea*, during the breeding-season, is of a bright blue color; and though generally peaceable, it attacked *S. ciris*, which has only the head blue, and completely scalped the unfortunate bird. Mr. Weir was also obliged to turn out a robin, as it fiercely attacked all the birds in his aviary with any red in their plumage, but no other kinds; it actually killed a red-breasted crossbill, and nearly killed a goldfinch. On the other band, he has observed that some birds, when first introduced, fly towards the species which resemble them most in color, and settle by their sides.

As male birds display their fine plumage and other ornaments with so much care before the females, it is obviously probable that these appreciate the beauty of their suitors. It is, however, difficult to obtain direct evidence of their capacity to appreciate beauty. When birds gaze at themselves in a looking-glass (of which many instances have been recorded) we cannot feel sure that it is not from jealousy of a supposed rival, though this is not the conclusion of some observers. In other cases it is difficult to distinguish between mere curiosity and admiration. It is perhaps the former feeling which, as stated by Lord Lilford, [15] attracts the ruff towards any bright object, so that, in the Ionian Islands, "it will dart down to a bright-colored handkerchief, regardless of repeated shots." The common lark is drawn down from the sky, and is caught in large numbers, by a small mirror made to move and glitter in the sun. Is it admiration or curiosity which leads the magpie, raven, and some other birds to steal and secrete bright objects, such as silver articles or jewels?

[15] The 'Ibis,' vol. ii. 1860, p. 344.

Mr. Gould states that certain humming-birds decorate the outsides of their nests "with the utmost taste; they instinctively fasten thereon beautiful pieces of flat lichen, the larger pieces in the middle, and the smaller on the part attached to the branch. Now and then a pretty feather is intertwined or fastened to the outer sides, the stem being always so placed that the feather stands out beyond the surface." The best evidence, however, of a taste for the beautiful is afforded by the three genera of Australian bower-birds already mentioned. Their bowers (Fig. 46), where the sexes congregate and play strange antics, are variously constructed, but what most concerns us is, that they are decorated by the several species in a different manner. The Satin bower-bird collects gaily-colored articles, such as the blue tail-feathers of parrakeets, bleached bones and shells, which it sticks between the twigs or arranges at the entrance. Mr. Gould found in one bower a neatly-worked stone tomahawk and a slip of blue cotton, evidently procured from a native encampment. These objects are continually re-arranged, and carried about by the birds whilst at play. The bower of the Spotted bower-bird "is beautifully lined with tall grasses, so disposed that the heads nearly meet, and the decorations are very profuse." Round stones are used to keep the grass-stems in their proper places, and to make divergent paths leading to the bower. The stones and shells are often brought from a great distance. The Regent bird, as described by Mr. Ramsay, ornaments its short bower with bleached land-shells belonging to five or six species, and with "berries of various colors, blue, red, and black, which give it when fresh a very pretty appearance. Besides these there were several newly-picked leaves and young shoots of a pinkish color, the whole showing a decided taste for the beautiful." Well may Mr. Gould say that "these highly decorated halls of assembly must be regarded as the most wonderful instances of bird-architecture yet discovered;" and the taste, as we see, of the several species certainly differs. [16]

[16] On the ornamented nests of humming-birds, Gould, 'Introduction to the Trochilidæ,' 1861, p. 19. On the bower-birds, Gould, 'Handbook to the Birds of Australia,' 1865, vol. i. pp. 444-461. Ramsay, in the 'Ibis,' 1867, p. 456.

PREFERENCE FOR PARTICULAR MALES BY THE FEMALES.

Having made these preliminary remarks on the discrimination and taste of birds, I will give all the facts known to me which bear on the preference shown by the female for particular males. It is certain that distinct species of birds occasionally pair in a state of nature and produce hybrids. Many instances could be given: thus Macgillivray relates how a male blackbird and female thrush "fell in love with each other," and produced offspring. [17] Several years ago eighteen cases had been recorded of the occurrence in Great Britain of hybrids between the black grouse and pheasant [18]; but most of these cases may perhaps be accounted for by solitary birds not finding one of their own species to pair with. With other birds, as Mr. Jenner Weir has reason to believe, hybrids are sometimes the result of the casual intercourse of birds building in close proximity. But these remarks do not apply to the many recorded instances of tamed or domestic birds, belonging to distinct species, which have become absolutely fascinated with each other, although living with their own species. Thus Waterton [19] states that out of a flock of twenty-three Canada geese, a female paired with a solitary Bernicle gander, although so different in appearance and size; and they produced hybrid offspring. A male wigeon

(*Mareca penelope*), living with females of the same species, has been known to pair with a pintail duck, *Querquedula acuta*. Lloyd describes the remarkable attachment between a shield-drake (*Tadorna vulpanser*) and a common duck. Many additional instances could be given; and the Rev. E.S. Dixon remarks that "those who have kept many different species of geese together well know what unaccountable attachments they are frequently forming, and that they are quite as likely to pair and rear young with individuals of a race (species) apparently the most alien to themselves as with their own stock."

[17] 'History of Brit. Birds,' vol. ii. p. 92.
[18] 'Zoologist,' 1853-1854, p. 3946.
[19] Waterton, 'Essays on Nat. Hist.' 2nd series, pp. 42 and 117. For the following statements see on the wigeon, 'Loudon's Mag. of Nat. Hist.' vol. ix. p. 616; L. Lloyd, 'Scandinavian Adventures,' vol. i. 1854, p. 452. Dixon, 'Ornamental and Domestic Poultry,' p. 137; Hewitt, in 'Journal of Horticulture,' Jan. 13, 1863, p. 40; Bechstein, 'Stubenvögel,' 1840, s. 230. Mr. J. Jenner Weir has lately given me an analogous case with ducks of two species.

The Rev. W.D. Fox informs me that he possessed at the same time a pair of Chinese geese (*Anser cygnoides*), and a common gander with three geese. The two lots kept quite separate, until the Chinese gander seduced one of the common geese to live with him. Moreover, of the young birds hatched from the eggs of the common geese, only four were pure, the other eighteen proving hybrids; so that the Chinese gander seems to have had prepotent charms over the common gander. I will give only one other case; Mr. Hewitt states that a wild duck, reared in captivity, "after breeding a couple of seasons with her own mallard, at once shook him off on my placing a male Pintail on the water. It was evidently a case of love at first sight, for she swam about the new-comer caressingly, though he appeared evidently alarmed and averse to her overtures of affection. From that hour she forgot her old partner. Winter passed by, and the next spring the pintail seemed to have become a convert to her blandishments, for they nested and produced seven or eight young ones."

What the charm may have been in these several cases, beyond mere novelty, we cannot even conjecture. Color, however, sometimes comes into play; for in order to raise hybrids from the siskin (*Fringilla spinus*) and the canary, it is much the best plan, according to Bechstein, to place birds of the same tint together. Mr. Jenner Weir turned a female canary into his aviary, where there were male linnets, goldfinches, siskins, greenfinches, chaffinches, and other birds, in order to see which she would choose; but there never was any doubt, and the greenfinch carried the day. They paired and produced hybrid offspring.

The fact of the female preferring to pair with one male rather than with another of the same species is not so likely to excite attention, as when this occurs, as we have just seen, between distinct species. The former cases can best be observed with domesticated or confined birds; but these are often pampered by high feeding, and sometimes have their instincts vitiated to an extreme degree. Of this latter fact I could give sufficient proofs with pigeons, and especially with fowls, but they cannot be here related. Vitiated instincts may also account for some of the hybrid unions above mentioned; but in many of these cases the birds were allowed to range freely over large ponds, and there is no reason to suppose that they were unnaturally stimulated by high feeding.

With respect to birds in a state of nature, the first and most obvious supposition which will occur to every one is that the female at the proper season accepts the first male whom she may encounter; but she has at least the opportunity for exerting a choice, as she is almost invariably pursued by many males. Audubon—and we must remember that he spent a long life in prowling about the forests of the United States and observing the birds—does not doubt that the female deliberately chooses her mate; thus, speaking of a woodpecker, he says the hen is followed by half-a-dozen gay suitors, who continue performing strange antics, "until a marked preference is shown for one." The female of the red-winged starling (*Agelæus phœniceus*) is likewise pursued by several males, "until, becoming fatigued, she alights, receives their addresses, and soon makes a choice." He describes also how several male night-jars repeatedly plunge through the air with astonishing rapidity, suddenly turning, and thus making a singular noise; "but no sooner has the female made her choice than the other males are driven away." With one of the vultures (*Cathartes aura*) of the United States, parties of eight, ten, or more males and females assemble on fallen logs, "exhibiting the strongest desire to please mutually," and after many caresses, each male leads off his partner on the wing. Audubon likewise carefully observed the wild flocks of Canada geese (*Anser canadensis*), and gives a graphic description of their love-antics; he says that the birds which had been previously mated "renewed their courtship as early as the month of January, while the others would be contending or coquetting for hours every day, until all seemed satisfied with the choice they had made, after which, although they remained together, any person could easily perceive that they were careful to keep in pairs. I have observed also that the older the birds the shorter were the preliminaries of their courtship. The bachelors and old maids whether in regret, or not caring to be disturbed by the bustle, quietly moved aside and lay down at some distance from the rest." [20] Many similar statements with respect to other birds could be cited from this same observer.

[20] Audubon, 'Ornithological Biography,' vol. i. pp. 191, 349; vol. ii. pp. 42, 275; vol. iii. p. 2.

Turning now to domesticated and confined birds, I will commence by giving what little I have learnt respecting the courtship of fowls. I have received long letters on this subject from Messrs. Hewitt and Tegetmeier, and almost an essay from the late Mr. Brent. It will be admitted by every one that these gentlemen, so well known from their published works, are careful and experienced observers. They do not believe that the females prefer certain males on account of the beauty of their plumage; but some allowance must be made for the artificial state under which these birds have long been kept. Mr. Tegetmeier is convinced that a gamecock, though disfigured by being dubbed and with his hackles trimmed, would be accepted as readily as a male retaining all his natural ornaments. Mr. Brent, however, admits that the beauty of the male probably aids in exciting the female; and her acquiescence is necessary. Mr. Hewitt is convinced that the union is by no means left to mere chance, for the female almost invariably prefers the most vigorous, defiant, and mettlesome male; hence it is almost useless, as he remarks, "to attempt true breeding if a game-cock in good health and condition runs the locality, for almost every hen on leaving the roosting-place will resort to the game-cock, even though that bird may not actually drive away the male of her own variety." Under ordinary circumstances the males and females of the fowl seem to come to a mutual understanding by means of certain gestures, described to me by Mr. Brent. But hens will often avoid the officious attentions of young males. Old hens, and hens of a pugnacious disposition, as the same writer informs me, dislike strange males, and will not yield until well beaten into compliance. Ferguson, however, describes how a quarrelsome hen was subdued by the gentle courtship of a Shanghai cock. [21]

[21] 'Rare and Prize Poultry,' 1854, p. 27.

There is reason to believe that pigeons of both sexes prefer pairing with birds of the same breed; and dovecot-pigeons dislike all the highly improved breeds. [22] Mr. Harrison Weir has lately heard from a trustworthy observer, who keeps blue pigeons, that these drive away all other colored varieties, such as white, red, and yellow; and from another observer, that a female dun carrier could not, after repeated trials, be matched with a black male, but immediately paired with a dun. Again, Mr. Tegetmeier had a female blue turbit that obstinately refused to pair with two males of the same breed, which were successively shut up with her for weeks; but on being let out she would have immediately accepted the first blue dragon that offered. As she was a valuable bird, she was then shut up for many weeks with a silver (*i.e.*, very pale blue) male, and at last mated with him. Nevertheless, as a general rule, color appears to have little influence on the pairing of pigeons. Mr. Tegetmeier, at my request, stained some of his birds with magenta, but they were not much noticed by the others.

[22] 'Variation of Animals and Plants under Domestication,' vol. ii. p. 103.

Female pigeons occasionally feel a strong antipathy towards certain males, without any assignable cause. Thus MM. Boitard and Corbié, whose experience extended over forty-five years, state: "Quand une femelle éprouve de l'antipathie pour un mâle avec lequel on veut l'accoupler, malgré tous les feux de l'amour, malgré l'alpiste et le chènevis dont on la nourrit pour augmenter son ardeur, malgré un emprisonnement de six mois et même d'un an, elle refuse constamment ses caresses; les avances empressées, les agaceries, les tournoiemens, les tendres roucoulemens, rien ne peut lui plaire ni l'émouvoir; gonflée, boudeuse, blottie dans un coin de sa prison, elle n'en sort que pour boire et manger, ou pour repousser avec une espèce de rage des caresses devenues trop pressantes." [23] On the other hand, Mr. Harrison Weir has himself observed, and has heard from several breeders, that a female pigeon will occasionally take a strong fancy for a particular male, and will desert her own mate for him. Some females, according to another experienced observer, Riedel [24], are of a profligate disposition, and prefer almost any stranger to their own mate. Some amorous males, called by our English fanciers "gay birds," are so successful in their gallantries, that, as Mr. H. Weir informs me, they must be shut up on account of the mischief which they cause.

[23] Boitard and Corbié, 'Les Pigeons,' etc., 1824, p. 12. Prosper Lucas ('Traité de l'Héréd. Nat.' tom. ii. 1850, p. 296) has himself observed nearly similar facts with pigeons.
[24] Die Taubenzucht, 1824, s. 86.

Wild turkeys in the United States, according to Audubon, "sometimes pay their addresses to the domesticated females, and are generally received by them with great pleasure." So that these females apparently prefer the wild to their own males. [25]

[25] 'Ornithological Biography,' vol. i. p. 13. See to the same effect, Dr. Bryant, in Allen's 'Mammals and Birds of Florida,' p. 344.

Here is a more curious case. Sir R. Heron during many years kept an account of the habits of the peafowl, which he bred in large numbers. He states that "the hens have frequently great preference to a particular peafowl. They were all so fond of an old pied cock, that one year, when he was confined, though still in view, they were constantly assembled close to the trellice-walls of his prison, and would not suffer a japanned peacock to touch them. On his being let out in the autumn, the oldest of the hens instantly courted him and was successful in her courtship. The next year he was shut up in a stable, and then the hens all courted his rival." [26] This rival was a japanned or black-winged peacock, to our eyes a more beautiful bird than the common kind.

[26] 'Proceedings, Zoological Society,' 1835, p. 54. The japanned peacock is considered by Mr. Sclater as a distinct species, and has been named *Pavo nigripennis*; but the evidence seems to me to show that it is only a variety.

Lichtenstein, who was a good observer and had excellent opportunities of observation at the Cape of Good Hope, assured Rudolphi that the female widow-bird (*Chera progne*) disowns the male when robbed of the long tail-feathers with which he is ornamented during the breeding-season. I presume that this observation must have been made on birds under confinement. [27] Here is an analogous case; Dr. Jaeger, [28] director of the Zoological Gardens of Vienna, states that a male silver-pheasant, who had been triumphant over all other males and was the accepted lover of the females, had his ornamental plumage spoiled. He was then immediately superseded by a rival, who got the upper hand and afterwards led the flock.

[27] Rudolphi, 'Beiträge zur Anthropologie,' 1812, s. 184.
[28] 'Die Darwin'sche Théorie, und ihre Stellung zu Moral und Religion,' 1869, s. 59.

It is a remarkable fact, as showing how important color is in the courtship of birds, that Mr. Boardman, a well-known collector and observer of birds for many years in the Northern United States, has never in his large experience seen an albino paired with another bird; yet he has had opportunities of observing many albinos belonging to several species. [29] It can hardly be maintained that albinos in a state of nature are incapable of breeding, as they can be raised with the greatest facility under confinement. It appears, therefore, that we must attribute the fact that they do not pair to their rejection by their normally colored comrades.

[29] This statement is given by Mr. A. Leith Adams, in his 'Field and Forest Rambles,' 1873, p. 76, and accords with his own experience.

Female birds not only exert a choice, but in some few cases they court the male, or even fight together for his possession. Sir R. Heron states that with peafowl, the first advances are always made by the female; something of the same kind takes place, according to Audubon, with the older females of the wild turkey. With the capercailzie, the females flit round the male whilst he is parading at one of the places of assemblage, and solicit his attention. [30]

[30] In regard to peafowl, see Sir R. Heron, 'Proc. Zoolog. Soc.' 1835, p. 54, and the Rev. E.S. Dixon, 'Ornamental Poultry,' 1848, p. 8. For the turkey, Audubon, ibid. p. 4. For the capercailzie, Lloyd, 'Game Birds of Sweden,' 1867, p. 23.

We have seen that a tame wild-duck seduced an unwilling pintail drake after a long courtship. Mr. Bartlett believes that the Lophophorus, like many other gallinaceous birds, is naturally polygamous, but two females cannot be placed in the same cage with a male, as they fight so much together. The following instance of rivalry is more surprising as it relates to bullfinches, which usually pair for life. Mr. Jenner Weir introduced a dull-colored and ugly female into his aviary, and she immediately attacked another mated female so unmercifully that the latter had to be separated. The new female did all the courtship, and was at last successful, for she paired with the male; but after a time she met with a just retribution, for, ceasing to be pugnacious, she was replaced by the old female, and the male then deserted his new and returned to his old love.

In all ordinary cases the male is so eager that he will accept any female, and does not, as far as we can judge, prefer one to the other; but, as we shall hereafter see, exceptions to this rule apparently occur in some few groups. With domesticated birds, I have heard of only one case of males showing any preference for certain females, namely, that of the domestic cock, who, according to the high authority of Mr. Hewitt, prefers the younger to the older hens. On the other hand, in effecting hybrid unions between the male pheasant and common hens, Mr. Hewitt is convinced that the pheasant invariably prefers the older birds. He does not appear to be in the least influenced by their color; but "is most capricious in his attachments" [31]: from some inexplicable cause he shows the most determined aversion to certain hens, which no care on the part of the breeder can overcome. Mr. Hewitt informs me that some hens are quite unattractive even to the males of their own species, so that they may be kept with several cocks during a whole season, and not one egg out of forty or fifty will prove fertile. On the other hand, with the long-tailed duck (*Harelda glacialis*), "it has been remarked," says M. Ekström, "that certain females are much more courted than the rest. Frequently, indeed, one sees an individual surrounded by six or eight amorous males." Whether this statement is credible, I know not; but the native sportsmen shoot these females in order to stuff them as decoys. [32]

[31] Mr. Hewitt, quoted in Tegetmeier's 'Poultry Book,' 1866, p. 165.
[32] Quoted in Lloyd's 'Game Birds of Sweden,' p. 345.

With respect to female birds feeling a preference for particular males, we must bear in mind that we can judge of choice being exerted only by analogy. If an inhabitant of another planet were to behold a number of young rustics at a fair courting a pretty girl, and quarrelling about her like birds at one of their places of assemblage, he would, by the eagerness of the wooers to please her and to display their finery, infer that she had the power of choice. Now with birds the evidence stands thus: they have acute powers of observation, and they seem to have some taste for the beautiful both in color and sound. It is certain that the females occasionally exhibit, from unknown causes, the strongest antipathies and preferences for particular males. When the sexes differ in color or in other ornaments the males with rare exceptions are the more decorated, either permanently or temporarily during the breeding-season. They sedulously display their various ornaments, exert their voices, and perform strange antics in the presence of the females. Even well-armed males, who, it might be thought, would altogether depend for success on the law of battle, are in most cases highly ornamented; and their ornaments have been acquired at the expense of some loss of power. In other cases ornaments have been acquired, at the cost of increased risk from birds and beasts of prey. With various species many individuals of both sexes congregate at the same spot, and their courtship is a prolonged affair. There is even reason to suspect that the males and females within the same district do not always succeed in pleasing each other and pairing.

What then are we to conclude from these facts and considerations? Does the male parade his charms with so much pomp and rivalry for no purpose? Are we not justified in believing that the female exerts a choice, and that she receives the addresses of the male who pleases her most? It is not probable that she consciously deliberates; but she is most excited or attracted by the most beautiful, or melodious, or gallant males. Nor need it be supposed that the female studies each stripe or spot of color; that the peahen, for instance, admires each detail in the gorgeous train of the peacock—she is probably struck only by the general effect. Nevertheless, after hearing how carefully the male Argus pheasant displays his elegant primary wing-feathers, and erects his ocellated plumes in the right position for their full effect; or again, how the male goldfinch alternately displays his gold-bespangled wings, we ought not to feel too sure that the female does not attend to each detail of beauty. We can judge, as already remarked, of choice being exerted, only from analogy; and the mental powers of birds do not differ fundamentally from ours. From these various considerations we may conclude that the pairing of birds is not left to chance; but that those males, which are best able by their various charms to please or excite the female, are under ordinary circumstances accepted. If this be admitted, there is not much difficulty in understanding how male birds have gradually acquired their ornamental characters. All animals present individual differences, and as man can modify his domesticated birds by selecting the individuals which appear to him the most beautiful, so the habitual or even occasional preference by the female of the more attractive males would almost certainly lead to their modification; and such modifications might in the course of time be augmented to almost any extent, compatible with the existence of the species.

VARIABILITY OF BIRDS, AND ESPECIALLY OF THEIR SECONDARY SEXUAL CHARACTERS.

Variability and inheritance are the foundations for the work of selection. That domesticated birds have varied greatly, their variations being inherited, is certain. That birds in a state of nature have been modified into distinct races is now universally admitted. [33] Variations may be divided into two classes; those which appear to our ignorance to arise spontaneously, and those which are directly related to the surrounding conditions, so that all or nearly all the individuals of the same species are similarly modified. Cases of the latter kind have recently been observed with care by Mr. J.A. Allen, [34] who shows that in the United States many species of birds gradually become more strongly colored in proceeding southward, and more lightly colored in proceeding westward to the arid plains of the interior. Both sexes seem generally to be affected in a like manner, but sometimes one sex more than the other. This result is not incompatible with the belief that the colors of birds are mainly due to the accumulation of successive variations through sexual selection; for even after the sexes have been greatly differentiated, climate might produce an equal effect on both sexes, or a greater effect on one sex than on the other, owing to some constitutional difference.

[33] According to Dr. Blasius ('Ibis,' vol. ii. 1860, p. 297), there are 425 indubitable species of birds which breed in Europe, besides sixty forms, which are frequently regarded as distinct species. Of the latter, Blasius thinks that only ten are really doubtful, and that the other fifty ought to be united with their nearest allies; but this shows that there must be a considerable amount of variation with some of our European birds. It is also an unsettled point with naturalists, whether several North American birds ought to be ranked as specifically distinct from the corresponding European species. So again many North American forms which until lately were named as distinct species, are now considered to be local races.

[34] 'Mammals and Birds of East Florida,' also an 'Ornithological Reconnaissance of Kansas,' etc. Notwithstanding the influence of climate on the colors of birds, it is difficult to account for the dull or dark tints of almost all the species inhabiting certain countries, for instance, the Galapagos Islands under the equator, the wide temperate plains of Patagonia, and, as it appears, Egypt (see Mr. Hartshorne in the 'American Naturalist,' 1873, p. 747). These countries are open, and afford little shelter to birds; but it seems doubtful whether the absence of brightly colored species can be explained on the principle of protection, for on the Pampas, which are equally open, though covered by green grass, and where the birds would be equally exposed to danger, many brilliant and conspicuously colored species are common. I have sometimes speculated whether the prevailing dull tints of the scenery in the above named countries may not have affected the appreciation of bright colors by the birds inhabiting them.

Individual differences between the members of the same species are admitted by every one to occur under a state of nature. Sudden and strongly marked variations are rare; it is also doubtful whether if beneficial they would often be preserved through selection and transmitted to succeeding generations. [35] Nevertheless, it may be worth while to give the few cases which I have been able to collect, relating chiefly to color,—simple albinism and melanism being excluded. Mr. Gould is well known to admit the existence of few varieties, for he esteems very slight differences as specific; yet he states [36] that near Bogota certain humming-birds belonging to the genus Cynanthus are divided into two or three races or varieties, which differ from each other in the coloring of the tail—"some having the whole of the feathers blue, while others have the eight central ones tipped with beautiful green." It does not appear that intermediate gradations have been observed in this or the following cases. In the males alone of one of the Australian parrakeets "the thighs in some are scarlet, in others grass-green." In another parrakeet of the same country "some individuals have the band across the wing-coverts bright-yellow, while in others the same part is tinged with red. [37] In the United States some few of the males of the scarlet tanager (*Tanagra rubra*) have "a beautiful transverse band of glowing red on the smaller wing-coverts" [38]; but this variation seems to be somewhat rare, so that its preservation through sexual selection would follow only under usually favourable circumstances. In Bengal the Honey buzzard (*Pernis cristata*) has either a small rudimental crest on its head, or none at all: so slight a difference, however, would not have been worth notice, had not this same species possessed in Southern India a well-marked occipital crest formed of several graduated feathers." [39]

[35] 'Origin of Species' fifth edit. 1869, p.104. I had always perceived, that rare and strongly-marked deviations of structure, deserving to be called monstrosities, could seldom be preserved through natural selection, and that the preservation of even highly-beneficial variations would depend to a certain extent on chance. I had also fully appreciated the importance of mere individual differences, and this led me to insist so strongly on the importance of that unconscious form of selection by man, which follows from the preservation of the most valued individuals of each breed, without any intention on his part to modify the characters of the breed. But until I read an able article in the 'North British Review' (March 1867, p. 289, et seq.), which has been of more use to me than any other Review, I did not see how great the chances were against the preservation of variations, whether slight or strongly pronounced, occurring only in single individuals.

[36] 'Introduction to the Trochlidæ,' p. 102.

[37] Gould, 'Handbook to Birds of Australia,' vol. ii. pp. 32 and 68.

[38] Audubon, 'Ornithological Biography,' 1838, vol. iv. p. 389.

[39] Jerdon, 'Birds of India,' vol. i. p. 108; and Mr. Blyth, in 'Land and Water,' 1868, p. 381.

The following case is in some respects more interesting. A pied variety of the raven, with the head, breast, abdomen, and parts of the wings and tail-feathers white, is confined to the Feroe Islands. It is not very rare there, for Graba saw during his visit from eight to ten living specimens. Although the characters of this variety are not quite constant, yet it has been named by several distinguished ornithologists as a distinct species. The fact of the pied birds being pursued and persecuted with much clamor by the other ravens of the island was the chief cause which led Brünnich to conclude that they were specifically distinct; but this is now known to be an error. [40] This case seems analogous to that lately given of albino birds not pairing from being rejected by their comrades.

[40] Graba, 'Tagebuch Reise nach Färo,' 1830, ss. 51-54. Macgillivray, 'History of British Birds,' vol. iii. p. 745, 'Ibis,' vol. v. 1863, p. 469.

In various parts of the northern seas a remarkable variety of the common Guillemot (*Uria troile*) is found; and in Feroe, one out of every five birds, according to Graba's estimation, presents this variation. It is characterized [41] by a pure white ring round the eye, with a curved narrow white line, an inch and a half in length, extending back from the ring. This conspicuous character has caused the bird to be ranked by several ornithologists as a distinct species under the name of *U. lacrymans*, but it is now known to be merely a variety. It often pairs with the common kind, yet intermediate gradations have never been seen; nor is this surprising, for variations which appear suddenly, are often, as I have elsewhere shown, [42] transmitted either unaltered or not at all. We thus see that two distinct forms of the same species may co-exist in the same district, and we cannot doubt that if the one had possessed any advantage over the other, it would soon have been multiplied to the exclusion of the latter. If, for instance, the male pied ravens, instead of being persecuted by their comrades, had been highly attractive (like the above pied peacock) to the black female ravens their numbers would have rapidly increased. And this would have been a case of sexual selection.

[41] Graba, ibid. s. 54. Macgillivray, ibid. vol. v. p. 327.
[42] 'Variation of Animals and Plants under Domestication,' vol. ii. p. 92.

With respect to the slight individual differences which are common, in a greater or less degree, to all the members of the same species, we have every reason to believe that they are by far the most important for the work of selection. Secondary sexual characters are eminently liable to vary, both with animals in a state of nature and under domestication. [43] There is also reason to believe, as we have seen in our eighth chapter, that variations are more apt to occur in the male than in the female sex. All these contingencies are highly favourable for sexual selection. Whether characters thus acquired are transmitted to one sex or to both sexes, depends, as we shall see in the following chapter, on the form of inheritance which prevails.

[43] On these points see also 'Variation of Animals and Plants under Domestication,' vol. i. p. 253; vol ii. pp. 73, 75.

It is sometimes difficult to form an opinion whether certain slight differences between the sexes of birds are simply the result of variability with sexually-limited inheritance, without the aid of sexual selection, or whether they have been augmented through this latter process. I do not here refer to the many instances where the male displays splendid colors or other ornaments, of which the female partakes to a slight degree; for these are almost certainly due to characters primarily acquired by the male having been more or less transferred to the female. But what are we to conclude with respect to certain birds in which, for instance, the eyes differ slightly in color in the two sexes? [44] In some cases the eyes differ conspicuously; thus with the storks of the genus *Xenorhynchus*, those of the male are blackish-hazel, whilst those of the females are gamboge-yellow; with many hornbills (Buceros), as I hear from Mr. Blyth, [45] the males have intense crimson eyes, and those of the females are white. In the *Buceros bicornis*, the hind margin of the casque and a stripe on the crest of the beak are black in the male, but not so in the female. Are we to suppose that these black marks and the crimson color of the eyes have been preserved or augmented through sexual selection in the males? This is very doubtful; for Mr. Bartlett showed me in the Zoological Gardens that the inside of the mouth of this Buceros is black in the male and flesh-colored in the female; and their external appearance or beauty would not be thus affected. I observed in Chile [46] that the iris in the condor, when about a year old, is dark-brown, but changes at maturity into yellowish-brown in the male, and into bright red in the female. The male has also a small, longitudinal, leaden-colored, fleshy crest or comb. The comb of many gallinaceous birds is highly ornamental, and assumes vivid colors during the act of courtship; but what are we to think of the dull-colored comb of the condor, which does not appear to us in the least ornamental? The same question may be asked in regard to various other characters, such as the knob on the base of the beak of the Chinese goose (*Anser cygnoides*), which is much larger in the male than in the female. No certain answer can be given to these questions; but we ought to be cautious in assuming that knobs and various fleshy appendages cannot be attractive to the female, when we remember that with savage races of man various hideous deformities—deep scars on the face with the flesh raised into protuberances, the septum of the nose pierced by sticks or bones, holes in the ears and lips stretched widely open—are all admired as ornamental.

[44] See, for instance, on the irides of a Podica and Gallicrex in 'Ibis,' vol. ii. 1860, p. 206; and vol. v. 1863, p. 426.
[45] See also Jerdon, 'Birds of India,' vol. i. pp. 243-245.
[46] 'Zoology of the Voyage of H.M.S. "Beagle,"' 1841, p. 6.

Whether or not unimportant differences between the sexes, such as those just specified, have been preserved through sexual selection, these differences, as well as all others, must primarily depend on the laws of variation. On the principle of correlated development, the plumage often varies on different parts of the body, or over the whole body, in the same manner. We see this well illustrated in certain breeds of the fowl. In all the breeds the feathers on the neck and loins of the males are elongated, and are called hackles; now when both sexes acquire a top-knot, which is a new character in the genus, the feathers on the head of the male become hackle-shaped, evidently on the principle of correlation; whilst those on the head of the female are of the ordinary shape. The color also of the hackles forming the top-knot of the male, is often correlated with that of the hackles on the neck and loins, as may be seen by comparing these feathers in the golden and silver-spangled Polish, the Houdans, and Cève-cœur breeds. In some natural species we may observe exactly the same correlation in the colors of these same feathers, as in the males of the splendid Gold and Amherst pheasants.

The structure of each individual feather generally causes any change in its coloring to be symmetrical; we see this in the various laced, spangled, and pencilled breeds of the fowl; and on the principle of correlation the feathers over the whole body are often colored in the same manner. We are thus enabled without much trouble to rear breeds with their plumage marked almost as symmetrically as in natural species. In laced and spangled fowls the colored margins of the feathers are abruptly defined; but in a mongrel raised by me from a black Spanish cock glossed with green, and a white game-hen, all the feathers were greenish-black, excepting towards their extremities, which were yellowish white; but between the white extremities and the black bases, there was on each feather a symmetrical, curved zone of dark-brown. In some instances the shaft of the feather determines the distribution of the tints; thus with the body-feathers of a mongrel from the same black Spanish cock and a silver-spangled Polish hen, the shaft, together with a narrow space on each side, was greenish-black, and this was surrounded by a regular zone of dark-brown, edged with brownish-white. In these cases we have feathers symmetrically shaded, like those which give so much elegance to the plumage of many natural species. I have also noticed a variety of the common pigeon with the wing-bars symmetrically zoned with three bright shades, instead of being simply black on a slaty-blue ground, as in the parent-species.

In many groups of birds the plumage is differently colored in the several species, yet certain spots, marks, or stripes are retained by all. Analogous cases occur with the breeds of the pigeon, which usually retain the two wing-bars, though they may be colored red, yellow, white, black, or blue, the rest of the plumage being of some wholly different tint. Here is a more curious case, in which certain marks are retained, though colored in a manner almost exactly the opposite of what is natural; the aboriginal pigeon has a blue tail, with the terminal halves of the outer webs of the two outer tail feathers white; now there is a sub-variety having a white instead of a blue tail, with precisely that part black which is white in the parent-species. [47]

[47] Bechstein, 'Naturgeschichte Deutschlands,' B. iv. 1795, s. 31, on a sub-variety of the

Monck pigeon.

FORMATION AND VARIABILITY OF THE OCELLI
OR EYE-LIKE SPOTS ON THE PLUMAGE OF BIRDS.

As no ornaments are more beautiful than the ocelli on the feathers of various birds, on the hairy coats of some mammals, on the scales of reptiles and fishes, on the skin of amphibians, on the wings of many Lepidoptera and other insects, they deserve to be especially noticed. An ocellus consists of a spot within a ring of another color, like the pupil within the iris, but the central spot is often surrounded by additional concentric zones. The ocelli on the tail-coverts of the peacock offer a familiar example, as well as those on the wings of the peacock-butterfly (Vanessa). Mr. Trimen has given me a description of a S. African moth (*Gynanisa isis*), allied to our Emperor moth, in which a magnificent ocellus occupies nearly the whole surface of each hinder wing; it consists of a black centre, including a semi-transparent crescent-shaped mark, surrounded by successive, ochre-yellow, black, ochre-yellow, pink, white, pink, brown, and whitish zones. Although we do not know the steps by which these wonderfully beautiful and complex ornaments have been developed, the process has probably been a simple one, at least with insects; for, as Mr. Trimen writes to me, "no characters of mere marking or coloration are so unstable in the Lepidoptera as the ocelli, both in number and size." Mr. Wallace, who first called my attention to this subject, showed me a series of specimens of our common meadow-brown butterfly (*Hipparchia janira*) exhibiting numerous gradations from a simple minute black spot to an elegantly-shaded ocellus. In a S. African butterfly (*Cyllo leda*, Linn.), belonging to the same family, the ocelli are even still more variable. In some specimens (A, Fig. 53) large spaces on the upper surface of the wings are colored black, and include irregular white marks; and from this state a complete gradation can be traced into a tolerably perfect ocellus (A[1]), and this results from the contraction of the irregular blotches of color. In another series of specimens a gradation can be followed from excessively minute white dots, surrounded by a scarcely visible black line (B), into perfectly symmetrical and large ocelli (B[1]). [48] In cases like these, the development of a perfect ocellus does not require a long course of variation and selection.

[48] This woodcut has been engraved from a beautiful drawing, most kindly made for me by Mr. Trimen; see also his description of the wonderful amount of variation in the coloration and shape of the wings of this butterfly, in his 'Rhopalocera Africae Australis,' p. 186.

With birds and many other animals, it seems to follow from the comparison of allied species that circular spots are often generated by the breaking up and contraction of stripes. In the Tragopan pheasant faint white lines in the female represent the beautiful white spots in the male; [49] and something of the same kind may be observed in the two sexes of the Argus pheasant. However this may be, appearances strongly favour the belief that on the one hand, a dark spot is often formed by the coloring matter being drawn towards a central point from a surrounding zone, which latter is thus rendered lighter; and, on the other hand, that a white spot is often formed by the color being driven away from a central point, so that it accumulates in a surrounding darker zone. In either case an ocellus is the result. The coloring matter seems to be a nearly constant quantity, but is redistributed, either centripetally or centrifugally. The feathers of the common guinea-fowl offer a good instance of white spots surrounded by darker zones; and wherever the white spots are large and stand near each other, the surrounding dark zones become confluent. In the same wing-feather of the Argus pheasant dark spots may be seen surrounded by a pale zone, and white spots by a dark zone. Thus the formation of an ocellus in its most elementary state appears to be a simple affair. By what further steps the more complex ocelli, which are surrounded by many successive zones of color, have been generated, I will not pretend to say. But the zoned feathers of the mongrels from differently colored fowls, and the extraordinary variability of the ocelli on many Lepidoptera, lead us to conclude that their formation is not a complex process, but depends on some slight and graduated change in the nature of the adjoining tissues.

[49] Jerdon, 'Birds of India,' vol. iii. p. 517.

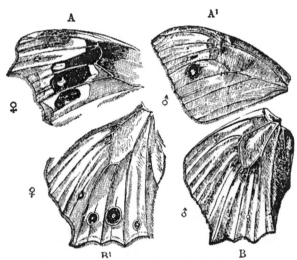

FIG. 53.—Cyllo leda, Linn., from a drawing by Mr. Trimen, showing the extreme range of variation in the ocelli.

A. Specimen, from Mauritius, upper surface of fore-wing.
A¹ Specimen, from Natal, ditto.

B. Specimen, from Java, upper surface of hind-wing.
B¹ Specimen, from Mauritius, ditto.

GRADATION OF SECONDARY SEXUAL CHARACTERS.

Cases of gradation are important, as showing us that highly complex ornaments may be acquired by small successive steps. In order to discover the actual steps by which the male of any existing bird has acquired his magnificent colors or other ornaments, we ought to behold the long line of his extinct progenitors; but this is obviously impossible. We may, however, generally gain a clue by comparing all the species of the same group, if it be a large one; for some of them will probably retain, at least partially, traces of their former characters. Instead of entering on tedious details respecting various groups, in which striking instances of gradation could be given, it seems the best plan to take one or two strongly marked cases, for instance that of the peacock, in order to see if light can be thrown on the steps by which this bird bas become so splendidly decorated. The peacock is chiefly remarkable from the extraordinary length of his tail-coverts; the tail itself not being much elongated. The barbs along nearly the whole length of these feathers stand separate or are decomposed; but this is the case with the feathers of many species, and with some varieties of the domestic fowl and pigeon. The barbs coalesce towards the extremity of the shaft forming the oval disc or ocellus, which is certainly one of the most beautiful objects in the world. It consists of an iridescent, intensely blue, indented centre, surrounded by a rich green zone, this by a broad coppery-brown zone, and this by five other narrow zones of slightly different iridescent shades. A trifling character in the disc deserves notice; the barbs, for a space along one of the concentric zones are more or less destitute of their barbules, so that a part of the disc is surrounded by an almost transparent zone, which gives it a highly finished aspect. But I have elsewhere described [50] an exactly analogous variation in the hackles of a sub-variety of the game-cock, in which the tips, having a metallic lustre, "are separated from the lower part of the feather by a symmetrically shaped transparent zone, composed of the naked portions of the barbs." The lower margin or base of the dark-blue centre of the ocellus is deeply indented on the line of the shaft. The surrounding zones likewise show traces, as may be seen in the drawing (Fig. 54), of indentations, or rather breaks. These indentations are common to the Indian and Javan peacocks (*Pavo cristatus* and *P. muticus*); and they seem to deserve particular attention, as probably connected with the development of the ocellus; but for a long time I could not conjecture their meaning.

[50] 'Variation of Animals and Plants under Domestication,' vol. i. p. 254.

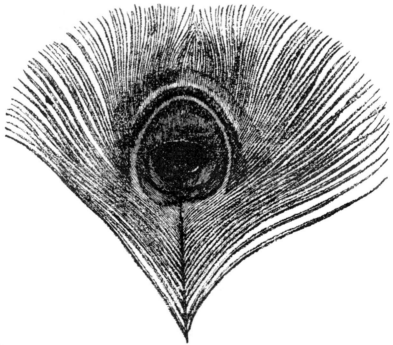

Fɪɢ. 54.—Feather of Peacock, about two-thirds of natural size, drawn by Mr. Ford.
The transparent zone is represented by the outermost white zone, confined to the upper end of the disc.

If we admit the principle of gradual evolution, there must formerly have existed many species which presented every successive step between the wonderfully elongated tail-coverts of the peacock and the short tail-coverts of all ordinary birds; and again between the magnificent ocelli of the former, and the simpler ocelli or mere colored spots on other birds; and so with all the other characters of the peacock. Let us look to the allied Gallinaceæ for any still-existing gradations. The species and sub-species of Polyplectron inhabit countries adjacent to the native land of the peacock; and they so far resemble this bird that they are sometimes called peacock-pheasants. I am also informed by Mr. Bartlett that they resemble the peacock in their voice and in some of their habits. During the spring the males, as previously described, strut about before the comparatively plain-colored females, expanding and erecting their tail and wing-feathers, which are ornamented with numerous ocelli. I request the reader to turn back to the drawing (Fig. 51) of a Polyplectron; In *P. napoleonis* the ocelli are confined to the tail, and the back is of a rich metallic blue; in which respects this species approaches the Java peacock. *P. hardwickii* possesses a peculiar top-knot, which is also somewhat like that of the Java peacock. In all the species the ocelli on the wings and tail are either circular or oval, and consist of a beautiful, iridescent, greenish-blue or greenish-purple disc, with a black border. This border in *P. chinquis* shades into brown, edged with cream color, so that the ocellus is here surrounded with variously shaded, though not bright, concentric zones. The unusual length of the tail-coverts is another remarkable character in Polyplectron; for in some of the species they are half, and in others two-thirds as long as the true tail-feathers. The tail-coverts are ocellated as in the peacock. Thus the several species of Polyplectron manifestly make a graduated approach to the peacock in the length of their tail-coverts, in the zoning of the ocelli, and in some other characters.

Notwithstanding this approach, the first species of Polyplectron which I examined almost made me give up the search; for I found not only that the true tail-feathers, which in the peacock are quite plain, were ornamented with ocelli, but that the ocelli on all the feathers differed fundamentally from those of the peacock, in there being two on the same feather (Fig. 55), one on each side of the shaft. Hence I concluded that the early progenitors of the peacock could not have resembled a Polyplectron. But on continuing my search, I observed that in some of the species the two ocelli stood very near each other; that in the tail-feathers of *P. hardwickii* they touched each other; and, finally, that on the tail-coverts of this same species as well as of *P. malaccense* (Fig. 56) they were actually confluent. As the central part alone is confluent, an indentation is left at both the upper and lower ends; and the surrounding colored zones are likewise indented. A single ocellus is thus formed on each tail-covert, though still plainly betraying its double origin. These confluent ocelli differ from the single ocelli of the peacock in having an indentation at both ends, instead of only at the lower or basal end. The explanation, however, of this difference is not difficult; in some species of Polyplectron the two oval ocelli on the same feather stand parallel to each other; in other species (as in *P. chinquis*) they converge towards one end; now the partial confluence of two convergent ocelli would manifestly leave a much deeper indentation at the divergent than at the convergent end. It is also manifest that if the convergence were strongly pronounced and the confluence complete, the indentation at the convergent end would tend to disappear.

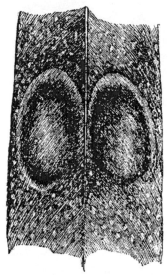

Fig. 55.— Part of a tail-covert of Polyplectron chinquis, with the two ocelli of nat. size.

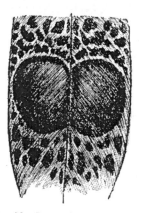

Fig. 56.—Part of a tail-covert of Polyplectron malaccense, with the two ocelli, partially confluent, of nat. size.

The tail-feathers in both species of the peacock are entirely destitute of ocelli, and this apparently is related to their being covered up and concealed by the long tail-coverts. In this respect they differ remarkably from the tail-feathers of Polyplectron, which in most of the species are ornamented with larger ocelli than those on the tail-coverts. Hence I was led carefully to examine the tail-feathers of the several species, in order to discover whether their ocelli showed any tendency to disappear; and to my great satisfaction, this appeared to be so. The central tail-feathers of *P. napoleonis* have the two ocelli on each side of the shaft perfectly developed; but the inner ocellus becomes less and less conspicuous on the more exterior tail-feathers, until a mere shadow or rudiment is left on the inner side of the outermost feather. Again, in *P. malaccense*, the ocelli on the tail-coverts are, as we have seen, confluent; and these feathers are of unusual length, being two-thirds of the length of the tail-feathers, so that in both these respects they approach the tail-coverts of the peacock. Now in *P. malaccense*, the two central tail-feathers alone are ornamented, each with two brightly-colored ocelli, the inner ocellus having completely disappeared from all the other tail-feathers. Consequently the tail-coverts and tail-feathers of this species of Polyplectron make a near approach in structure and ornamentation to the corresponding feathers of the peacock.

As far, then, as gradation throws light on the steps by which the magnificent train of the peacock has been acquired, hardly anything more is needed. If we picture to ourselves a progenitor of the peacock in an almost exactly intermediate condition between the existing peacock, with his enormously elongated tail-coverts, ornamented with single ocelli, and an ordinary gallinaceous bird with short tail-coverts, merely spotted with some color, we shall see a bird allied to Polyplectron—that is, with tail-coverts, capable of erection and expansion, ornamented with two partially confluent ocelli, and long enough almost to conceal the tail-feathers, the latter having already partially lost their ocelli. The indentation of the central disc and of the surrounding zones of the ocellus, in both species of peacock, speaks plainly in favour of this view, and is otherwise inexplicable. The males of Polyplectron are no doubt beautiful birds, but their beauty, when viewed from a little distance, cannot be compared with that of the peacock. Many female progenitors of the peacock must, during a long line of descent, have appreciated this superiority; for they have unconsciously, by the continued preference for the most beautiful males, rendered the peacock the most splendid of living birds.

ARGUS PHEASANT.

Another excellent case for investigation is offered by the ocelli on the wing-feathers of the Argus pheasant, which are shaded in so wonderful a manner as to resemble balls lying loose within sockets, and consequently differ from ordinary ocelli. No one, I presume, will attribute the shading, which has excited the admiration of many experienced artists, to chance—to the fortuitous concourse of atoms of coloring matter. That these ornaments should have been formed through the selection of many successive variations, not one of which was originally intended to produce the ball-and-socket effect, seems as incredible as that one of Raphael's Madonnas should have been formed by the selection of chance daubs of paint made by a long succession of young artists, not one of whom intended at first to draw the human figure. In order to discover how the ocelli have been developed, we cannot look to a long line of progenitors, nor to many closely-allied forms, for such do not now exist. But fortunately the several feathers on the wing suffice to give us a clue to the problem, and they prove to demonstration that a gradation is at least possible from a mere spot to a finished ball-and-socket ocellus.

The wing-feathers, bearing the ocelli, are covered with dark stripes (Fig. 57) or with rows of dark spots (Fig. 59), each stripe or row of spots running obliquely down the outer side of the shaft to one of the ocelli. The spots are generally elongated in a line transverse to the row in which they stand. They often become confluent either in the line of the row—and then they form a longitudinal stripe—or transversely, that is, with the spots in the adjoining rows, and then they form transverse stripes. A spot sometimes breaks up into smaller spots, which still stand in their proper places.

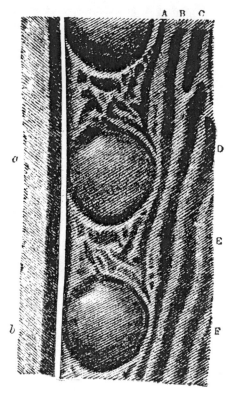

It will be convenient first to describe a perfect ball-and-socket ocellus. This consists of an intensely black circular ring, surrounding a space shaded so as exactly to resemble a ball. The figure here given has been admirably drawn by Mr. Ford and well engraved, but a woodcut cannot exhibit the exquisite shading of the original. The ring is almost always slightly broken or interrupted (Fig. 57) at a point in the upper half, a little to the right of and above the white shade on the enclosed ball; it is also sometimes broken towards the base on the right hand. These little breaks have an important meaning. The ring is always much thickened, with the edges ill-defined towards the left-hand upper corner, the feather being held erect, in the position in which it is here drawn. Beneath this thickened part there is on the surface of the ball an oblique almost pure-white mark, which shades off downwards into a pale-leaden hue, and this into yellowish and brown tints, which insensibly become darker and darker towards the lower part of the ball. It is this shading which gives so admirably the effect of light shining on a convex surface. If one of the balls be examined, it will be seen that the lower part is of a brown tint and is indistinctly separated by a curved oblique line from the upper part, which is yellower and more leaden; this curved oblique line runs at right angles to the longer axis of the white patch of light, and indeed of all the shading; but this difference in color, which cannot of course be shown in the woodcut, does not in the least interfere with the perfect shading of the ball. It should be particularly observed that each ocellus stands in obvious connection either with a dark stripe, or with a longitudinal row of dark spots, for both occur indifferently on the same feather. Thus in Fig. 57 stripe A runs to ocellus a; B runs to ocellus b; stripe C is broken in the upper part, and runs down to the next succeeding ocellus, not represented in the woodcut; D to the next lower one, and so with the stripes E and F. Lastly, the several ocelli are separated from each other by a pale surface bearing irregular black marks.

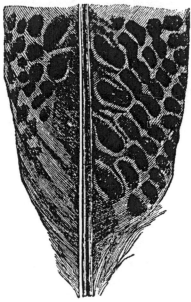

Fig. 58.—Basal part of the secondary
wing-feather, nearest to the body.

I will next describe the other extreme of the series, namely, the first trace of an ocellus. The short secondary wing-feather (Fig. 58), nearest to the body, is marked like the other feathers, with oblique, longitudinal, rather irregular, rows of very dark spots. The basal spot, or that nearest the shaft, in the five lower rows (excluding the lowest one) is a little larger than the other spots of the same row, and a little more elongated in a transverse direction. It differs also from the other spots by being bordered on its upper side with some dull fulvous shading. But this spot is not in any way more remarkable than those on the plumage of many birds, and might easily be overlooked. The next higher spot does not differ at all from the upper ones in the same row. The larger basal spots occupy exactly the same relative position on these feathers as do the perfect ocelli on the longer wing-feathers.

352

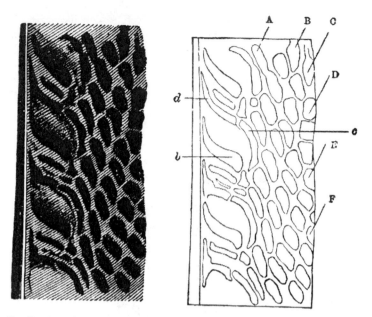

F₁ɢ. 59.—Portion of one of the secondary wing-feathers near to the body, showing the so-called elliptic ornaments. The right-hand figure is given merely as a diagram for the sake of the letters of reference.

A, B, C, D, etc. Rows of spots running down to and forming the elliptic ornaments.

b. Lowest spot or mark in row B.

c. The next succeeding spot or mark in the same row.

d. Apparently a broken prolongation of the spot *c.* in the same row B.

By looking to the next two or three succeeding wing-feathers, an absolutely insensible gradation can be traced from one of the last-described basal spots, together with the next higher one in the same row, to a curious ornament, which cannot be called an ocellus, and which I will name, from the want of a better term, an "elliptic ornament." These are shown in the accompanying figure (Fig. 59). We here see several oblique rows, A, B, C, D, etc. (see the lettered diagram on the right hand), of dark spots of the usual character. Each row of spots runs down to and is connected with one of the elliptic ornaments, in exactly the same manner as each stripe in Fig. 57 runs down to and is connected with one of the ball-and-socket ocelli. Looking to any one row, for instance, B, in Fig. 59, the lowest mark (*b*) is thicker and considerably longer than the upper spots, and has its left extremity pointed and curved upwards. This black mark is abruptly bordered on its upper side by a rather broad space of richly shaded tints, beginning with a narrow brown zone, which passes into orange, and this into a pale leaden tint, with the end towards the shaft much paler. These shaded tints together fill up the whole inner space of the elliptic ornament. The mark (*b*) corresponds in every respect with the basal shaded spot of the simple feather described in the last paragraph (Fig. 58), but is more highly developed and more brightly colored. Above and to the right of this spot (*b*, Fig. 59), with its bright shading, there is a long narrow, black mark (*c*), belonging to the same row, and which is arched a little downwards so as to face (*b*). This mark is sometimes broken into two portions. It is also narrowly edged on the lower side with a fulvous tint. To the left of and above c, in the same oblique direction, but always more or less distinct from it, there is another black mark (*d*). This mark is generally sub-triangular and irregular in shape, but in the one lettered in the diagram it is unusually narrow, elongated, and regular. It apparently consists of a lateral and broken prolongation of the mark (c), together with its confluence with a broken and prolonged part of the next spot above; but I do not feel sure of this. These three marks, b, c, and d, with the intervening bright shades, form together the so-called elliptic ornament. These ornaments placed parallel to the shaft, manifestly correspond in position with the ball-and-socket ocelli. Their extremely elegant appearance cannot be appreciated in the drawing, as the orange and leaden tints, contrasting so well with the black marks, cannot be shown.

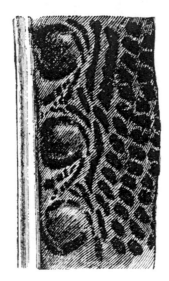

FIG. 60.—An ocellus in an intermediate condition between the elliptic ornament and the perfect ball-and-socket ocellus.

Between one of the elliptic ornaments and a perfect ball-and-socket ocellus, the gradation is so perfect that it is scarcely possible to decide when the latter term ought to be used. The passage from the one into the other is effected by the elongation and greater curvature in opposite directions of the lower black mark (b, Fig. 59), and more especially of the upper one (c), together with the contraction of the elongated sub-triangular or narrow mark (d), so that at last these three marks become confluent, forming an irregular elliptic ring. This ring is gradually rendered more and more circular and regular, increasing at the same time in diameter. I have here given a drawing (Fig. 60) of the natural size of an ocellus not as yet quite perfect. The lower part of the black ring is much more curved than is the lower mark in the elliptic ornament (b, Fig. 59). The upper part of the ring consists of two or three separate portions; and there is only a trace of the thickening of the portion which forms the black mark above the white shade. This white shade itself is not as yet much concentrated; and beneath it the surface is brighter colored than in a perfect ball-and-socket ocellus. Even in the most perfect ocelli traces of the junction of three or four elongated black marks, by which the ring has been formed, may often be detected. The irregular sub-triangular or narrow mark (d, Fig. 59), manifestly forms, by its contraction and equalization, the thickened portion of the ring above the white shade on a perfect ball-and-socket ocellus. The lower part of the ring is invariably a little thicker than the other parts (Fig. 57), and this follows from the lower black mark of the elliptic ornament (b, Fig. 59) having originally been thicker than the upper mark (c). Every step can be followed in the process of confluence and modification; and the black ring which surrounds the ball of the ocellus is unquestionably formed by the union and modification of the three black marks, b, c, d, of the elliptic ornament. The irregular zigzag black marks between the successive ocelli (Fig. 57) are plainly due to the breaking up of the somewhat more regular but similar marks between the elliptic ornaments.

The successive steps in the shading of the ball-and-socket ocelli can be followed out with equal clearness. The brown, orange, and pale-leadened narrow zones, which border the lower black mark of the elliptic ornament, can be seen gradually to become more and more softened and shaded into each other, with the upper lighter part towards the left-hand corner rendered still lighter, so as to become almost white, and at the same time more contracted. But even in the most perfect ball-and-socket ocelli a slight difference in the tints, though not in the shading, between the upper and lower parts of the ball can be perceived, as before noticed; and the line of separation is oblique, in the same direction as the bright colored shades of the elliptic ornaments. Thus almost every minute detail in the shape and coloring of the ball-and-socket ocelli can be shown to follow from gradual changes in the elliptic ornaments; and the development of the latter can be traced by equally small steps from the union of two almost simple spots, the lower one (Fig. 58) having some dull fulvous shading on its upper side.

The extremities of the longer secondary feathers which bear the perfect ball-and-socket ocelli, are peculiarly ornamented (Fig. 61). The oblique longitudinal stripes suddenly cease upwards and become confused; and above this limit the whole upper end of the feather (a) is covered with white dots, surrounded by little black rings, standing on a dark ground. The oblique stripe belonging to the uppermost ocellus (b) is barely represented by a very short irregular black mark with the usual, curved, transverse base. As this stripe is thus abruptly cut off, we can perhaps understand from what has gone before, how it is that the upper thickened part of the ring is here absent; for, as before stated, this thickened part apparently stands in some relation with a broken prolongation from the next higher spot. From the absence of the upper and thickened part of the ring, the uppermost ocellus, though perfect in all other respects, appears as if its top had been obliquely sliced off. It would, I think, perplex any one, who believes that the plumage of the Argus pheasant was created as we now see it, to account for the imperfect condition of the uppermost ocellus. I should add that on the secondary wing-feather farthest from the body all the ocelli are smaller and less perfect than on the other feathers, and have the upper part of the ring deficient, as in the case just mentioned. The imperfection here seems to be connected with the fact that the spots on this feather show less tendency than usual to become confluent into stripes; they are, on the contrary, often broken up into smaller spots, so that two or three rows run down to the same ocellus.

Fig. 61.—Portion near summit of one of the secondary wing-feathers, bearing perfect ball-and-socket ocelli.

a, Ornamented upper part.
b. Uppermost, imperfect ball-and-socket ocellus. (The shading above the white mark on the summit of the ocellus is here a little too dark.)
c. Perfect ocellus.

There still remains another very curious point, first observed by Mr. T.W. Wood, [51] which deserves attention. In a photograph, given me by Mr. Ward, of a specimen mounted as in the act of display, it may be seen that on the feathers which are held perpendicularly, the white marks on the ocelli, representing light reflected from a convex surface, are at the upper or further end, that is, are directed upwards; and the bird whilst displaying himself on the ground would naturally be illuminated from above. But here comes the curious point; the outer feathers are held almost horizontally, and their ocelli ought likewise to appear as if illuminated from above, and consequently the white marks ought to be placed on the upper sides of the ocelli; and, wonderful as is the fact, they are thus placed! Hence the ocelli on the several feathers, though occupying very different positions with respect to the light, all appear as if illuminated from above, just as an artist would have shaded them. Nevertheless they are not illuminated from strictly the same point as they ought to be; for the white marks on the ocelli of the feathers which are held almost horizontally, are placed rather too much towards the further end; that is, they are not sufficiently lateral. We have, however, no right to expect absolute perfection in a part rendered ornamental through sexual selection, any more than we have in a part modified through natural selection for real use; for instance, in that wondrous organ the human eye. And we know what Helmholtz, the highest authority in Europe on the subject, has said about the human eye; that if an optician had sold him an instrument so carelessly made, he would have thought himself fully justified in returning it. [52]

[51] The 'Field,' May 28, 1870.
[52] 'Popular Lectures on Scientific Subjects,' Eng. trans. 1873, pp. 219, 227, 269, 390.

We have now seen that a perfect series can be followed, from simple spots to the wonderful ball-and-socket ornaments. Mr. Gould, who kindly gave me some of these feathers, fully agrees with me in the completeness of the gradation. It is obvious that the stages in development exhibited by the feathers on the same bird do not at all necessarily show us the steps passed through by the extinct progenitors of the species; but they probably give us the clue to the actual steps, and they at least prove to demonstration that a gradation is possible. Bearing in mind how carefully the male Argus pheasant displays his plumes before the female, as well as the many facts rendering it probable that female birds prefer the more attractive males, no one who admits the agency of sexual selection in any case will deny that a simple dark spot with some fulvous shading might be converted, through the approximation and modification of two adjoining spots, together with some slight increase of color, into one of the so-called elliptic ornaments. These latter ornaments have been shown to many persons, and all have admitted that they are beautiful, some thinking them even more so than the ball-and-socket ocelli. As the secondary plumes became lengthened through sexual selection, and as the elliptic ornaments increased in diameter, their colors apparently became less bright; and then the ornamentation of the plumes had to be gained by an improvement in the pattern and shading; and this process was carried on until the wonderful ball-and-socket ocelli were finally developed. Thus we can understand—and in no other way as it seems to me—the present condition and origin of the ornaments on the wing-feathers of the Argus pheasant.

From the light afforded by the principle of gradation—from what we know of the laws of variation—from the changes which have taken place in many of our domesticated birds—and, lastly, from the character (as we shall hereafter see more clearly) of the immature plumage of young birds—we can sometimes indicate, with a certain amount of confidence, the probable steps by which the males have acquired their brilliant plumage and various ornaments; yet in many cases we are involved in complete darkness. Mr. Gould several years ago pointed out to me a humming-bird, the *Urosticte benjamini*, remarkable for the curious differences between the sexes. The male, besides a splendid gorget, has greenish-black tail-feathers, with the four *central* ones tipped with white; in the female, as with most of the allied species, the three *outer* tail-feathers on each side are tipped with white, so that the male has the four central, whilst the female has the six exterior feathers ornamented with white tips. What makes the case more curious is that, although the coloring of the tail differs remarkably in both sexes of many kinds of humming-birds, Mr. Gould does not know a single species, besides the Urosticte, in which the male has the four central feathers tipped with white.

The Duke of Argyll, in commenting on this case, [53] passes over sexual selection, and asks, "What explanation does the law of natural selection give of such specific varieties as these?" He answers "none whatever"; and I quite agree with him. But can this be so confidently said of sexual selection? Seeing in how many ways the tail-feathers of humming-birds differ, why should not the four central feathers have varied in this one species alone, so as to have acquired white tips? The variations may have been gradual, or somewhat abrupt as in the case recently given of the humming-birds near Bogota, in which certain individuals alone have the "central tail-feathers tipped with beautiful green." In the female of the Urosticte I noticed extremely minute or rudimental white tips to the two outer of the four central black tail-feathers; so that here we have an indication of change of some kind in the plumage of this species. If we grant the possibility of the central tail-feathers of the male varying in whiteness, there is nothing strange in such variations having been sexually selected. The white tips, together with the small white ear-tufts, certainly add, as the Duke of Argyll admits, to the beauty of the male; and whiteness is apparently appreciated by other birds, as may be inferred from such cases as the snow-white male of the Bell-bird. The statement made by Sir R. Heron should not be forgotten, namely, that his peahens, when debarred from access to the pied peacock, would not unite with any other male, and during that season produced no offspring. Nor is it strange that variations in the tail-feathers of the Urosticte should have been specially selected for the sake of ornament, for the next succeeding genus in the family takes its name of Metallura from the splendor of these feathers. We have, moreover, good evidence that humming-birds take especial pains in displaying their tail-feathers; Mr. Belt, [54] after describing the beauty of the *Florisuga mellivora*, says, "I have seen the female sitting on a branch, and two males displaying their charms in front of her. One would shoot up like a rocket, then suddenly expanding the snow-white tail, like an inverted parachute, slowly descend in front of her, turning round gradually to show off back and front...The expanded white tail covered more space than all the rest of the bird, and was evidently the grand feature in the performance. Whilst one male was descending, the other would shoot up and come slowly down expanded. The entertainment would end in a fight between the two performers; but whether the most beautiful or the most pugnacious was the accepted suitor, I know not." Mr. Gould, after describing the peculiar plumage of the Urosticte, adds, "that ornament and variety is the sole object, I have myself but little doubt." [55] If this be admitted, we can perceive that the males which during former times were decked in the most elegant and novel manner would have gained an advantage, not in the ordinary struggle for life, but in rivalry with other males, and would have left a larger number of offspring to inherit their newly-acquired beauty.

[53] 'The Reign of Law,' 1867, p. 247.
[54] 'The Naturalist in Nicaragua,' 1874, p. 112.
[55] 'Introduction to the Trochilidæ,' 1861, p. 110.

CHAPTER XV.

BIRDS—continued.

Discussion as to why the males alone of some species, and both sexes of others, are brightly colored—On sexually-limited inheritance, as applied to various structures and to brightly-colored plumage—Nidification in relation to color—Loss of nuptial plumage during the winter.

We have in this chapter to consider why the females of many birds have not acquired the same ornaments as the male; and why, on the other hand, both sexes of many other birds are equally, or almost equally, ornamented? In the following chapter we shall consider the few cases in which the female is more conspicuously colored than the male.

In my 'Origin of Species' [1] I briefly suggested that the long tail of the peacock would be inconvenient and the conspicuous black color of the male capercailzie dangerous, to the female during the period of incubation: and consequently that the transmission of these characters from the male to the female offspring had been checked through natural selection. I still think that this may have occurred in some few instances: but after mature reflection on all the facts which I have been able to collect, I am now inclined to believe that when the sexes differ, the successive variations have generally been from the first limited in their transmission to the same sex in which they first arose. Since my remarks appeared, the subject of sexual coloration has been discussed in some very interesting papers by Mr. Wallace, [2] who believes that in almost all cases the successive variations tended at first to be transmitted equally to both sexes; but that the female was saved, through natural selection, from acquiring the conspicuous colors of the male, owing to the danger which she would thus have incurred during incubation.

[1] Fourth edition, 1866, p. 241.
[2] 'Westminster Review,' July 1867. 'Journal of Travel,' vol. i. 1868, p. 73.

This view necessitates a tedious discussion on a difficult point, namely, whether the transmission of a character, which is at first inherited by both sexes can be subsequently limited in its transmission to one sex alone by means of natural selection. We must bear in mind, as shown in the preliminary chapter on sexual selection, that characters which are limited in their development to one sex are always latent in the other. An imaginary illustration will best aid us in seeing the difficulty of the case; we may suppose that a fancier wished to make a breed of pigeons, in which the males alone should be colored of a pale blue, whilst the females retained their former slaty tint. As with pigeons characters of all kinds are usually transmitted to both sexes equally, the fancier would have to try to convert this latter form of inheritance into sexually-limited transmission. All that he could do would be to persevere in selecting every male pigeon which was in the least degree of a paler blue; and the natural result of this process, if steadily carried on for a long time, and if the pale variations were strongly inherited or often recurred, would be to make his whole stock of a lighter blue. But our fancier would be compelled to match, generation after generation, his pale blue males with slaty females, for he wishes to keep the latter of this color. The result would generally be the production either of a mongrel piebald lot, or more probably the speedy and complete loss of the pale-blue tint; for the primordial slaty color would be transmitted with prepotent force. Supposing, however, that some pale-blue males and slaty females were produced during each successive generation, and were always crossed together, then the slaty females would have, if I may use the expression, much blue blood in their veins, for their fathers, grandfathers, etc., will all have been blue birds. Under these circumstances it is conceivable (though I know of no distinct facts rendering it probable) that the slaty females might acquire so strong a latent tendency to pale-blueness, that they would not destroy this color in their male offspring, their female offspring still inheriting the slaty tint. If so, the desired end of making a breed with the two sexes permanently different in color might be gained.

The extreme importance, or rather necessity in the above case of the desired character, namely, pale-blueness, being present though in a latent state in the female, so that the male offspring should not be deteriorated, will be best appreciated as follows: the male of Sœmmerring's pheasant has a tail thirty-seven inches in length, whilst that of the female is only eight inches; the tail of the male common pheasant is about twenty inches, and that of the female twelve inches long. Now if the female Sœmmerring pheasant with her *short* tail were crossed with the male common pheasant, there can be no doubt that the male hybrid offspring would have a much *longer* tail than that of the pure offspring of the common pheasant. On the other hand, if the female common pheasant, with a tail much longer than that of the female Sœmmerring pheasant, were crossed with the male of the latter, the male hybrid offspring would have a much *shorter* tail than that of the pure offspring of Sœmmerring's pheasant. [3]

[3] Temminck says that the tail of the female *Phasianus Sœmmerringii* is only six inches long, 'Planches coloriées,' vol. v. 1838, pp. 487 and 488: the measurements above given were made for me by Mr. Sclater. For the common pheasant, see Macgillivray, 'History of British Birds,' vol. i. pp. 118-121.

Our fancier, in order to make his new breed with the males of a pale-blue tint, and the females unchanged, would have to continue selecting the males during many generations; and each stage of paleness would have to be fixed in the males, and rendered latent in the females. The task would be an extremely difficult one, and has never been tried, but might possibly be successfully carried out. The chief obstacle would be the early and complete loss of the pale-blue tint, from the necessity of reiterated crosses with the slaty female, the latter not having at first any *latent* tendency to produce pale-blue offspring.

On the other hand, if one or two males were to vary ever so slightly in paleness, and the variations were from the first limited in their transmission to the male sex, the task of making a new breed of the desired kind would be easy, for such males would simply have to be selected and matched with ordinary females. An analogous case has actually occurred, for there are breeds of the pigeon in Belgium [4] in which the males alone are marked with black striæ. So again Mr. Tegetmeier has recently shown [5] that dragons not rarely produce silver-colored birds, which are almost always hens; and he himself has bred ten such females. It is on the other hand a very unusual event when a silver male is produced; so that nothing would be easier, if desired, than to make a breed of dragons with blue males and silver females. This tendency is indeed so strong that when Mr. Tegetmeier at last got a silver male and matched him with one of the silver females, he expected to get a breed with both sexes thus colored; he was however disappointed, for the young male reverted to the blue color of his grandfather, the young female alone being silver. No doubt with patience this tendency to reversion in the males, reared from an occasional silver male matched with a silver hen, might be eliminated, and then both sexes would be colored alike; and this very process has been followed with success by Mr. Esquilant in the case of silver turbits.

[4] Dr. Chapuis, 'Le Pigeon Voyageur Belge,' 1865, p. 87.
[5] The 'Field,' Sept. 1872.

With fowls, variations of color, limited in their transmission to the male sex, habitually occur. When this form of inheritance prevails, it might well happen that some of the successive variations would be transferred to the female, who would then slightly resemble the male, as actually occurs in some breeds. Or again, the greater number, but not all, of the successive steps might be transferred to both sexes, and the female would then closely resemble the male. There can hardly be a doubt that this is the cause of the male pouter pigeon having a somewhat larger crop, and of the male carrier pigeon having somewhat larger wattles, than their respective females; for fanciers have not selected one sex more than the other, and have had no wish that these characters should be more strongly displayed in the male than in the female, yet this is the case with both breeds.

The same process would have to be followed, and the same difficulties encountered, if it were desired to make a breed with the females alone of some new color.

Lastly, our fancier might wish to make a breed with the two sexes differing from each other, and both from the parent species. Here the difficulty would be extreme, unless the successive variations were from the first sexually limited on both sides, and then there would be no difficulty. We see this with the fowl; thus the two sexes of the pencilled Hamburghs differ greatly from each other, and from the two sexes of the aboriginal *Gallus bankiva*; and both are now kept constant to their standard of excellence by continued selection, which would be impossible unless the distinctive characters of both were limited in their transmission.

The Spanish fowl offers a more curious case; the male has an immense comb, but some of the successive variations, by the accumulation of which it was acquired, appear to have been transferred to the female; for she has a comb many times larger than that of the females of the parent species. But the comb of the female differs in one respect from that of the male, for it is apt to lop over; and within a recent period it has been ordered by the fancy that this should always be the case, and success has quickly followed the order. Now the lopping of the comb must be sexually limited in its transmission, otherwise it would prevent the comb of the male from being perfectly upright, which would be abhorrent to every fancier. On the other hand, the uprightness of the comb in the male must likewise be a sexually-limited character, otherwise it would prevent the comb of the female from lopping over.

From the foregoing illustrations, we see that even with almost unlimited time at command, it would be an extremely difficult and complex, perhaps an impossible process, to change one form of transmission into the other through selection. Therefore, without distinct evidence in each case, I am unwilling to admit that this has been effected in natural species. On the other hand, by means of successive variations, which were from the first sexually limited in their transmission, there would not be the least difficulty in rendering a male bird widely different in color or in any other character from the female; the latter being left unaltered, or slightly altered, or specially modified for the sake of protection.

As bright colors are of service to the males in their rivalry with other males, such colors would be selected whether or not they were transmitted exclusively to the same sex. Consequently the females might be expected often to partake of the brightness of the males to a greater or less degree; and this occurs with a host of species. If all the successive variations were transmitted equally to both sexes, the females would be indistinguishable from the males; and this likewise occurs with many birds. If, however, dull colors were of high importance for the safety of the female during incubation, as with many ground birds, the females which varied in brightness, or which received through inheritance from the males any marked accession of brightness, would sooner or later be destroyed. But the tendency in the males to continue for an indefinite period transmitting to their female offspring their own brightness, would have to be eliminated by a change in the form of inheritance; and this, as shown by our previous illustration, would be extremely difficult. The more probable result of the long-continued destruction of the more brightly-colored females, supposing the equal form of transmission to prevail, would be the lessening or annihilation of the bright colors of the males, owing to their continual crossing with the duller females. It would be tedious to follow out all the other possible results; but I may remind the reader that if sexually-limited variations in brightness occurred in the females, even if they were not in the least injurious to them and consequently were not eliminated, yet they would not be favoured or selected, for the male usually accepts any female, and does not select the more attractive individuals; consequently these variations would be liable to be lost, and would have little influence on the character of the race; and this will aid in accounting for the females being commonly duller-colored than the males.

In the eighth chapter instances were given, to which many might here be added, of variations occurring at various ages, and inherited at the corresponding age. It was also shown that variations which occur late in life are commonly transmitted to the same sex in which they first appear; whilst variations occurring early in life are apt to be transmitted to both sexes; not that all the cases of sexually-limited transmission can thus be accounted for. It was further shown that if a male bird varied by becoming brighter whilst young, such variations would be of no service until the age for reproduction had arrived, and there was competition between rival males. But in the case of birds living on the ground and commonly in need of the protection of dull colors, bright tints would be far more dangerous to the young and inexperienced than to the adult males. Consequently the males which varied in brightness whilst young would suffer much destruction and be eliminated through natural selection; on the other hand, the males which varied in this manner when nearly mature, notwithstanding that they were exposed to some additional danger, might survive, and from being favoured through sexual selection, would procreate their kind. As a relation often exists between the period of variation and the form of transmission, if the bright-colored young males were destroyed and the mature ones were successful in their courtship, the males alone would acquire brilliant colors and would transmit them exclusively to their male offspring. But I by no means wish to maintain that the influence of age on the form of transmission, is the sole cause of the great difference in brilliancy between the sexes of many birds.

When the sexes of birds differ in color, it is interesting to determine whether the males alone have been modified by sexual selection, the females having been left unchanged, or only partially and indirectly thus changed; or whether the females have been specially modified through natural selection for the sake of protection. I will therefore discuss this question at some length, even more fully than its intrinsic importance deserves; for various curious collateral points may thus be conveniently considered.

Before we enter on the subject of color, more especially in reference to Mr. Wallace's conclusions, it may be useful to discuss some other sexual differences under a similar point of view. A breed of fowls formerly existed in Germany [6] in which the hens were furnished with spurs; they were good layers, but they so greatly disturbed their nests with their spurs that they could not be allowed to sit on their own eggs. Hence at one time it appeared to me probable that with the females of the wild Gallineæ the development of spurs had been checked through natural selection, from the injury thus caused to their nests. This seemed all the more probable, as wing-spurs, which would not be injurious during incubation, are often as well-developed in the female as in the male; though in not a few cases they are rather larger in the male. When the male is furnished with leg-spurs the female almost always exhibits rudiments of them,—the rudiment sometimes consisting of a mere scale, as in Gallus. Hence it might be argued that the females had aboriginally been furnished with well-developed spurs, but that these had subsequently been lost through disuse or natural selection. But if this view be admitted, it would have to be extended to innumerable other cases; and it implies that the female progenitors of the existing spur-bearing species were once encumbered with an injurious appendage.

[6] Bechstein, 'Naturgeschichte Deutschlands,' 1793, B. iii. 339.

In some few genera and species, as in Galloperdix, Acomus, and the Javan peacock (*Pavo muticus*), the females, as well as the males, possess well-developed leg-spurs. Are we to infer from this fact that they construct a different sort of nest from that made by their nearest allies, and not liable to be injured by their spurs; so that the spurs have not been removed? Or are we to suppose that the females of these several species especially require spurs for their defence? It is a more probable conclusion that both the presence and absence of spurs in the females result from different laws of inheritance having prevailed, independently of natural selection. With the many females in which spurs appear as rudiments, we may conclude that some few of the successive variations, through which they were developed in the males, occurred very early in life, and were consequently transferred to the females. In the other and much rarer cases, in which the females possess fully developed spurs, we may conclude that all the successive variations were transferred to them; and that they gradually acquired and inherited the habit of not disturbing their nests.

The vocal organs and the feathers variously modified for producing sound, as well as the proper instincts for using them, often differ in the two sexes, but are sometimes the same in both. Can such differences be accounted for by the males having acquired these organs and instincts, whilst the females have been saved from inheriting them, on account of the danger to which they would have been exposed by attracting the attention of birds or beasts of prey? This does not seem to me probable, when we think of the multitude of birds which with impunity gladden the country with their voices during the spring. [7] It is a safer conclusion that, as vocal and instrumental organs are of special service only to the males during their courtship, these organs were developed through sexual selection and their constant use in that sex alone—the successive variations and the effects of use having been from the first more or less limited in transmission to the male offspring.

[7] Daines Barrington, however, thought it probable ('Philosophical Transactions,' 1773, p. 164) that few female birds sing, because the talent would have been dangerous to them during incubation. He adds, that a similar view may possibly account for the inferiority of the female to the male in plumage.

Many analogous cases could be adduced; those for instance of the plumes on the head being generally longer in the male than in the female, sometimes of equal length in both sexes, and occasionally absent in the female,—these several cases occurring in the same group of birds. It would be difficult to account for such a difference between the sexes by the female having been benefited by possessing a slightly shorter crest than the male, and its consequent diminution or complete suppression through natural selection. But I will take a more favourable case, namely the length of the tail. The long train of the peacock would have been not only inconvenient but dangerous to the peahen during the period of incubation and whilst accompanying her young. Hence there is not the least *à priori* improbability in the development of her tail having been checked through natural selection. But the females of various pheasants, which apparently are exposed on their open nests to as much danger as the peahen, have tails of considerable length. The females as well as the males of the *Menura superba* have long tails, and they build a domed nest, which is a great anomaly in so large a bird. Naturalists have wondered how the female Menura could manage her tail during incubation; but it is now known [8] that she "enters the nest head first, and then turns round with her tail sometimes over her back, but more often bent round by her side. Thus in time the tail becomes quite askew, and is a tolerable guide to the length of time the bird has been sitting." Both sexes of an Australian kingfisher (*Tanysiptera sylvia*) have the middle tail-feathers greatly lengthened, and the female makes her nest in a hole; and as I am informed by Mr. R.B. Sharpe these feathers become much crumpled during incubation.

[8] Mr. Ramsay, in 'Proc. Zoolog. Soc.' 1868, p. 50.

In these two latter cases the great length of the tail-feathers must be in some degree inconvenient to the female; and as in both species the tail-feathers of the female are somewhat shorter than those of the male, it might be argued that their full development had been prevented through natural selection. But if the development of the tail of the peahen had been checked only when it became inconveniently or dangerously great, she would have retained a much longer tail than she actually possesses; for her tail is not nearly so long, relatively to the size of her body, as that of many female pheasants, nor longer than that of the female turkey. It must also be borne in mind that, in accordance with this view, as soon as the tail of the peahen became dangerously long, and its development was consequently checked, she would have continually reacted on her male progeny, and thus have prevented the peacock from acquiring his present magnificent train. We may therefore infer that the length of the tail in the peacock and its shortness in the peahen are the result of the requisite variations in the male having been from the first transmitted to the male offspring alone.

We are led to a nearly similar conclusion with respect to the length of the tail in the various species of pheasants. In the Eared pheasant (*Crossoptilon auritum*) the tail is of equal length in both sexes, namely sixteen or seventeen inches; in the common pheasant it is about twenty inches long in the male and twelve in the female; in Sœmmerring's pheasant, thirty-seven inches in the male and only eight in the female; and lastly in Reeve's pheasant it is sometimes actually seventy-two inches long in the male and sixteen in the female. Thus in the several species, the tail of the female differs much in length, irrespectively of that of the male; and this can be accounted for, as it seems to me, with much more probability, by the laws of inheritance,—that is by the successive variations having been from the first more or less closely limited in their transmission to the male sex than by the agency of natural selection, resulting from the length of tail being more or less injurious to the females of these several allied species.

We may now consider Mr. Wallace's arguments in regard to the sexual coloration of birds. He believes that the bright tints originally acquired through sexual selection by the males would in all, or almost all cases, have been transmitted to the females, unless the transference had been checked through natural selection. I may here remind the reader that various facts opposed to this view have already been given under reptiles, amphibians, fishes and lepidoptera. Mr. Wallace rests his belief chiefly, but not exclusively, as we shall see in the next chapter, on the following statement, [9] that when both sexes are colored in a very conspicuous manner, the nest is of such a nature as to conceal the sitting bird; but when there is a marked contrast of color between the sexes, the male being gay and the female dull-colored, the nest is open and exposes the sitting bird to view. This coincidence, as far as it goes, certainly seems to favour the belief that the females which sit on open nests have been specially modified for the sake of protection; but we shall presently see that there is another and more probable explanation, namely, that conspicuous females have acquired the instinct of building domed nests oftener than dull-colored birds. Mr. Wallace admits that there are, as might have been expected, some exceptions to his two rules, but it is a question whether the exceptions are not so numerous as seriously to invalidate them.

[9] 'Journal of Travel,' edited by A. Murray, vol. i. 1868, p. 78.

There is in the first place much truth in the Duke of Argyll's remark [10] that a large domed nest is more conspicuous to an enemy, especially to all tree-haunting carnivorous animals, than a smaller open nest. Nor must we forget that with many birds which build open nests, the male sits on the eggs and aids the female in feeding the young: this is the case, for instance, with *Pyranga æstiva*, [11] one of the most splendid birds in the United States, the male being vermilion, and the female light brownish-green. Now if brilliant colors had been extremely dangerous to birds whilst sitting on their open nests, the males in these cases would have suffered greatly. It might, however, be of such paramount importance to the male to be brilliantly colored, in order to beat his rivals, that this may have more than compensated some additional danger.

[10] 'Journal of Travel,' edited by A. Murray, vol. i. 1868, p. 281.
[11] Audubon, 'Ornithological Biography,' vol. i. p. 233.

Mr. Wallace admits that with the King-crows (Dicrurus), Orioles, and Pittidæ, the females are conspicuously colored, yet build open nests; but he urges that the birds of the first group are highly pugnacious and could defend themselves; that those of the second group take extreme care in concealing their open nests, but this does not invariably hold good; [12] and that with the birds of the third group the females are brightly colored chiefly on the under surface. Besides these cases, pigeons which are sometimes brightly, and almost always conspicuously colored, and which are notoriously liable to the attacks of birds of prey, offer a serious exception to the rule, for they almost always build open and exposed nests. In another large family, that of the humming-birds, all the species build open nests, yet with some of the most gorgeous species the sexes are alike; and in the majority, the females, though less brilliant than the males, are brightly colored. Nor can it be maintained that all female humming-birds, which are brightly colored, escape detection by their tints being green, for some display on their upper surfaces red, blue, and other colors. [13]

[12] Jerdon, 'Birds of India,' vol. ii. p. 108. Gould's 'Handbook of the Birds of Australia,'

vol. i. p. 463.

[13] For instance, the female *Eupetomena macroura* has the head and tail dark blue with reddish loins; the female *Lampornis porphyrurus* is blackish-green on the upper surface, with the lores and sides of the throat crimson; the female *Eulampis jugularis* has the top of the head and back green, but the loins and the tail are crimson. Many other instances of highly conspicuous females could be given. See Mr. Gould's magnificent work on this family.

In regard to birds which build in holes or construct domed nests, other advantages, as Mr. Wallace remarks, besides concealment are gained, such as shelter from the rain, greater warmth, and in hot countries protection from the sun; [14] so that it is no valid objection to his view that many birds having both sexes obscurely colored build concealed nests. [15] The female Horn-bill (*Buceros*), for instance, of India and Africa is protected during incubation with extraordinary care, for she plasters up with her own excrement the orifice of the hole in which she sits on her eggs, leaving only a small orifice through which the male feeds her; she is thus kept a close prisoner during the whole period of incubation; [16] yet female horn-bills are not more conspicuously colored than many other birds of equal size which build open nests. It is a more serious objection to Mr. Wallace's view, as is admitted by him, that in some few groups the males are brilliantly colored and the females obscure, and yet the latter hatch their eggs in domed nests. This is the case with the Grallinæ of Australia, the Superb Warblers (Maluridæ) of the same country, the Sun-birds (Nectariniæ), and with several of the Australian Honey-suckers or Meliphagidæ. [17]

[14] Mr. Salvin noticed in Guatemala ('Ibis,' 1864, p. 375) that humming-birds were much more unwilling to leave their nests during very hot weather, when the sun was shining brightly, as if their eggs would be thus injured, than during cool, cloudy, or rainy weather.

[15] I may specify, as instances of dull-colored birds building concealed nests, the species belonging to eight Australian genera described in Gould's 'Handbook of the Birds of Australia,' vol. i. pp. 340, 362, 365, 383, 387, 389, 391, 414.

[16] Mr. C. Horne, 'Proc. Zoolog. Soc.' 1869. p. 243.

[17] On the nidification and colors of these latter species, see Gould's 'Handbook to the Birds of Australia,' vol. i. pp. 504, 527.

If we look to the birds of England we shall see that there is no close and general relation between the colors of the female and the nature of the nest which is constructed. About forty of our British birds (excluding those of large size which could defend themselves) build in holes in banks, rocks, or trees, or construct domed nests. If we take the colors of the female goldfinch, bullfinch, or blackbird, as a standard of the degree of conspicuousness, which is not highly dangerous to the sitting female, then out of the above forty birds the females of only twelve can be considered as conspicuous to a dangerous degree, the remaining twenty-eight being inconspicuous. [18] Nor is there any close relation within the same genus between a well-pronounced difference in color between the sexes, and the nature of the nest constructed. Thus the male house sparrow (*Passer domesticus*) differs much from the female, the male tree-sparrow (*P. montanus*) hardly at all, and yet both build well-concealed nests. The two sexes of the common fly-catcher (*Muscicapa grisola*) can hardly be distinguished, whilst the sexes of the pied fly-catcher (*M. luctuosa*) differ considerably, and both species build in holes or conceal their nests. The female blackbird (*Turdus merula*) differs much, the female ring-ouzel (*T. torquatus*) differs less, and the female common thrush (*T. musicus*) hardly at all from their respective males; yet all build open nests. On the other hand, the not very distantly-allied water-ouzel (*Cinclus aquaticus*) builds a domed nest, and the sexes differ about as much as in the ring-ouzel. The black and red grouse (*Tetrao tetrix* and *T. scoticus*) build open nests in equally well-concealed spots, but in the one species the sexes differ greatly, and in the other very little.

[18] I have consulted, on this subject, Macgillivray's 'British Birds,' and though doubts may be entertained in some cases in regard to the degree of concealment of the nest, and to the degree of conspicuousness of the female, yet the following birds, which all lay their eggs in holes or in domed nests, can hardly be considered, by the above standard, as conspicuous: Passer, 2 species; Sturnus, of which the female is considerably less brilliant than the male; Cinclus; Motallica boarula (?); Erithacus (?); Fruticola, 2 sp.; Saxicola; Ruticilla, 2 sp.; Sylvia, 3 sp.; Parus, 3 sp.; Mecistura; Anorthura; Certhia; Sitta; Yunx; Muscicapa, 2 sp.; Hirundo, 3 sp.; and Cypselus. The females of the following 12 birds may be considered as conspicuous according to the same standard, viz., Pastor, Motacilla alba, Parus major and P. cæruleus, Upupa, Picus, 4 sp., Coracias, Alcedo, and Merops.

Notwithstanding the foregoing objections, I cannot doubt, after reading Mr. Wallace's excellent essay, that looking to the birds of the world, a large majority of the species in which the females are conspicuously colored (and in this case the males with rare exceptions are equally conspicuous), build concealed nests for the sake of protection. Mr. Wallace enumerates [19] a long series of groups in which this rule holds good; but it will suffice here to give, as instances, the more familiar groups of kingfishers, toucans, trogons, puff-birds (Capitonidæ), plantain-eaters (Musophagæ, woodpeckers, and parrots. Mr. Wallace believes that in these groups, as the males gradually acquired through sexual selection their brilliant colors, these were transferred to the females and were not eliminated by natural selection, owing to the protection which they already enjoyed from their manner of nidification. According to this view, their present manner of nesting was acquired before their present colors. But it seems to me much more probable that in most cases, as the females were gradually rendered more and more brilliant from partaking of the colors of the male, they were gradually led to change their instincts (supposing that they originally built open nests), and to seek protection by building domed or concealed nests. No one who studies, for instance, Audubon's account of the differences in the nests of the same species in the Northern and Southern United States, [20] will feel any great difficulty in admitting that birds, either by a change (in the strict sense of the word) of their habits, or through the natural selection of so-called spontaneous variations of instinct, might readily be led to modify their manner of nesting.

[19] 'Journal of Travel,' edited by A. Murray, vol. i. p. 78.
[20] See many statements in the 'Ornithological Biography.' See also some curious observations on the nests of Italian birds by Eugenio Bettoni, in the 'Atti della Società Italiana,' vol. xi. 1869, p. 487.

This way of viewing the relation, as far as it holds good, between the bright colors of female birds and their manner of nesting, receives some support from certain cases occurring in the Sahara Desert. Here, as in most other deserts, various birds, and many other animals, have had their colors adapted in a wonderful manner to the tints of the surrounding surface. Nevertheless there are, as I am informed by the Rev. Mr. Tristram, some curious exceptions to the rule; thus the male of the *Monticola cyanea* is conspicuous from his bright blue color, and the female almost equally conspicuous from her mottled brown and white plumage; both sexes of two species of Dromolæa are of a lustrous black; so that these three species are far from receiving protection from their colors, yet they are able to survive, for they have acquired the habit of taking refuge from danger in holes or crevices in the rocks.

With respect to the above groups in which the females are conspicuously colored and build concealed nests, it is not necessary to suppose that each separate species had its nidifying instinct specially modified; but only that the early progenitors of each group were gradually led to build domed or concealed nests, and afterwards transmitted this instinct, together with their bright colors, to their modified descendants. As far as it can be trusted, the conclusion is interesting, that sexual selection together with equal or nearly equal inheritance by both sexes, have indirectly determined the manner of nidification of whole groups of birds.

According to Mr. Wallace, even in the groups in which the females, from being protected in domed nests during incubation, have not had their bright colors eliminated through natural selection, the males often differ in a slight, and occasionally in a considerable degree from the females. This is a significant fact, for such differences in color must be accounted for by some of the variations in the males having been from the first limited in transmission to the same sex; as it can hardly be maintained that these differences, especially when very slight, serve as a protection to the female. Thus all the species in the splendid group of the Trogons build in holes; and Mr. Gould gives figures [21] of both sexes of twenty-five species, in all of which, with one partial exception, the sexes differ sometimes slightly, sometimes conspicuously, in color,—the males being always finer than the females, though the latter are likewise beautiful. All the species of kingfishers build in holes, and with most of the species the sexes are equally brilliant, and thus far Mr. Wallace's rule holds good; but in some of the Australian species the colors of the females are rather less vivid than those of the male; and in one splendidly-colored species, the sexes differ so much that they were at first thought to be specifically distinct. [22] Mr. R.B. Sharpe, who has especially studied this group, has shown me some American species (Ceryle) in which the breast of the male is belted with black. Again, in Carcineutes, the difference between the sexes is conspicuous: in the male the upper surface is dull-blue banded with black, the lower surface being partly fawn-colored, and there is much red about the head; in the female the upper surface is reddish-brown banded with black, and the lower surface white with black markings. It is an interesting fact, as showing how the same peculiar style of sexual coloring often characterizes allied forms, that in three species of Dacelo the male differs from the female only in the tail being dull-blue banded with black, whilst that of the female is brown with blackish bars; so that here the tail differs in color in the two sexes in exactly the same manner as the whole upper surface in the two sexes of Carcineutes.

[21] See his Monograph of the Trogonidæ, 1st edition.
[22] Namely, Cyanalcyon, Gould's 'Handbook to the Birds of Australia,' vol. i. p. 133; see, also, pp. 130, 136.

With parrots, which likewise build in holes, we find analogous cases: in most of the species, both sexes are brilliantly colored and indistinguishable, but in not a few species the males are colored rather more vividly than the females, or even very differently from them. Thus, besides other strongly-marked differences, the whole under surface of the male King Lory (*Aprosmictus scapulatus*) is scarlet, whilst the throat and chest of the female is green tinged with red: in the *Euphema splendida* there is a similar difference, the face and wing coverts moreover of the female being of a paler blue than in the male. [23] In the family of the tits (*Parinæ*), which build concealed nests, the female of our common blue tomtit (*Parus cæruleus*), is "much less brightly colored" than the male: and in the magnificent Sultan yellow tit of India the difference is greater. [24]

[23] Every gradation of difference between the sexes may be followed in the parrots of Australia. See Gould's 'Handbook,' etc., vol. ii. pp. 14-102.
[24] Macgillivray's 'British Birds,' vol. ii. p. 433. Jerdon, 'Birds of India,' vol. ii. p. 282.

Again, in the great group of the woodpeckers, [25] the sexes are generally nearly alike, but in the *Megapicus validus* all those parts of the head, neck, and breast, which are crimson in the male are pale brown in the female. As in several woodpeckers the head of the male is bright crimson, whilst that of the female is plain, it occurred to me that this color might possibly make the female dangerously conspicuous, whenever she put her head out of the hole containing her nest, and consequently that this color, in accordance with Mr. Wallace's belief, had been eliminated. This view is strengthened by what Malherbe states with respect to *Indopicus carlotta*; namely, that the young females, like the young males, have some crimson about their heads, but that this color disappears in the adult female, whilst it is intensified in the adult male. Nevertheless the following considerations render this view extremely doubtful: the male takes a fair share in incubation, [26] and would be thus almost equally exposed to danger; both sexes of many species have their heads of an equally bright crimson; in other species the difference between the sexes in the amount of scarlet is so slight that it can hardly make any appreciable difference in the danger incurred; and lastly, the coloring of the head in the two sexes often differs slightly in other ways.

[25] All the following facts are taken from M. Malherbe's magnificent 'Monographie des Picidées,' 1861.
[26] Audubon's 'Ornithological Biography,' vol. ii. p. 75; see also the 'Ibis,' vol. i. p. 268.

The cases, as yet given, of slight and graduated differences in color between the males and females in the groups, in which as a general rule the sexes resemble each other, all relate to species which build domed or concealed nests. But similar gradations may likewise be observed in groups in which the sexes as a general rule resemble each other, but which build open nests.

As I have before instanced the Australian parrots, so I may here instance, without giving any details, the Australian pigeons. [27] It deserves especial notice that in all these cases the slight differences in plumage between the sexes are of the same general nature as the occasionally greater differences. A good illustration of this fact has already been afforded by those kingfishers in which either the tail alone or the whole upper surface of the plumage differs in the same manner in the two sexes. Similar cases may be observed with parrots and pigeons. The differences in color between the sexes of the same species are, also, of the same general nature as the differences in color between the distinct species of the same group. For when in a group in which the sexes are usually alike, the male differs considerably from the female, he is not colored in a quite new style. Hence we may infer that within the same group the special colors of both sexes when they are alike, and the colors of the male, when he differs slightly or even considerably from the female, have been in most cases determined by the same general cause; this being sexual selection.

[27] Gould's 'Handbook to the Birds of Australia,' vol. ii. pp. 109-149.

It is not probable, as has already been remarked, that differences in color between the sexes, when very slight, can be of service to the female as a protection. Assuming, however, that they are of service, they might be thought to be cases of transition; but we have no reason to believe that many species at any one time are undergoing change. Therefore we can hardly admit that the numerous females which differ very slightly in color from their males are now all commencing to become obscure for the sake of protection. Even if we consider somewhat more marked sexual differences, is it probable, for instance, that the head of the female chaffinch,—the crimson on the breast of the female bullfinch,—the green of the female greenfinch,—the crest of the female golden-crested wren, have all been rendered less bright by the slow process of selection for the sake of protection? I cannot think so; and still less with the slight differences between the sexes of those birds which build concealed nests. On the other hand, the differences in color between the sexes, whether great or small, may to a large extent be explained on the principle of the successive variations, acquired by the males through sexual selection, having been from the first more or less limited in their transmission to the females. That the degree of limitation should differ in different species of the same group will not surprise any one who has studied the laws of inheritance, for they are so complex that they appear to us in our ignorance to be capricious in their action. [28]

[28] See remarks to this effect in 'Variation of Animals and Plants under Domestication,' vol. ii. chap. xii.

As far as I can discover there are few large groups of birds in which all the species have both sexes alike and brilliantly colored, but I hear from Mr. Sclater, that this appears to be the case with the Musophagæ or plantain-eaters. Nor do I believe that any large group exists in which the sexes of all the species are widely dissimilar in color: Mr. Wallace informs me that the chatterers of S. America (*Cotingidæ*) offer one of the best instances; but with some of the species, in which the male has a splendid red breast, the female exhibits some red on her breast; and the females of other species show traces of the green and other colors of the males. Nevertheless we have a near approach to close sexual similarity or dissimilarity throughout several groups: and this, from what has just been said of the fluctuating nature of inheritance, is a somewhat surprising circumstance. But that the same laws should largely prevail with allied animals is not surprising. The domestic fowl has produced a great number of breeds and sub-breeds, and in these the sexes generally differ in plumage; so that it has been noticed as an unusual circumstance when in certain sub-breeds they resemble each other. On the other hand, the domestic pigeon has likewise produced a vast number of distinct breeds and sub-breeds, and in these, with rare exceptions, the two sexes are identically alike.

Therefore if other species of Gallus and Columba were domesticated and varied, it would not be rash to predict that similar rules of sexual similarity and dissimilarity, depending on the form of transmission, would hold good in both cases. In like manner the same form of transmission has generally prevailed under nature throughout the same groups, although marked exceptions to this rule occur. Thus within the same family or even genus, the sexes may be identically alike, or very different in color. Instances have already been given in the same genus, as with sparrows, fly-catchers, thrushes and grouse. In the family of pheasants the sexes of almost all the species are wonderfully dissimilar, but are quite alike in the eared pheasant or *Crossoptilon auritum*. In two species of Chloephaga, a genus of geese, the male cannot be distinguished from the females, except by size; whilst in two others, the sexes are so unlike that they might easily be mistaken for distinct species. [29]

[29] The 'Ibis,' vol. vi. 1864, p. 122.

The laws of inheritance can alone account for the following cases, in which the female acquires, late in life, certain characters proper to the male, and ultimately comes to resemble him more or less completely. Here protection can hardly have come into play. Mr. Blyth informs me that the females of *Oriolus melanocephalus* and of some allied species, when sufficiently mature to breed, differ considerably in plumage from the adult males; but after the second or third moults they differ only in their beaks having a slight greenish tinge. In the dwarf bitterns (Ardetta), according to the same authority, "the male acquires his final livery at the first moult, the female not before the third or fourth moult; in the meanwhile she presents an intermediate garb, which is ultimately exchanged for the same livery as that of the male." So again the female *Falco peregrinus* acquires her blue plumage more slowly than the male. Mr. Swinhoe states that with one of the Drongo shrikes (*Dicrurus macrocercus*) the male, whilst almost a nestling, moults his soft brown plumage and becomes of a uniform glossy greenish-black; but the female retains for a long time the white striæ and spots on the axillary feathers; and does not completely assume the uniform black color of the male for three years. The same excellent observer remarks that in the spring of the second year the female spoon-bill (Platalea) of China resembles the male of the first year, and that apparently it is not until the third spring that she acquires the same adult plumage as that possessed by the male at a much earlier age. The female *Bombycilla carolinensis* differs very little from the male, but the appendages, which like beads of red sealing-wax ornament the wing-feathers, [30] are not developed in her so early in life as in the male. In the male of an Indian parrakeet (*Palæornis javanicus*) the upper mandible is coral-red from his earliest youth, but in the female, as Mr. Blyth has observed with caged and wild birds, it is at first black and does not become red until the bird is at least a year old, at which age the sexes resemble each other in all respects. Both sexes of the wild turkey are ultimately furnished with a tuft of bristles on the breast, but in two-year-old birds the tuft is about four inches long in the male and hardly apparent in the female; when, however, the latter has reached her fourth year, it is from four to five inches in length. [31]

[30] When the male courts the female, these ornaments are vibrated, and "are shown off to great advantage," on the outstretched wings: A. Leith Adams, 'Field and Forest Rambles,' 1873, p. 153.

[31] On Ardetta, Translation of Cuvier's 'Règne Animal,' by Mr. Blyth, footnote, p. 159. On the Peregrine Falcon, Mr. Blyth, in Charlesworth's 'Mag. of Nat. Hist.' vol. i. 1837, p. 304. On Dicrurus, 'Ibis,' 1863, p. 44. On the Platalea, 'Ibis,' vol. vi. 1864, p. 366. On the Bombycilla, Audubon's 'Ornitholog. Biography,' vol. i. p. 229. On the Palæornis, see, also, Jerdon, 'Birds of India,' vol. i. p. 263. On the wild turkey, Audubon, ibid. vol. i. p. 15; but I hear from Judge Caton that in Illinois the female very rarely acquires a tuft. Analogous cases with the females of Petrocossyphus are given by Mr. R. Sharpe, 'Proceedings of the Zoological Society,' 1872, p. 496.

These cases must not be confounded with those where diseased or old females abnormally assume masculine characters, nor with those where fertile females, whilst young, acquire the characters of the male, through variation or some unknown cause. [32] But all these cases have so much in common that they depend, according to the hypothesis of pangenesis, on gemmules derived from each part of the male being present, though latent, in the female; their development following on some slight change in the elective affinities of her constituent tissues.

[32] Of these latter cases Mr. Blyth has recorded (Translation of Cuvier's 'Règne Animal,' p. 158) various instances with Lanius, Ruticilla, Linaria, and Anas. Audubon has also recorded a similar case ('Ornitholog. Biography,' vol. v. p. 519) with *Pyranga œstiva*.

A few words must be added on changes of plumage in relation to the season of the year. From reasons formerly assigned there can be little doubt that the elegant plumes, long pendant feathers, crests, etc., of egrets, herons, and many other birds, which are developed and retained only during the summer, serve for ornamental and nuptial purposes, though common to both sexes. The female is thus rendered more conspicuous during the period of incubation than during the winter; but such birds as herons and egrets would be able to defend themselves. As, however, plumes would probably be inconvenient and certainly of no use during the winter, it is possible that the habit of moulting twice in the year may have been gradually acquired through natural selection for the sake of casting off inconvenient ornaments during the winter. But this view cannot be extended to the many waders, whose summer and winter plumages differ very little in color. With defenceless species, in which both sexes, or the males alone, become extremely conspicuous during the breeding-season,—or when the males acquire at this season such long wing or tail-feathers as to impede their flight, as with Cosmetornis and Vidua,—it certainly at first appears highly probable that the second moult has been gained for the special purpose of throwing off these ornaments. We must, however, remember that many birds, such as some of the Birds of Paradise, the Argus pheasant and peacock, do not cast their plumes during the winter; and it can hardly be maintained that the constitution of these birds, at least of the Gallinaceæ, renders a double moult impossible, for the ptarmigan moults thrice in the year. [33] Hence it must be considered as doubtful whether the many species which moult their ornamental plumes or lose their bright colors during the winter, have acquired this habit on account of the inconvenience or danger which they would otherwise have suffered.

[33] See Gould's 'Birds of Great Britain.'

I conclude, therefore, that the habit of moulting twice in the year was in most or all cases first acquired for some distinct purpose, perhaps for gaining a warmer winter covering; and that variations in the plumage occurring during the summer were accumulated through sexual selection, and transmitted to the offspring at the same season of the year; that such variations were inherited either by both sexes or by the males alone, according to the form of inheritance which prevailed. This appears more probable than that the species in all cases originally tended to retain their ornamental plumage during the winter, but were saved from this through natural selection, resulting from the inconvenience or danger thus caused.

I have endeavoured in this chapter to show that the arguments are not trustworthy in favour of the view that weapons, bright colors, and various ornaments, are now confined to the males owing to the conversion, by natural selection, of the equal transmission of characters to both sexes, into transmission to the male sex alone. It is also doubtful whether the colors of many female birds are due to the preservation, for the sake of protection, of variations which were from the first limited in their transmission to the female sex. But it will be convenient to defer any further discussion on this subject until I treat, in the following chapter, of the differences in plumage between the young and old.

CHAPTER XVI.

BIRDS—concluded.

The immature plumage in relation to the character of the plumage in both sexes when adult—Six classes of cases—Sexual differences between the males of closely-allied or representative species—The female assuming the characters of the male—Plumage of the young in relation to the summer and winter plumage of the adults—On the increase of beauty in the birds of the world—Protective coloring—Conspicuously colored birds—Novelty appreciated—Summary of the four chapters on Birds.

We must now consider the transmission of characters, as limited by age, in reference to sexual selection. The truth and importance of the principle of inheritance at corresponding ages need not here be discussed, as enough has already been said on the subject. Before giving the several rather complex rules or classes of cases, under which the differences in plumage between the young and the old, as far as known to me, may be included, it will be well to make a few preliminary remarks.

With animals of all kinds when the adults differ in color from the young, and the colors of the latter are not, as far as we can see, of any special service, they may generally be attributed, like various embryological structures, to the retention of a former character. But this view can be maintained with confidence, only when the young of several species resemble each other closely, and likewise resemble other adult species belonging to the same group; for the latter are the living proofs that such a state of things was formerly possible. Young lions and pumas are marked with feeble stripes or rows of spots, and as many allied species both young and old are similarly marked, no believer in evolution will doubt that the progenitor of the lion and puma was a striped animal, and that the young have retained vestiges of the stripes, like the kittens of black cats, which are not in the least striped when grown up. Many species of deer, which when mature are not spotted, are whilst young covered with white spots, as are likewise some few species in the adult state. So again the young in the whole family of pigs (Suidæ), and in certain rather distantly allied animals, such as the tapir, are marked with dark longitudinal stripes; but here we have a character apparently derived from an extinct progenitor, and now preserved by the young alone. In all such cases the old have had their colors changed in the course of time, whilst the young have remained but little altered, and this has been effected through the principle of inheritance at corresponding ages.

This same principle applies to many birds belonging to various groups, in which the young closely resemble each other, and differ much from their respective adult parents. The young of almost all the Gallinaceæ, and of some distantly allied birds such as ostriches, are covered with longitudinally striped down; but this character points back to a state of things so remote that it hardly concerns us. Young cross-bills (Loxia) have at first straight beaks like those of other finches, and in their immature striated plumage they resemble the mature red-pole and female siskin, as well as the young of the goldfinch, greenfinch, and some other allied species. The young of many kinds of buntings (Emberiza) resemble one another, and likewise the adult state of the common bunting, *E. miliaria*. In almost the whole large group of thrushes the young have their breasts spotted—a character which is retained throughout life by many species, but is quite lost by others, as by the *Turdus migratorius*. So again with many thrushes, the feathers on the back are mottled before they are moulted for the first time, and this character is retained for life by certain eastern species. The young of many species of shrikes (Lanius), of some woodpeckers, and of an Indian pigeon (*Chalcophaps indicus*), are transversely striped on the under surface; and certain allied species or whole genera are similarly marked when adult. In some closely-allied and resplendent Indian cuckoos (Chrysococcyx), the mature species differ considerably from one another in color, but the young cannot be distinguished. The young of an Indian goose (*Sarkidiornis melanonotus*) closely resemble in plumage an allied genus, Dendrocygna, when mature. [1] Similar facts will hereafter be given in regard to certain herons. Young black-grouse (*Tetrao tetrix*) resemble the young as well as the old of certain other species, for instance the red-grouse or *T. scoticus*. Finally, as Mr. Blyth, who has attended closely to this subject, has well remarked, the natural affinities of many species are best exhibited in their immature plumage; and as the true affinities of all organic beings depend on their descent from a common progenitor, this remark strongly confirms the belief that the immature plumage approximately shows us the former or ancestral condition of the species.

[1] In regard to thrushes, shrikes, and woodpeckers, see Mr. Blyth, in Charlesworth's 'Mag. of Nat. Hist.' vol. i. 1837, p. 304; also footnote to his translation of Cuvier's 'Règne Animal,' p. 159. I give the case of Loxia on Mr. Blyth's information. On thrushes, see also Audubon, 'Ornith. Biog.' vol. ii. p. 195. On Chrysococcyx and Chalcophaps, Blyth, as quoted in Jerdon's 'Birds of India,' vol. iii. p. 485. On Sarkidiornis, Blyth, in 'Ibis,' 1867, p. 175.

Although many young birds, belonging to various families, thus give us a glimpse of the plumage of their remote progenitors, yet there are many other birds, both dull-colored and bright-colored, in which the young closely resemble their parents. In such cases the young of the different species cannot resemble each other more closely than do the parents; nor can they strikingly resemble allied forms when adult. They give us but little insight into the plumage of their progenitors, excepting in so far that, when the young and the old are colored in the same general manner throughout a whole group of species, it is probable that their progenitors were similarly colored.

We may now consider the classes of cases, under which the differences and resemblances between the plumage of the young and the old, in both sexes or in one sex alone, may be grouped. Rules of this kind were first enounced by Cuvier; but with the progress of knowledge they require some modification and amplification. This I have attempted to do, as far as the extreme complexity of the subject permits, from information derived from various sources; but a full essay on this subject by some competent ornithologist is much needed. In order to ascertain to what extent each rule prevails, I have tabulated the facts given in four great works, namely, by Macgillivray on the birds of Britain, Audubon on those of North America, Jerdon on those of India, and Gould on those of Australia. I may here premise, first, that the several cases or rules graduate into each other; and secondly, that when the young are said to resemble their parents, it is not meant that they are identically alike, for their colors are almost always less vivid, and the feathers are softer and often of a different shape.

RULES OR CLASSES OF CASES.

I. When the adult male is more beautiful or conspicuous than the adult female, the young of both sexes in their first plumage closely resemble the adult female, as with the common fowl and peacock; or, as occasionally occurs, they resemble her much more closely than they do the adult male.

II. When the adult female is more conspicuous than the adult male, as sometimes though rarely occurs, the young of both sexes in their first plumage resemble the adult male.

III. When the adult male resembles the adult female, the young of both sexes have a peculiar first plumage of their own, as with the robin.

IV. When the adult male resembles the adult female, the young of both sexes in their first plumage resemble the adults, as with the kingfisher, many parrots, crows, hedge-warblers.

V. When the adults of both sexes have a distinct winter and summer plumage, whether or not the male differs from the female, the young resemble the adults of both sexes in their winter dress, or much more rarely in their summer dress, or they resemble the females alone. Or the young may have an intermediate character; or again they may differ greatly from the adults in both their seasonal plumages.

VI. In some few cases the young in their first plumage differ from each other according to sex; the young males resembling more or less closely the adult males, and the young females more or less closely the adult females.

CLASS I.

In this class, the young of both sexes more or less closely resemble the adult female, whilst the adult male differs from the adult female, often in the most conspicuous manner. Innumerable instances in all Orders could be given; it will suffice to call to mind the common pheasant, duck, and house-sparrow. The cases under this class graduate into others. Thus the two sexes when adult may differ so slightly, and the young so slightly from the adults, that it is doubtful whether such cases ought to come under the present, or under the third or fourth classes. So again the young of the two sexes, instead of being quite alike, may differ in a slight degree from each other, as in our sixth class. These transitional cases, however, are few, or at least are not strongly pronounced, in comparison with those which come strictly under the present class.

The force of the present law is well shown in those groups, in which, as a general rule, the two sexes and the young are all alike; for when in these groups the male does differ from the female, as with certain parrots, kingfishers, pigeons, etc., the young of both sexes resemble the adult female. [2] We see the same fact exhibited still more clearly in certain anomalous cases; thus the male of *Heliothrix auriculata* (one of the humming-birds) differs conspicuously from the female in having a splendid gorget and fine ear-tufts, but the female is remarkable from having a much longer tail than that of the male; now the young of both sexes resemble (with the exception of the breast being spotted with bronze) the adult female in all other respects, including the length of her tail, so that the tail of the male actually becomes shorter as he reaches maturity, which is a most unusual circumstance. [3] Again, the plumage of the male goosander (*Mergus merganser*) is more conspicuously colored than that of the female, with the scapular and secondary wing-feathers much longer; but differently from what occurs, as far as I know, in any other bird, the crest of the adult male, though broader than that of the female, is considerably shorter, being only a little above an inch in length; the crest of the female being two and a half inches long. Now the young of both sexes entirely resemble the adult female, so that their crests are actually of greater length, though narrower, than in the adult male. [4]

[2] See, for instance, Mr. Gould's account ('Handbook to the Birds of Australia,' vol. i. p. 133) of Cyanalcyon (one of the Kingfishers), in which, however, the young male, though resembling the adult female, is less brilliantly colored. In some species of Dacelo the males have blue tails, and the females brown ones; and Mr. R.B. Sharpe informs me that the tail of the young male of *D. gaudichaudi* is at first brown. Mr. Gould has described (ibid. vol. ii. pp. 14, 20, 37) the sexes and the young of certain black Cockatoos and of the King Lory, with which the same rule prevails. Also Jerdon ('Birds of India,' vol. i. p. 260) on the *Palæornis rosa*, in which the young are more like the female than the male. See Audubon ('Ornithological Biography,' vol. ii. p. 475) on the two sexes and the young of *Columba passerina*.
[3] I owe this information to Mr. Gould, who showed me the specimens; see also his 'Introduction to the *Trochilidæ*,' 1861, p. 120.
[4] Macgillivray, 'Hist. Brit. Birds,' vol. v. pp. 207-214.

When the young and the females closely resemble each other and both differ from the males, the most obvious conclusion is that the males alone have been modified. Even in the anomalous cases of the Heliothrix and Mergus, it is probable that originally both adult sexes were furnished—the one species with a much elongated tail, and the other with a much elongated crest—these characters having since been partially lost by the adult males from some unexplained cause, and transmitted in their diminished state to their male offspring alone, when arrived at the corresponding age of maturity. The belief that in the present class the male alone has been modified, as far as the differences between the male and the female together with her young are concerned, is strongly supported by some remarkable facts recorded by Mr. Blyth, [5] with respect to closely-allied species which represent each other in distinct countries. For with several of these representative species the adult males have undergone a certain amount of change and can be distinguished; the females and the young from the distinct countries being indistinguishable, and therefore absolutely unchanged. This is the case with certain

Indian chats (Thamnobia), with certain honey-suckers (Nectarinia), shrikes (Tephrodornis), certain kingfishers (Tanysiptera), Kalij pheasants (Gallophasis), and tree-partridges (Arboricola).

[5] See his admirable paper in the 'Journal of the Asiatic Soc. of Bengal,' vol. xix. 1850, p. 223; see also Jerdon, 'Birds of India,' vol. i. introduction, p. xxix. In regard to Tanysiptera, Prof. Schlegel told Mr. Blyth that he could distinguish several distinct races, solely by comparing the adult males.

In some analogous cases, namely with birds having a different summer and winter plumage, but with the two sexes nearly alike, certain closely-allied species can easily be distinguished in their summer or nuptial plumage, yet are indistinguishable in their winter as well as in their immature plumage. This is the case with some of the closely-allied Indian wagtails or Motacillæ. Mr. Swinhoe [6] informs me that three species of Ardeola, a genus of herons, which represent one another on separate continents, are "most strikingly different" when ornamented with their summer plumes, but are hardly, if at all, distinguishable during the winter. The young also of these three species in their immature plumage closely resemble the adults in their winter dress. This case is all the more interesting, because with two other species of Ardeola both sexes retain, during the winter and summer, nearly the same plumage as that possessed by the three first species during the winter and in their immature state; and this plumage, which is common to several distinct species at different ages and seasons, probably shows us how the progenitors of the genus were colored. In all these cases, the nuptial plumage which we may assume was originally acquired by the adult males during the breeding-season, and transmitted to the adults of both sexes at the corresponding season, has been modified, whilst the winter and immature plumages have been left unchanged.

[6] See also Mr. Swinhoe, in 'Ibis,' July 1863, p. 131; and a previous paper, with an extract from a note by Mr. Blyth, in 'Ibis,' January, 1861, p. 25.

The question naturally arises, how is it that in these latter cases the winter plumage of both sexes, and in the former cases the plumage of the adult females, as well as the immature plumage of the young, have not been at all affected? The species which represent each other in distinct countries will almost always have been exposed to somewhat different conditions, but we can hardly attribute to this action the modification of the plumage in the males alone, seeing that the females and the young, though similarly exposed, have not been affected. Hardly any fact shows us more clearly how subordinate in importance is the direct action of the conditions of life, in comparison with the accumulation through selection of indefinite variations, than the surprising difference between the sexes of many birds; for both will have consumed the same food, and have been exposed to the same climate. Nevertheless we are not precluded from believing that in the course of time new conditions may produce some direct effect either on both sexes, or from their constitutional differences chiefly on one sex. We see only that this is subordinate in importance to the accumulated results of selection. Judging, however, from a wide-spread analogy, when a species migrates into a new country (and this must precede the formation of representative species), the changed conditions to which they will almost always have been exposed will cause them to undergo a certain amount of fluctuating variability. In this case sexual selection, which depends on an element liable to change—the taste or admiration of the female—will have had new shades of color or other differences to act on and accumulate; and as sexual selection is always at work, it would (from what we know of the results on domestic animals of man's unintentional selection), be surprising if animals inhabiting separate districts, which can never cross and thus blend their newly-acquired characters, were not, after a sufficient lapse of time, differently modified. These remarks likewise apply to the nuptial or summer plumage, whether confined to the males, or common to both sexes.

Although the females of the above closely-allied or representative species, together with their young, differ hardly at all from one another, so that the males alone can be distinguished, yet the females of most species within the same genus obviously differ from each other. The differences, however, are rarely as great as between the males. We see this clearly in the whole family of the Gallinaceæ: the females, for instance, of the common and Japan pheasant, and especially of the gold and Amherst pheasant —of the silver pheasant and the wild fowl—resemble one another very closely in color, whilst the males differ to an extraordinary degree. So it is with the females of most of the Cotingidæ, Fringillidæ, and many other families. There can indeed be no doubt that, as a general rule, the females have been less modified than the males. Some few birds, however, offer a singular and inexplicable exception; thus the females of *Paradisea apoda* and *P. papuana* differ from each other more than do their respective males; [7] the female of the latter species having the under surface pure white, whilst the female *P. apoda* is deep brown beneath. So, again, as I hear from Professor Newton, the males of two species of Oxynotus (shrikes), which represent each other in the islands of Mauritius and Bourbon, [8] differ but little in color, whilst the females differ much. In the Bourbon species the female appears to have partially retained an immature condition of plumage, for at first sight she "might be taken for the young of the Mauritian species." These differences may be compared with those inexplicable ones, which occur independently of man's selection in certain sub-breeds of the game-fowl, in which the females are very different, whilst the males can hardly be distinguished. [9]

[7] Wallace, 'The Malay Archipelago,' vol. ii. 1869, p. 394.
[8] These species are described with colored figures, by M. F. Pollen, in 'Ibis,' 1866, p. 275.
[9] 'Variation of Animals,' etc., vol. i. p. 251.

As I account so largely by sexual selection for the differences between the males of allied species, how can the differences between the females be accounted for in all ordinary cases? We need not here consider the species which belong to distinct genera; for with these, adaptation to different habits of life, and other agencies, will have come into play. In regard to the differences between the females within the same genus, it appears to me almost certain, after looking through various large groups, that the chief agent has been the greater or less transference to the female of the characters acquired by the males through sexual selection. In the several British finches, the two sexes differ either very slightly or considerably; and if we compare the females of the greenfinch, chaffinch, goldfinch, bullfinch, crossbill, sparrow, etc., we shall see that they differ from one another chiefly in the points in which they partially resemble their respective males; and the colors of the males may safely be attributed to sexual selection. With many gallinaceous species the sexes differ to an extreme degree, as with the peacock, pheasant, and fowl, whilst with other species there has been a partial or even complete transference of character from the male to the female. The females of the several species of Polyplectron exhibit in a dim condition, and chiefly on the tail, the splendid ocelli of their males. The female partridge differs from the male only in the red mark on her breast being smaller; and the female wild turkey only in her colors being much duller. In the guinea-fowl the two sexes are indistinguishable. There is no improbability in the plain, though peculiarly spotted plumage of this latter bird having been acquired through sexual selection by the males, and then transmitted to both sexes; for it is not essentially different from the much more beautifully spotted plumage, characteristic of the males alone of the Tragopan pheasants.

It should be observed that, in some instances, the transference of characters from the male to the female has been effected apparently at a remote period, the male having subsequently undergone great changes, without transferring to the female any of his later-gained characters. For instance, the female and the young of the black-grouse (*Tetrao tetrix*) resemble pretty closely both sexes and the young of the red-grouse (*T. scoticus*); and we may consequently infer that the black-grouse is descended from some ancient species, of which both sexes were colored in nearly the same manner as the red-grouse. As both sexes of this latter species are more distinctly barred during the breeding-season than at any other time, and as the male differs slightly from the female in his more strongly-pronounced red and brown tints, [10] we may conclude that his plumage has been influenced by sexual selection, at least to a certain extent. If so, we may further infer that nearly similar plumage of the female black-grouse was similarly produced at some former period. But since this period the male black-grouse has acquired his fine black plumage, with his forked and outwardly-curled tail-feathers; but of these characters there has hardly been any transference to the female, excepting that she shows in her tail a trace of the curved fork.

[10] Macgillivray, 'History of British Birds,' vol. i. pp. 172-174.

We may therefore conclude that the females of distinct though allied species have often had their plumage rendered more or less different by the transference in various degrees of characters acquired by the males through sexual selection, both during former and recent times. But it deserves especial attention that brilliant colors have been transferred much more rarely than other tints. For instance, the male of the red-throated blue-breast (*Cyanecula suecica*) has a rich blue breast, including a sub-triangular red mark; now marks of nearly the same shape have been transferred to the female, but the central space is fulvous instead of red, and is surrounded by mottled instead of blue feathers. The Gallinaceæ offer many analogous cases; for none of the species, such as partridges, quails, guinea-fowls, etc., in which the colors of the plumage have been largely transferred from the male to the female, are brilliantly colored. This is well exemplified with the pheasants, in which the male is generally so much more brilliant than the female; but with the Eared and Cheer pheasants (*Crossoptilon auritum* and *Phasianus wallichii*) the sexes closely resemble each other and their colors are dull. We may go so far as to believe that if any part of the plumage in the males of these two pheasants had been brilliantly colored, it would not have been transferred to the females. These facts strongly support Mr. Wallace's view that with birds which are exposed to much danger during incubation, the transference of bright colors from the male to the female has been checked through natural selection. We must not, however, forget that another explanation, before given, is possible; namely, that the males which varied and became bright, whilst they were young and inexperienced, would have been exposed to much danger, and would generally have been destroyed; the older and more cautious males, on the other hand, if they varied in a like manner, would not only have been able to survive, but would have been favoured in their rivalry with other males. Now variations occurring late in life tend to be transmitted exclusively to the same sex, so that in this case extremely bright tints would not have been transmitted to the females. On the other hand, ornaments of a less conspicuous kind, such as those possessed by the Eared and Cheer pheasants, would not have been dangerous, and if they appeared during early youth, would generally have been transmitted to both sexes.

In addition to the effects of the partial transference of characters from the males to the females, some of the differences between the females of closely allied species may be attributed to the direct or definite action of the conditions of life. [11] With the males, any such action would generally have been masked by the brilliant colors gained through sexual selection; but not so with the females. Each of the endless diversities in plumage which we see in our domesticated birds is, of course, the result of some definite cause; and under natural and more uniform conditions, some one tint, assuming that it was in no way injurious, would almost certainly sooner or later prevail. The free intercrossing of the many individuals belonging to the same species would ultimately tend to make any change of color, thus induced, uniform in character.

[11] See, on this subject, chap. xxiii. in the 'Variation of Animals and Plants under Domestication.'

No one doubts that both sexes of many birds have had their colors adapted for the sake of protection; and it is possible that the females alone of some species may have been modified for this end. Although it would be a difficult, perhaps an impossible process, as shown in the last chapter, to convert one form of transmission into another through selection, there would not be the least difficulty in adapting the colors of the female, independently of those of the male, to surrounding objects, through the accumulation of variations which were from the first limited in their transmission to the female sex. If the variations were not thus limited, the bright tints of the male would be deteriorated or destroyed. Whether the females alone of many species have been thus specially modified, is at present very doubtful. I wish I could follow Mr. Wallace to the full extent; for the admission would remove some difficulties. Any variations which were of no service to the female as a protection would be at once obliterated, instead of being lost simply by not being selected, or from free intercrossing, or from being eliminated when transferred to the male and in any way injurious to him. Thus the plumage of the female would be kept constant in character. It would also be a relief if we could admit that the obscure tints of both sexes of many birds had been acquired and preserved for the sake of protection,—for example, of the hedge-warbler or kitty-wren (*Accentor modularis* and *Troglodytes vulgaris*), with respect to which we have no sufficient evidence of the action of sexual selection. We ought, however, to be cautious in concluding that colors which appear to us dull, are not attractive to the females of certain species; we should bear in mind such cases as that of the common house-sparrow, in which the male differs much from the female, but does not exhibit any bright tints. No one probably will dispute that many gallinaceous birds which live on the open ground, have acquired their present colors, at least in part, for the sake of protection. We know how well they are thus concealed; we know that ptarmigans, whilst changing from their winter to their summer plumage, both of which are protective, suffer greatly from birds of prey. But can we believe that the very slight differences in tints and markings between, for instance, the female black-grouse and red-grouse serve as a protection? Are partridges, as they are now colored, better protected than if they had resembled quails? Do the slight differences between the females of the common pheasant, the Japan and gold pheasants, serve as a protection, or might not their plumages have been interchanged with impunity? From what Mr. Wallace has observed of the habits of certain gallinaceous birds in the East, he thinks that such slight differences are beneficial. For myself, I will only say that I am not convinced.

Formerly when I was inclined to lay much stress on protection as accounting for the duller colors of female birds, it occurred to me that possibly both sexes and the young might aboriginally have been equally bright colored; but that subsequently, the females from the danger incurred during incubation, and the young from being inexperienced, had been rendered dull as a protection. But this view is not supported by any evidence, and is not probable; for we thus in imagination expose during past times the females and the young to danger, from which it has subsequently been necessary to shield their modified descendants. We have, also, to reduce, through a gradual process of selection, the females and the young to almost exactly the same tints and markings, and to transmit them to the corresponding sex and period of life. On the supposition that the females and the young have partaken during each stage of the process of modification of a tendency to be as brightly colored as the males, it is also a somewhat strange fact that the females have never been rendered dull-colored without the young participating in the same change; for there are no instances, as far as I can discover, of species with the females dull and the young bright colored. A partial exception, however, is offered by the young of certain woodpeckers, for they have "the whole upper part of the head tinged with red," which afterwards either decreases into a mere circular red line in the adults of both sexes, or quite disappears in the adult females. [12]

[12] Audubon, 'Ornith. Biography,' vol. i. p. 193. Macgillivray, 'History of British Birds,' vol. iii. p. 85. See also the case before given of *Indopicus carlotta*.

Finally, with respect to our present class of cases, the most probable view appears to be that successive variations in brightness or in other ornamental characters, occurring in the males at a rather late period of life have alone been preserved; and that most or all of these variations, owing to the late period of life at which they appeared, have been from the first transmitted only to the adult male offspring. Any variations in brightness occurring in the females or in the young, would have been of no service to them, and would not have been selected; and moreover, if dangerous, would have been eliminated. Thus the females and the young will either have been left unmodified, or (as is much more common) will have been partially modified by receiving through transference from the males some of his successive variations. Both sexes have perhaps been directly acted on by the conditions of life to which they have long been exposed: but the females from not being otherwise much modified, will best exhibit any such effects. These changes and all others will have been kept uniform by the free intercrossing of many individuals. In some cases, especially with ground birds, the females and the young may possibly have been modified, independently of the males, for the sake of protection, so as to have acquired the same dull-colored plumage.

CLASS II.

WHEN THE ADULT FEMALE IS MORE CONSPICUOUS THAN THE ADULT MALE, THE YOUNG OF BOTH SEXES IN THEIR FIRST PLUMAGE RESEMBLE THE ADULT MALE.

This class is exactly the reverse of the last, for the females are here brighter colored or more conspicuous than the males; and the young, as far as they are known, resemble the adult males instead of the adult females. But the difference between the sexes is never nearly so great as with many birds in the first class, and the cases are comparatively rare. Mr. Wallace, who first called attention to the singular relation which exists between the less bright colors of the males and their performing the duties of incubation, lays great stress on this point, [13] as a crucial test that obscure colors have been acquired for the sake of protection during the period of nesting. A different view seems to me more probable. As the cases are curious and not numerous, I will briefly give all that I have been able to find.

[13] 'Westminster Review,' July 1867, and A. Murray, 'Journal of Travel,' 1868, p. 83.

In one section of the genus Turnix, quail-like birds, the female is invariably larger than the male (being nearly twice as large in one of the Australian species), and this is an unusual circumstance with the Gallinaceæ. In most of the species the female is more distinctly colored and brighter than the male, [14] but in some few species the sexes are alike. In *Turnix taigoor* of India the male "wants the black on the throat and neck, and the whole tone of the plumage is lighter and less pronounced than that of the female." The female appears to be noisier, and is certainly much more pugnacious than the male; so that the females and not the males are often kept by the natives for fighting, like game-cocks. As male birds are exposed by the English bird-catchers for a decoy near a trap, in order to catch other males by exciting their rivalry, so the females of this Turnix are employed in India. When thus exposed the females soon begin their "loud purring call, which can be heard a long way off, and any females within ear-shot run rapidly to the spot, and commence fighting with the caged bird." In this way from twelve to twenty birds, all breeding females, may be caught in the course of a single day. The natives assert that the females after laying their eggs associate in flocks, and leave the males to sit on them. There is no reason to doubt the truth of this assertion, which is supported by some observations made in China by Mr. Swinhoe. [15] Mr. Blyth believes, that the young of both sexes resemble the adult male.

[14] For the Australian species, see Gould's 'Handbook,' etc., vol. ii. pp. 178, 180, 186, and 188. In the British Museum specimens of the Australian Plain-wanderer (*Pedionomus torquatus*) may be seen, showing similar sexual differences.

[15] Jerdon, 'Birds of India,' vol. iii. p. 596. Mr. Swinhoe, in 'Ibis,' 1865, p. 542; 1866, pp. 131, 405.

FIG. 62.—Rhynchæa capensis (from Brehm).

The females of the three species of Painted Snipes (Rhynchæa, Fig. 62) "are not only larger but much more richly colored than the males." [16] With all other birds in which the trachea differs in structure in the two sexes it is more developed and complex in the male than in the female; but in the *Rhynchæa australis* it is simple in the male, whilst in the female it makes four distinct convolutions before entering the lungs. [17] The female therefore of this species has acquired an eminently masculine character. Mr. Blyth ascertained, by examining many specimens, that the trachea is not convoluted in either sex of *R. bengalensis*, which species resembles *R. australis* so closely, that it can hardly be distinguished except by its shorter toes. This fact is another striking instance of the law that secondary sexual characters are often widely different in closely-allied forms, though it is a very rare circumstance when such differences relate to the female sex. The young of both sexes of *R. bengalensis* in their first plumage are said to resemble the mature male. [18] There is also reason to believe that the male undertakes the duty of incubation, for Mr. Swinhoe [19] found the females before the close of the summer associated in flocks, as occurs with the females of the Turnix.

[16] Jerdon, 'Birds of India,' vol. iii. p. 677.
[17] Gould's 'Handbook to the Birds of Australia,' vol. ii. p. 275.
[18] 'The Indian Field,' Sept. 1858, p. 3.
[19] 'Ibis,' 1866, p. 298.

The females of *Phalaropus fulicarius* and *P. hyperboreus* are larger, and in their summer plumage "more gaily attired than the males." But the difference in color between the sexes is far from conspicuous. According to Professor Steenstrup, the male alone of P. fulicarius undertakes the duty of incubation; this is likewise shown by the state of his breast-feathers during the breeding-season. The female of the dotterel plover (*Eudromias morinellus*) is larger than the male, and has the red and black tints on the lower surface, the white crescent on the breast, and the stripes over the eyes, more strongly pronounced. The male also takes at least a share in hatching the eggs; but the female likewise attends to the young. [20] I have not been able to discover whether with these species the young resemble the adult males more closely than the adult females; for the comparison is somewhat difficult to make on account of the double moult.

[20] For these several statements, see Mr. Gould's 'Birds of Great Britain.' Prof. Newton informs me that he has long been convinced, from his own observations and from those of others, that the males of the above-named species take either the whole or a large share of the duties of incubation, and that they "show much greater devotion towards their young, when in danger, than do the females." So it is, as he informs me, with *Limosa lapponica* and some few other Waders, in which the females are larger and have more strongly contrasted colors than the males.

Turning now to the ostrich Order: the male of the common cassowary (*Casuarius galeatus*) would be thought by any one to be the female, from his smaller size and from the appendages and naked skin about his head being much less brightly colored; and I am informed by Mr. Bartlett that in the Zoological Gardens, it is certainly the male alone who sits on the eggs and takes care of the young. [21] The female is said by Mr. T.W. Wood [22] to exhibit during the breeding-season a most pugnacious disposition; and her wattles then become enlarged and more brilliantly colored. So again the female of one of the emus (*Dromæus irroratus*) is considerably larger than the male, and she possesses a slight top-knot, but is otherwise indistinguishable in plumage. She appears, however, "to have greater power, when angry or otherwise excited, of erecting, like a turkey-cock, the feathers of her neck and breast. She is usually the more courageous and pugilistic. She makes a deep hollow guttural boom especially at night, sounding like a small gong. The male has a slenderer frame and is more docile, with no voice beyond a suppressed hiss when angry, or a croak." He not only performs the whole duty of incubation, but has to defend the young from their mother; "for as soon as she catches sight of her progeny she becomes violently agitated, and notwithstanding the resistance of the father appears to use her utmost endeavours to destroy them. For months afterwards it is unsafe to put the parents together, violent quarrels being the inevitable result, in which the female generally comes off conqueror." [23] So that with this emu we have a complete reversal not only of the parental and incubating instincts, but of the usual moral qualities of the two sexes; the females being savage, quarrelsome, and noisy, the males gentle and good. The case is very different with the African ostrich, for the male is somewhat larger than the female and has finer plumes with more strongly contrasted colors; nevertheless he undertakes the whole duty of incubation. [24]

[21] The natives of Ceram (Wallace, 'Malay Archipelago,' vol. ii. p. 150) assert that the male and female sit alternately on the eggs; but this assertion, as Mr. Bartlett thinks,

may be accounted for by the female visiting the nest to lay her eggs.

[22] The 'Student,' April 1870, p. 124.

[23] See the excellent account of the habits of this bird under confinement, by Mr. A.W. Bennett, in 'Land and Water,' May 1868, p. 233.

[24] Mr. Sclater, on the incubation of the Struthiones, 'Proc. Zool. Soc.' June 9, 1863. So it is with the *Rhea darwinii*: Captain Musters says ('At Home with the Patagonians,' 1871, p. 128), that the male is larger, stronger and swifter than the female, and of slightly darker colors; yet he takes sole charge of the eggs and of the young, just as does the male of the common species of Rhea.

I will specify the few other cases known to me, in which the female is more conspicuously colored than the male, although nothing is known about the manner of incubation. With the carrion-hawk of the Falkland Islands (*Milvago leucurus*) I was much surprised to find by dissection that the individuals, which had all their tints strongly pronounced, with the cere and legs orange-colored, were the adult females; whilst those with duller plumage and grey legs were the males or the young. In an Australian tree-creeper (*Climacteris erythrops*) the female differs from the male in "being adorned with beautiful, radiated, rufous markings on the throat, the male having this part quite plain." Lastly, in an Australian night-jar "the female always exceeds the male in size and in the brilliance of her tints; the males, on the other hand, have two white spots on the primaries more conspicuous than in the female." [25]

[25] For the Milvago, see 'Zoology of the Voyage of the "Beagle," Birds,' 1841, p. 16. For the Climacteris and night-jar (Eurostopodus), see Gould's 'Handbook to the Birds of Australia,' vol. i. pp. 602 and 97. The New Zealand shieldrake (*Tadorna variegata*) offers a quite anomalous case; the head of the female is pure white, and her back is redder than that of the male; the head of the male is of a rich dark bronzed color, and his back is clothed with finely pencilled slate-colored feathers, so that altogether he may be considered as the more beautiful of the two. He is larger and more pugnacious than the female, and does not sit on the eggs. So that in all these respects this species comes under our first class of cases; but Mr. Sclater ('Proceedings of the Zoological Society,' 1866, p. 150) was much surprised to observe that the young of both sexes, when about three months old, resembled in their dark heads and necks the adult males, instead of the adult females; so that it would appear in this case that the females have been modified, whilst the males and the young have retained a former state of plumage.

We thus see that the cases in which female birds are more conspicuously colored than the males, with the young in their immature plumage resembling the adult males instead of the adult females, as in the previous class, are not numerous, though they are distributed in various Orders. The amount of difference, also, between the sexes is incomparably less than that which frequently occurs in the last class; so that the cause of the difference, whatever it may have been, has here acted on the females either less energetically or less persistently than on the males in the last class. Mr. Wallace believes that the males have had their colors rendered less conspicuous for the sake of protection during the period of incubation; but the difference between the sexes in hardly any of the foregoing cases appears sufficiently great for this view to be safely accepted. In some of the cases, the brighter tints of the female are almost confined to the lower surface, and the males, if thus colored, would not have been exposed to danger whilst sitting on the eggs. It should also be borne in mind that the males are not only in a slight degree less conspicuously colored than the females, but are smaller and weaker. They have, moreover, not only acquired the maternal instinct of incubation, but are less pugnacious and vociferous than the females, and in one instance have simpler vocal organs. Thus an almost complete transposition of the instincts, habits, disposition, color, size, and of some points of structure, has been effected between the two sexes.

Now if we might assume that the males in the present class have lost some of that ardour which is usual to their sex, so that they no longer search eagerly for the females; or, if we might assume that the females have become much more numerous than the males—and in the case of one Indian Turnix the females are said to be "much more commonly met with than the males" [26]—then it is not improbable that the females would have been led to court the males, instead of being courted by them. This indeed is the case to a certain extent with some birds, as we have seen with the peahen, wild turkey, and certain kinds of grouse. Taking as our guide the habits of most male birds, the greater size and strength as well as the extraordinary pugnacity of the females of the Turnix and emu, must mean that they endeavour to drive away rival females, in order to gain possession of the male; and on this view all the facts become clear; for the males would probably be most charmed or excited by the females which were the most attractive to them by their bright colors, other ornaments, or vocal powers. Sexual selection would then do its work, steadily adding to the attractions of the females; the males and the young being left not at all, or but little modified.

[26] Jerdon, 'Birds of India,' vol. iii. p. 598.

CLASS III.

WHEN THE ADULT MALE RESEMBLES THE ADULT FEMALE, THE YOUNG OF BOTH SEXES HAVE A PECULIAR FIRST PLUMAGE OF THEIR OWN.

In this class the sexes when adult resemble each other, and differ from the young. This occurs with many birds of many kinds. The male robin can hardly be distinguished from the female, but the young are widely different, with their mottled dusky-olive and brown plumage. The male and female of the splendid scarlet ibis are alike, whilst the young are brown; and the scarlet color, though common to both sexes, is apparently a sexual character, for it is not well developed in either sex under confinement; and a loss of color often occurs with brilliant males when they are confined. With many species of herons the young differ greatly from the adults; and the summer plumage of the latter, though common to both sexes, clearly has a nuptial character. Young swans are slate-colored, whilst the mature birds are pure white; but it would be superfluous to give additional instances. These differences between the young and the old apparently depend, as in the last two classes, on the young having retained a former or ancient state of plumage, whilst the old of both sexes have acquired a new one. When the adults are bright colored, we may conclude from the remarks just made in relation to the scarlet ibis and to many herons, and from the analogy of the species in the first class, that such colors have been acquired through sexual selection by the nearly mature males; but that, differently from what occurs in the first two classes, the transmission, though limited to the same age, has not been limited to the same sex. Consequently, the sexes when mature resemble each other and differ from the young.

CLASS IV.

WHEN THE ADULT MALE RESEMBLES THE ADULT FEMALE, THE YOUNG OF BOTH SEXES IN THEIR FIRST PLUMAGE RESEMBLE THE ADULTS.

In this class the young and the adults of both sexes, whether brilliantly or obscurely colored, resemble each other. Such cases are, I think, more common than those in the last class. We have in England instances in the kingfisher, some woodpeckers, the jay, magpie, crow, and many small dull-colored birds, such as the hedge-warbler or kitty-wren. But the similarity in plumage between the young and the old is never complete, and graduates away into dissimilarity. Thus the young of some members of the kingfisher family are not only less vividly colored than the adults, but many of the feathers on the lower surface are edged with brown, [27]—a vestige probably of a former state of the plumage. Frequently in the same group of birds, even within the same genus, for instance in an Australian genus of parrakeets (Platycercus), the young of some species closely resemble, whilst the young of other species differ considerably, from their parents of both sexes, which are alike. [28] Both sexes and the young of the common jay are closely similar; but in the Canada jay (*Perisoreus canadensis*) the young differ so much from their parents that they were formerly described as distinct species. [29]

[27] Jerdon, 'Birds of India,' vol. i. pp. 222, 228. Gould's 'Handbook to the Birds of Australia,' vol. i. pp. 124, 130.
[28] Gould, ibid. vol. ii. pp. 37, 46, 56.
[29] Audubon, 'Ornith. Biography,' vol. ii. p. 55.

I may remark before proceeding that, under the present and next two classes of cases, the facts are so complex and the conclusions so doubtful, that any one who feels no especial interest in the subject had better pass them over.

The brilliant or conspicuous colors which characterise many birds in the present class, can rarely or never be of service to them as a protection; so that they have probably been gained by the males through sexual selection, and then transferred to the females and the young. It is, however, possible that the males may have selected the more attractive females; and if these transmitted their characters to their offspring of both sexes, the same results would follow as from the selection of the more attractive males by the females. But there is evidence that this contingency has rarely, if ever, occurred in any of those groups of birds in which the sexes are generally alike; for, if even a few of the successive variations had failed to be transmitted to both sexes, the females would have slightly exceeded the males in beauty. Exactly the reverse occurs under nature; for, in almost every large group in which the sexes generally resemble each other, the males of some few species are in a slight degree more brightly colored than the females. It is again possible that the females may have selected the more beautiful males, these males having reciprocally selected the more beautiful females; but it is doubtful whether this double process of selection would be likely to occur, owing to the greater eagerness of one sex than the other, and whether it would be more efficient than selection on one side alone. It is, therefore, the most probable view that sexual selection has acted, in the present class, as far as ornamental characters are concerned, in accordance with the general rule throughout the animal kingdom, that is, on the males; and that these have transmitted their gradually-acquired colors, either equally or almost equally, to their offspring of both sexes.

Another point is more doubtful, namely, whether the successive variations first appeared in the males after had become nearly mature, or whilst quite young. In either case sexual selection must have acted on the male when he had to compete with rivals for the possession of the female; and in both cases the characters thus acquired have been transmitted to both sexes and all ages. But these characters if acquired by the males when adult, may have been transmitted at first to the adults alone, and at some subsequent period transferred to the young. For it is known that, when the law of inheritance at corresponding ages fails, the offspring often inherit characters at an earlier age than that at which they first appeared in their parents. [30] Cases apparently of this kind have been observed with birds in a state of nature. For instance Mr. Blyth has seen specimens of *Lanius rufus* and of *Colymbus glacialis* which had assumed whilst young, in a quite anomalous manner, the adult plumage of their parents. [31] Again, the young of the common swan (*Cygnus olor*) do not cast off their dark feathers and become white until eighteen months or two years old; but Dr. F. Forel has described the case of three vigorous young birds, out of a brood of four, which were born pure white. These young birds were not albinos, as shown by the color of their beaks and legs, which nearly resembled the same parts in the adults. [32]

[30] 'Variation of Animals and Plants under Domestication,' vol. ii. p. 79.
[31] 'Charlesworth's Magazine of Natural History,' vol. i. 1837, pp. 305, 306.
[32] 'Bulletin de la Soc. Vaudoise des Sc. Nat.' vol. x. 1869, p. 132. The young of the Polish swan, *Cygnus immutabilis* of Yarrell, are always white; but this species, as Mr. Sclater informs me, is believed to be nothing more than a variety of the domestic swan (*Cygnus olor*).

It may be worth while to illustrate the above three modes by which, in the present class, the two sexes and the young may have come to resemble each other, by the curious case of the genus Passer. [33] In the house-sparrow (*P. domesticus*) the male differs much from the female and from the young. The young and the females are alike, and resemble to a large extent both sexes and the young of the sparrow of Palestine (*P. brachydactylus*), as well as of some allied species. We may therefore assume that the female and young of the house-sparrow approximately show us the plumage of the progenitor of the genus. Now with the tree-sparrow (*P. montanus*) both sexes and the young closely resemble the male of the house-sparrow; so that they have all been modified in the same manner, and all depart from the typical coloring of their early progenitor. This may have been effected by a male ancestor of the tree-sparrow having varied, firstly, when nearly mature; or, secondly, whilst quite young, and by having in either case transmitted his modified plumage to the females and the young; or, thirdly, he may have varied when adult and transmitted his plumage to both adult sexes, and, owing to the failure of the law of inheritance at corresponding ages, at some subsequent period to his young.

[33] I am indebted to Mr. Blyth for information in regard to this genus. The sparrow of Palestine belongs to the sub-genus Petronia.

It is impossible to decide which of these three modes has generally prevailed throughout the present class of cases. That the males varied whilst young, and transmitted their variations to their offspring of both sexes, is the most probable. I may here add that I have, with little success, endeavoured, by consulting various works, to decide how far the period of variation in birds has generally determined the transmission of characters to one sex or to both. The two rules, often referred to (namely, that variations occurring late in life are transmitted to one and the same sex, whilst those which occur early in life are transmitted to both sexes), apparently hold good in the first, [34] second, and fourth classes of cases; but they fail in the third, often in the fifth [35], and in the sixth small class. They apply, however, as far as I can judge, to a considerable majority of the species; and we must not forget the striking generalization by Dr. W. Marshall with respect to the protuberances on the heads of birds. Whether or not the two rules generally hold good, we may conclude from the facts given in the eighth chapter, that the period of variation is one important element in determining the form of transmission.

[34] For instance, the males of *Tanagra æstiva* and *Fringilla cyanea* require three years, the male of *Fringilla ciris* four years, to complete their beautiful plumage. (See Audubon, 'Ornith. Biography,' vol. i. pp. 233, 280, 378). The Harlequin duck takes three years (ibid. vol. iii. p. 614). The male of the Gold pheasant, as I hear from Mr. Jenner Weir, can be distinguished from the female when about three months old, but he does not acquire his full splendour until the end of the September in the following year.

[35] Thus the *Ibis tantalus* and *Grus americanus* take four years, the Flamingo several years, and the *Ardea ludovicana* two years, before they acquire their perfect plumage. See Audubon, ibid. vol. i. p. 221; vol. iii. pp. 133, 139, 211.

With birds it is difficult to decide by what standard we ought to judge of the earliness or lateness of the period of variation, whether by the age in reference to the duration of life, or to the power of reproduction, or to the number of moults through which the species passes. The moulting of birds, even within the same family, sometimes differs much without any assignable cause. Some birds moult so early, that nearly all the body feathers are cast off before the first wing-feathers are fully grown; and we cannot believe that this was the primordial state of things. When the period of moulting has been accelerated, the age at which the colors of the adult plumage are first developed will falsely appear to us to be earlier than it really is. This may be illustrated by the practice followed by some bird-fanciers, who pull out a few feathers from the breast of nestling bullfinches, and from the head or neck of young gold-pheasants, in order to ascertain their sex; for in the males, these feathers are immediately replaced by colored ones. [36] The actual duration of life is known in but few birds, so that we can hardly judge by this standard. And, with reference to the period at which the power of reproduction is gained, it is a remarkable fact that various birds occasionally breed whilst retaining their immature plumage. [37].

[36] Mr. Blyth, in Charlesworth's 'Magazine of Natural History,' vol. i. 1837, p. 300. Mr. Bartlett has informed me in regard to gold pheasants.

[37] I have noticed the following cases in Audubon's 'Ornith. Biography.' The redstart of America (Muscapica ruticilla, vol. i. p. 203) The *Ibis tantalus* takes four years to

come to full maturity, but sometimes breeds in the second year (vol. iii. p. 133). The *Grus americanus* takes the same time, but breeds before acquiring its full plumage (vol. iii. p. 211). The adults of *Ardea cærulea* are blue, and the young white; and white, mottled, and mature blue birds may all be seen breeding together (vol. iv. p. 58): but Mr. Blyth informs me that certain herons apparently are dimorphic, for white and colored individuals of the same age may be observed. The Harlequin duck (*Anas histrionica*, Linn.) takes three years to acquire its full plumage, though many birds breed in the second year (vol. iii. p. 614). The White-headed Eagle (*Falco leucocephalus*, vol. iii. p. 210) is likewise known to breed in its immature state. Some species of Oriolus (according to Mr. Blyth and Mr. Swinhoe, in 'Ibis,' July 1863, p. 68) likewise breed before they attain their full plumage.

The fact of birds breeding in their immature plumage seems opposed to the belief that sexual selection has played as important a part, as I believe it has, in giving ornamental colors, plumes, etc., to the males, and, by means of equal transmission, to the females of many species. The objection would be a valid one, if the younger and less ornamented males were as successful in winning females and propagating their kind, as the older and more beautiful males. But we have no reason to suppose that this is the case. Audubon speaks of the breeding of the immature males of *Ibis tantalus* as a rare event, as does Mr. Swinhoe, in regard to the immature males of Oriolus. [38] If the young of any species in their immature plumage were more successful in winning partners than the adults, the adult plumage would probably soon be lost, as the males would prevail, which retained their immature dress for the longest period, and thus the character of the species would ultimately be modified. [39] If, on the other hand, the young never succeeded in obtaining a female, the habit of early reproduction would perhaps be sooner or later eliminated, from being superfluous and entailing waste of power.

[38] See footnote 37 above.
[39] Other animals, belonging to quite distinct classes, are either habitually or occasionally capable of breeding before they have fully acquired their adult characters. This is the case with the young males of the salmon. Several amphibians have been known to breed whilst retaining their larval structure. Fritz Müller has shown ('Facts and arguments for Darwin,' Eng. trans. 1869, p. 79) that the males of several amphipod crustaceans become sexually mature whilst young; and I infer that this is a case of premature breeding, because they have not as yet acquired their fully-developed claspers. All such facts are highly interesting, as bearing on one means by which species may undergo great modifications of character.

The plumage of certain birds goes on increasing in beauty during many years after they are fully mature; this is the case with the train of the peacock, with some of the birds of paradise, and with the crest and plumes of certain herons, for instance, the *Ardea ludovicana*. [40] But it is doubtful whether the continued development of such feathers is the result of the selection of successive beneficial variations (though this is the most probable view with birds of paradise) or merely of continuous growth. Most fishes continue increasing in size, as long as they are in good health and have plenty of food; and a somewhat similar law may prevail with the plumes of birds.

[40] Jerdon, 'Birds of India,' vol. iii. p. 507, on the peacock. Dr. Marshall thinks that the

older and more brilliant males of birds of paradise, have an advantage over the younger males; see 'Archives Néerlandaises,' tom. vi. 1871.—On Ardea, Audubon, ibid. vol. iii. p. 139.

CLASS V.

WHEN THE ADULTS OF BOTH SEXES HAVE A DISTINCT WINTER AND SUMMER PLUMAGE, WHETHER OR NOT THE MALE DIFFERS FROM THE FEMALE, THE YOUNG RESEMBLE THE ADULTS OF BOTH SEXES IN THEIR WINTER DRESS, OR MUCH MORE RARELY IN THEIR SUMMER DRESS, OR THEY RESEMBLE THE FEMALES ALONE. OR THE YOUNG MAY HAVE AN INTERMEDIATE CHARACTER; OR, AGAIN, THEY MAY DIFFER GREATLY FROM THE ADULTS IN BOTH THEIR SEASONAL PLUMAGES.

The cases in this class are singularly complex; nor is this surprising, as they depend on inheritance, limited in a greater or less degree in three different ways, namely, by sex, age, and the season of the year. In some cases the individuals of the same species pass through at least five distinct states of plumage. With the species, in which the male differs from the female during the summer season alone, or, which is rarer, during both seasons, [41] the young generally resemble the females,—as with the so-called goldfinch of North America, and apparently with the splendid Maluri of Australia. [42] With those species, the sexes of which are alike during both the summer and winter, the young may resemble the adults, firstly, in their winter dress; secondly, and this is of much rarer occurrence, in their summer dress; thirdly, they may be intermediate between these two states; and, fourthly, they may differ greatly from the adults at all seasons. We have an instance of the first of these four cases in one of the egrets of India (*Buphus coromandus*), in which the young and the adults of both sexes are white during the winter, the adults becoming golden-buff during the summer. With the gaper (*Anastomus oscitans*) of India we have a similar case, but the colors are reversed: for the young and the adults of both sexes are grey and black during the winter, the adults becoming white during the summer. [43] As an instance of the second case, the young of the razor-bill (*Alca torda*, Linn.), in an early state of plumage, are colored like the adults during the summer; and the young of the white-crowned sparrow of North America (*Fringilla leucophrys*), as soon as fledged, have elegant white stripes on their heads, which are lost by the young and the old during the winter. [44] With respect to the third case, namely, that of the young having an intermediate character between the summer and winter adult plumages, Yarrell [45] insists that this occurs with many waders. Lastly, in regard to the young differing greatly from both sexes in their adult summer and winter plumages, this occurs with some herons and egrets of North America and India,—the young alone being white.

[41] For illustrative cases, see vol. iv. of Macgillivray's 'History of British Birds;' on Tringa, etc., pp. 229, 271; on the Machetes, p. 172; on the *Charadrius hiaticula*, p. 118; on the *Charadrius pluvialis*, p. 94.
[42] For the goldfinch of N. America, *Fringilla tristis*, Linn., see Audubon, 'Ornithological Biography,' vol. i. p. 172. For the Maluri, Gould's 'Handbook of the Birds of Australia,' vol. i. p. 318.
[43] I am indebted to Mr. Blyth for information as to the Buphus; see also Jerdon, 'Birds

of India,' vol. iii. p. 749. On the Anastomus, see Blyth, in 'Ibis,' 1867, p. 173.

[44] On the Alca, see Macgillivray, 'Hist. Brit. Birds,' vol. v. p. 347. On the *Fringilla leucophrys*, Audubon, ibid. vol. ii. p. 89. I shall have hereafter to refer to the young of certain herons and egrets being white.

[45] 'History of British Birds,' vol. i. 1839, p. 159.

I will make only a few remarks on these complicated cases. When the young resemble the females in their summer dress, or the adults of both sexes in their winter dress, the cases differ from those given under Classes I. and III. only in the characters originally acquired by the males during the breeding-season, having been limited in their transmission to the corresponding season. When the adults have a distinct summer and winter plumage, and the young differ from both, the case is more difficult to understand. We may admit as probable that the young have retained an ancient state of plumage; we can account by sexual selection for the summer or nuptial plumage of the adults, but how are we to account for their distinct winter plumage? If we could admit that this plumage serves in all cases as a protection, its acquirement would be a simple affair; but there seems no good reason for this admission. It may be suggested that the widely different conditions of life during the winter and summer have acted in a direct manner on the plumage; this may have had some effect, but I have not much confidence in so great a difference as we sometimes see between the two plumages, having been thus caused. A more probable explanation is, that an ancient style of plumage, partially modified through the transference of some characters from the summer plumage, has been retained by the adults during the winter. Finally, all the cases in our present class apparently depend on characters acquired by the adult males, having been variously limited in their transmission according to age, season, and sex; but it would not be worth while to attempt to follow out these complex relations.

CLASS VI.

THE YOUNG IN THEIR FIRST PLUMAGE DIFFER FROM EACH OTHER ACCORDING TO SEX; THE YOUNG MALES RESEMBLING MORE OR LESS CLOSELY THE ADULT MALES, AND THE YOUNG FEMALES MORE OR LESS CLOSELY THE ADULT FEMALES.

The cases in the present class, though occurring in various groups, are not numerous; yet it seems the most natural thing that the young should at first somewhat resemble the adults of the same sex, and gradually become more and more like them. The adult male blackcap (*Sylvia atricapilla*) has a black head, that of the female being reddish-brown; and I am informed by Mr. Blyth, that the young of both sexes can be distinguished by this character even as nestlings. In the family of thrushes an unusual number of similar cases have been noticed; thus, the male blackbird (*Turdus merula*) can be distinguished in the nest from the female. The two sexes of the mocking bird (*Turdus polyglottus*, Linn.) differ very little from each other, yet the males can easily be distinguished at a very early age from the females by showing more pure white. [46] The males of a forest-thrush and of a rock-thrush (*Orocetes erythrogastra* and *Petrocincla cyanea*) have much of their plumage of a fine blue, whilst the females are brown; and the nestling males of both species have their main wing and tail-feathers edged with blue whilst those of the female are edged with brown. [47] In the young blackbird the wing-feathers assume their mature character and become black after the others; on the other hand, in the two species just named the wing-feathers become blue before the others. The most probable view with reference to the cases in the present class is that the males, differently from what occurs in Class I., have transmitted their colors to their male offspring at an earlier age than that at which they were first acquired; for, if the males had varied whilst quite young, their characters would probably have been transmitted to both sexes. [48]

[46] Audubon, 'Ornith. Biography,' vol. i. p. 113.

[47] Mr. C.A. Wright, in 'Ibis,' vol. vi. 1864, p. 65. Jerdon, 'Birds of India,' vol. i. p. 515. See also on the blackbird, Blyth in Charlesworth's 'Magazine of Natural History,' vol. i. 1837, p. 113.

[48] The following additional cases may be mentioned; the young males of *Tanagra rubra* can be distinguished from the young females (Audubon, 'Ornith. Biography,' vol. iv. p. 392), and so it is within the nestlings of a blue nuthatch, *Dendrophila frontalis* of India (Jerdon, 'Birds of India,' vol. i. p. 389). Mr. Blyth also informs me that the sexes of the stonechat, *Saxicola rubicola*, are distinguishable at a very early age. Mr. Salvin gives ('Proc. Zoolog. Soc.' 1870, p. 206) the case of a humming-bird, like the following one of Eustephanus.

In *Aïthurus polytmus*, a humming-bird, the male is splendidly colored black and green, and two of the tail-feathers are immensely lengthened; the female has an ordinary tail and inconspicuous colors; now the young males, instead of resembling the adult female, in accordance with the common rule, begin from the first to assume the colors proper to their sex, and their tail-feathers soon become elongated. I owe this information to Mr. Gould, who has given me the following more striking and as yet unpublished case. Two humming-birds belonging to the genus Eustephanus, both beautifully colored, inhabit the small island of Juan Fernandez, and have always been ranked as specifically distinct. But it has lately been ascertained that the one which is of a rich chestnut-brown color with a golden-red head, is the male, whilst the other which is elegantly variegated with green and white with a metallic green head is the female. Now the young from the first somewhat resemble the adults of the corresponding sex, the resemblance gradually becoming more and more complete.

In considering this last case, if as before we take the plumage of the young as our guide, it would appear that both sexes have been rendered beautiful independently; and not that one sex has partially transferred its beauty to the other. The male apparently has acquired his bright colors through sexual selection in the same manner as, for instance, the peacock or pheasant in our first class of cases; and the female in the same manner as the female Rhynchæa or Turnix in our second class of cases. But there is much difficulty in understanding how this could have been effected at the same time with the two sexes of the same species. Mr. Salvin states, as we have seen in the eighth chapter, that with certain humming-birds the males greatly exceed the females in number, whilst with other species inhabiting the same country the females greatly exceed the males. If, then, we might assume that during some former lengthened period the males of the Juan Fernandez species had greatly exceeded the females in number, but that during another lengthened period the females had far exceeded the males, we could understand how the males at one time, and the females at another, might have been rendered beautiful by the selection of the brighter colored individuals of either sex; both sexes transmitting their characters to their young at a rather earlier age than usual. Whether this is the true explanation I will not pretend to say; but the case is too remarkable to be passed over without notice.

We have now seen in all six classes, that an intimate relation exists between the plumage of the young and the adults, either of one sex or both. These relations are fairly well explained on the principle that one sex—this being in the great majority of cases the male—first acquired through variation and sexual selection bright colors or other ornaments, and transmitted them in various ways, in accordance with the recognized laws of inheritance. Why variations have occurred at different periods of life, even sometimes with species of the same group, we do not know, but with respect to the form of transmission, one important determining cause seems to be the age at which the variations first appear.

From the principle of inheritance at corresponding ages, and from any variations in color which occurred in the males at an early age not being then selected—on the contrary being often eliminated as dangerous—whilst similar variations occurring at or near the period of reproduction have been preserved, it follows that the plumage of the young will often have been left unmodified, or but little modified. We thus get some insight into the coloring of the progenitors of our existing species. In a vast number of species in five out of our six classes of cases, the adults of one sex or of both are bright colored, at least during the breeding-season, whilst the young are invariably less brightly colored than the adults, or are quite dull colored; for no instance is known, as far as I can discover, of the young of dull-colored species displaying bright colors, or of the young of bright-colored species being more brilliant than their parents. In the fourth class, however, in which the young and the old resemble each other, there are many species (though by no means all), of which the young are bright-colored, and as these form old groups, we may infer that their early progenitors were likewise bright. With this exception, if we look to the birds of the world, it appears that their beauty has been much increased since that period, of which their immature plumage gives us a partial record.

ON THE COLOR OF THE PLUMAGE IN RELATION TO PROTECTION.

It will have been seen that I cannot follow Mr. Wallace in the belief that dull colors, when confined to the females, have been in most cases specially gained for the sake of protection. There can, however, be no doubt, as formerly remarked, that both sexes of many birds have had their colors modified, so as to escape the notice of their enemies; or in some instances, so as to approach their prey unobserved, just as owls have had their plumage rendered soft, that their flight may not be overheard. Mr. Wallace remarks [49] that "it is only in the tropics, among forests which never lose their foliage, that we find whole groups of birds, whose chief color is green." It will be admitted by every one, who has ever tried, how difficult it is to distinguish parrots in a leaf-covered tree. Nevertheless, we must remember that many parrots are ornamented with crimson, blue, and orange tints, which can hardly be protective. Woodpeckers are eminently arboreal, but besides green species, there are many black, and black-and-white kinds—all the species being apparently exposed to nearly the same dangers. It is therefore probable that with tree-haunting birds, strongly-pronounced colors have been acquired through sexual selection, but that a green tint has been acquired oftener than any other, from the additional advantage of protection.

[49] 'Westminster Review,' July 1867, p. 5.

In regard to birds which live on the ground, every one admits that they are colored so as to imitate the surrounding surface. How difficult it is to see a partridge, snipe, woodcock, certain plovers, larks, and night-jars when crouched on ground. Animals inhabiting deserts offer the most striking cases, for the bare surface affords no concealment, and nearly all the smaller quadrupeds, reptiles, and birds depend for safety on their colors. Mr. Tristram has remarked in regard to the inhabitants of the Sahara, that all are protected by their "isabelline or sand-color." [50] Calling to my recollection the desert-birds of South America, as well as most of the ground-birds of Great Britain, it appeared to me that both sexes in such cases are generally colored nearly alike. Accordingly, I applied to Mr. Tristram with respect to the birds of the Sahara, and he has kindly given me the following information. There are twenty-six species belonging to fifteen genera, which manifestly have their plumage colored in a protective manner; and this coloring is all the more striking, as with most of these birds it differs from that of their congeners. Both sexes of thirteen out of the twenty-six species are colored in the same manner; but these belong to genera in which this rule commonly prevails, so that they tell us nothing about the protective colors being the same in both sexes of desert-birds. Of the other thirteen species, three belong to genera in which the sexes usually differ from each other, yet here they have the sexes alike. In the remaining ten species, the male differs from the female; but the difference is confined chiefly to the under surface of the plumage, which is concealed when the bird crouches on the ground; the head and back being of the same sand-colored hue in the two sexes. So that in these ten species the upper surfaces of both sexes have been acted on and rendered alike, through natural selection, for the sake of protection; whilst the lower surfaces of the males alone have been diversified, through sexual selection, for the sake of ornament. Here, as both sexes are equally well protected, we clearly see that the females have not been prevented by natural selection from inheriting the colors of their male parents; so that we must look to the law of sexually-limited transmission.

[50] 'Ibis,' 1859, vol. i. p. 429, *et seq.* Dr. Rohlfs, however, remarks to me in a letter that according to his experience of the Sahara, this statement is too strong.

In all parts of the world both sexes of many soft-billed birds, especially those which frequent reeds or sedges, are obscurely colored. No doubt if their colors had been brilliant, they would have been much more conspicuous to their enemies; but whether their dull tints have been specially gained for the sake of protection seems, as far as I can judge, rather doubtful. It is still more doubtful whether such dull tints can have been gained for the sake of ornament. We must, however, bear in mind that male birds, though dull-colored, often differ much from their females (as with the common sparrow), and this leads to the belief that such colors have been gained through sexual selection, from being attractive. Many of the soft-billed birds are songsters; and a discussion in a former chapter should not be forgotten, in which it was shown that the best songsters are rarely ornamented with bright tints. It would appear that female birds, as a general rule, have selected their mates either for their sweet voices or gay colors, but not for both charms combined. Some species, which are manifestly colored for the sake of protection, such as the jack-snipe, woodcock, and night-jar, are likewise marked and shaded, according to our standard of taste, with extreme elegance. In such cases we may conclude that both natural and sexual selection have acted conjointly for protection and ornament. Whether any bird exists which does not possess some special attraction, by which to charm the opposite sex, may be doubted. When both sexes are so obscurely colored that it would be rash to assume the agency of sexual selection, and when no direct evidence can be advanced showing that such colors serve as a protection, it is best to own complete ignorance of the cause, or, which comes to nearly the same thing, to attribute the result to the direct action of the conditions of life.

Both sexes of many birds are conspicuously, though not brilliantly colored, such as the numerous black, white, or piebald species; and these colors are probably the result of sexual selection. With the common blackbird, capercailzie, blackcock, black scoter-duck (Oidemia), and even with one of the birds of paradise (*Lophorina atra*), the males alone are black, whilst the females are brown or mottled; and there can hardly be a doubt that blackness in these cases has been a sexually selected character. Therefore it is in some degree probable that the complete or partial blackness of both sexes in such birds as crows, certain cockatoos, storks, and swans, and many marine birds, is likewise the result of sexual selection, accompanied by equal transmission to both sexes; for blackness can hardly serve in any case as a protection. With several birds, in which the male alone is black, and in others in which both sexes are black, the beak or skin about the head is brightly colored, and the contrast thus afforded adds much to their beauty; we see this in the bright yellow beak of the male blackbird, in the crimson skin over the eyes of the blackcock and capercailzie, in the brightly and variously colored beak of the scoter-drake (Oidemia), in the red beak of the chough (*Corvus graculus*, Linn.), of the black swan, and the black stork. This leads me to remark that it is not incredible that toucans may owe the enormous size of their beaks to sexual selection, for the sake of displaying the diversified and vivid stripes of color, with which these organs are ornamented. [51] The naked skin, also, at the base of the beak and round the eyes is likewise often brilliantly colored; and Mr. Gould, in speaking of one species, [52] says that the colors of the beak "are doubtless in the finest and most brilliant state during the time of pairing." There is no greater improbability that toucans should be encumbered with immense beaks, though rendered as light as possible by their cancellated structure, for the display of fine colors (an object falsely appearing to us unimportant), than that the male Argus pheasant and some other birds should be encumbered with plumes so long as to impede their flight.

[51] No satisfactory explanation has ever been offered of the immense size, and still less of the bright colors, of the toucan's beak. Mr. Bates ('The Naturalist on the Amazons,' vol. ii. 1863, p. 341) states that they use their beaks for reaching fruit at the extreme tips of the branches; and likewise, as stated by other authors, for extracting eggs and young birds from the nests of other birds. But, as Mr. Bates admits, the beak "can scarcely be considered a very perfectly-formed instrument for the end to which it is applied." The great bulk of the beak, as shown by its breadth, depth, as well as length, is not intelligible on the view, that it serves merely as an organ of prehension. Mr. Belt believes ('The Naturalist in Nicaragua.' p. 197) that the principal use of the beak is as a defence against enemies, especially to the female whilst nesting in a hole in a tree.

[52] *Rhamphastos carinatus*, Gould's 'Monograph of Ramphastidæ.'

In the same manner, as the males alone of various species are black, the females being dull-colored; so in a few cases the males alone are either wholly or partially white, as with the several bell-birds of South America (Chasmorhynchus), the Antarctic goose (*Bernicla antarctica*), the silver pheasant, etc., whilst the females are brown or obscurely mottled. Therefore, on the same principle as before, it is probable that both sexes of many birds, such as white cockatoos, several egrets with their beautiful plumes, certain ibises, gulls, terns, etc., have acquired their more or less completely white plumage through sexual selection. In some of these cases the plumage becomes white only at maturity. This is the case with certain gannets, tropic-birds, etc., and with the snow-goose (*Anser hyperboreus*). As the latter breeds on the "barren grounds," when not covered with snow, and as it migrates southward during the winter, there is no reason to suppose that its snow-white adult plumage serves as a protection. In the *Anastomus oscitans*, we have still better evidence that the white plumage is nuptial character, for it is developed only during the summer; the young in their immature state, and the adults in their winter dress, being grey and black. With many kinds of gulls (Larus), the head and neck become pure white during the summer, being grey or mottled during the winter and in the young state. On the other hand, with the smaller gulls, or sea-mews (Gavia), and with some terns (Sterna), exactly the reverse occurs; for the heads of the young birds during the first year, and of the adults during the winter, are either pure white, or much paler colored than during the breeding-season. These latter cases offer another instance of the capricious manner in which sexual selection appears often to have acted. [53]

[53] On Larus, Gavia, and Sterna, see Macgillivray, 'History of British Birds,' vol. v. pp. 515, 584, 626. On the Anser hyperboreus, Audubon, 'Ornithological Biography,' vol. iv. p. 562. On the Anastomus, Mr. Blyth, in 'Ibis,' 1867, p. 173.

That aquatic birds have acquired a white plumage so much oftener than terrestrial birds, probably depends on their large size and strong powers of flight, so that they can easily defend themselves or escape from birds of prey, to which moreover they are not much exposed. Consequently, sexual selection has not here been interfered with or guided for the sake of protection. No doubt with birds which roam over the open ocean, the males and females could find each other much more easily, when made conspicuous either by being perfectly white or intensely black; so that these colors may possible serve the same end as the call-notes of many land-birds. [54] A white or black bird when it discovers and flies down to a carcase floating on the sea or cast up on the beach, will be seen from a great distance, and will guide other birds of the same and other species, to the prey; but as this would be a disadvantage to the first finders, the individuals which were the whitest or blackest would not thus procure more food than the less strongly colored individuals. Hence conspicuous colors cannot have been gradually acquired for this purpose through natural selection.

[54] It may be noticed that with vultures, which roam far and wide high in the air, like marine birds over the ocean, three or four species are almost wholly or largely white, and that many others are black. So that here again conspicuous colors may possibly aid the sexes in finding each other during the breeding-season.

As sexual selection depends on so fluctuating an element as taste, we can understand how it is that, within the same group of birds having nearly the same habits, there should exist white or nearly white, as well as black, or nearly black species,—for instance, both white and black cockatoos, storks, ibises, swans, terns, and petrels. Piebald birds likewise sometimes occur in the same groups together with black and white species; for instance, the black-necked swan, certain terns, and the common magpie. That a strong contrast in color is agreeable to birds, we may conclude by looking through any large collection, for the sexes often differ from each other in the male having the pale parts of a purer white, and the variously colored dark parts of still darker tints than the female.

It would even appear that mere novelty, or slight changes for the sake of change, have sometimes acted on female birds as a charm, like changes of fashion with us. Thus the males of some parrots can hardly be said to be more beautiful than the females, at least according to our taste, but they differ in such points, as in having a rose-colored collar instead of "a bright emeraldine narrow green collar"; or in the male having a black collar instead of "a yellow demi-collar in front," with a pale roseate instead of a plumb-blue head. [55] As so many male birds have elongated tail-feathers or elongated crests for their chief ornament, the shortened tail, formerly described in the male of a humming-bird, and the shortened crest of the male goosander, seem like one of the many changes of fashion which we admire in our own dresses.

[55] See Jerdon on the genus Palæornis, 'Birds of India,' vol. i. pp. 258-260.

Some members of the heron family offer a still more curious case of novelty in coloring having, as it appears, been appreciated for the sake of novelty. The young of the *Ardea asha* are white, the adults being dark slate-colored; and not only the young, but the adults in their winter plumage, of the allied *Buphus coromandus* are white, this color changing into a rich golden-buff during the breeding-season. It is incredible that the young of these two species, as well as of some other members of the same family, [56] should for any special purpose have been rendered pure white and thus made conspicuous to their enemies; or that the adults of one of these two species should have been specially rendered white during the winter in a country which is never covered with snow. On the other hand we have good reason to believe that whiteness has been gained by many birds as a sexual ornament. We may therefore conclude that some early progenitor of the *Ardea asha* and the Buphus acquired a white plumage for nuptial purposes, and transmitted this color to their young; so that the young and the old became white like certain existing egrets; and that the whiteness was afterwards retained by the young, whilst it was exchanged by the adults for more strongly-pronounced tints. But if we could look still further back to the still earlier progenitors of these two species, we should probably see the adults dark-colored. I infer that this would be the case, from the analogy of many other birds, which are dark whilst young, and when adult are white; and more especially from the case of the *Ardea gularis*, the colors of which are the reverse of those of *A. asha*, for the young are dark-colored and the adults white, the young having retained a former state of plumage. It appears therefore that, during a long line of descent, the adult progenitors of the *Ardea asha*, the Buphus, and of some allies, have undergone the following changes of color: first, a dark shade; secondly, pure white; and thirdly, owing to another change of fashion (if I may so express myself), their present slaty, reddish, or golden-buff tints. These successive changes are intelligible only on the principle of novelty having been admired by birds for its own sake.

[56] The young of *Ardea rufescens* and *A. cærulea* of the United States are likewise white, the adults being colored in accordance with their specific names. Audubon ('Ornithological Biography,' vol. iii. p. 416; vol. iv. p. 58) seems rather pleased at the

thought that this remarkable change of plumage will greatly "disconcert the systematists."

Several writers have objected to the whole theory of sexual selection, by assuming that with animals and savages the taste of the female for certain colors or other ornaments would not remain constant for many generations; that first one color and then another would be admired, and consequently that no permanent effect could be produced. We may admit that taste is fluctuating, but it is not quite arbitrary. It depends much on habit, as we see in mankind; and we may infer that this would hold good with birds and other animals. Even in our own dress, the general character lasts long, and the changes are to a certain extent graduated. Abundant evidence will be given in two places in a future chapter, that savages of many races have admired for many generations the same cicatrices on the skin, the same hideously perforated lips, nostrils, or ears, distorted heads, etc.; and these deformities present some analogy to the natural ornaments of various animals. Nevertheless, with savages such fashions do not endure for ever, as we may infer from the differences in this respect between allied tribes on the same continent. So again the raisers of fancy animals certainly have admired for many generations and still admire the same breeds; they earnestly desire slight changes, which are considered as improvements, but any great or sudden change is looked at as the greatest blemish. With birds in a state of nature we have no reason to suppose that they would admire an entirely new style of coloration, even if great and sudden variations often occurred, which is far from being the case. We know that dovecot pigeons do not willingly associate with the variously colored fancy breeds; that albino birds do not commonly get partners in marriage; and that the black ravens of the Feroe Islands chase away their piebald brethren. But this dislike of a sudden change would not preclude their appreciating slight changes, any more than it does in the case of man. Hence with respect to taste, which depends on many elements, but partly on habit and partly on a love of novelty, there seems no improbability in animals admiring for a very long period the same general style of ornamentation or other attractions, and yet appreciating slight changes in colors, form, or sound.

SUMMARY OF THE FOUR CHAPTERS ON BIRDS.

Most male birds are highly pugnacious during the breeding-season, and some possess weapons adapted for fighting with their rivals. But the most pugnacious and the best armed males rarely or never depend for success solely on their power to drive away or kill their rivals, but have special means for charming the female. With some it is the power of song, or of giving forth strange cries, or instrumental music, and the males in consequence differ from the females in their vocal organs, or in the structure of certain feathers. From the curiously diversified means for producing various sounds, we gain a high idea of the importance of this means of courtship. Many birds endeavour to charm the females by love-dances or antics, performed on the ground or in the air, and sometimes at prepared places. But ornaments of many kinds, the most brilliant tints, combs and wattles, beautiful plumes, elongated feathers, top-knots, and so forth, are by far the commonest means. In some cases mere novelty appears to have acted as a charm. The ornaments of the males must be highly important to them, for they have been acquired in not a few cases at the cost of increased danger from enemies, and even at some loss of power in fighting with their rivals. The males of very many species do not assume their ornamental dress until they arrive at maturity, or they assume it only during the breeding-season, or the tints then become more vivid. Certain ornamental appendages become enlarged, turgid, and brightly colored during the act of courtship. The males display their charms with elaborate care and to the best effect; and this is done in the presence of the females. The courtship is sometimes a prolonged affair, and many males and females congregate at an appointed place. To suppose that the females do not appreciate the beauty of the males, is to admit that their splendid decorations, all their pomp and display, are useless; and this is incredible. Birds have fine powers of discrimination, and in some few instances it can be shown that they have a taste for the beautiful. The females, moreover, are known occasionally to exhibit a marked preference or antipathy for certain individual males.

If it be admitted that the females prefer, or are unconsciously excited by the more beautiful males, then the males would slowly but surely be rendered more and more attractive through sexual selection. That it is this sex which has been chiefly modified, we may infer from the fact that, in almost every genus where the sexes differ, the males differ much more from one another than do the females; this is well shown in certain closely-allied representative species, in which the females can hardly be distinguished, whilst the males are quite distinct. Birds in a state of nature offer individual differences which would amply suffice for the work of sexual selection; but we have seen that they occasionally present more strongly marked variations which recur so frequently that they would immediately be fixed, if they served to allure the female. The laws of variation must determine the nature of the initial changes, and will have largely influenced the final result. The gradations, which may be observed between the males of allied species, indicate the nature of the steps through which they have passed. They explain also in the most interesting manner how certain characters have originated, such as the indented ocelli on the tail-feathers of the peacock, and the ball-and-socket ocelli on the wing-feathers of the Argus pheasant. It is evident that the brilliant colors, top-knots, fine plumes, etc., of many male birds cannot have been acquired as a protection; indeed, they sometimes lead to danger. That they are not due to the direct and definite action of the conditions of life, we may feel assured, because the females have been exposed to the same conditions, and yet often differ from the males to an extreme degree. Although it is probable that changed conditions acting during a lengthened period have in some cases produced a definite effect on both sexes, or sometimes on one sex alone, the more important result will have been an increased tendency to vary or to present more strongly-marked individual differences; and such differences will have afforded an excellent ground-work for the action of sexual selection.

The laws of inheritance, irrespectively of selection, appear to have determined whether the characters acquired by the males for the sake of ornament, for producing various sounds, and for fighting together, have been transmitted to the males alone or to both sexes, either permanently, or periodically during certain seasons of the year. Why various characters should have been transmitted sometimes in one way and sometimes in another, is not in most cases known; but the period of variability seems often to have been the determining cause. When the two sexes have inherited all characters in common they necessarily resemble each other; but as the successive variations may be differently transmitted, every possible gradation may be found, even within the same genus, from the closest similarity to the widest dissimilarity between the sexes. With many closely-allied species, following nearly the same habits of life, the males have come to differ from each other chiefly through the action of sexual selection; whilst the females have come to differ chiefly from partaking more or less of the characters thus acquired by the males. The effects, moreover, of the definite action of the conditions of life, will not have been masked in the females, as in the males, by the accumulation through sexual selection of strongly-pronounced colors and other ornaments. The individuals of both sexes, however affected, will have been kept at each successive period nearly uniform by the free intercrossing of many individuals.

With species, in which the sexes differ in color, it is possible or probable that some of the successive variations often tended to be transmitted equally to both sexes; but that when this occurred the females were prevented from acquiring the bright colors of the males, by the destruction which they suffered during incubation. There is no evidence that it is possible by natural selection to convert one form of transmission into another. But there would not be the least difficulty in rendering a female dull-colored, the male being still kept bright-colored, by the selection of successive variations, which were from the first limited in their transmission to the same sex. Whether the females of many species have actually been thus modified, must at present remain doubtful. When, through the law of the equal transmission of characters to both sexes, the females were rendered as conspicuously colored as the males, their instincts appear often to have been modified so that they were led to build domed or concealed nests.

In one small and curious class of cases the characters and habits of the two sexes have been completely transposed, for the females are larger, stronger, more vociferous and brighter colored than the males. They have, also, become so quarrelsome that they often fight together for the possession of the males, like the males of other pugnacious species for the possession of the females. If, as seems probable, such females habitually drive away their rivals, and by the display of their bright colors or other charms endeavour to attract the males, we can understand how it is that they have gradually been rendered, by sexual selection and sexually-limited transmission, more beautiful than the males—the latter being left unmodified or only slightly modified.

Whenever the law of inheritance at corresponding ages prevails but not that of sexually-limited transmission, then if the parents vary late in life—and we know that this constantly occurs with our poultry, and occasionally with other birds—the young will be left unaffected, whilst the adults of both sexes will be modified. If both these laws of inheritance prevail and either sex varies late in life, that sex alone will be modified, the other sex and the young being unaffected. When variations in brightness or in other conspicuous characters occur early in life, as no doubt often happens, they will not be acted on through sexual selection until the period of reproduction arrives; consequently if dangerous to the young, they will be eliminated through natural selection. Thus we can understand how it is that variations arising late in life have so often been preserved for the ornamentation of the males; the females and the young being left almost unaffected, and therefore like each other. With species having a distinct summer and winter plumage, the males of which either resemble or differ from the females during both seasons or during the summer alone, the degrees and kinds of resemblance between the young and the old are exceedingly complex; and this complexity apparently depends on characters, first acquired by the males, being transmitted in various ways and degrees, as limited by age, sex, and season.

As the young of so many species have been but little modified in color and in other ornaments, we are enabled to form some judgment with respect to the plumage of their early progenitors; and we may infer that the beauty of our existing species, if we look to the whole class, has been largely increased since that period, of which the immature plumage gives us an indirect record. Many birds, especially those which live much on the ground, have undoubtedly been obscurely colored for the sake of protection. In some instances the upper exposed surface of the plumage has been thus colored in both sexes, whilst the lower surface in the males alone has been variously ornamented through sexual selection. Finally, from the facts given in these four chapters, we may conclude that weapons for battle, organs for producing sound, ornaments of many kinds, bright and conspicuous colors, have generally been acquired by the males through variation and sexual selection, and have been transmitted in various ways according to the several laws of inheritance—the females and the young being left comparatively but little modified. [57]

[57] I am greatly indebted to the kindness of Mr. Sclater for having looked over these four chapters on birds, and the two following ones on mammals. In this way I have been saved from making mistakes about the names of the species, and from stating anything as a fact which is known to this distinguished naturalist to be erroneous. But, of course, he is not at all answerable for the accuracy of the statements quoted by me from various authorities.

CHAPTER XVII.

SECONDARY SEXUAL CHARACTERS OF MAMMALS.

The law of battle—Special weapons, confined to the males—Cause of absence of weapons in the female—Weapons common to both sexes, yet primarily acquired by the male—Other uses of such weapons—Their high importance—Greater size of the male—Means of defence—On the preference shown by either sex in the pairing of quadrupeds.

With mammals the male appears to win the female much more through the law of battle than through the display of his charms. The most timid animals, not provided with any special weapons for fighting, engage in desperate conflicts during the season of love. Two male hares have been seen to fight together until one was killed; male moles often fight, and sometimes with fatal results; male squirrels engage in frequent contests, "and often wound each other severely"; as do male beavers, so that "hardly a skin is without scars." [1] I observed the same fact with the hides of the guanacoes in Patagonia; and on one occasion several were so absorbed in fighting that they fearlessly rushed close by me. Livingstone speaks of the males of the many animals in Southern Africa as almost invariably showing the scars received in former contests.

[1] See Waterton's account of two hares fighting, 'Zoologist,' vol. i. 1843, p. 211. On moles, Bell, 'Hist. of British Quadrupeds,' 1st ed., p. 100. On squirrels, Audubon and Bachman, Viviparous Quadrupeds of N. America, 1846, p. 269. On beavers, Mr. A.H. Green, in 'Journal of Lin. Society, Zoology,' vol. x. 1869, p. 362.

The law of battle prevails with aquatic as with terrestrial mammals. It is notorious how desperately male seals fight, both with their teeth and claws, during the breeding-season; and their hides are likewise often covered with scars. Male sperm-whales are very jealous at this season; and in their battles "they often lock their jaws together, and turn on their sides and twist about"; so that their lower jaws often become distorted. [2]

[2] On the battles of seals, see Capt. C. Abbott in 'Proc. Zool. Soc.' 1868, p. 191; Mr. R. Brown, ibid. 1868, p. 436; also L. Lloyd, 'Game Birds of Sweden,' 1867, p. 412; also Pennant. On the sperm-whale see Mr. J.H. Thompson, in 'Proc. Zool. Soc.' 1867, p. 246.

All male animals which are furnished with special weapons for fighting, are well known to engage in fierce battles. The courage and the desperate conflicts of stags have often been described; their skeletons have been found in various parts of the world, with the horns inextricably locked together, showing how miserably the victor and vanquished had perished. [3] No animal in the world is so dangerous as an elephant in must. Lord Tankerville has given me a graphic description of the battles between the wild bulls in Chillingham Park, the descendants, degenerated in size but not in courage, of the gigantic *Bos primigenius*. In 1861 several contended for mastery; and it was observed that two of the younger bulls attacked in concert the old leader of the herd, overthrew and disabled him, so that he was believed by the keepers to be lying mortally wounded in a neighbouring wood. But a few days afterwards one of the young bulls approached the wood alone; and then the "monarch of the chase," who had been lashing himself up for vengeance, came out and, in a short time, killed his antagonist. He then quietly joined the herd, and long held undisputed sway. Admiral Sir B.J. Sulivan informs me that, when he lived in the Falkland Islands, he imported a young English stallion, which frequented the hills near Port William with eight mares. On these hills there were two wild stallions, each with a small troop of mares; "and it is certain that these stallions would never have approached each other without fighting. Both had tried singly to fight the English horse and drive away his mares, but had failed. One day they came in *together* and attacked him. This was seen by the captain who had charge of the horses, and who, on riding to the spot, found one of the two stallions engaged with the English horse, whilst the other was driving away the mares, and had already separated four from the rest. The captain settled the matter by driving the whole party into the corral, for the wild stallions would not leave the mares."

[3] See Scrope ('Art of Deer-stalking,' p. 17) on the locking of the horns with the *Cervus elaphus*. Richardson, in 'Fauna Bor. Americana,' 1829, p. 252, says that the wapiti, moose, and reindeer have been found thus locked together. Sir. A. Smith found at the Cape of Good Hope the skeletons of two gnus in the same condition.

Male animals which are provided with efficient cutting or tearing teeth for the ordinary purposes of life, such as the carnivora, insectivora, and rodents, are seldom furnished with weapons especially adapted for fighting with their rivals. The case is very different with the males of many other animals. We see this in the horns of stags and of certain kinds of antelopes in which the females are hornless. With many animals the canine teeth in the upper or lower jaw, or in both, are much larger in the males than in the females, or are absent in the latter, with the exception sometimes of a hidden rudiment. Certain antelopes, the musk-deer, camel, horse, boar, various apes, seals, and the walrus, offer instances. In the females of the walrus the tusks are sometimes quite absent. [4] In the male elephant of India and in the male dugong [5] the upper incisors form offensive weapons. In the male narwhal the left canine alone is developed into the well-known, spirally-twisted, so-called horn, which is sometimes from nine to ten feet in length. It is believed that the males use these horns for fighting together; for "an unbroken one can rarely be got, and occasionally one may be found with the point of another jammed into the broken place." [6] The tooth on the opposite side of the head in the male consists of a rudiment about ten inches in length, which is embedded in the jaw; but sometimes, though rarely, both are equally developed on the two sides. In the female both are always rudimentary. The male cachalot has a larger head than that of the female, and it no doubt aids him in his aquatic battles. Lastly, the adult male ornithorhynchus is provided with a remarkable apparatus, namely a spur on the foreleg, closely resembling the poison-fang of a venomous snake; but according to Harting, the secretion from the gland is not poisonous; and on the leg of the female there is a hollow, apparently for the reception of the spur. [7]

[4] Mr. Lamont ('Seasons with the Sea-Horses,' 1861, p. 143) says that a good tusk of the male walrus weighs 4 pounds, and is longer than that of the female, which weighs about 3 pounds. The males are described as fighting ferociously. On the occasional absence of the tusks in the female, see Mr. R. Brown, 'Proceedings, Zoological Society,' 1868, p. 429.

[5] Owen, 'Anatomy of Vertebrates,' vol. iii. p. 283.

[6] Mr. R. Brown, in 'Proc. Zool. Soc.' 1869, p. 553. See Prof. Turner, in 'Journal of Anat. and Phys.' 1872, p. 76, on the homological nature of these tusks. Also Mr. J.W. Clarke on two tusks being developed in the males, in 'Proceedings of the Zoological Society,' 1871, p. 42.

[7] Owen on the cachalot and Ornithorhynchus, ibid. vol. iii. pp. 638, 641. Harting is quoted by Dr. Zouteveen in the Dutch translation of this work, vol. ii. p. 292.

When the males are provided with weapons which in the females are absent, there can be hardly a doubt that these serve for fighting with other males; and that they were acquired through sexual selection, and were transmitted to the male sex alone. It is not probable, at least in most cases, that the females have been prevented from acquiring such weapons, on account of their being useless, superfluous, or in some way injurious. On the contrary, as they are often used by the males for various purposes, more especially as a defence against their enemies, it is a surprising fact that they are so poorly developed, or quite absent, in the females of so many animals. With female deer the development during each recurrent season of great branching horns, and with female elephants the development of immense tusks, would be a great waste of vital power, supposing that they were of no use to the females. Consequently, they would have tended to be eliminated in the female through natural selection; that is, if the successive variations were limited in their transmission to the female sex, for otherwise the weapons of the males would have been injuriously affected, and this would have been a greater evil. On the whole, and from the consideration of the following facts, it seems probable that when the various weapons differ in the two sexes, this has generally depended on the kind of transmission which has prevailed.

As the reindeer is the one species in the whole family of Deer, in which the female is furnished with horns, though they are somewhat smaller, thinner, and less branched than in the male, it might naturally be thought that, at least in this case, they must be of some special service to her. The female retains her horns from the time when they are fully developed, namely, in September, throughout the winter until April or May, when she brings forth her young. Mr. Crotch made particular enquiries for me in Norway, and it appears that the females at this season conceal themselves for about a fortnight in order to bring forth their young, and then reappear, generally hornless. In Nova Scotia, however, as I hear from Mr. H. Reeks, the female sometimes retains her horns longer. The male on the other hand casts his horns much earlier, towards the end of November. As both sexes have the same requirements and follow the same habits of life, and as the male is destitute of horns during the winter, it is improbable that they can be of any special service to the female during this season, which includes the larger part of the time during which she is horned. Nor is it probable that she can have inherited horns from some ancient progenitor of the family of deer, for, from the fact of the females of so many species in all quarters of the globe not having horns, we may conclude that this was the primordial character of the group. [8]

[8] On the structure and shedding of the horns of the reindeer, Hoffberg, 'Amœnitates Acad.' vol. iv. 1788, p. 149. See Richardson, 'Fauna Bor. Americana,' p. 241, in regard to the American variety or species: also Major W. Ross King, 'The Sportsman in Canada,' 1866, p. 80.

The horns of the reindeer are developed at a most unusually early age; but what the cause of this may be is not known. The effect has apparently been the transference of the horns to both sexes. We should bear in mind that horns are always transmitted through the female, and that she has a latent capacity for their development, as we see in old or diseased females. [9] Moreover the females of some other species of deer exhibit, either normally or occasionally, rudiments of horns; thus the female of Cervulus moschatus has "bristly tufts, ending in a knob, instead of a horn"; and "in most specimens of the female wapiti (Cervus canadensis) there is a sharp bony protuberance in the place of the horn." [10] From these several considerations we may conclude that the possession of fairly well-developed horns by the female reindeer, is due to the males having first acquired them as weapons for fighting with other males; and secondarily to their development from some unknown cause at an unusually early age in the males, and their consequent transference to both sexes.

[9] Isidore Geoffroy St.-Hilaire, 'Essais de Zoolog. Générale,' 1841, p. 513. Other masculine characters, besides the horns, are sometimes similarly transferred to the female; thus Mr. Boner, in speaking of an old female chamois ('Chamois Hunting in the Mountains of Bavaria,' 1860, 2nd ed., p. 363), says, "not only was the head very male-looking, but along the back there was a ridge of long hair, usually to be found only in bucks."

[10] On the Cervulus, Dr. Gray, 'Catalogue of Mammalia in the British Museum,' part iii. p. 220. On the Cervus canadensis or wapiti, see Hon. J.D. Caton, 'Ottawa Academy of Nat. Sciences,' May 1868, p. 9.

Turning to the sheath-horned ruminants: with antelopes a graduated series can be formed, beginning with species, the females of which are completely destitute of horns—passing on to those which have horns so small as to be almost rudimentary (as with the Antilocapra americana, in which species they are present in only one out of four or five females [11])—to those which have fairly developed horns, but manifestly smaller and thinner than in the male and sometimes of a different shape [12],—and ending with those in which both sexes have horns of equal size. As with the reindeer, so with antelopes, there exists, as previously shown, a relation between the period of the development of the horns and their transmission to one or both sexes; it is therefore probable that their presence or absence in the females of some species, and their more or less perfect condition in the females of other species, depends, not on their being of any special use, but simply on inheritance. It accords with this view that even in the same restricted genus both sexes of some species, and the males alone of others, are thus provided. It is also a remarkable fact that, although the females of *Antilope bezoartica* are normally destitute of horns, Mr. Blyth has seen no less than three females thus furnished; and there was no reason to suppose that they were old or diseased.

[11] I am indebted to Dr. Canfield for this information; see also his paper in the 'Proceedings of the Zoological Society,' 1866, p. 105.

[12] For instance the horns of the female *Ant. euchore* resemble those of a distinct species, viz. the *Ant. dorcas* var. *Corine*, see Desmarest, 'Mammalogie,' p. 455.

In all the wild species of goats and sheep the horns are larger in the male than in the female, and are sometimes quite absent in the latter. [13] In several domestic breeds of these two animals, the males alone are furnished with horns; and in some breeds, for instance, in the sheep of North Wales, though both sexes are properly horned, the ewes are very liable to be hornless. I have been informed by a trustworthy witness, who purposely inspected a flock of these same sheep during the lambing season, that the horns at birth are generally more fully developed in the male than in the female. Mr. J. Peel crossed his Lonk sheep, both sexes of which always bear horns, with hornless Leicesters and hornless Shropshire Downs; and the result was that the male offspring had their horns considerably reduced, whilst the females were wholly destitute of them. These several facts indicate that, with sheep, the horns are a much less firmly fixed character in the females than in the males; and this leads us to look at the horns as properly of masculine origin.

[13] Gray, 'Catalogue of Mammalia, the British Museum,' part iii. 1852, p. 160.

With the adult musk-ox (*Ovibos moschatus*) the horns of the male are larger than those of the female, and in the latter the bases do not touch. [14] In regard to ordinary cattle Mr. Blyth remarks: "In most of the wild bovine animals the horns are both longer and thicker in the bull than in the cow, and in the cow-banteng (*Bos sondaicus*) the horns are remarkably small, and inclined much backwards. In the domestic races of cattle, both of the humped and humpless types, the horns are short and thick in the bull, longer and more slender in the cow and ox; and in the Indian buffalo, they are shorter and thicker in the bull, longer and more slender in the cow. In the wild gaour (*B. gaurus*) the horns are mostly both longer and thicker in the bull than in the cow." [15] Dr. Forsyth Major also informs me that a fossil skull, believed to be that of the female *Bos etruscus*, has been found in Val d'Arno, which is wholly without horns. In the *Rhinoceros simus*, as I may add, the horns of the female are generally longer but less powerful than in the male; and in some other species of rhinoceros they are said to be shorter in the female. [16] From these various facts we may infer as probable that horns of all kinds, even when they are equally developed in the two sexes, were primarily acquired by the male in order to conquer other males, and have been transferred more or less completely to the female.

[14] Richardson, 'Fauna Bor. Americana,' p. 278.

[15] 'Land and Water,' 1867, p. 346.

[16] Sir Andrew Smith, 'Zoology of S. Africa,' pl. xix. Owen, 'Anatomy of Vertebrates,' vol. iii. p. 624.

The effects of castration deserve notice, as throwing light on this same point. Stags after the operation never renew their horns. The male reindeer, however, must be excepted, as after castration he does renew them. This fact, as well as the possession of horns by both sexes, seems at first to prove that the horns in this species do not constitute a sexual character; [17] but as they are developed at a very early age, before the sexes differ in constitution, it is not surprising that they should be unaffected by castration, even if they were aboriginally acquired by the male. With sheep both sexes properly bear horns; and I am informed that with Welch sheep the horns of the males are considerably reduced by castration; but the degree depends much on the age at which the operation is performed, as is likewise the case with other animals. Merino rams have large horns, whilst the ewes "generally speaking are without horns"; and in this breed castration seems to produce a somewhat greater effect, so that if performed at an early age the horns "remain almost undeveloped." [18] On the Guinea coast there is a breed in which the females never bear horns, and, as Mr. Winwood Reade informs me, the rams after castration are quite destitute of them. With cattle, the horns of the males are much altered by castration; for instead of being short and thick, they become longer than those of the cow, but otherwise resemble them. The *Antilope bezoartica* offers a somewhat analogous case: the males have long straight spiral horns, nearly parallel to each other, and directed backwards; the females occasionally bear horns, but these when present are of a very different shape, for they are not spiral, and spreading widely, bend round with the points forwards. Now it is a remarkable fact that, in the castrated male, as Mr. Blyth informs me, the horns are of the same peculiar shape as in the female, but longer and thicker. If we may judge from analogy, the female probably shows us, in these two cases of cattle and the antelope, the former condition of the horns in some early progenitor of each species. But why castration should lead to the reappearance of an early condition of the horns cannot be explained with any certainty. Nevertheless, it seems probable, that in nearly the same manner as the constitutional disturbance in the offspring, caused by a cross between two distinct species or races, often leads to the reappearance of long-lost characters; [19] so here, the disturbance in the constitution of the individual, resulting from castration, produces the same effect.

[17] This is the conclusion of Seidlitz, 'Die Darwinsche Theorie,' 1871, p. 47.

[18] I am much obliged to Prof. Victor Carus, for having made enquiries for me in Saxony on this subject. H. von Nathusius ('Viehzucht,' 1872, p. 64) says that the horns of sheep castrated at an early period, either altogether disappear or remain as mere rudiments; but I do not know whether he refers to merinos or to ordinary breeds.

[19] I have given various experiments and other evidence proving that this is the case, in my 'Variation of Animals and Plants under Domestication,' vol. ii. 1868, pp. 39-47.

The tusks of the elephant, in the different species or races, differ according to sex, nearly as do the horns of ruminants. In India and Malacca the males alone are provided with well-developed tusks. The elephant of Ceylon is considered by most naturalists as a distinct race, but by some as a distinct species, and here "not one in a hundred is found with tusks, the few that possess them being exclusively males." [20] The African elephant is undoubtedly distinct, and the female has large well-developed tusks, though not so large as those of the male.

[20] Sir J. Emerson Tennent, 'Ceylon,' 1859, vol. ii. p. 274. For Malacca, 'Journal of Indian Archipelago,' vol. iv. p. 357.

These differences in the tusks of the several races and species of elephants—the great variability of the horns of deer, as notably in the wild reindeer—the occasional presence of horns in the female *Antilope Bezoartica*, and their frequent absence in the female of *Antilocapra americana*—the presence of two tusks in some few male narwhals—the complete absence of tusks in some female walruses—are all instances of the extreme variability of secondary sexual characters, and of their liability to differ in closely-allied forms.

Although tusks and horns appear in all cases to have been primarily developed as sexual weapons, they often serve other purposes. The elephant uses his tusks in attacking the tiger; according to Bruce, he scores the trunks of trees until they can be thrown down easily, and he likewise thus extracts the farinaceous cores of palms; in Africa he often uses one tusk, always the same, to probe the ground and thus ascertain whether it will bear his weight. The common bull defends the herd with his horns; and the elk in Sweden has been known, according to Lloyd, to strike a wolf dead with a single blow of his great horns. Many similar facts could be given. One of the most curious secondary uses to which the horns of an animal may be occasionally put is that observed by Captain Hutton [21] with the wild goat (*Capra ægagrus*) of the Himalayas and, as it is also said with the ibex, namely that when the male accidentally falls from a height he bends inwards his head, and by alighting on his massive horns, breaks the shock. The female cannot thus use her horns, which are smaller, but from her more quiet disposition she does not need this strange kind of shield so much.

[21] 'Calcutta Journal of Natural History,' vol. ii, 1843, p. 526.

Each male animal uses his weapons in his own peculiar fashion. The common ram makes a charge and butts with such force with the bases of his horns, that I have seen a powerful man knocked over like a child. Goats and certain species of sheep, for instance the *Ovis cycloceros* of Afghanistan, [22] rear on their hind legs, and then not only butt, but "make a cut down and a jerk up, with the ribbed front of their scimitar-shaped horn, as with a sabre. When the *O. cycloceros* attacked a large domestic ram, who was a noted bruiser, he conquered him by the sheer novelty of his mode of fighting, always closing at once with his adversary, and catching him across the face and nose with a sharp drawing jerk of the head, and then bounding out of the way before the blow could be returned." In Pembrokeshire a male goat, the master of a flock which during several generations had run wild, was known to have killed several males in single combat; this goat possessed enormous horns, measuring thirty-nine inches in a straight line from tip to tip. The common bull, as every one knows, gores and tosses his opponent; but the Italian buffalo is said never to use his horns: he gives a tremendous blow with his convex forehead, and then tramples on his fallen enemy with his knees—an instinct which the common bull does not possess. [23] Hence a dog who pins a buffalo by the nose is immediately crushed. We must, however, remember that the Italian buffalo has been long domesticated, and it is by no means certain that the wild parent-form had similar horns. Mr. Bartlett informs me that when a female Cape buffalo (*Bubalus caffer*) was turned into an enclosure with a bull of the same species, she attacked him, and he in return pushed her about with great violence. But it was manifest to Mr. Bartlett that, had not the bull shown dignified forbearance, he could easily have killed her by a single lateral thrust with his immense horns. The giraffe uses his short, hair-covered horns, which are rather longer in the male than in the female, in a curious manner; for, with his long neck, he swings his head to either side, almost upside down, with such force that I have seen a hard plank deeply indented by a single blow.

[22] Mr. Blyth, in 'Land and Water,' March, 1867, p. 134, on the authority of Capt. Hutton and others. For the wild Pembrokeshire goats, see the 'Field,' 1869, p. 150.
[23] M. E.M. Bailly, "Sur l'usage des Cornes," etc., .Annal des Sciences Nat.' tom. ii. 1824, p. 369.

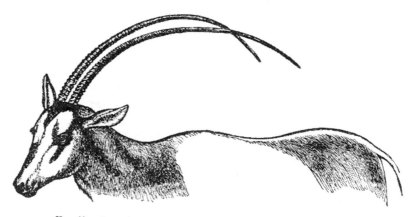

FIG. 63.—Oryx leucoryx, male (from the Knowsley Menagerie).

With antelopes it is sometimes difficult to imagine how they can possibly use their curiously-shaped horns; thus the springboc (*Ant. euchore*) has rather short upright horns, with the sharp points bent inwards almost at right angles, so as to face each other; Mr. Bartlett does not know how they are used, but suggests that they would inflict a fearful wound down each side of the face of an antagonist. The slightly-curved horns of the *Oryx leucoryx* (Fig. 63) are directed backwards, and are of such length that their points reach beyond the middle of the back, over which they extend in almost parallel lines. Thus they seem singularly ill-fitted for fighting; but Mr. Bartlett informs me that when two of these animals prepare for battle, they kneel down, with their beads between their fore legs, and in this attitude the horns stand nearly parallel and close to the ground, with the points directed forwards and a little upwards. The combatants then gradually approach each other, and each endeavours to get the upturned points under the body of the other; if one succeeds in doing this, he suddenly springs up, throwing up his head at the same time, and can thus wound or perhaps even transfix his antagonist. Both animals always kneel down, so as to guard as far as possible against this manœuvre. It has been recorded that one of these antelopes has used his horn with effect even against a lion; yet from being forced to place his head between the forelegs in order to bring the points of the horns forward, he would generally be under a great disadvantage when attacked by any other animal. It is, therefore, not probable that the horns have been modified into their present great length and peculiar position, as a protection against beasts of prey. We can however see that, as soon as some ancient male progenitor of the Oryx acquired moderately long horns, directed a little backwards, he would be compelled, in his battles with rival males, to bend his head somewhat inwards or downwards, as is now done by certain stags; and it is not improbable that he might have acquired the habit of at first occasionally and afterwards of regularly kneeling down. In this case it is almost certain that the males which possessed the longest horns would have had a great advantage over others with shorter horns; and then the horns would gradually have been rendered longer and longer, through sexual selection, until they acquired their present extraordinary length and position.

With stags of many kinds the branches of the horns offer a curious case of difficulty; for certainly a single straight point would inflict a much more serious wound than several diverging ones. In Sir Philip Egerton's museum there is a horn of the red-deer (*Cervus elaphus*), thirty inches in length, with "not fewer than fifteen snags or branches"; and at Moritzburg there is still preserved a pair of antlers of a red-deer, shot in 1699 by Frederick I., one of which bears the astonishing number of thirty-three branches and the other twenty-seven, making altogether sixty branches. Richardson figures a pair of antlers of the wild reindeer with twenty-nine points. [24] From the manner in which the horns are branched, and more especially from deer being known occasionally to fight together by kicking with their fore-feet, [25] M. Bailly actually comes to the conclusion that their horns are more injurious than useful to them. But this author overlooks the pitched battles between rival males. As I felt much perplexed about the use or advantage of the branches, I applied to Mr. McNeill of Colonsay, who has long and carefully observed the habits of red-deer, and he informs me that he has never seen some of the branches brought into use, but that the brow antlers, from inclining downwards, are a great protection to the forehead, and their points are likewise used in attack. Sir Philip Egerton also informs me both as to red-deer and fallow-deer that, in fighting, they suddenly dash together, and getting their horns fixed against each other's bodies, a desperate struggle ensues. When one is at last forced to yield and turn round, the victor endeavours to plunge his brow antlers into his defeated foe. It thus appears that the upper branches are used chiefly or exclusively for pushing and fencing. Nevertheless in some species the upper branches are used as weapons of offence; when a man was attacked by a wapiti deer (*Cervus canadensis*) in Judge Caton's park in Ottawa, and several men tried to rescue him, the stag "never raised his head from the ground; in fact he kept his face almost flat on the ground, with his nose nearly between his fore feet, except when he rolled his head to one side to take a new observation preparatory to a plunge." In this position the ends of the horns were directed against his adversaries. "In rolling his head he necessarily raised it somewhat, because his antlers were so long that he could not roll his head without raising them on one side, while, on the other side they touched the ground." The stag by this procedure gradually drove the party of rescuers backwards to a distance of 150 or 200 feet; and the attacked man was killed. [26]

[24] On the horns of red-deer, Owen, 'British Fossil Mammals,' 1846, p. 478; Richardson on the horns of the reindeer, 'Fauna Bor. Americana,' 1829, p. 240. I am indebted to Prof. Victor Carus, for the Moritzburg case.

410

[25] Hon. J.D. Caton ('Ottawa Acad. of Nat. Science,' May 1868, p. 9) says that the American deer fight with their fore-feet, after "the question of superiority has been once settled and acknowledged in the herd." Bailly, 'Sur l'usage des Cornes,' 'Annales des Sciences Nat.' tom. ii. 1824, p. 371.

[26] See a most interesting account in the Appendix to Hon. J.D. Caton's paper, as above quoted.

Fɪɢ. 64.—Strepsiceros Kudu (from Sir Andrew Smith's ' Zoology of South Africa ').

Although the horns of stags are efficient weapons, there can, I think be no doubt that a single point would have been much more dangerous than a branched antler; and Judge Caton, who has had large experience with deer, fully concurs in this conclusion. Nor do the branching horns, though highly important as a means of defence against rival stags, appear perfectly well adapted for this purpose, as they are liable to become interlocked. The suspicion has therefore crossed my mind that they may serve in part as ornaments. That the branched antlers of stags as well as the elegant lyrated horns of certain antelopes, with their graceful double curvature (Fig. 64), are ornamental in our eyes, no one will dispute. If, then, the horns, like the splendid accoutrements of the knights of old, add to the noble appearance of stags and antelopes, they may have been modified partly for this purpose, though mainly for actual service in battle; but I have no evidence in favour of this belief.

An interesting case has lately been published, from which it appears that the horns of a deer in one district in the United States are now being modified through sexual and natural selection. A writer in an excellent American Journal [27] says, that he has hunted for the last twenty-one years in the Adirondacks, where the *Cervus virginianus* abounds. About fourteen years ago he first heard of *spike-horn bucks*. These became from year to year more common; about five years ago he shot one, and afterwards another, and now they are frequently killed. "The spike-horn differs greatly from the common antler of the *C. virginianus*. It consists of a single spike, more slender than the antler, and scarcely half so long, projecting forward from the brow, and terminating in a very sharp point. It gives a considerable advantage to its possessor over the common buck. Besides enabling him to run more swiftly through the thick woods and underbrush (every hunter knows that does and yearling bucks run much more rapidly than the large bucks when armed with their cumbrous antlers), the spike-horn is a more effective weapon than the common antler. With this advantage the spike-horn bucks are gaining upon the common bucks, and may, in time, entirely supersede them in the Adirondacks. Undoubtedly, the first spike-horn buck was merely an accidental freak of nature. But his spike-horns gave him an advantage, and enabled him to propagate his peculiarity. His descendants having a like advantage, have propagated the peculiarity in a constantly increasing ratio, till they are slowly crowding the antlered deer from the region they inhabit." A critic has well objected to this account by asking, why, if the simple horns are now so advantageous, were the branched antlers of the parent-form ever developed? To this I can only answer by remarking, that a new mode of attack with new weapons might be a great advantage, as shown by the case of the *Ovis cycloceros*, who thus conquered a domestic ram famous for his fighting power. Though the branched antlers of a stag are well adapted for fighting with his rivals, and though it might be an advantage to the prong-horned variety slowly to acquire long and branched horns, if he had to fight only with others of the same kind, yet it by no means follows that branched horns would be the best fitted for conquering a foe differently armed. In the foregoing case of the *Oryx leucoryx*, it is almost certain that the victory would rest with an antelope having short horns, and who therefore did not need to kneel down, though an oryx might profit by having still longer horns, if he fought only with his proper rivals.

[27] The 'American Naturalist,' Dec. 1869, p. 552.

Male quadrupeds, which are furnished with tusks, use them in various ways, as in the case of horns. The boar strikes laterally and upwards; the musk-deer downwards with serious effect. [28] The walrus, though having so short a neck and so unwieldy a body, "can strike either upwards, or downwards, or sideways, with equal dexterity." [29] I was informed by the late Dr. Falconer, that the Indian elephant fights in a different manner according to the position and curvature of his tusks. When they are directed forwards and upwards he is able to fling a tiger to a great distance—it is said to even thirty feet; when they are short and turned downwards he endeavours suddenly to pin the tiger to the ground and, in consequence, is dangerous to the rider, who is liable to be jerked off the howdah. [30]

[28] Pallas, 'Spicilegia Zoologica,' fasc. xiii. 1779, p. 18.

[29] Lamont, 'Seasons with the Sea-Horses,' 1861, p. 141.

[30] See also Corse ('Philosophical Transactions,' 1799, p. 212) on the manner in which the short-tusked Mooknah variety attacks other elephants.

Very few male quadrupeds possess weapons of two distinct kinds specially adapted for fighting with rival males. The male muntjac-deer (Cervulus), however, offers an exception, as he is provided with horns and exserted canine teeth. But we may infer from what follows that one form of weapon has often been replaced in the course of ages by another. With ruminants the development of horns generally stands in an inverse relation with that of even moderately developed canine teeth. Thus camels, guanacoes, chevrotains, and musk-deer, are hornless, and they have efficient canines; these teeth being "always of smaller size in the females than in the males." The Camelidæ have, in addition to their true canines, a pair of canine-shaped incisors in their upper jaws. [31] Male deer and antelopes, on the other hand, possess horns, and they rarely have canine teeth; and these, when present, are always of small size, so that it is doubtful whether they are of any service in their battles. In *Antilope montana* they exist only as rudiments in the young male, disappearing as he grows old; and they are absent in the female at all ages; but the females of certain other antelopes and of certain deer have been known occasionally to exhibit rudiments of these teeth. [32] Stallions have small canine teeth, which are either quite absent or rudimentary in the mare; but they do not appear to be used in fighting, for stallions bite with their incisors, and do not open their mouths wide like camels and guanacoes. Whenever the adult male possesses canines, now inefficient, whilst the female has either none or mere rudiments, we may conclude that the early male progenitor of the species was provided with efficient canines, which have been partially transferred to the females. The reduction of these teeth in the males seems to have followed from some change in their manner of fighting, often (but not in the horse) caused by the development of new weapons.

[31] Owen, 'Anatomy of Vertebrates,' vol. iii. p. 349.

[32] See Rüppell (in 'Proc. Zoolog. Soc.' Jan. 12, 1836, p. 3) on the canines in deer and antelopes, with a note by Mr. Martin on a female American deer. See also Falconer ('Palæont. Memoirs and Notes,' vol. i. 1868, p. 576) on canines in an adult female deer. In old males of the musk-deer the canines (Pallas, 'Spic. Zoolog.' fasc. xiii. 1779, p. 18) sometimes grow to the length of three inches, whilst in old females a rudiment projects scarcely half an inch above the gums.

Tusks and horns are manifestly of high importance to their possessors, for their development consumes much organized matter. A single tusk of the Asiatic elephant—one of the extinct woolly species—and of the African elephant, have been known to weigh respectively 150, 160, and 180 pounds; and even greater weights have been given by some authors. [33] With deer, in which the horns are periodically renewed, the drain on the constitution must be greater; the horns, for instance, of the moose weigh from fifty to sixty pounds, and those of the extinct Irish elk from sixty to seventy pounds—the skull of the latter weighing on an average only five pounds and a quarter. Although the horns are not periodically renewed in sheep, yet their development, in the opinion of many agriculturists, entails a sensible loss to the breeder. Stags, moreover, in escaping from beasts of prey are loaded with an additional weight for the race, and are greatly retarded in passing through a woody country. The moose, for instance, with horns extending five and a half feet from tip to tip, although so skilful in their use that he will not touch or break a twig when walking quietly, cannot act so dexterously whilst rushing away from a pack of wolves. "During his progress he holds his nose up, so as to lay the horns horizontally back; and in this attitude cannot see the ground distinctly." [34] The tips of the horns of the great Irish elk were actually eight feet apart! Whilst the horns are covered with velvet, which lasts with red-deer for about twelve weeks, they are extremely sensitive to a blow; so that in Germany the stags at this time somewhat change their habits, and avoiding dense forests, frequent young woods and low thickets. [35] These facts remind us that male birds have acquired ornamental plumes at the cost of retarded flight, and other ornaments at the cost of some loss of power in their battles with rival males.

[33] Emerson Tennent, 'Ceylon,' 1859, vol. ii. p. 275; Owen, 'British Fossil Mammals,' 1846, p. 245.
[34] Richardson, 'Fauna Bor. Americana,' on the moose, *Alces palmata*, pp. 236, 237; on the expanse of the horns, 'Land and Water,' 1869, p. 143. See also Owen, 'British Fossil Mammals,' on the Irish elk, pp. 447, 455.
[35] 'Forest Creatures,' by C. Boner, 1861, p. 60.

With mammals, when, as is often the case, the sexes differ in size, the males are almost always larger and stronger. I am informed by Mr. Gould that this holds good in a marked manner with the marsupials of Australia, the males of which appear to continue growing until an unusually late age. But the most extraordinary case is that of one of the seals (*Callorhinus ursinus*), a full-grown female weighing less than one-sixth of a full-grown male. [36] Dr. Gill remarks that it is with the polygamous seals, the males of which are well known to fight savagely together, that the sexes differ much in size; the monogamous species differing but little. Whales also afford evidence of the relation existing between the pugnacity of the males and their large size compared with that of the female; the males of the right-whales do not fight together, and they are not larger, but rather smaller, than their females; on the other hand, male sperm-whales fight much together, and their bodies are "often found scarred with the imprint of their rival's teeth," and they are double the size of the females. The greater strength of the male, as Hunter long ago remarked, [37] is invariably displayed in those parts of the body which are brought into action in fighting with rival males—for instance, in the massive neck of the bull. Male quadrupeds are also more courageous and pugnacious than the females. There can be little doubt that these characters have been gained, partly through sexual selection, owing to a long series of victories, by the stronger and more courageous males over the weaker, and partly through the inherited effects of use. It is probable that the successive variations in strength, size, and courage, whether due to mere variability or to the effects of use, by the accumulation of which male quadrupeds have acquired these characteristic qualities, occurred rather late in life, and were consequently to a large extent limited in their transmission to the same sex.

[36] See the very interesting paper by Mr. J.A. Allen in 'Bull. Mus. Comp. Zoology of Cambridge, United States,' vol. ii. No. 1, p. 82. The weights were ascertained by a careful observer, Capt. Bryant. Dr. Gill in 'The American Naturalist,' January, 1871,

Prof. Shaler on the relative size of the sexes of whales, 'American Naturalist,' January, 1873.

[37] 'Animal Economy,' p. 45.

From these considerations I was anxious to obtain information as to the Scotch deer-hound, the sexes of which differ more in size than those of any other breed (though blood-hounds differ considerably), or than in any wild canine species known to me. Accordingly, I applied to Mr. Cupples, well known for his success with this breed, who has weighed and measured many of his own dogs, and who has with great kindness collected for me the following facts from various sources. Fine male dogs, measured at the shoulder, range from 28 inches, which is low, to 33 or even 34 inches in height; and in weight from 80 pounds, which is light, to 120 pounds, or even more. The females range in height from 23 to 27, or even to 28 inches; and in weight from 50 to 70, or even 80 pounds. [38] Mr. Cupples concludes that from 95 to 100 pounds for the male, and 70 for the female, would be a safe average; but there is reason to believe that formerly both sexes attained a greater weight. Mr. Cupples has weighed puppies when a fortnight old; in one litter the average weight of four males exceeded that of two females by six and a half ounces; in another litter the average weight of four males exceeded that of one female by less than one ounce; the same males when three weeks old, exceeded the female by seven and a half ounces, and at the age of six weeks by nearly fourteen ounces. Mr. Wright of Yeldersley House, in a letter to Mr. Cupples, says: "I have taken notes on the sizes and weights of puppies of many litters, and as far as my experience goes, dog-puppies as a rule differ very little from bitches till they arrive at about five or six months old; and then the dogs begin to increase, gaining upon the bitches both in weight and size. At birth, and for several weeks afterwards, a bitch-puppy will occasionally be larger than any of the dogs, but they are invariably beaten by them later." Mr. McNeill, of Colonsay, concludes that "the males do not attain their full growth till over two years old, though the females attain it sooner." According to Mr. Cupples' experience, male dogs go on growing in stature till they are from twelve to eighteen months old, and in weight till from eighteen to twenty-four months old; whilst the females cease increasing in stature at the age of from nine to fourteen or fifteen months, and in weight at the age of from twelve to fifteen months. From these various statements it is clear that the full difference in size between the male and female Scotch deer-hound is not acquired until rather late in life. The males almost exclusively are used for coursing, for, as Mr. McNeill informs me, the females have not sufficient strength and weight to pull down a full-grown deer. From the names used in old legends, it appears, as I hear from Mr. Cupples, that, at a very ancient period, the males were the most celebrated, the females being mentioned only as the mothers of famous dogs. Hence, during many generations, it is the male which has been chiefly tested for strength, size, speed, and courage, and the best will have been bred from. As, however, the males do not attain their full dimensions until rather late in life, they will have tended, in accordance with the law often indicated, to transmit their characters to their male offspring alone; and thus the great inequality in size between the sexes of the Scotch deer-hound may probably be accounted for.

[38] See also Richardson's 'Manual on the Dog,' p. 59. Much valuable information on the Scottish deer-hound is given by Mr. McNeill, who first called attention to the inequality in size between the sexes, in Scrope's 'Art of Deer-Stalking.' I hope that Mr. Cupples will keep to his intention of publishing a full account and history of this

famous breed.

Fig. 65.--Head of common wild boar, in prime of life, (from Brehm).

The males of some few quadrupeds possess organs or parts developed solely as a means of defence against the attacks of other males. Some kinds of deer use, as we have seen, the upper branches of their horns chiefly or exclusively for defending themselves; and the Oryx antelope, as I am informed by Mr. Bartlett, fences most skilfully with his long, gently curved horns; but these are likewise used as organs of offence. The same observer remarks that rhinoceroses in fighting, parry each other's sidelong blows with their horns, which clatter loudly together, as do the tusks of boars. Although wild boars fight desperately, they seldom, according to Brehm, receive fatal wounds, as the blows fall on each other's tusks, or on the layer of gristly skin covering the shoulder, called by the German hunters, the shield; and here we have a part specially modified for defence. With boars in the prime of life (Fig. 65) the tusks in the lower jaw are used for fighting, but they become in old age, as Brehm states, so much curved inwards and upwards over the snout that they can no longer be used in this way. They may, however, still serve, and even more effectively, as a means of defence. In compensation for the loss of the lower tusks as weapons of offence, those in the upper jaw, which always project a little laterally, increase in old age so much in length and curve so much upwards that they can be used for attack. Nevertheless, an old boar is not so dangerous to man as one at the age of six or seven years. [39]

[39] Brehm, 'Thierleben,' B. ii. ss. 729-732.

In the full-grown male Babirusa pig of Celebes (Fig. 66), the lower tusks are formidable weapons, like those of the European boar in the prime of life, whilst the upper tusks are so long and have their points so much curled inwards, sometimes even touching the forehead, that they are utterly useless as weapons of attack. They more nearly resemble horns than teeth, and are so manifestly useless as teeth that the animal was formerly supposed to rest his head by hooking them on to a branch! Their convex surfaces, however, if the head were held a little laterally, would serve as an excellent guard; and hence, perhaps, it is that in old animals they "are generally broken off, as if by fighting." [40] Here, then, we have the curious case of the upper tusks of the Babirusa regularly assuming during the prime of life a structure which apparently renders them fitted only for defence; whilst in the European boar the lower tusks assume in a less degree and only during old age nearly the same form, and then serve in like manner solely for defence.

[40] See Mr. Wallace's interesting account of this animal, 'The Malay Archipelago,' 1869, vol. i. p. 435.

FIG. 66.—Skull of the Babirusa Pig (from Wallace's 'Malay Archipelago').

In the wart-hog (see *Phacochoerus æthiopicus*, Fig. 67) the tusks in the upper jaw of the male curve upwards during the prime of life, and from being pointed serve as formidable weapons. The tusks in the lower jaw are sharper than those in the upper, but from their shortness it seems hardly possible that they can be used as weapons of attack. They must, however, greatly strengthen those in the upper jaw, from being ground so as to fit closely against their bases. Neither the upper nor the lower tusks appear to have been specially modified to act as guards, though no doubt they are to a certain extent used for this purpose. But the wart-hog is not destitute of other special means of protection, for it has, on each side of the face, beneath the eyes, a rather stiff, yet flexible, cartilaginous, oblong pad (Fig. 67), which projects two or three inches outwards; and it appeared to Mr. Bartlett and myself, when viewing the living animal, that these pads, when struck from beneath by the tusks of an opponent, would be turned upwards, and would thus admirably protect the somewhat prominent eyes. I may add, on the authority of Mr. Bartlett, that these boars when fighting stand directly face to face.

FIG. 67.--Head of female Æthiopian wart-hog, from 'Proc. Zool. Soc.' 1869, showing the same characters as the male, though on a reduced scale.

N.B. When the engraving was first made, I was under the impression that it
represented the male.

Lastly, the African river-hog (*Potomochoerus penicillatus*) has a hard cartilaginous knob on each side of the face beneath the eyes, which answers to the flexible pad of the wart-hog; it has also two bony prominences on the upper jaw above the nostrils. A boar of this species in the Zoological Gardens recently broke into the cage of the wart-hog. They fought all night long, and were found in the morning much exhausted, but not seriously wounded. It is a significant fact, as showing the purposes of the above-described projections and excrescences, that these were covered with blood, and were scored and abraded in an extraordinary manner.

Although the males of so many members of the pig family are provided with weapons, and as we have just seen with means of defence, these weapons seem to have been acquired within a rather late geological period. Dr. Forsyth Major specifies [41] several miocene species, in none of which do the tusks appear to have been largely developed in the males; and Professor Rütimeyer was formerly struck with this same fact.

[41] 'Atti della Soc. Italiana di Sc. Nat.' 1873, vol. xv. fasc. iv.

The mane of the lion forms a good defence against the attacks of rival lions, the one danger to which he is liable; for the males, as Sir A. Smith informs me, engage in terrible battles, and a young lion dares not approach an old one. In 1857 a tiger at Bromwich broke into the cage of a lion and a fearful scene ensued: "the lion's mane saved his neck and head from being much injured, but the tiger at last succeeded in ripping up his belly, and in a few minutes he was dead." [42] The broad ruff round the throat and chin of the Canadian lynx (*Felis canadensis*) is much longer in the male than in the female; but whether it serves as a defence I do not know. Male seals are well known to fight desperately together, and the males of certain kinds (*Otaria jubata*) [43] have great manes, whilst the females have small ones or none. The male baboon of the Cape of Good Hope (*Cynocephalus porcarius*) has a much longer mane and larger canine teeth than the female; and the mane probably serves as a protection, for, on asking the keepers in the Zoological Gardens, without giving them any clue to my object, whether any of the monkeys especially attacked each other by the nape of the neck, I was answered that this was not the case, except with the above baboon. In the Hamadryas baboon, Ehrenberg compares the mane of the adult male to that of a young lion, whilst in the young of both sexes and in the female the mane is almost absent.

[42] 'The Times,' Nov. 10, 1857. In regard to the Canada lynx, see Audubon and Bachman, 'Quadrupeds of North America,' 1846, p. 139.
[43] Dr. Murie, on Otaria, 'Proc. Zoolog. Soc.' 1869, p. 109. Mr. J.A. Allen, in the paper above quoted (p. 75), doubts whether the hair, which is longer on the neck in the male than in the female, deserves to be called a mane.

It appeared to me probable that the immense woolly mane of the male American bison, which reaches almost to the ground, and is much more developed in the males than in the females, served as a protection to them in their terrible battles; but an experienced hunter told Judge Caton that he had never observed anything which favoured this belief. The stallion has a thicker and fuller mane than the mare; and I have made particular inquiries of two great trainers and breeders, who have had charge of many entire horses, and am assured that they "invariably endeavour to seize one another by the neck." It does not, however, follow from the foregoing statements, that when the hair on the neck serves as a defence, that it was originally developed for this purpose, though this is probable in some cases, as in that of the lion. I am informed by Mr. McNeill that the long hairs on the throat of the stag (*Cervus elaphus*) serve as a great protection to him when hunted, for the dogs generally endeavour to seize him by the throat; but it is not probable that these hairs were specially developed for this purpose; otherwise the young and the females would have been equally protected.

CHOICE IN PAIRING BY EITHER SEX OF QUADRUPEDS.

Before describing in the next chapter, the differences between the sexes in voice, odors emitted, and ornaments, it will be convenient here to consider whether the sexes exert any choice in their unions. Does the female prefer any particular male, either before or after the males may have fought together for supremacy; or does the male, when not a polygamist, select any particular female? The general impression amongst breeders seems to be that the male accepts any female; and this owing to his eagerness, is, in most cases, probably the truth. Whether the female as a general rule indifferently accepts any male is much more doubtful. In the fourteenth chapter, on Birds, a considerable body of direct and indirect evidence was advanced, showing that the female selects her partner; and it would be a strange anomaly if female quadrupeds, which stand higher in the scale and have higher mental powers, did not generally, or at least often, exert some choice. The female could in most cases escape, if wooed by a male that did not please or excite her; and when pursued by several males, as commonly occurs, she would often have the opportunity, whilst they were fighting together, of escaping with some one male, or at least of temporarily pairing with him. This latter contingency has often been observed in Scotland with female red-deer, as I am informed by Sir Philip Egerton and others. [44]

[44] Mr. Boner, in his excellent description of the habits of the red-deer in Germany ('Forest Creatures,' 1861, p. 81) says, "while the stag is defending his rights against one intruder, another invades the sanctuary of his harem, and carries off trophy after trophy." Exactly the same thing occurs with seals; see Mr. J.A. Allen, ibid. p. 100.

It is scarcely possible that much should be known about female quadrupeds in a state of nature making any choice in their marriage unions. The following curious details on the courtship of one of the eared seals (*Callorhinus ursinus*) are given [45] on the authority of Capt. Bryant, who had ample opportunities for observation. He says, "Many of the females on their arrival at the island where they breed appear desirous of returning to some particular male, and frequently climb the outlying rocks to overlook the rookeries, calling out and listening as if for a familiar voice. Then changing to another place they do the same again...As soon as a female reaches the shore, the nearest male goes down to meet her, making meanwhile a noise like the clucking of a hen to her chickens. He bows to her and coaxes her until he gets between her and the water so that she cannot escape him. Then his manner changes, and with a harsh growl he drives her to a place in his harem. This continues until the lower row of harems is nearly full. Then the males higher up select the time when their more fortunate neighbours are off their guard to steal their wives. This they do by taking them in their mouths and lifting them over the heads of the other females, and carefully placing them in their own harem, carrying them as cats do their kittens. Those still higher up pursue the same method until the whole space is occupied. Frequently a struggle ensues between two males for the possession of the same female, and both seizing her at once pull her in two or terribly lacerate her with their teeth. When the space is all filled, the old male walks around complacently reviewing his family, scolding those who crowd or disturb the others, and fiercely driving off all intruders. This surveillance always keeps him actively occupied."

[45] Mr. J.A. Allen in 'Bull. Mus. Comp. Zoolog. of Cambridge, United States,' vol. ii. No. 1, p. 99.

As so little is known about the courtship of animals in a state of nature, I have endeavoured to discover how far our domesticated quadrupeds evince any choice in their unions. Dogs offer the best opportunity for observation, as they are carefully attended to and well understood. Many breeders have expressed a strong opinion on this head. Thus, Mr. Mayhew remarks, "The females are able to bestow their affections; and tender recollections are as potent over them as they are known to be in other cases, where higher animals are concerned. Bitches are not always prudent in their loves, but are apt to fling themselves away on curs of low degree. If reared with a companion of vulgar appearance, there often springs up between the pair a devotion which no time can afterwards subdue. The passion, for such it really is, becomes of a more than romantic endurance." Mr. Mayhew, who attended chiefly to the smaller breeds, is convinced that the females are strongly attracted by males of a large size. [46] The well-known veterinary Blaine states [47] that his own female pug dog became so attached to a spaniel, and a female setter to a cur, that in neither case would they pair with a dog of their own breed until several weeks had elapsed. Two similar and trustworthy accounts have been given me in regard to a female retriever and a spaniel, both of which became enamoured with terrier-dogs.

[46] 'Dogs: their Management,' by E. Mayhew, M.R.C.V.S., 2nd ed., 1864, pp. 187-192.
[47] Quoted by Alex. Walker, 'On Intermarriage,' 1838, p. 276; see also p. 244.

Mr. Cupples informs me that he can personally vouch for the accuracy of the following more remarkable case, in which a valuable and wonderfully-intelligent female terrier loved a retriever belonging to a neighbour to such a degree, that she had often to be dragged away from him. After their permanent separation, although repeatedly showing milk in her teats, she would never acknowledge the courtship of any other dog, and to the regret of her owner never bore puppies. Mr. Cupples also states, that in 1868, a female deerhound in his kennel thrice produced puppies, and on each occasion showed a marked preference for one of the largest and handsomest, but not the most eager, of four deerhounds living with her, all in the prime of life. Mr. Cupples has observed that the female generally favours a dog whom she has associated with and knows; her shyness and timidity at first incline her against a strange dog. The male, on the contrary, seems rather inclined towards strange females. It appears to be rare when the male refuses any particular female, but Mr. Wright, of Yeldersley House, a great breeder of dogs, informs me that he has known some instances; he cites the case of one of his own deerhounds, who would not take any notice of a particular female mastiff, so that another deerhound had to be employed. It would be superfluous to give, as I could, other instances, and I will only add that Mr. Barr, who has carefully bred many bloodhounds, states that in almost every instance particular individuals of opposite sexes show a decided preference for each other. Finally, Mr. Cupples, after attending to this subject for another year, has written to me, "I have had full confirmation of my former statement, that dogs in breeding form decided preferences for each other, being often influenced by size, bright color, and individual characters, as well as by the degree of their previous familiarity."

In regard to horses, Mr. Blenkiron, the greatest breeder of race-horses in the world, informs me that stallions are so frequently capricious in their choice, rejecting one mare and without any apparent cause taking to another, that various artifices have to be habitually used. The famous Monarque, for instance, would never consciously look at the dam of Gladiateur, and a trick had to be practised. We can partly see the reason why valuable race-horse stallions, which are in such demand as to be exhausted, should be so particular in their choice. Mr. Blenkiron has never known a mare reject a horse; but this has occurred in Mr. Wright's stable, so that the mare had to be cheated. Prosper Lucas [48] quotes various statements from French authorities, and remarks, "On voit des étalons qui s'éprennent d'une jument, et négligent toutes les autres." He gives, on the authority of Baëlen, similar facts in regard to bulls; and Mr. H. Reeks assures me that a famous short-horn bull belonging to his father "invariably refused to be matched with a black cow." Hoffberg, in describing the domesticated reindeer of Lapland says, "Fœminæ majores et fortiores mares præ cæteris admittunt, ad eos confugiunt, a junioribus agitatæ, qui hos in fugam conjiciunt." [49] A clergyman, who has bred many pigs, asserts that sows often reject one boar and immediately accept another.

[48] 'Traité de l'Héréd. Nat.' tom. ii. 1850, p. 296.
[49] 'Amœnitates Acad.' vol. iv. 1788, p. 160.

From these facts there can be no doubt that, with most of our domesticated quadrupeds, strong individual antipathies and preferences are frequently exhibited, and much more commonly by the female than by the male. This being the case, it is improbable that the unions of quadrupeds in a state of nature should be left to mere chance. It is much more probable that the females are allured or excited by particular males, who possess certain characters in a higher degree than other males; but what these characters are, we can seldom or never discover with certainty.

CHAPTER XVIII.

SECONDARY SEXUAL CHARACTERS OF MAMMALS—continued.

Voice—Remarkable sexual peculiarities in seals—Odor—Development of the hair—Color of the hair and skin—Anomalous case of the female being more ornamented than the male—Color and ornaments due to sexual selection—Color acquired for the sake of protection—Color, though common to both sexes, often due to sexual selection—On the disappearance of spots and stripes in adult quadrupeds—On the colors and ornaments of the Quadrumana—Summary.

Quadrupeds use their voices for various purposes, as a signal of danger, as a call from one member of a troop to another, or from the mother to her lost offspring, or from the latter for protection to their mother; but such uses need not here be considered. We are concerned only with the difference between the voices of the sexes, for instance between that of the lion and lioness, or of the bull and cow. Almost all male animals use their voices much more during the rutting-season than at any other time; and some, as the giraffe and porcupine, [1] are said to be completely mute excepting at this season. As the throats (i.e. the larynx and thyroid bodies [2]) of stags periodically become enlarged at the beginning of the breeding-season, it might be thought that their powerful voices must be somehow of high importance to them; but this is very doubtful. From information given to me by two experienced observers, Mr. McNeill and Sir P. Egerton, it seems that young stags under three years old do not roar or bellow; and that the old ones begin bellowing at the commencement of the breeding-season, at first only occasionally and moderately, whilst they restlessly wander about in search of the females. Their battles are prefaced by loud and prolonged bellowing, but during the actual conflict they are silent. Animals of all kinds which habitually use their voices utter various noises under any strong emotion, as when enraged and preparing to fight; but this may merely be the result of nervous excitement, which leads to the spasmodic contraction of almost all the muscles of the body, as when a man grinds his teeth and clenches his fists in rage or agony. No doubt stags challenge each other to mortal combat by bellowing; but those with the more powerful voices, unless at the same time the stronger, better-armed, and more courageous, would not gain any advantage over their rivals.

[1] Owen, 'Anatomy of Vertebrates,' vol. iii. p. 585.
[2] Ibid. p. 595.

It is possible that the roaring of the lion may be of some service to him by striking terror into his adversary; for when enraged he likewise erects his mane and thus instinctively tries to make himself appear as terrible as possible. But it can hardly be supposed that the bellowing of the stag, even if it be of service to him in this way, can have been important enough to have led to the periodical enlargement of the throat. Some writers suggest that the bellowing serves as a call to the female; but the experienced observers above quoted inform me that female deer do not search for the male, though the males search eagerly for the females, as indeed might be expected from what we know of the habits of other male quadrupeds. The voice of the female, on the other hand, quickly brings to her one or more stags, [3] as is well known to the hunters who in wild countries imitate her cry. If we could believe that the male had the power to excite or allure the female by his voice, the periodical enlargement of his vocal organs would be intelligible on the principle of sexual selection, together with inheritance limited to the same sex and season; but we have no evidence in favour of this view. As the case stands, the loud voice of the stag during the breeding-season does not seem to be of any special service to him, either during his courtship or battles, or in any other way. But may we not believe that the frequent use of the voice, under the strong excitement of love, jealousy, and rage, continued during many generations, may at last have produced an inherited effect on the vocal organs of the stag, as well as of other male animals? This appears to me, in our present state of knowledge, the most probable view.

[3] See, for instance, Major W. Ross King ('The Sportsman in Canada,' 1866, pp. 53, 131) on the habits of the moose and wild reindeer.

The voice of the adult male gorilla is tremendous, and he is furnished with a laryngeal sack, as is the adult male orang. [4] The gibbons rank among the noisiest of monkeys, and the Sumatra species (*Hylobates syndactylus*) is also furnished with an air sack; but Mr. Blyth, who has had opportunities for observation, does not believe that the male is noisier than the female. Hence, these latter monkeys probably use their voices as a mutual call; and this is certainly the case with some quadrupeds, for instance the beaver. [5] Another gibbon, the *H. agilis*, is remarkable, from having the power of giving a complete and correct octave of musical notes, [6] which we may reasonably suspect serves as a sexual charm; but I shall have to recur to this subject in the next chapter. The vocal organs of the American *Mycetes caraya* are one-third larger in the male than in the female, and are wonderfully powerful. These monkeys in warm weather make the forests resound at morning and evening with their overwhelming voices. The males begin the dreadful concert, and often continue it during many hours, the females sometimes joining in with their less powerful voices. An excellent observer, Rengger, [7] could not perceive that they were excited to begin by any special cause; he thinks that, like many birds, they delight in their own music, and try to excel each other. Whether most of the foregoing monkeys have acquired their powerful voices in order to beat their rivals and charm the females—or whether the vocal organs have been strengthened and enlarged through the inherited effects of long-continued use without any particular good being thus gained—I will not pretend to say; but the former view, at least in the case of the *Hylobates agilis*, seems the most probable.

[4] Owen 'Anatomy of Vertebrates,' vol. iii. p. 600.
[5] Mr. Green, in 'Journal of Linnean Society,' vol. x. 'Zoology,' 1869, note 362.
[6] C.L. Martin, 'General Introduction to the Natural History of Mamm. Animals,' 1841, p. 431.
[7] 'Naturgeschichte der Säugethiere von Paraguay,' 1830, ss. 15, 21.

I may here mention two very curious sexual peculiarities occurring in seals, because they have been supposed by some writers to affect the voice. The nose of the male sea-elephant (*Macrorhinus proboscideus*) becomes greatly elongated during the breeding-season, and can then be erected. In this state it is sometimes a foot in length. The female is not thus provided at any period of life. The male makes a wild, hoarse, gurgling noise, which is audible at a great distance and is believed to be strengthened by the proboscis;

the voice of the female being different. Lesson compares the erection of the proboscis, with the swelling of the wattles of male gallinaceous birds whilst courting the females. In another allied kind of seal, the bladder-nose (*Cystophora cristata*), the head is covered by a great hood or bladder. This is supported by the septum of the nose, which is produced far backwards and rises into an internal crest seven inches in height. The hood is clothed with short hair, and is muscular; can be inflated until it more than equals the whole head in size! The males when rutting, fight furiously on the ice, and their roaring "is said to be sometimes so loud as to be heard four miles off." When attacked they likewise roar or bellow; and whenever irritated the bladder is inflated and quivers. Some naturalists believe that the voice is thus strengthened, but various other uses have been assigned to this extraordinary structure. Mr. R. Brown thinks that it serves as a protection against accidents of all kinds; but this is not probable, for, as I am assured by Mr. Lamont who killed 600 of these animals, the hood is rudimentary in the females, and it is not developed in the males during youth. [8]

[8] On the sea-elephant, see an article by Lesson, in 'Dict. Class. Hist. Nat.' tom. xiii. p. 418. For the Cystophora, or Stemmatopus, see Dr. Dekay, 'Annals of Lyceum of Nat. Hist.' New York, vol. i. 1824, p. 94. Pennant has also collected information from the sealers on this animal. The fullest account is given by Mr. Brown, in 'Proc. Zoolog. Soc.' 1868, p. 435.

ODOR.

With some animals, as with the notorious skunk of America, the overwhelming odor which they emit appears to serve exclusively as a defence. With shrew-mice (Sorex) both sexes possess abdominal scent-glands, and there can be little doubt, from the rejection of their bodies by birds and beasts of prey, that the odor is protective; nevertheless, the glands become enlarged in the males during the breeding-season. In many other quadrupeds the glands are of the same size in both sexes [9], but their uses are not known. In other species the glands are confined to the males, or are more developed than in the females; and they almost always become more active during the rutting-season. At this period the glands on the sides of the face of the male elephant enlarge, and emit a secretion having a strong musky odor. The males, and rarely the females, of many kinds of bats have glands and protrudable sacks situated in various parts; and it is believed that these are odoriferous.

[9] As with the castoreum of the beaver, see Mr. L.H. Morgan's most interesting work, 'The American Beaver,' 1868, p. 300. Pallas ('Spic. Zoolog.' fasc. viii. 1779, p. 23) has well discussed the odoriferous glands of mammals. Owen ('Anat. of Vertebrates,' vol. iii. p. 634) also gives an account of these glands, including those of the elephant, and (p. 763) those of shrew-mice. On bats, Mr. Dobson in 'Proceedings of the Zoological Society' 1873, p. 241.

The rank effluvium of the male goat is well known, and that of certain male deer is wonderfully strong and persistent. On the banks of the Plata I perceived the air tainted with the odor of the male *Cervus campestris*, at half a mile to leeward of a herd; and a silk handkerchief, in which I carried home a skin, though often used and washed, retained, when first unfolded, traces of the odor for one year and seven months. This animal does not emit its strong odor until more than a year old, and if castrated whilst young never emits it. [10] Besides the general odor, permeating the whole body of certain ruminants (for instance, *Bos moschatus*) in the breeding-season, many deer, antelopes, sheep, and goats possess odoriferous glands in various situations, more especially on their faces. The so-called tear-sacks, or suborbital pits, come under this head. These glands secrete a semi-fluid fetid matter which is sometimes so copious as to stain the whole face, as I have myself seen in an antelope. They are "usually larger in the male than in the female, and their development is checked by castration." [11] According to Desmarest they are altogether absent in the female of *Antilope subgutturosa*. Hence, there can be no doubt that they stand in close relation with the reproductive functions. They are also sometimes present, and sometimes absent, in nearly allied forms. In the adult male musk-deer (*Moschus moschiferus*), a naked space round the tail is bedewed with an odoriferous fluid, whilst in the adult female, and in the male until two years old, this space is covered with hair and is not odoriferous. The proper musk-sack of this deer is from its position necessarily confined to the male, and forms an additional scent-organ. It is a singular fact that the matter secreted by this latter gland, does not, according to Pallas, change in consistence, or increase in quantity, during the rutting-season; nevertheless this naturalist admits that its presence is in some way connected with the act of reproduction. He gives, however, only a conjectural and unsatisfactory explanation of its use. [12]

[10] Rengger, 'Naturgeschichte der Säugethiere von Paraguay,' 1830, s. 355. This observer also gives some curious particulars in regard to the odor.
[11] Owen, 'Anatomy of Vertebrates,' vol. iii. p. 632. See also Dr. Murie's observations on those glands in the 'Proc. Zoolog. Soc.' 1870, p. 340. Desmarest, 'On the *Antilope subgutturosa*, 'Mammalogie,' 1820, p. 455.
[12] Pallas, 'Spicilegia Zoolog.' fasc. xiii. 1799, p. 24; Desmoulins, 'Dict. Class. d'Hist. Nat.' tom. iii. p. 586.

In most cases, when only the male emits a strong odor during the breeding-season, it probably serves to excite or allure the female. We must not judge on this head by our own taste, for it is well known that rats are enticed by certain essential oils, and cats by valerian, substances far from agreeable to us; and that dogs, though they will not eat carrion, sniff and roll on it. From the reasons given when discussing the voice of the stag, we may reject the idea that the odor serves to bring the females from a distance to the males. Active and long-continued use cannot here have come into play, as in the case of the vocal organs. The odor emitted must be of considerable importance to the male, inasmuch as large and complex glands, furnished with muscles for everting the sack, and for closing or opening the orifice, have in some cases been developed. The development of these organs is intelligible through sexual selection, if the most odoriferous males are the most successful in winning the females, and in leaving offspring to inherit their gradually perfected glands and odors.

DEVELOPMENT OF THE HAIR.

We have seen that male quadrupeds often have the hair on their necks and shoulders much more developed than the females; and many additional instances could be given. This sometimes serves as a defence to the male during his battles; but whether the hair in most cases has been specially developed for this purpose, is very doubtful. We may feel almost certain that this is not the case, when only a thin and narrow crest runs along the back; for a crest of this kind would afford scarcely any protection, and the ridge of the back is not a place likely to be injured; nevertheless such crests are sometimes confined to the males, or are much more developed in them than in the females. Two antelopes, the *Tragelaphus scriptus* [13] (Fig. 70) and *Portax picta* may be given as instances. When stags, and the males of the wild goat, are enraged or terrified, these crests stand erect; [14] but it cannot be supposed that they have been developed merely for the sake of exciting fear in their enemies. One of the above-named antelopes, the *Portax picta*, has a large well-defined brush of black hair on the throat, and this is much larger in the male than in the female. In the *Ammotragus tragelaphus* of North Africa, a member of the sheep-family, the fore-legs are almost concealed by an extraordinary growth of hair, which depends from the neck and upper halves of the legs; but Mr. Bartlett does not believe that this mantle is of the least use to the male, in whom it is much more developed than in the female.

[13] Dr. Gray, 'Gleanings from the Menagerie at Knowsley,' pl. 28.
[14] Judge Caton on the Wapiti, 'Transact. Ottawa Acad. Nat. Sciences,' 1868, pp. 36, 40; Blyth, 'Land and Water,' on *Capra ægagrus* 1867, p. 37.

Male quadrupeds of many kinds differ from the females in having more hair, or hair of a different character, on certain parts of their faces. Thus the bull alone has curled hair on the forehead. [15] In three closely-allied sub-genera of the goat family, only the males possess beards, sometimes of large size; in two other sub-genera both sexes have a beard, but it disappears in some of the domestic breeds of the common goat; and neither sex of the Hemitragus has a beard. In the ibex the beard is not developed during the summer, and is so small at other times that it may be called rudimentary. [16] With some monkeys the beard is confined to the male, as in the orang; or is much larger in the male than in the female, as in the *Mycetes caraya* and *Pithecia satanas* (Fig. 68). So it is with the whiskers of some species of Macaous, [17] and, as we have seen, with the manes of some species of baboons. But with most kinds of monkeys the various tufts of hair about the face and head are alike in both sexes.

[15] Hunter's 'Essays and Observations,' edited by Owen, 1861. vol. i. p. 236.
[16] See Dr. Gray's 'Catalogue of Mammalia in the British Museum,' part iii. 1852, p. 144.
[17] Rengger, 'Säugethiere,' etc., s. 14; Desmarest, 'Mammalogie,' p. 86.

The males of various members of the ox family (Bovidæ), and of certain antelopes, are furnished with a dewlap, or great fold of skin on the neck, which is much less developed in the female.

Fᴵɢ. 68.—Pithecia satanas, male (from Brehm).

Now, what must we conclude with respect to such sexual differences as these? No one will pretend that the beards of certain male goats, or the dewlaps of the bull, or the crests of hair along the backs of certain male antelopes, are of any use to them in their ordinary habits. It is possible that the immense beard of the male Pithecia, and the large beard of the male orang, may protect their throats when fighting; for the keepers in the Zoological Gardens inform me that many monkeys attack each other by the throat; but it is not probable that the beard has been developed for a distinct purpose from that served by the whiskers, moustache, and other tufts of hair on the face; and no one will suppose that these are useful as a protection. Must we attribute all these appendages of hair or skin to mere purposeless variability in the male? It cannot be denied that this is possible; for in many domesticated quadrupeds, certain characters, apparently not derived through reversion from any wild parent form, are confined to the males, or are more developed in them than in the females—for instance, the hump on the male zebu-cattle of India, the tail of fat-tailed rams, the arched outline of the forehead in the males of several breeds of sheep, and lastly, the mane, the long hairs on the hind legs, and the dewlap of the male of the Berbura goat. [18] The mane, which occurs only in the rams of an African breed of sheep, is a true secondary sexual character, for, as I hear from Mr. Winwood Reade, it is not developed if the animal be castrated. Although we ought to be extremely cautious, as shown in my work on 'Variation under Domestication,' in concluding that any character, even with animals kept by semi-civilized people, has not been subjected to selection by man, and thus augmented, yet in the cases just specified this is improbable; more especially as the characters are confined to the males, or are more strongly developed in them than in the females. If it were positively known that the above African ram is a descendant of the same primitive stock as the other breeds of sheep, and if the Berbura male-goat with his mane, dewlap, etc., is descended from the same stock as other goats, then, assuming that selection has not been applied to these characters, they must be due to simple variability, together with sexually-limited inheritance.

[18] See the chapters on these several animals in vol. i. of my 'Variation of Animals under Domestication;' also vol. ii. p. 73; also chap. xx. on the practice of selection by semi-civilized people. For the Berbura goat, see Dr. Gray, 'Catalogue,' ibid. p. 157.

Hence it appears reasonable to extend this same view to all analogous cases with animals in a state of nature. Nevertheless I cannot persuade myself that it generally holds good, as in the case of the extraordinary development of hair on the throat and fore-legs of the male Ammotragus, or in that of the immense beard of the male Pithecia. Such study as I have been able to give to nature makes me believe that parts or organs which are highly developed, were acquired at some period for a special purpose. With those antelopes in which the adult male is more strongly-colored than the female, and with those monkeys in which the hair on the face is elegantly arranged and colored in a diversified manner, it seems probable that the crests and tufts of hair were gained as ornaments; and this I know is the opinion of some naturalists. If this be correct, there can be little doubt that they were gained or at least modified through sexual selection; but how far the same view may be extended to other mammals is doubtful.

COLOR OF THE HAIR AND OF THE NAKED SKIN.

I will first give briefly all the cases known to me of male quadrupeds differing in color from the females. With Marsupials, as I am informed by Mr. Gould, the sexes rarely differ in this respect; but the great red kangaroo offers a striking exception, "delicate blue being the prevailing tint in those parts of the female which in the male are red." [19] In the *Didelphis opossum* of Cayenne the female is said to be a little more red than the male. Of the Rodents, Dr. Gray remarks: "African squirrels, especially those found in the tropical regions, have the fur much brighter and more vivid at some seasons of the year than at others, and the fur of the male is generally brighter than that of the female." [20] Dr. Gray informs me that he specified the African squirrels, because, from their unusually bright colors, they best exhibit this difference. The female of the *Mus minutus* of Russia is of a paler and dirtier tint than the male. In a large number of bats the fur of the male is lighter than in the female. [21] Mr. Dobson also remarks, with respect to these animals: "Differences, depending partly or entirely on the possession by the male of fur of a much more brilliant hue, or distinguished by different markings or by the greater length of certain portions, are met only, to any appreciable extent, in the frugivorous bats in which the sense of sight is well developed." This last remark deserves attention, as bearing on the question whether bright colors are serviceable to male animals from being ornamental. In one genus of sloths, it is now established, as Dr. Gray states, "that the males are ornamented differently from the females—that is to say, that they have a patch of soft short hair between the shoulders, which is generally of a more or less orange color, and in one species pure white. The females, on the contrary, are destitute of this mark."

[19] *Osphranter rufus*, Gould, 'Mammals of Australia,' 1863, vol. ii. On the Didelphis, Desmarest, 'Mammalogie,' p. 256.

[20] 'Annals and Magazine of Natural History,' Nov. 1867, p. 325. On the *Mus minutus*, Desmarest, 'Mammalogie,' p. 304.

[21] J.A. Allen, in 'Bulletin of Mus. Comp. Zoolog. of Cambridge, United States,' 1869, p. 207. Mr. Dobson on sexual characters in the Chiroptera, 'Proceedings of the Zoological Society,' 1873, p. 241. Dr. Gray on Sloths, ibid. 1871, p. 436.

The terrestrial Carnivora and Insectivora rarely exhibit sexual differences of any kind, including color. The ocelot (*Felis pardalis*), however, is exceptional, for the colors of the female, compared with those of the male, are "moins apparentes, le fauve, étant plus terne, le blanc moins pur, les raies ayant moins de largeur et les taches moins de diametre." [22] The sexes of the allied *Felis mitis* also differ, but in a less degree; the general hues of the female being rather paler than in the male, with the spots less black. The marine Carnivora or seals, on the other hand, sometimes differ considerably in color, and they present, as we have already seen, other remarkable sexual differences. Thus the male of the *Otaria nigrescens* of the southern hemisphere is of a rich brown shade above; whilst the female, who acquires her adult tints earlier in life than the male, is dark-grey above, the young of both sexes being of a deep chocolate color. The male of the northern *Phoca grœnlandica* is tawny grey, with a curious saddle-shaped dark mark on the back; the female is much smaller, and has a very different appearance, being "dull white or yellowish straw-color, with a tawny hue on the back"; the young at first are pure white, and can "hardly be distinguished among the icy hummocks and snow, their color thus acting as a protection." [23]

[22] Desmarest, 'Mammalogie,' 1820, p. 220. On *Felis Mitis*, Rengger, ibid. s. 194.

[23] Dr. Murie on the Otaria, 'Proceedings Zoological Society,' 1869, p. 108. Mr. R. Brown on the *P. groenlandica*, ibid. 1868, p. 417. See also on the colors of seals, Desmarest, ibid. pp. 243, 249.

With Ruminants sexual differences of color occur more commonly than in any other order. A difference of this kind is general in the Strepsicerene antelopes; thus the male nilghau (*Portax picta*) is bluish-grey and much darker than the female, with the square white patch on the throat, the white marks on the fetlocks, and the black spots on the ears all much more distinct. We have seen that in this species the crests and tufts of hair are likewise more developed in the male than in the hornless female. I am informed by Mr. Blyth that the male, without shedding his hair, periodically becomes darker during the breeding-season. Young males cannot be distinguished from young females until about twelve months old; and if the male is emasculated before this period, he never, according to the same authority, changes color. The importance of this latter fact, as evidence that the coloring of the Portax is of sexual origin, becomes obvious, when we hear [24] that neither the red summer-coat nor the blue winter-coat of the Virginian deer is at all affected by emasculation. With most or all of the highly-ornamented species of Tragelaphus the males are darker than the hornless females, and their crests of hair are more fully developed. In the male of that magnificent antelope, the Derbyan eland, the body is redder, the whole neck much blacker, and the white band which separates these colors broader than in the female. In the Cape eland, also, the male is slightly darker than the female. [25]

[24] Judge Caton, in 'Transactions of the Ottawa Academy of Natural Sciences,' 1868, p.

4.

[25] Dr. Gray, 'Cat. of Mamm. in Brit. Mus.' part iii. 1852, pp. 134-142; also Dr. Gray, 'Gleanings from the Menagerie of Knowsley,' in which there is a splendid drawing of the *Oreas derbianus*: see the text on Tragelaphus. For the Cape eland (*Oreas canna*), see Andrew Smith, 'Zoology of S. Africa,' pl. 41 and 42. There are also many of these Antelopes in the Zoological Gardens.

In the Indian black-buck (*A. bezoartica*), which belongs to another tribe of antelopes, the male is very dark, almost black; whilst the hornless female is fawn-colored. We meet in this species, as Mr. Blyth informs me, with an exactly similar series of facts, as in the *Portax picta*, namely, in the male periodically changing color during the breeding-season, in the effects of emasculation on this change, and in the young of both sexes being indistinguishable from each other. In the *Antilope niger* the male is black, the female, as well as the young of both sexes, being brown; in *A. sing-sing* the male is much brighter colored than the hornless female, and his chest and belly are blacker; in the male *A. caama*, the marks and lines which occur on various parts of the body are black, instead of brown as in the female; in the brindled gnu (*A. gorgon*) "the colors of the male are nearly the same as those of the female, only deeper and of a brighter hue." [26] Other analogous cases could be added.

[26] On the *Ant. niger*, see 'Proc. Zool. Soc.' 1850, p. 133. With respect to an allied species, in which there is an equal sexual difference in color, see Sir S. Baker, 'The Albert Nyanza,' 1866, vol. ii. p. 627. For the *A. sing-sing*, Gray, 'Cat. B. Mus.' p. 100. Desmarest, 'Mammalogie,' p. 468, on the *A. caama*. Andrew Smith, 'Zoology of S. Africa,' on the Gnu.

The Banteng bull (*Bos sondaicus*) of the Malayan Archipelago is almost black, with white legs and buttocks; the cow is of a bright dun, as are the young males until about the age of three years, when they rapidly change color. The emasculated bull reverts to the color of the female. The female Kemas goat is paler, and both it and the female *Capra ægagrus* are said to be more uniformly tinted than their males. Deer rarely present any sexual differences in color. Judge Caton, however, informs me that in the males of the wapiti deer (*Cervus canadensis*) the neck, belly, and legs are much darker than in the female; but during the winter the darker tints gradually fade away and disappear. I may here mention that Judge Caton has in his park three races of the Virginian deer, which differ slightly in color, but the differences are almost exclusively confined to the blue winter or breeding-coat; so that this case may be compared with those given in a previous chapter of closely-allied or representative species of birds, which differ from each other only in their breeding plumage. [27] The females of *Cervus paludosus* of S. America, as well as the young of both sexes, do not possess the black stripes on the nose and the blackish-brown line on the breast, which are characteristic of the adult males. [28] Lastly, as I am informed by Mr. Blyth, the mature male of the beautifully colored and spotted axis deer is considerably darker than the female: and this hue the castrated male never acquires.

[27] 'Ottawa Academy of Sciences,' May 21, 1868, pp. 3, 5.
[28] S. Müller, on the Banteng, 'Zoog. Indischen Archipel.' 1839-1844, tab. 35; see also Raffles, as quoted by Mr. Blyth, in 'Land and Water,' 1867, p. 476. On goats, Dr.

Gray, 'Catalogue of the British Museum,' p. 146; Desmarest, 'Mammalogie,' p. 482. On the *Cervus paludosus*, Rengger, ibid. s. 345.

The last Order which we need consider is that of the Primates. The male of the *Lemur macaco* is generally coal-black, whilst the female is brown. [29] Of the Quadrumana of the New World, the females and young of *Mycetes caraya* are grayish-yellow and like each other; in the second year the young male becomes reddish-brown; in the third, black, excepting the stomach, which, however, becomes quite black in the fourth or fifth year. There is also a strongly-marked difference in color between the sexes of *Mycetes seniculus* and *Cebus capucinus*; the young of the former, and I believe of the latter species, resembling the females. With *Pithecia leucocephala* the young likewise resemble the females, which are brownish-black above and light rusty-red beneath, the adult males being black. The ruff of hair round the face of *Ateles marginatus* is tinted yellow in the male and white in the female. Turning to the Old World, the males of *Hylobates hoolock* are always black, with the exception of a white band over the brows; the females vary from whity-brown to a dark tint mixed with black, but are never wholly black. [30] In the beautiful *Cercopithecus diana*, the head of the adult male is of an intense black, whilst that of the female is dark grey; in the former the fur between the thighs is of an elegant fawn-color, in the latter it is paler. In the beautiful and curious moustache monkey (*Cercopithecus cephus*) the only difference between the sexes is that the tail of the male is chestnut and that of the female grey; but Mr. Bartlett informs me that all the hues become more pronounced in the male when adult, whilst in the female they remain as they were during youth. According to the colored figures given by Solomon Müller, the male of *Semnopithecus chrysomelas* is nearly black, the female being pale brown. In the *Cercopithecus cynosurus* and *griseoviridis* one part of the body, which is confined to the male sex, is of the most brilliant blue or green, and contrasts strikingly with the naked skin on the hinder part of the body, which is vivid red.

[29] Sclater, 'Proc. Zool. Soc.' 1866, p. i. The same fact has also been fully ascertained by MM. Pollen and van Dam. See, also, Dr. Gray in 'Annals and Magazine of Natural History,' May 1871, p. 340.
[30] On Mycetes, Rengger, ibid. s. 14; and Brehm, 'Thierleben,' B. i. s. 96, 107. On Ateles Desmarest, 'Mammalogie,' p. 75. On Hylobates, Blyth, 'Land and Water,' 1867, p. 135. On the Semnopithecus, S. Müller, 'Zoog. Indischen Archipel.' tab. x.

Lastly, in the baboon family, the adult male of *Cynocephalus hamadryas* differs from the female not only by his immense mane, but slightly in the color of the hair and of the naked callosities. In the drill (*C. leucophœus*) the females and young are much paler-colored, with less green, than the adult males. No other member in the whole class of mammals is colored in so extraordinary a manner as the adult male mandrill (*C. mormon*). The face at this age becomes of a fine blue, with the ridge and tip of the nose of the most brilliant red. According to some authors, the face is also marked with whitish stripes, and is shaded in parts with black, but the colors appear to be variable. On the forehead there is a crest of hair, and on the chin a yellow beard. "Toutes les parties supérieures de leurs cuisses et le grand espace nu de leurs fesses sont également colorés du rouge le plus vif, avec un mélange de bleu qui ne manque *réellement pas d'élégance*." [31] When the animal is excited all the naked parts become much more vividly tinted. Several authors have used the strongest expressions in describing these resplendent colors, which they compare with those of the most brilliant birds. Another remarkable peculiarity is that when the great canine teeth are fully developed, immense protuberances of bone are formed on each cheek, which are deeply furrowed longitudinally, and the naked skin over them is brilliantly-colored, as just-described. (Fig. 69.) In the adult females and in the young of both sexes these protuberances are scarcely perceptible; and the naked parts are much less bright colored, the face being almost black, tinged with blue. In the adult female, however, the nose at certain regular intervals of time becomes tinted with red.

[31] Gervais, 'Hist. Nat. des Mammifères,' 1854, p. 103. Figures are given of the skull of the male. Also Desmarest, 'Mammalogie,' p. 70. Geoffroy St.-Hilaire and F. Cuvier, 'Hist. Nat. des Mamm.' 1824, tom. i.

FIG. 69.—Head of male Mandrill (from Gervais, ' Hist. Nat. des Mammifères ').

In all the cases hitherto given the male is more strongly or brighter colored than the female, and differs from the young of both sexes. But as with some few birds it is the female which is brighter colored than the male, so with the Rhesus monkey (*Macacus rhesus*), the female has a large surface of naked skin round the tail, of a brilliant carmine red, which, as I was assured by the keepers in the Zoological Gardens, periodically becomes even yet more vivid, and her face also is pale red. On the other hand, in the adult male and in the young of both sexes (as I saw in the Gardens), neither the naked skin at the posterior end of the body, nor the face, show a trace of red. It appears, however, from some published accounts, that the male does occasionally, or during certain seasons, exhibit some traces of the red. Although he is thus less ornamented than the female, yet in the larger size of his body larger canine teeth, more developed whiskers, more prominent superciliary ridges, he follows the common rule of the male excelling the female.

I have now given all the cases known to me of a difference in color between the sexes of mammals. Some of these may be the result of variations confined to one sex and transmitted to the same sex, without any good being gained, and therefore without the aid of selection. We have instances of this with our domesticated animals, as in the males of certain cats being rusty-red, whilst the females are tortoise-shell colored. Analogous cases occur in nature: Mr. Bartlett has seen many black varieties of the jaguar, leopard, vulpine phalanger, and wombat; and he is certain that all, or nearly all these animals, were males. On the other hand, with wolves, foxes, and apparently American squirrels, both sexes are occasionally born black. Hence it is quite possible that with some mammals a difference in color between the sexes, especially when this is congenital, may simply be the result, without the aid of selection, of the occurrence of one or more variations, which from the first were sexually limited in their transmission. Nevertheless it is improbable that the diversified, vivid, and contrasted colors of certain quadrupeds, for instance, of the above monkeys and antelopes, can thus be accounted for. We should bear in mind that these colors do not appear in the male at birth, but only at or near maturity; and that unlike ordinary variations, they are lost if the male be emasculated. It is on the whole probable that the strongly-marked colors and other ornamental characters of male quadrupeds are beneficial to them in their rivalry with other males, and have consequently been acquired through sexual selection. This view is strengthened by the differences in color between the sexes occurring almost exclusively, as may be collected from the previous details, in those groups and sub-groups of mammals which present other and strongly-marked secondary sexual characters; these being likewise due to sexual selection.

Quadrupeds manifestly take notice of color. Sir S. Baker repeatedly observed that the African elephant and rhinoceros attacked white or grey horses with special fury. I have elsewhere shown [32] that half-wild horses apparently prefer to pair with those of the same color, and that herds of fallow-deer of different colors, though living together, have long kept distinct. It is a more significant fact that a female zebra would not admit the addresses of a male ass until he was painted so as to resemble a zebra, and then, as John Hunter remarks, "she received him very readily. In this curious fact, we have instinct excited by mere color, which had so strong an effect as to get the better of everything else. But the male did not require this, the female being an animal somewhat similar to himself, was sufficient to rouse him." [33]

[32] The 'Variation of Animals and Plants under Domestication,' 1868, vol. ii. pp. 102, 103.
[33] 'Essays and Observations,' by J. Hunter, edited by Owen, 1861, vol. i. p. 194.

In an earlier chapter we have seen that the mental powers of the higher animals do not differ in kind, though greatly in degree, from the corresponding powers of man, especially of the lower and barbarous races; and it would appear that even their taste for the beautiful is not widely different from that of the Quadrumana. As the negro of Africa raises the flesh on his face into parallel ridges "or cicatrices, high above the natural surface, which unsightly deformities are considered great personal attractions"; [34]—as negroes and savages in many parts of the world paint their faces with red, blue, white, or black bars,—so the male mandrill of Africa appears to have acquired his deeply-furrowed and gaudily-colored face from having been thus rendered attractive to the female. No doubt it is to us a most grotesque notion that the posterior end of the body should be colored for the sake of ornament even more brilliantly than the face; but this is not more strange than that the tails of many birds should be especially decorated.

[34] Sir S. Baker, 'The Nile Tributaries of Abyssinia,' 1867.

With mammals we do not at present possess any evidence that the males take pains to display their charms before the female; and the elaborate manner in which this is performed by male birds and other animals is the strongest argument in favour of the belief that the females admire, or are excited by, the ornaments and colors displayed before them. There is, however, a striking parallelism between mammals and birds in all their secondary sexual characters, namely in their weapons for fighting with rival males, in their ornamental appendages, and in their colors. In both classes, when the male differs from the female, the young of both sexes almost always resemble each other, and in a large majority of cases resemble the adult female. In both classes the male assumes the characters proper to his sex shortly before the age of reproduction; and if emasculated at an early period, loses them. In both classes the change of color is sometimes seasonal, and the tints of the naked parts sometimes become more vivid during the act of courtship. In both classes the male is almost always more vividly or strongly colored than the female, and is ornamented with larger crests of hair or feathers, or other such appendages. In a few exceptional cases the female in both classes is more highly ornamented than the male. With many mammals, and at least in the case of one bird, the male is more odoriferous than the female. In both classes the voice of the male is more powerful than that of the female. Considering this parallelism, there can be little doubt that the same cause, whatever it may be, has acted on mammals and birds; and the result, as far as ornamental characters are concerned, may be attributed, as it appears to me, to the long-continued preference of the individuals of one sex for certain individuals of the opposite sex, combined with their success in leaving a larger number of offspring to inherit their superior attractions.

EQUAL TRANSMISSION OF ORNAMENTAL CHARACTERS TO BOTH SEXES.

With many birds, ornaments, which analogy leads us to believe were primarily acquired by the males, have been transmitted equally, or almost equally, to both sexes; and we may now enquire how far this view applies to mammals. With a considerable number of species, especially of the smaller kinds, both sexes have been colored, independently of sexual selection, for the sake of protection; but not, as far as I can judge, in so many cases, nor in so striking a manner, as in most of the lower classes. Audubon remarks that he often mistook the musk-rat, [35] whilst sitting on the banks of a muddy stream, for a clod of earth, so complete was the resemblance. The hare on her form is a familiar instance of concealment through color; yet this principle partly fails in a closely-allied species, the rabbit, for when running to its burrow, it is made conspicuous to the sportsman, and no doubt to all beasts of prey, by its upturned white tail. No one doubts that the quadrupeds inhabiting snow-clad regions have been rendered white to protect them from their enemies, or to favour their stealing on their prey. In regions where snow never lies for long, a white coat would be injurious; consequently, species of this color are extremely rare in the hotter parts of the world. It deserves notice that many quadrupeds inhabiting moderately cold regions, although they do not assume a white winter dress, become paler during this season; and this apparently is the direct result of the conditions to which they have long been exposed. Pallas [36] states that in Siberia a change of this nature occurs with the wolf, two species of Mustela, the domestic horse, the *Equus hemionus*, the domestic cow, two species of antelopes, the musk-deer, the roe, elk, and reindeer. The roe, for instance, has a red summer and a grayish-white winter coat; and the latter may perhaps serve as a protection to the animal whilst wandering through the leafless thickets, sprinkled with snow and hoar-frost. If the above-named animals were gradually to extend their range into regions perpetually covered with snow, their pale winter-coats would probably be rendered through natural selection, whiter and whiter, until they became as white as snow.

[35] *Fiber zibethicus*, Audubon and Bachman, 'The Quadrupeds of North America,' 1846, p. 109.

[36] 'Novæ species Quadrupedum e Glirium ordine,' 1778, p. 7. What I have called the roe is the *Capreolus sibiricus subecaudatus* of Pallas.

Mr. Reeks has given me a curious instance of an animal profiting by being peculiarly colored. He raised from fifty to sixty white and brown piebald rabbits in a large walled orchard; and he had at the same time some similarly colored cats in his house. Such cats, as I have often noticed, are very conspicuous during day; but as they used to lie in watch during the dusk at the mouths of the burrows, the rabbits apparently did not distinguish them from their parti-colored brethren. The result was that, within eighteen months, every one of these parti-colored rabbits was destroyed; and there was evidence that this was effected by the cats. Color seems to be advantageous to another animal, the skunk, in a manner of which we have had many instances in other classes. No animal will voluntarily attack one of these creatures on account of the dreadful odor which it emits when irritated; but during the dusk it would not easily be recognized and might be attacked by a beast of prey. Hence it is, as Mr. Belt believes, [37] that the skunk is provided with a great white bushy tail, which serves as a conspicuous warning.

[37] 'The Naturalist in Nicaragua,' p. 249.

Fɪɢ. 70.--Tragelaphus scriptus, male (from the Knowsley Menagerie).

Although we must admit that many quadrupeds have received their present tints either as a protection, or as an aid in procuring prey, yet with a host of species, the colors are far too conspicuous and too singularly arranged to allow us to suppose that they serve for these purposes. We may take as an illustration certain antelopes; when we see the square white patch on the throat, the white marks on the fetlocks, and the round black spots on the ears, all more distinct in the male of the *Portax picta*, than in the female;—when we see that the colors are more vivid, that the narrow white lines on the flank and the broad white bar on the shoulder are more distinct in the male *Oreas derbyanus* than in the female;—when we see a similar difference between the sexes of the curiously-ornamented *Tragelaphus scriptus* (Fig. 70),—we cannot believe that differences of this kind are of any service to either sex in their daily habits of life. It seems a much more probable conclusion that the various marks were first acquired by the males and their colors intensified through sexual selection, and then partially transferred to the females. If this view be admitted, there can be little doubt that the equally singular colors and marks of many other antelopes, though common to both sexes, have been gained and transmitted in a like manner. Both sexes, for instance, of the koodoo (*Strepsiceros kudu*) (Fig. 64) have narrow white vertical lines on their hind flanks, and an elegant angular white mark on their foreheads. Both sexes in the genus Damalis are very oddly colored; in *D. pygarga* the back and neck are purplish-red, shading on the flanks into black; and these colors are abruptly separated from the white belly and from a large white space on the buttocks; the head is still more oddly colored, a large oblong white mask, narrowly-edged with black, covers the face up to the eyes (Fig. 71); there are three white stripes on the forehead, and the ears are marked with white. The fawns of this species are of a uniform pale yellowish-brown. In *Damalis albifrons* the coloring of the head differs from that in the last species in a single white stripe replacing the three stripes, and in the ears being almost wholly white. [38] After having studied to the best of my ability the sexual differences of animals belonging to all classes, I cannot avoid the conclusion that the curiously-arranged colors of many antelopes, though common to both sexes, are the result of sexual selection primarily applied to the male.

[38] See the fine plates in A. Smith's 'Zoology of South Africa,' and Dr. Gray's 'Gleanings from the Menagerie of Knowsley.'

FIG. 71.—Damalis pygarga, male (from the Knowsley menagerie).

The same conclusion may perhaps be extended to the tiger, one of the most beautiful animals in the world, the sexes of which cannot be distinguished by color, even by the dealers in wild beasts. Mr. Wallace believes [39] that the striped coat of the tiger "so assimilates with the vertical stems of the bamboo, as to assist greatly in concealing him from his approaching prey." But this view does not appear to me satisfactory. We have some slight evidence that his beauty may be due to sexual selection, for in two species of Felis the analogous marks and colors are rather brighter in the male than in the female. The zebra is conspicuously striped, and stripes cannot afford any protection in the open plains of South Africa. Burchell [40] in describing a herd says, "their sleek ribs glistened in the sun, and the brightness and regularity of their striped coats presented a picture of extraordinary beauty, in which probably they are not surpassed by any other quadruped." But as throughout the whole group of the Equidæ the sexes are identical in color, we have here no evidence of sexual selection. Nevertheless he who attributes the white and dark vertical stripes on the flanks of various antelopes to this process, will probably extend the same view to the Royal Tiger and beautiful Zebra.

[39] 'Westminster Review,' July 1, 1867, p. 5.
[40] 'Travels in South Africa,' 1824, vol. ii. p. 315.

We have seen in a former chapter that when young animals belonging to any class follow nearly the same habits of life as their parents, and yet are colored in a different manner, it may be inferred that they have retained the coloring of some ancient and extinct progenitor. In the family of pigs, and in the tapirs, the young are marked with longitudinal stripes, and thus differ from all the existing adult species in these two groups. With many kinds of deer the young are marked with elegant white spots, of which their parents exhibit not a trace. A graduated series can be followed from the axis deer, both sexes of which at all ages and during all seasons are beautifully spotted (the male being rather more strongly colored than the female), to species in which neither the old nor the young are spotted. I will specify some of the steps in this series. The Mantchurian deer (*Cervus mantchuricus*) is spotted during the whole year, but, as I have seen in the Zoological Gardens, the spots are much plainer during the summer, when the general color of the coat is lighter, than during the winter, when the general color is darker and the horns are fully developed. In the hog-deer (*Hyelaphus porcinus*) the spots are extremely conspicuous during the summer when the coat is reddish-brown, but quite disappear during the winter when the coat is brown. [41] In both these species the young are spotted. In the Virginian deer the young are likewise spotted, and about five per cent. of the adult animals living in Judge Caton's park, as I am informed by him, temporarily exhibit at the period when the red summer coat is being replaced by the bluish winter coat, a row of spots on each flank, which are always the same in number, though very variable in distinctness. From this condition there is but a very small step to the complete absence of spots in the adults at all seasons; and, lastly, to their absence at all ages and seasons, as occurs with certain species. From the existence of this perfect series, and more especially from the fawns of so many species being spotted, we may conclude that the now living members of the deer family are the descendants of some ancient species which, like the axis deer, was spotted at all ages and seasons. A still more ancient progenitor probably somewhat resembled the *Hyomoschus aquaticus*—for this animal is spotted, and the hornless males have large exserted canine teeth, of which some few true deer still retain rudiments. Hyomoschus, also, offers one of those interesting cases of a form linking together two groups, for it is intermediate in certain osteological characters between the pachyderms and ruminants, which were formerly thought to be quite distinct. [42]

[41] Dr. Gray, 'Gleanings from the Menagerie of Knowsley,' p. 64. Mr. Blyth, in speaking ('Land and Water,' 1869, p. 42) of the hog-deer of Ceylon, says it is more brightly spotted with white than the common hog-deer, at the season when it renews its horns.
[42] Falconer and Cautley, 'Proc. Geolog. Soc.' 1843; and Falconer's 'Pal. Memoirs,' vol. i. p. 196.

A curious difficulty here arises. If we admit that colored spots and stripes were first acquired as ornaments, how comes it that so many existing deer, the descendants of an aboriginally spotted animal, and all the species of pigs and tapirs, the descendants of an aboriginally striped animal, have lost in their adult state their former ornaments? I cannot satisfactorily answer this question. We may feel almost sure that the spots and stripes disappeared at or near maturity in the progenitors of our existing species, so that they were still retained by the young; and, owing to the law of inheritance at corresponding ages, were transmitted to the young of all succeeding generations. It may have been a great advantage to the lion and puma, from the open nature of their usual haunts, to have lost their stripes, and to have been thus rendered less conspicuous to their prey; and if the successive variations, by which this end was gained, occurred rather late in life, the young would have retained their stripes, as is now the case. As to deer, pigs, and tapirs, Fritz Müller has suggested to me that these animals, by the removal of their spots or stripes through natural selection, would have been less easily seen by their enemies; and that they would have especially required this protection, as soon as the carnivora increased in size and number during the tertiary periods. This may be the true explanation, but it is rather strange that the young should not have been thus protected, and still more so that the adults of some species should have retained their spots, either partially or completely, during part of the year. We know that, when the domestic ass varies and becomes reddish-brown, grey, or black, the stripes on the shoulders and even on the spine frequently disappear, though we cannot explain the cause. Very few horses, except dun-colored kinds, have stripes on any part of their bodies, yet we have good reason to believe that the aboriginal horse was striped on the legs and spine, and probably on the shoulders. [43] Hence the disappearance of the spots and stripes in our adult existing deer, pigs, and tapirs, may be due to a change in the general color of their coats; but whether this change was effected through sexual or natural selection, or was due to the direct action of the conditions of life, or to some other unknown cause, it is impossible to decide. An observation made by Mr. Sclater well illustrates our ignorance of the laws which regulate the appearance and disappearance of stripes; the species of Asinus which inhabit the Asiatic continent are destitute of stripes, not having even the cross shoulder-stripe, whilst those which inhabit Africa are conspicuously striped, with the partial exception of A. tæniopus, which has only the cross shoulder-stripe and generally some faint bars on the legs; and this species inhabits the almost intermediate region of Upper Egypt and Abyssinia. [44]

[43] The 'Variation of Animals and Plants under Domestication,' 1868, vol. i. pp. 61-64.
[44] 'Proc. Zool. Soc.' 1862, p. 164. See, also, Dr. Hartmann, 'Ann. d. Landw.' Bd. xliii. s. 222.

FIG. 72.—Head of Semnopithecus rubicundus. This and the following figures (from Prof. Gervais) are given to show the odd arrangement and development of the hair on the head.

QUADRUMANA.

Before we conclude, it will be well to add a few remarks on the ornaments of monkeys. In most of the species the sexes resemble each other in color, but in some, as we have seen, the males differ from the females, especially in the color of the naked parts of the skin, in the development of the beard, whiskers, and mane. Many species are colored either in so extraordinary or so beautiful a manner, and are furnished with such curious and elegant crests of hair, that we can hardly avoid looking at these characters as having been gained for the sake of ornament. The accompanying figures (Figs. 72 to 76) serve to show the arrangement of the hair on the face and head in several species. It is scarcely conceivable that these crests of hair, and the strongly contrasted colors of the fur and skin, can be the result of mere variability without the aid of selection; and it is inconceivable that they can be of use in any ordinary way to these animals. If so, they have probably been gained through sexual selection, though transmitted equally, or almost equally, to both sexes. With many of the Quadrumana, we have additional evidence of the action of sexual selection in the greater size and strength of the males, and in the greater development of their canine teeth, in comparison with the females.

FIG. 73.—Head of Semnopithecus comatus.   FIG. 74.—Head of Cebus capucinus.

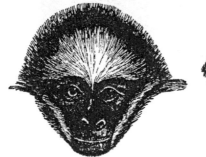

FIG. 75. –Head of Ateles marginatus.   FIG. 76.—Head of Cebus vellerosus

A few instances will suffice of the strange manner in which both sexes of some species are colored, and of the beauty of others. The face of the *Cercopithecus petaurista* (Fig. 77) is black, the whiskers and beard being white, with a defined, round, white spot on the nose, covered with short white hair, which gives to the animal an almost ludicrous aspect. The *Semnopithecus frontatus* likewise has a blackish face with a long black beard, and a large naked spot on the forehead of a bluish-white color. The face of *Macacus lasiotus* is dirty flesh-colored, with a defined red spot on each cheek. The appearance of *Cercocebus æthiops* is grotesque, with its black face, white whiskers and collar, chestnut head, and a large naked white spot over each eyelid. In very many species, the beard, whiskers, and crests of hair round the face are of a different color from the rest of the head, and when different, are always of a lighter tint, [45] being often pure white, sometimes bright yellow, or reddish. The whole face of the South American *Brachyurus calvus* is of a "glowing scarlet hue"; but this color does not appear until the animal is nearly mature. [46] The naked skin of the face differs wonderfully in color in the various species. It is often brown or flesh-color, with parts perfectly white, and often as black as that of the most sooty negro. In the Brachyurus the scarlet tint is brighter than that of the most blushing Caucasian damsel. It is sometimes more distinctly orange than in any Mongolian, and in several species it is blue, passing into violet or grey. In all the species known to Mr. Bartlett, in which the adults of both sexes have strongly-colored faces, the colors are dull or absent during early youth. This likewise holds good with the mandrill and Rhesus, in which the face and the posterior parts of the body are brilliantly colored in one sex alone. In these latter cases we have reason to believe that the colors were acquired through sexual selection; and we are naturally led to extend the same view to the foregoing species, though both sexes when adult have their faces colored in the same manner.

[45] I observed this fact in the Zoological Gardens; and many cases may be seen in the colored plates in Geoffroy St.-Hilaire and F. Cuvier, 'Histoire Nat. des Mammifères,' tom. i. 1824.

[46] Bates, 'The Naturalist on the Amazons,' 1863, vol. ii. p. 310.

FIG. 77.--Cercopithecus petaurista (from Brehm).

FIG. 78.—Cercopithecus diana (from Brehm).

Although many kinds of monkeys are far from beautiful according to our taste, other species are universally admired for their elegant appearance and bright colors. The *Semnopithecus nemæus*, though peculiarly colored, is described as extremely pretty; the orange-tinted face is surrounded by long whiskers of glossy whiteness, with a line of chestnut-red over the eyebrows; the fur on the back is of a delicate grey, with a square patch on the loins, the tail and the fore-arms being of a pure white; a gorget of chestnut surmounts the chest; the thighs are black, with the legs chestnut-red. I will mention only two other monkeys for their beauty; and I have selected these as presenting slight sexual differences in color, which renders it in some degree probable that both sexes owe their elegant appearance to sexual selection. In the moustache-monkey (*Cercopithecus cephus*) the general color of the fur is mottled-greenish with the throat white; in the male the end of the tail is chestnut, but the face is the most ornamented part, the skin being chiefly bluish-grey, shading into a blackish tint beneath the eyes, with the upper lip of a delicate blue, clothed on the lower edge with a thin black moustache; the whiskers are orange-colored, with the upper part black, forming a band which extends backwards to the ears, the latter being clothed with whitish hairs. In the Zoological Society's Gardens I have often overheard visitors admiring the beauty of another monkey, deservedly called

Cercopithecus diana (Fig. 78); the general color of the fur is grey; the chest and inner surface of the forelegs are white; a large triangular defined space on the hinder part of the back is rich chestnut; in the male the inner sides of the thighs and the abdomen are delicate fawn-colored, and the top of the head is black; the face and ears are intensely black, contrasting finely with a white transverse crest over the eyebrows and a long white peaked beard, of which the basal portion is black. [47]

[47] I have seen most of the above monkeys in the Zoological Society's Gardens. The description of the Semnopithecus nemaeus is taken from Mr. W.C. Martin's 'Natural History of Mammalia,' 1841, p. 460; see also pp. 475, 523.

In these and many other monkeys, the beauty and singular arrangement of their colors, and still more the diversified and elegant arrangement of the crests and tufts of hair on their heads, force the conviction on my mind that these characters have been acquired through sexual selection exclusively as ornaments.

SUMMARY.

The law of battle for the possession of the female appears to prevail throughout the whole great class of mammals. Most naturalists will admit that the greater size, strength, courage, and pugnacity of the male, his special weapons of offence, as well as his special means of defence, have been acquired or modified through that form of selection which I have called sexual. This does not depend on any superiority in the general struggle for life, but on certain individuals of one sex, generally the male, being successful in conquering other males, and leaving a larger number of offspring to inherit their superiority than do the less successful males.

There is another and more peaceful kind of contest, in which the males endeavour to excite or allure the females by various charms. This is probably carried on in some cases by the powerful odors emitted by the males during the breeding-season; the odoriferous glands having been acquired through sexual selection. Whether the same view can be extended to the voice is doubtful, for the vocal organs of the males must have been strengthened by use during maturity, under the powerful excitements of love, jealousy or rage, and will consequently have been transmitted to the same sex. Various crests, tufts, and mantles of hair, which are either confined to the male, or are more developed in this sex than in the female, seem in most cases to be merely ornamental, though they sometimes serve as a defence against rival males. There is even reason to suspect that the branching horns of stags, and the elegant horns of certain antelopes, though properly serving as weapons of offence or defence, have been partly modified for ornament.

When the male differs in color from the female, he generally exhibits darker and more strongly-contrasted tints. We do not in this class meet with the splendid red, blue, yellow, and green tints, so common with male birds and many other animals. The naked parts, however, of certain Quadrumana must be excepted; for such parts, often oddly situated, are brilliantly colored in some species. The colors of the male in other cases may be due to simple variation, without the aid of selection. But when the colors are diversified and strongly pronounced, when they are not developed until near maturity, and when they are lost after emasculation, we can hardly avoid the conclusion that they have been acquired through sexual selection for the sake of ornament, and have been transmitted exclusively, or almost exclusively, to the same sex. When both sexes are colored in the same manner, and the colors are conspicuous or curiously arranged, without being of the least apparent use as a protection, and especially when they are associated with various other ornamental appendages, we are led by analogy to the same conclusion, namely, that they have been acquired through sexual selection, although transmitted to both sexes. That conspicuous and diversified colors, whether confined to the males or common to both sexes, are as a general rule associated in the same groups and sub-groups with other secondary sexual characters serving for war or for ornament, will be found to hold good, if we look back to the various cases given in this and the last chapter.

The law of the equal transmission of characters to both sexes, as far as color and other ornaments are concerned, has prevailed far more extensively with mammals than with birds; but weapons, such as horns and tusks, have often been transmitted either exclusively or much more perfectly to the males than to the females. This is surprising, for, as the males generally use their weapons for defence against enemies of all kinds, their weapons would have been of service to the females. As far as we can see, their absence in this sex can be accounted for only by the form of inheritance which has prevailed. Finally, with quadrupeds the contest between the individuals of the same sex, whether peaceful or bloody, has, with the rarest exceptions, been confined to the males; so that the latter have been modified through sexual selection, far more commonly than the females, either for fighting with each other or for alluring the opposite sex.

PART III.

SEXUAL SELECTION IN RELATION TO MAN, AND CONCLUSION.

CHAPTER XIX.

SECONDARY SEXUAL CHARACTERS OF MAN.

Differences between man and woman—Causes of such differences and of certain characters common to both sexes—Law of battle—Differences in mental powers, and voice—On the influence of beauty in determining the marriages of mankind—Attention paid by savages to ornaments—Their ideas of beauty in woman—The tendency to exaggerate each natural peculiarity.

With mankind the differences between the sexes are greater than in most of the Quadrumana, but not so great as in some, for instance, the mandrill. Man on an average is considerably taller, heavier, and stronger than woman, with squarer shoulders and more plainly-pronounced muscles. Owing to the relation which exists between muscular development and the projection of the brows, [1] the superciliary ridge is generally more marked in man than in woman. His body, and especially his face, is more hairy, and his voice has a different and more powerful tone. In certain races the women are said to differ slightly in tint from the men. For instance, Schweinfurth, in speaking of a negress belonging to the Monbuttoos, who inhabit the interior of Africa a few degrees north of the equator, says, "Like all her race, she had a skin several shades lighter than her husband's, being something of the color of half-roasted coffee." [2] As the women labor in the fields and are quite unclothed, it is not likely that they differ in color from the men owing to less exposure to the weather. European women are perhaps the brighter colored of the two sexes, as may be seen when both have been equally exposed.

[1] Schaaffhausen, translation in 'Anthropological Review,' Oct. 1868, pp. 419, 420, 427.
[2] 'The Heart of Africa,' English transl. 1873, vol i. p. 544.

Man is more courageous, pugnacious and energetic than woman, and has a more inventive genius. His brain is absolutely larger, but whether or not proportionately to his larger body, has not, I believe, been fully ascertained. In woman the face is rounder; the jaws and the base of the skull smaller; the outlines of the body rounder, in parts more prominent; and her pelvis is broader than in man; [3] but this latter character may perhaps be considered rather as a primary than a secondary sexual character. She comes to maturity at an earlier age than man.

[3] Ecker, translation, in 'Anthropological Review,' Oct. 1868, pp. 351-356. The comparison of the form of the skull in men and women has been followed out with much care by Welcker.

As with animals of all classes, so with man, the distinctive characters of the male sex are not fully developed until he is nearly mature; and if emasculated they never appear. The beard, for instance, is a secondary sexual character, and male children are beardless, though at an early age they have abundant hair on the head. It is probably due to the rather late appearance in life of the successive variations whereby man has acquired his masculine characters, that they are transmitted to the male sex alone. Male and female children resemble each other closely, like the young of so many other animals in which the adult sexes differ widely; they likewise resemble the mature female much more closely than the mature male. The female, however, ultimately assumes certain distinctive characters, and in the formation of her skull, is said to be intermediate between the child and the man. [4] Again, as the young of closely allied though distinct species do not differ nearly so much from each other as do the adults, so it is with the children of the different races of man. Some have even maintained that race-differences cannot be detected in the infantile skull. [5] In regard to color, the new-born negro child is reddish nut-brown, which soon becomes slaty-grey; the black color being fully developed within a year in the Soudan, but not until three years in Egypt. The eyes of the negro are at first blue, and the hair chestnut-brown rather than black, being curled only at the ends. The children of the Australians immediately after birth are yellowish-brown, and become dark at a later age. Those of the Guaranys of Paraguay are whitish-yellow, but they acquire in the course of a few weeks the yellowish-brown tint of their parents. Similar observations have been made in other parts of America. [6]

[4] Ecker and Welcker, ibid. pp. 352, 355; Vogt, 'Lectures on Man,' Eng. translat. p. 81.
[5] Schaaffhausen, 'Anthropolog. Review,' ibid. p. 429.
[6] Pruner-Bey, on negro infants as quoted by Vogt, 'Lectures on Man,' Eng. translat. 1864, p. 189: for further facts on negro infants, as quoted from Winterbottom and Camper, see Lawrence, 'Lectures on Physiology,' etc. 1822, p. 451. For the infants of the Guaranys, see Rengger, 'Säugethiere,' etc. s. 3. See also Godron, 'De l'Espèce,' tom. ii. 1859, p. 253. For the Australians, Waitz, 'Introduction to Anthropology,' Eng. translat. 1863, p. 99.

I have specified the foregoing differences between the male and female sex in mankind, because they are curiously like those of the Quadrumana. With these animals the female is mature at an earlier age than the male; at least this is certainly the case in *Cebus azaræ*. [7] The males of most species are larger and stronger than the females, of which fact the gorilla affords a well-known instance. Even in so trifling a character as the greater prominence of the superciliary ridge, the males of certain monkeys differ from the females, [8] and agree in this respect with mankind. In the gorilla and certain other monkeys, the cranium of the adult male presents a strongly-marked sagittal crest, which is absent in the female; and Ecker found a trace of a similar difference between the two sexes in the Australians. [9] With monkeys when there is any difference in the voice, that of the male is the more powerful. We have seen that certain male monkeys have a well-developed beard, which is quite deficient, or much less developed in the female. No instance is known of the beard, whiskers, or moustache being larger in the female than in the male monkey. Even in the color of the beard there is a curious parallelism between man and the Quadrumana, for with man when the beard differs in color from the hair of the head, as is commonly the case, it is, I believe, almost always of a lighter tint, being often reddish. I have repeatedly observed this fact in England; but two gentlemen have lately written to me, saying that they form an exception to the rule. One of these gentlemen accounts for the fact by the wide difference in color of the hair on the paternal and maternal sides of his family. Both had been long aware of this peculiarity (one of them having often been accused of dyeing his beard), and had been thus led to observe other men, and were convinced that the exceptions were very rare. Dr. Hooker attended to this little point for me in Russia, and found no exception to the rule. In Calcutta, Mr. J. Scott, of the Botanic Gardens, was so kind as to observe the many races of men to be seen there, as well as in some other parts of India, namely, two races of Sikhim, the Bhoteas, Hindoos, Burmese, and Chinese, most of which races have very little hair on the face; and he always found that when there was any difference in color between the hair of the head and the beard, the latter was invariably lighter. Now with monkeys, as has already been stated, the beard frequently differs strikingly in color from the hair of the head, and in such cases it is always of a lighter hue, being often pure white, sometimes yellow or reddish. [10]

[7] Rengger, 'Säugethiere,' etc., 1830, s. 49.

[8] As in *Macacus cynomolgus* (Desmarest, 'Mammalogie,' p. 65), and in *Hylobates agilis* (Geoffroy St.-Hilaire and F. Cuvier, 'Histoire Nat. des Mamm.' 1824, tom. i. p. 2).

[9] 'Anthropological Review,' Oct. 1868, p. 353.

[10] Mr. Blyth informs me that he has only seen one instance of the beard, whiskers, etc., in a monkey becoming white with old age, as is so commonly the case with us. This, however, occurred in an aged *Macacus cynomolgus*, kept in confinement whose moustaches were "remarkably long and human-like." Altogether this old monkey presented a ludicrous resemblance to one of the reigning monarchs of Europe, after whom he was universally nick-named. In certain races of man the hair on the head hardly ever becomes grey; thus Mr. D. Forbes has never, as he informs me, seen an instance with the Aymaras and Quichuas of South America.

In regard to the general hairiness of the body, the women in all races are less hairy than the men; and in some few Quadrumana the under side of the body of the female is less hairy than that of the male. [11] Lastly, male monkeys, like men, are bolder and fiercer than the females. They lead the troop, and when there is danger, come to the front. We thus see how close is the parallelism between the sexual differences of man and the Quadrumana. With some few species, however, as with certain baboons, the orang and the gorilla, there is a considerably greater difference between the sexes, as in the size of the canine teeth, in the development and color of the hair, and especially in the color of the naked parts of the skin, than in mankind.

[11] This is the case with the females of several species of Hylobates; see Geoffroy St.-Hilaire and F. Cuvier, 'Hist. Nat. des Mamm.' tom. i. See also, on *H. lar*, 'Penny Cyclopedia,' vol. ii. pp. 149, 150.

All the secondary sexual characters of man are highly variable, even within the limits of the same race; and they differ much in the several races. These two rules hold good generally throughout the animal kingdom. In the excellent observations made on board the *Novara*, [12] the male Australians were found to exceed the females by only 65 millim. in height, whilst with the Javans the average excess was 218 millim.; so that in this latter race the difference in height between the sexes is more than thrice as great as with the Australians. Numerous measurements were carefully made of the stature, the circumference of the neck and chest, the length of the back-bone and of the arms, in various races; and nearly all these measurements show that the males differ much more from one another than do the females. This fact indicates that, as far as these characters are concerned, it is the male which has been chiefly modified, since the several races diverged from their common stock.

[12] The results were deduced by Dr. Weisbach from the measurements made by Drs. K. Scherzer and Schwarz, see 'Reise der Novara: Anthropolog. Theil,' 1867, ss. 216, 231, 234, 236, 239, 269.

The development of the beard and the hairiness of the body differ remarkably in the men of distinct races, and even in different tribes or families of the same race. We Europeans see this amongst ourselves. In the Island of St. Kilda, according to Martin, [13] the men do not acquire beards until the age of thirty or upwards, and even then the beards are very thin. On the Europæo-Asiatic continent, beards prevail until we pass beyond India; though with the natives of Ceylon they are often absent, as was noticed in ancient times by Diodorus. [14] Eastward of India beards disappear, as with the Siamese, Malays, Kalmucks, Chinese, and Japanese; nevertheless, the Ainos, [15] who inhabit the northernmost islands of the Japan Archipelago, are the hairiest men in the world. With negroes the beard is scanty or wanting, and they rarely have whiskers; in both sexes the body is frequently almost destitute of fine down. [16] On the other hand, the Papuans of the Malay Archipelago, who are nearly as black as negroes, possess well-developed beards. [17] In the Pacific Ocean the inhabitants of the Fiji Archipelago have large bushy beards, whilst those of the not distant archipelagoes of Tonga and Samoa are beardless; but these men belong to distinct races. In the Ellice group all the inhabitants belong to the same race; yet on one island alone, namely Nunemaya, "the men have splendid beards"; whilst on the other islands "they have, as a rule, a dozen straggling hairs for a beard." [18]

[13] 'Voyage to St. Kilda' (3rd ed. 1753), p. 37.

[14] Sir J.E. Tennent, 'Ceylon,' vol. ii. 1859, p. 107.

[15] Quatrefages, 'Revue des Cours Scientifiques,' Aug. 29, 1868, p. 630; Vogt, 'Lectures on Man,' Eng. trans. p. 127.

[16] On the beards of negroes, Vogt, 'Lectures,' etc. p. 127; Waitz, 'Introduct. to Anthropology,' Engl. translat. 1863, vol. i. p. 96. It is remarkable that in the United States ('Investigations in Military and Anthropological Statistics of American Soldiers,' 1869, p. 569) the pure negroes and their crossed offspring seem to have bodies almost as hairy as Europeans.

[17] Wallace, 'The Malay Arch.' vol. ii. 1869, p. 178.

[18] Dr. J. Barnard Davis on Oceanic Races, in 'Anthropological Review,' April 1870, pp. 185, 191.

Throughout the great American continent the men may be said to be beardless; but in almost all the tribes a few short hairs are apt to appear on the face, especially in old age. With the tribes of North America, Catlin estimates that eighteen out of twenty men are completely destitute by nature of a beard; but occasionally there may be seen a man, who has neglected to pluck out the hairs at puberty, with a soft beard an inch or two in length. The Guaranys of Paraguay differ from all the surrounding tribes in having a small beard, and even some hair on the body, but no whiskers. [19] I am informed by Mr. D. Forbes, who particularly attended to this point, that the Aymaras and Quichuas of the Cordillera are remarkably hairless, yet in old age a few straggling hairs occasionally appear on the chin. The men of these two tribes have very little hair on the various parts of the body where hair grows abundantly in Europeans, and the women have none on the corresponding parts. The hair on the head, however, attains an extraordinary length in both sexes, often reaching almost to the ground; and this is likewise the case with some of the N. American tribes. In the amount of hair, and in the general shape of the body, the sexes of the American aborigines do not differ so much from each other, as in most other races. [20] This fact is analogous with what occurs with some closely allied monkeys; thus the sexes of the chimpanzee are not as different as those of the orang or gorilla. [21]

[19] Catlin, 'North American Indians,' 3rd. ed. 1842, vol. ii. p. 227. On the Guaranys, see Azara, 'Voyages dans l'Amérique Mérid.' tom. ii. 1809, p. 85; also Rengger, 'Säugethiere von Paraguay,' s. 3.

[20] Prof. and Mrs. Agassiz ('Journey in Brazil,' p. 530) remark that the sexes of the American Indians differ less than those of the negroes and of the higher races. See also Rengger, ibid. p. 3, on the Guaranys.

[21] Rütimeyer, 'Die Grenzen der Thierwelt; eine Betrachtung zu Darwin's Lehre,' 1868, s. 54.

In the previous chapters we have seen that with mammals, birds, fishes, insects, etc., many characters, which there is every reason to believe were primarily gained through sexual selection by one sex, have been transferred to the other. As this same form of transmission has apparently prevailed much with mankind, it will save useless repetition if we discuss the origin of characters peculiar to the male sex together with certain other characters common to both sexes.

LAW OF BATTLE.

With savages, for instance, the Australians, the women are the constant cause of war both between members of the same tribe and between distinct tribes. So no doubt it was in ancient times; "nam fuit ante Helenam mulier teterrima belli causa." With some of the North American Indians, the contest is reduced to a system. That excellent observer, Hearne, [22] says:—"It has ever been the custom among these people for the men to wrestle for any woman to whom they are attached; and, of course, the strongest party always carries off the prize. A weak man, unless he be a good hunter, and well-beloved, is seldom permitted to keep a wife that a stronger man thinks worth his notice. This custom prevails throughout all the tribes, and causes a great spirit of emulation among their youth, who are upon all occasions, from their childhood, trying their strength and skill in wrestling." With the Guanas of South America, Azara states that the men rarely marry till twenty years old or more, as before that age they cannot conquer their rivals.

[22] 'A Journey from Prince of Wales Fort,' 8vo. ed. Dublin, 1796, p. 104. Sir J. Lubbock ('Origin of Civilization,' 1870, p. 69) gives other and similar cases in North America. For the Guanas of South America see Azara, 'Voyages,' etc. tom. ii. p. 94.

Other similar facts could be given; but even if we had no evidence on this head, we might feel almost sure, from the analogy of the higher Quadrumana, [23] that the law of battle had prevailed with man during the early stages of his development. The occasional appearance at the present day of canine teeth which project above the others, with traces of a diastema or open space for the reception of the opposite canines, is in all probability a case of reversion to a former state, when the progenitors of man were provided with these weapons, like so many existing male Quadrumana. It was remarked in a former chapter that as man gradually became erect, and continually used his hands and arms for fighting with sticks and stones, as well as for the other purposes of life, he would have used his jaws and teeth less and less. The jaws, together with their muscles, would then have been reduced through disuse, as would the teeth through the not well understood principles of correlation and economy of growth; for we everywhere see that parts, which are no longer of service, are reduced in size. By such steps the original inequality between the jaws and teeth in the two sexes of mankind would ultimately have been obliterated. The case is almost parallel with that of many male Ruminants, in which the canine teeth have been reduced to mere rudiments, or have disappeared, apparently in consequence of the development of horns. As the prodigious difference between the skulls of the two sexes in the orang and gorilla stands in close relation with the development of the immense canine teeth in the males, we may infer that the reduction of the jaws and teeth in the early male progenitors of man must have led to a most striking and favourable change in his appearance.

[23] On the fighting of the male gorillas, see Dr. Savage, in 'Boston Journal of Natural History,' vol. v. 1847, p. 423. On *Presbytis entellus*, see the 'Indian Field,' 1859, p. 146.

There can be little doubt that the greater size and strength of man, in comparison with woman, together with his broader shoulders, more developed muscles, rugged outline of body, his greater courage and pugnacity, are all due in chief part to inheritance from his half-human male ancestors. These characters would, however, have been preserved or even augmented during the long ages of man's savagery, by the success of the strongest and boldest men, both in the general struggle for life and in their contests for wives; a success which would have ensured their leaving a more numerous progeny than their less favoured brethren. It is not probable that the greater strength of man was primarily acquired through the inherited effects of his having worked harder than woman for his own subsistence and that of his family; for the women in all barbarous nations are compelled to work at least as hard as the men. With civilized people the arbitrament of battle for the possession of the women has long ceased; on the other hand, the men, as a general rule, have to work harder than the women for their joint subsistence, and thus their greater strength will have been kept up.

DIFFERENCE IN THE MENTAL POWERS OF THE TWO SEXES.

With respect to differences of this nature between man and woman, it is probable that sexual selection has played a highly important part. I am aware that some writers doubt whether there is any such inherent difference; but this is at least probable from the analogy of the lower animals which present other secondary sexual characters. No one disputes that the bull differs in disposition from the cow, the wild-boar from the sow, the stallion from the mare, and, as is well known to the keepers of menageries, the males of the larger apes from the females. Woman seems to differ from man in mental disposition, chiefly in her greater tenderness and less selfishness; and this holds good even with savages, as shown by a well-known passage in Mungo Park's Travels, and by statements made by many other travellers. Woman, owing to her maternal instincts, displays these qualities towards her infants in an eminent degree; therefore it is likely that she would often extend them towards her fellow-creatures. Man is the rival of other men; he delights in competition, and this leads to ambition which passes too easily into selfishness. These latter qualities seem to be his natural and unfortunate birthright. It is generally admitted that with woman the powers of intuition, of rapid perception, and perhaps of imitation, are more strongly marked than in man; but some, at least, of these faculties are characteristic of the lower races, and therefore of a past and lower state of civilization.

The chief distinction in the intellectual powers of the two sexes is shown by man's attaining to a higher eminence, in whatever he takes up, than can woman—whether requiring deep thought, reason, or imagination, or merely the use of the senses and hands. If two lists were made of the most eminent men and women in poetry, painting, sculpture, music (inclusive both of composition and performance), history, science, and philosophy, with half-a-dozen names under each subject, the two lists would not bear comparison. We may also infer, from the law of the deviation from averages, so well illustrated by Mr. Galton, in his work on 'Hereditary Genius,' that if men are capable of a decided pre-eminence over women in many subjects, the average of mental power in man must be above that of woman.

Amongst the half-human progenitors of man, and amongst savages, there have been struggles between the males during many generations for the possession of the females. But mere bodily strength and size would do little for victory, unless associated with courage, perseverance, and determined energy. With social animals, the young males have to pass through many a contest before they win a female, and the older males have to retain their females by renewed battles. They have, also, in the case of mankind, to defend their females, as well as their young, from enemies of all kinds, and to hunt for their joint subsistence. But to avoid enemies or to attack them with success, to capture wild animals, and to fashion weapons, requires the aid of the higher mental faculties, namely, observation, reason, invention, or imagination. These various faculties will thus have been continually put to the test and selected during manhood; they will, moreover, have been strengthened by use during this same period of life. Consequently in accordance with the principle often alluded to, we might expect that they would at least tend to be transmitted chiefly to the male offspring at the corresponding period of manhood.

Now, when two men are put into competition, or a man with a woman, both possessed of every mental quality in equal perfection, save that one has higher energy, perseverance, and courage, the latter will generally become more eminent in every pursuit, and will gain the ascendancy. [24] He may be said to possess genius—for genius has been declared by a great authority to be patience; and patience, in this sense, means unflinching, undaunted perseverance. But this view of genius is perhaps deficient; for without the higher powers of the imagination and reason, no eminent success can be gained in many subjects. These latter faculties, as well as the former, will have been developed in man, partly through sexual selection,—that is, through the contest of rival males, and partly through natural selection, that is, from success in the general struggle for life; and as in both cases the struggle will have been during maturity, the characters gained will have been transmitted more fully to the male than to the female offspring. It accords in a striking manner with this view of the modification and re-inforcement of many of our mental faculties by sexual selection, that, firstly, they notoriously undergo a considerable change at puberty, [25] and, secondly, that eunuchs remain throughout life inferior in these same qualities. Thus, man has ultimately become superior to woman. It is, indeed, fortunate that the law of the equal transmission of characters to both sexes prevails with mammals; otherwise, it is probable that man would have become as superior in mental endowment to woman, as the peacock is in ornamental plumage to the peahen.

[24] J. Stuart Mill remarks ('The Subjection of Women,' 1869, p. 122), "The things in which man most excels woman are those which require most plodding, and long hammering at single thoughts." What is this but energy and perseverance?
[25] Maudsley, 'Mind and Body,' p. 31.

It must be borne in mind that the tendency in characters acquired by either sex late in life, to be transmitted to the same sex at the same age, and of early acquired characters to be transmitted to both sexes, are rules which, though general, do not always hold. If they always held good, we might conclude (but I here exceed my proper bounds) that the inherited effects of the early education of boys and girls would be transmitted equally to both sexes; so that the present inequality in mental power between the sexes would not be effaced by a similar course of early training; nor can it have been caused by their dissimilar early training. In order that woman should reach the same standard as man, she ought, when nearly adult, to be trained to energy and perseverance, and to have her reason and imagination exercised to the highest point; and then she would probably transmit these qualities chiefly to her adult daughters. All women, however, could not be thus raised, unless during many generations those who excelled in the above robust virtues were married, and produced offspring in larger numbers than other women. As before remarked of bodily strength, although men do not now fight for their wives, and this form of selection has passed away, yet during manhood, they generally undergo a severe struggle in order to maintain themselves and their families; and this will tend to keep up or even increase their mental powers, and, as a consequence, the present inequality between the sexes. [26]

[26] An observation by Vogt bears on this subject: he says, "It is a remarkable circumstance, that the difference between the sexes, as regards the cranial cavity, increases with the development of the race, so that the male European excels much more the female, than the negro the negress. Welcker confirms this statement of Huschke from his measurements of negro and German skulls." But Vogt admits ('Lectures on Man,' Eng. translat. 1864, p. 81) that more observations are requisite on this point.

VOICE AND MUSICAL POWERS.

In some species of Quadrumana there is a great difference between the adult sexes, in the power of their voices and in the development of the vocal organs; and man appears to have inherited this difference from his early progenitors. His vocal cords are about one-third longer than in woman, or than in boys; and emasculation produces the same effect on him as on the lower animals, for it "arrests that prominent growth of the thyroid, etc., which accompanies the elongation of the cords." [27] With respect to the cause of this difference between the sexes, I have nothing to add to the remarks in the last chapter on the probable effects of the long-continued use of the vocal organs by the male under the excitement of love, rage and jealousy. According to Sir Duncan Gibb, [28] the voice and the form of the larynx differ in the different races of mankind; but with the Tartars, Chinese, etc., the voice of the male is said not to differ so much from that of the female, as in most other races.

[27] Owen, 'Anatomy of Vertebrates,' vol. iii. p. 603.
[28] 'Journal of the Anthropological Society,' April 1869, p. lvii. and lxvi.

The capacity and love for singing or music, though not a sexual character in man, must not here be passed over. Although the sounds emitted by animals of all kinds serve many purposes, a strong case can be made out, that the vocal organs were primarily used and perfected in relation to the propagation of the species. Insects and some few spiders are the lowest animals which voluntarily produce any sound; and this is generally effected by the aid of beautifully constructed stridulating organs, which are often confined to the males. The sounds thus produced consist, I believe in all cases, of the same note, repeated rhythmically [29]; and this is sometimes pleasing even to the ears of man. The chief and, in some cases, exclusive purpose appears to be either to call or charm the opposite sex.

[29] Dr. Scudder, 'Notes on Stridulation,' in 'Proc. Boston Soc. of Nat. Hist.' vol. xi. April 1868.

The sounds produced by fishes are said in some cases to be made only by the males during the breeding-season. All the air-breathing Vertebrata necessarily possess an apparatus for inhaling and expelling air, with a pipe capable of being closed at one end. Hence when the primeval members of this class were strongly excited and their muscles violently contracted, purposeless sounds would almost certainly have been produced; and these, if they proved in any way serviceable, might readily have been modified or intensified by the preservation of properly adapted variations. The lowest Vertebrates which breathe air are Amphibians; and of these, frogs and toads possess vocal organs, which are incessantly used during the breeding-season, and which are often more highly developed in the male than in the female. The male alone of the tortoise utters a noise, and this only during the season of love. Male alligators roar or bellow during the same season. Every one knows how much birds use their vocal organs as a means of courtship; and some species likewise perform what may be called instrumental music.

In the class of Mammals, with which we are here more particularly concerned, the males of almost all the species use their voices during the breeding-season much more than at any other time; and some are absolutely mute excepting at this season. With other species both sexes, or only the females, use their voices as a love-call. Considering these facts, and that the vocal organs of some quadrupeds are much more largely developed in the male than in the female, either permanently or temporarily during the breeding-season; and considering that in most of the lower classes the sounds produced by the males, serve not only to call but to excite or allure the female, it is a surprising fact that we have not as yet any good evidence that these organs are used by male mammals to charm the females. The American *Mycetes caraya* perhaps forms an exception, as does the *Hylobates agilis*, an ape allied to man. This gibbon has an extremely loud but musical voice. Mr. Waterhouse states, [30]"It appeared to me that in ascending and descending the scale, the intervals were always exactly half-tones; and I am sure that the highest note was the exact octave to the lowest. The quality of the notes is very musical; and I do not doubt that a good violinist would be able to give a correct idea of the gibbon's composition, excepting as regards its loudness." Mr. Waterhouse then gives the notes. Professor Owen, who is a musician, confirms the foregoing statement, and remarks, though erroneously, that this gibbon "alone of brute mammals may be said to sing." It appears to be much excited after its performance. Unfortunately, its habits have never been closely observed in a state of nature; but from the analogy of other animals, it is probable that it uses its musical powers more especially during the season of courtship.

[30] Given in W.C.L. Martin's 'General Introduction to Natural History of Mamm. Animals,' 1841, p. 432; Owen, 'Anatomy of Vertebrates,' vol. iii, p. 600.

This gibbon is not the only species in the genus which sings, for my son, Francis Darwin, attentively listened in the Zoological Gardens to *H. leuciscus* whilst singing a cadence of three notes, in true musical intervals and with a clear musical tone. It is a more surprising fact that certain rodents utter musical sounds. Singing mice have often been mentioned and exhibited, but imposture has commonly been suspected. We have, however, at last a clear account by a well-known observer, the Rev. S. Lockwood, [31] of the musical powers of an American species, the *Hesperomys cognatus*, belonging to a genus distinct from that of the English mouse. This little animal was kept in confinement, and the performance was repeatedly heard. In one of the two chief songs, "the last bar would frequently be prolonged to two or three; and she would sometimes change from C sharp and D, to C natural and D, then warble on these two notes awhile, and wind up with a quick chirp on C sharp and D. The distinctness between the semitones was very marked, and easily appreciable to a good ear." Mr. Lockwood gives both songs in musical notation; and adds that though this little mouse "had no ear for time, yet she would keep to the key of B (two flats) and strictly in a major key."..."Her soft clear voice falls an octave with all the precision possible; then at the wind up, it rises again into a very quick trill on C sharp and D."

[31] The 'American Naturalist,' 1871, p. 761.

A critic has asked how the ears of man, and he ought to have added of other animals, could have been adapted by selection so as to distinguish musical notes. But this question shows some confusion on the subject; a noise is the sensation resulting from the co-existence of several aerial "simple vibrations" of various periods, each of which intermits so frequently that its separate existence cannot be perceived. It is only in the want of continuity of such vibrations, and in their want of harmony *inter se*, that a noise differs from a musical note. Thus an ear to be capable of discriminating noises—and the high importance of this power to all animals is admitted by every one—must be sensitive to musical notes. We have evidence of this capacity even low down in the animal scale: thus Crustaceans are provided with auditory hairs of different lengths, which have been seen to vibrate when the proper musical notes are struck. [32] As stated in a previous chapter, similar observations have been made on the hairs of the antennæ of gnats. It has been positively asserted by good observers that spiders are attracted by music. It is also well known that some dogs howl when hearing particular tones. [33] Seals apparently appreciate music, and their fondness for it "was well known to the ancients, and is often taken advantage of by the hunters at the present day." [34]

[32] Helmholtz, 'Théorie Phys. de la Musique,' 1868, p. 187.
[33] Several accounts have been published to this effect. Mr. Peach writes to me that an old dog of his howls when B flat is sounded on the flute, and to no other note. I may add another instance of a dog always whining, when one note on a concertina, which was out of tune, was played.
[34] Mr. R. Brown, in 'Proc. Zool. Soc.' 1868, p. 410.

Therefore, as far as the mere perception of musical notes is concerned, there seems no special difficulty in the case of man or of any other animal. Helmholtz has explained on physiological principles why concords are agreeable, and discords disagreeable to the human ear; but we are little concerned with these, as music in harmony is a late invention. We are more concerned with melody, and here again, according to Helmholtz, it is intelligible why the notes of our musical scale are used. The ear analyses all sounds into their component "simple vibrations," although we are not conscious of this analysis. In a musical note the lowest in pitch of these is generally predominant, and the others which are less marked are the octave, the twelfth, the second octave, etc., all harmonies of the fundamental predominant note; any two notes of our scale have many of these harmonic over-tones in common. It seems pretty clear then, that if an animal always wished to sing precisely the same song, he would guide himself by sounding those notes in succession, which possess many over-tones in common—that is, he would choose for his song, notes which belong to our musical scale.

But if it be further asked why musical tones in a certain order and rhythm give man and other animals pleasure, we can no more give the reason than for the pleasantness of certain tastes and smells. That they do give pleasure of some kind to animals, we may infer from their being produced during the season of courtship by many insects, spiders, fishes, amphibians, and birds; for unless the females were able to appreciate such sounds and were excited or charmed by them, the persevering efforts of the males, and the complex structures often possessed by them alone, would be useless; and this it is impossible to believe.

Human song is generally admitted to be the basis or origin of instrumental music. As neither the enjoyment nor the capacity of producing musical notes are faculties of the least use to man in reference to his daily habits of life, they must be ranked amongst the most mysterious with which he is endowed. They are present, though in a very rude condition, in men of all races, even the most savage; but so different is the taste of the several races, that our music gives no pleasure to savages, and their music is to us in most cases hideous and unmeaning. Dr. Seemann, in some interesting remarks on this subject [35], "doubts whether even amongst the nations of Western Europe, intimately connected as they are by close and frequent intercourse, the music of the one is interpreted in the same sense by the others. By travelling eastwards we find that there is certainly a different language of music. Songs of joy and dance-accompaniments are no longer, as with us, in the major keys, but always in the minor." Whether or not the half-human progenitors of man possessed, like the singing gibbons, the capacity of producing, and therefore no doubt of appreciating, musical notes, we know that man possessed these faculties at a very remote period. M. Lartet has described two flutes made out of the bones and horns of the reindeer, found in caves together with flint tools and the remains of extinct animals. The arts of singing and of dancing are also very ancient, and are now practised by all or nearly all the lowest races of man. Poetry, which may be considered as the offspring of song, is likewise so ancient, that many persons have felt astonished that it should have arisen during the earliest ages of which we have any record.

[35] 'Journal of Anthropological Society,' Oct. 1870, p. clv. See also the several later chapters in Sir John Lubbock's 'Prehistoric Times,' 2nd ed. 1869, which contain an admirable account of the habits of savages.

We see that the musical faculties, which are not wholly deficient in any race, are capable of prompt and high development, for Hottentots and Negroes have become excellent musicians, although in their native countries they rarely practise anything that we should consider music. Schweinfurth, however, was pleased with some of the simple melodies which he heard in the interior of Africa. But there is nothing anomalous in the musical faculties lying dormant in man: some species of birds which never naturally sing, can without much difficulty be taught to do so; thus a house-sparrow has learnt the song of a linnet. As these two species are closely allied, and belong to the order of Insessores, which includes nearly all the singing-birds in the world, it is possible that a progenitor of the sparrow may have been a songster. It is more remarkable that parrots, belonging to a group distinct from the Insessores, and having differently constructed vocal organs, can be taught not only to speak, but to pipe or whistle tunes invented by man, so that they must have some musical capacity. Nevertheless it would be very rash to assume that parrots are descended from some ancient form which was a songster. Many cases could be advanced of organs and instincts originally adapted for one purpose, having been utilized for some distinct purpose. [36] Hence the capacity for high musical development which the savage races of man possess, may be due either to the practice by our semi-human progenitors of some rude form of music, or simply to their having acquired the proper vocal organs for a different purpose. But in this latter case we must assume, as in the above instance of parrots, and as seems to occur with many animals, that they already possessed some sense of melody.

[36] Since this chapter was printed, I have seen a valuable article by Mr. Chauncey Wright ('North American Review,' Oct. 1870, page 293), who, in discussing the above subject, remarks, "There are many consequences of the ultimate laws or uniformities of nature, through which the acquisition of one useful power will bring with it many resulting advantages as well as limiting disadvantages, actual or possible, which the principle of utility may not have comprehended in its action." As I have attempted to show in an early chapter of this work, this principle has an important bearing on the acquisition by man of some of his mental characteristics.

Music arouses in us various emotions, but not the more terrible ones of horror, fear, rage, etc. It awakens the gentler feelings of tenderness and love, which readily pass into devotion. In the Chinese annals it is said, "Music hath the power of making heaven descend upon earth." It likewise stirs up in us the sense of triumph and the glorious ardour for war. These powerful and mingled feelings may well give rise to the sense of sublimity. We can concentrate, as Dr. Seemann observes, greater intensity of feeling in a single musical note than in pages of writing. It is probable that nearly the same emotions, but much weaker and far less complex, are felt by birds when the male pours forth his full volume of song, in rivalry with other males, to captivate the female. Love is still the commonest theme of our songs. As Herbert Spencer remarks, "music arouses dormant sentiments of which we had not conceived the possibility, and do not know the meaning; or, as Richter says, tells us of things we have not seen and shall not see." Conversely, when vivid emotions are felt and expressed by the orator, or even in common speech, musical cadences and rhythm are instinctively used. The negro in Africa when excited often bursts forth in song; "another will reply in song, whilst the company, as if touched by a musical wave, murmur a chorus in perfect unison." [37] Even monkeys express strong feelings in different tones—anger and impatience by low,—fear and pain by high notes. [38] The sensations and ideas thus excited in us by music, or expressed by the cadences of oratory, appear from their vagueness, yet depth, like mental reversions to the emotions and thoughts of a long-past age.

[37] Winwood Reade, 'The Martyrdom of Man,' 1872, p. 441, and 'African Sketch Book,' 1873, vol. ii. p. 313.
[38] Rengger, 'Säugethiere von Paraguay,' s. 49.

All these facts with respect to music and impassioned speech become intelligible to a certain extent, if we may assume that musical tones and rhythm were used by our half-human ancestors, during the season of courtship, when animals of all kinds are excited not only by love, but by the strong passions of jealousy, rivalry, and triumph. From the deeply-laid principle of inherited associations, musical tones in this case would be likely to call up vaguely and indefinitely the strong emotions of a long-past age. As we have every reason to suppose that articulate speech is one of the latest, as it certainly is the highest, of the arts acquired by man, and as the instinctive power of producing musical notes and rhythms is developed low down in the animal series, it would be altogether opposed to the principle of evolution, if we were to admit that man's musical capacity has been developed from the tones used in impassioned speech. We must suppose that the rhythms and cadences of oratory are derived from previously developed musical powers. [39] We can thus understand how it is that music, dancing, song, and poetry are such very ancient arts. We may go even further than this, and, as remarked in a former chapter, believe that musical sounds afforded one of the bases for the development of language. [40]

[39] See the very interesting discussion on the 'Origin and Function of Music,' by Mr. Herbert Spencer, in his collected 'Essays,' 1858, p. 359. Mr. Spencer comes to an exactly opposite conclusion to that at which I have arrived. He concludes, as did Diderot formerly, that the cadences used in emotional speech afford the foundation from which music has been developed; whilst I conclude that musical notes and rhythm were first acquired by the male or female progenitors of mankind for the sake of charming the opposite sex. Thus musical tones became firmly associated with some of the strongest passions an animal is capable of feeling, and are consequently used instinctively, or through association when strong emotions are expressed in speech. Mr. Spencer does not offer any satisfactory explanation, nor can I, why high or deep notes should be expressive, both with man and the lower animals, of certain emotions. Mr. Spencer gives also an interesting discussion on the relations between poetry, recitative and song.

[40] I find in Lord Monboddo's 'Origin of Language,' vol. i. 1774, p. 469, that Dr. Blacklock likewise thought "that the first language among men was music, and that before our ideas were expressed by articulate sounds, they were communicated by tones varied according to different degrees of gravity and acuteness."

As the males of several quadrumana animals have their vocal organs much more developed than in the females, and as a gibbon, one of the anthropomorphous apes, pours forth a whole octave of musical notes and may be said to sing, it appears probable that the progenitors of man, either the males or females or both sexes, before acquiring the power of expressing their mutual love in articulate language, endeavoured to charm each other with musical notes and rhythm. So little is known about the use of the voice by the Quadrumana during the season of love, that we have no means of judging whether the habit of singing was first acquired by our male or female ancestors. Women are generally thought to possess sweeter voices than men, and as far as this serves as any guide, we may infer that they first acquired musical powers in order to attract the other sex. [41] But if so, this must have occurred long ago, before our ancestors had become sufficiently human to treat and value their women merely as useful slaves. The impassioned orator, bard, or musician, when with his varied tones and cadences he excites the strongest emotions in his hearers, little suspects that he uses the same means by which his half-human ancestors long ago aroused each other's ardent passions, during their courtship and rivalry.

[41] See an interesting discussion on this subject by Häckel, 'Generelle Morph.' B. ii. 1866, s. 246.

THE INFLUENCE OF BEAUTY IN
DETERMINING THE MARRIAGES OF MANKIND.

In civilized life man is largely, but by no means exclusively, influenced in the choice of his wife by external appearance; but we are chiefly concerned with primeval times, and our only means of forming a judgment on this subject is to study the habits of existing semi-civilized and savage nations. If it can be shown that the men of different races prefer women having various characteristics, or conversely with the women, we have then to enquire whether such choice, continued during many generations, would produce any sensible effect on the race, either on one sex or both according to the form of inheritance which has prevailed.

It will be well first to show in some detail that savages pay the greatest attention to their personal appearance. [42] That they have a passion for ornament is notorious; and an English philosopher goes so far as to maintain, that clothes were first made for ornament and not for warmth. As Professor Waitz remarks, "however poor and miserable man is, he finds a pleasure in adorning himself." The extravagance of the naked Indians of South America in decorating themselves is shown "by a man of large stature gaining with difficulty enough by the labor of a fortnight to procure in exchange the *chica* necessary to paint himself red." [43] The ancient barbarians of Europe during the Reindeer period brought to their caves any brilliant or singular objects which they happened to find. Savages at the present day everywhere deck themselves with plumes, necklaces, armlets, ear-rings, etc. They paint themselves in the most diversified manner. "If painted nations," as Humboldt observes, "had been examined with the same attention as clothed nations, it would have been perceived that the most fertile imagination and the most mutable caprice have created the fashions of painting, as well as those of garments."

[42] A full and excellent account of the manner in which savages in all parts of the world ornament themselves, is given by the Italian traveller, Professor Mantegazza, 'Rio de la Plata, Viaggi e Studi,' 1867, pp. 525-545; all the following statements, when other references are not given, are taken from this work. See, also, Waitz, 'Introduction to Anthropology,' Eng. translat. vol. i. 1863, p. 275, *et passim*. Lawrence also gives very full details in his 'Lectures on Physiology,' 1822. Since this chapter was written Sir J. Lubbock has published his 'Origin of Civilization,' 1870, in which there is an interesting chapter on the present subject, and from which (pp. 42, 48) I have taken some facts about savages dyeing their teeth and hair, and piercing their teeth.

[43] Humboldt, 'Personal Narrative,' Eng. translat. vol. iv. p. 515; on the imagination shown in painting the body, p. 522; on modifying the form of the calf of the leg, p. 466.

In one part of Africa the eyelids are colored black; in another the nails are colored yellow or purple. In many places the hair is dyed of various tints. In different countries the teeth are stained black, red, blue, etc., and in the Malay Archipelago it is thought shameful to have white teeth "like those of a dog." Not one great country can be named, from the polar regions in the north to New Zealand in the south, in which the aborigines do not tattoo themselves. This practice was followed by the Jews of old, and by the ancient Britons. In Africa some of the natives tattoo themselves, but it is a much more common practice to raise protuberances by rubbing salt into incisions made in various parts of the body; and these are considered by the inhabitants of Kordofan and Darfur "to be great personal attractions." In the Arab countries no beauty can be perfect until the cheeks "or temples have been gashed." [44] In South America, as Humboldt remarks, "a mother would be accused of culpable indifference towards her children, if she did not employ artificial means to shape the calf of the leg after the fashion of the country." In the Old and New Worlds the shape of the skull was formerly modified during infancy in the most extraordinary manner, as is still the case in many places, and such deformities are considered ornamental. For instance, the savages of Colombia [45] deem a much flattened head "an essential point of beauty."

[44] 'The Nile Tributaries,' 1867; 'The Albert N'yanza,' 1866, vol. i. p. 218.
[45] Quoted by Prichard, 'Physical History of Mankind,' 4th ed. vol. i. 1851, p. 321.

The hair is treated with especial care in various countries; it is allowed to grow to full length, so as to reach to the ground, or is combed into "a compact frizzled mop, which is the Papuan's pride and glory." [46] In northern Africa "a man requires a period of from eight to ten years to perfect his coiffure." With other nations the head is shaved, and in parts of South America and Africa even the eyebrows and eyelashes are eradicated. The natives of the Upper Nile knock out the four front teeth, saying that they do not wish to resemble brutes. Further south, the Batokas knock out only the two upper incisors, which, as Livingstone [47] remarks, gives the face a hideous appearance, owing to the prominence of the lower jaw; but these people think the presence of the incisors most unsightly, and on beholding some Europeans, cried out, "Look at the great teeth!" The chief Sebituani tried in vain to alter this fashion. In various parts of Africa and in the Malay Archipelago the natives file the incisors into points like those of a saw, or pierce them with holes, into which they insert studs.

[46] On the Papuans, Wallace, 'The Malay Archipelago,' vol. ii. p. 445. On the coiffure of the Africans, Sir S. Baker, 'The Albert N'yanza,' vol. i. p. 210.
[47] 'Travels,' p. 533.

As the face with us is chiefly admired for its beauty, so with savages it is the chief seat of mutilation. In all quarters of the world the septum, and more rarely the wings of the nose are pierced; rings, sticks, feathers, and other ornaments being inserted into the holes. The ears are everywhere pierced and similarly ornamented, and with the Botocudos and Lenguas of South America the hole is gradually so much enlarged that the lower edge touches the shoulder. In North and South America and in Africa either the upper or lower lip is pierced; and with the Botocudos the hole in the lower lip is so large that a disc of wood, four inches in diameter, is placed in it. Mantegazza gives a curious account of the shame felt by a South American native, and of the ridicule which he excited, when he sold his *tembeta*,—the large colored piece of wood which is passed through the hole. In Central Africa the women perforate the lower lip and wear a crystal, which, from the movement of the tongue, has "a wriggling motion, indescribably ludicrous during conversation." The wife of the chief of Latooka told Sir S. Baker [48] that Lady Baker "would be much improved if she would extract her four front teeth from the lower jaw, and wear the long pointed polished crystal in her under lip." Further south with the Makalolo, the upper lip is perforated, and a large metal and bamboo ring, called a *pelelé*, is worn in the hole. "This caused the lip in one case to project two inches beyond the tip of the nose; and when the lady smiled, the contraction of the muscles elevated it over the eyes. 'Why do the women wear these things?' the venerable chief, Chinsurdi, was asked. Evidently surprised at such a stupid question, he replied, 'For beauty! They are the only beautiful things women have; men have beards, women have none. What kind of a person would she be without the pelelé? She would not be a woman at all with a mouth like a man, but no beard.'" [49]

[48] 'The Albert N'yanza,' 1866, vol. i. p. 217.
[49] Livingstone, 'British Association,' 1860; report given in the 'Athenaeum,' July 7, 1860, p. 29.

Hardly any part of the body, which can be unnaturally modified, has escaped. The amount of suffering thus caused must have been extreme, for many of the operations require several years for their completion, so that the idea of their necessity must be imperative. The motives are various; the men paint their bodies to make themselves appear terrible in battle; certain mutilations are connected with religious rites, or they mark the age of puberty, or the rank of the man, or they serve to distinguish the tribes. Amongst savages the same fashions prevail for long periods, [50] and thus mutilations, from whatever cause first made, soon come to be valued as distinctive marks. But self-adornment, vanity, and the admiration of others, seem to be the commonest motives. In regard to tattooing, I was told by the missionaries in New Zealand that when they tried to persuade some girls to give up the practice, they answered, "We must just have a few lines on our lips; else when we grow old we shall be so very ugly." With the men of New Zealand, a most capable judge [51] says, "to have fine tattooed faces was the great ambition of the young, both to render themselves attractive to the ladies, and conspicuous in war." A star tattooed on the forehead and a spot on the chin are thought by the women in one part of Africa to be irresistible attractions. [52] In most, but not all parts of the world, the men are more ornamented than the women, and often in a different manner; sometimes, though rarely, the women are hardly at all ornamented. As the women are made by savages to perform the greatest share of the work, and as they are not allowed to eat the best kinds of food, so it accords with the characteristic selfishness of man that they should not be allowed to obtain, or use the finest ornaments. Lastly, it is a remarkable fact, as proved by the foregoing quotations, that the same fashions in modifying the shape of the head, in ornamenting the hair, in painting, tattooing, in perforating the nose, lips, or ears, in removing or filing the teeth, etc., now prevail, and have long prevailed, in the most distant quarters of the world. It is extremely improbable that these practices, followed by so many distinct nations, should be due to tradition from any common source. They indicate the close similarity of the mind of man, to whatever race he may belong, just as do the almost universal habits of dancing, masquerading, and making rude pictures.

[50] Sir S. Baker (ibid. vol. i. p. 210) speaking of the natives of Central Africa says, "every tribe has a distinct and unchanging fashion for dressing the hair." See Agassiz ('Journey in Brazil,' 1868, p. 318) on invariability of the tattooing of Amazonian Indians.
[51] Rev. R. Taylor, 'New Zealand and its Inhabitants,' 1855, p. 152.
[52] Mantegazza, 'Viaggi e Studi,' p. 542.

Having made these preliminary remarks on the admiration felt by savages for various ornaments, and for deformities most unsightly in our eyes, let us see how far the men are attracted by the appearance of their women, and what are their ideas of beauty. I have heard it maintained that savages are quite indifferent about the beauty of their women, valuing them solely as slaves; it may therefore be well to observe that this conclusion does not at all agree with the care which the women take in ornamenting themselves, or with their vanity. Burchell [53] gives an amusing account of a Bush-woman who used as much grease, red ochre, and shining powder "as would have ruined any but a very rich husband." She displayed also "much vanity and too evident a consciousness of her superiority." Mr. Winwood Reade informs me that the negroes of the West Coast often discuss the beauty of their women. Some competent observers have attributed the fearfully common practice of infanticide partly to the desire felt by the women to retain their good looks. [54] In several regions the women wear charms and use love-philters to gain the affections of the men; and Mr. Brown enumerates four plants used for this purpose by the women of North-Western America. [55]

[53] 'Travels in South Africa,' 1824, vol. i. p. 414.
[54] See, for references, Gerland, 'Ueber das Aussterben der Naturvölker,' 1868, ss. 51, 53, 55; also Azara, 'Voyages,' etc., tom. ii. p. 116.
[55] On the vegetable productions used by the North-Western American Indians, see 'Pharmaceutical Journal,' vol. x.

Hearne, [56] an excellent observer, who lived many years with the American Indians, says, in speaking of the women, "Ask a Northern Indian what is beauty, and he will answer, a broad flat face, small eyes, high cheek-bones, three or four broad black lines across each cheek, a low forehead, a large broad chin, a clumsy hook nose, a tawny hide, and breasts hanging down to the belt." Pallas, who visited the northern parts of the Chinese empire, says, "those women are preferred who have the Mandschú form; that is to say, a broad face, high cheek-bones, very broad noses, and enormous ears"[57]; and Vogt remarks that the obliquity of the eye, which is proper to the Chinese and Japanese, is exaggerated in their pictures for the purpose, as it "seems, of exhibiting its beauty, as contrasted with the eye of the red-haired barbarians." It is well known, as Huc repeatedly remarks, that the Chinese of the interior think Europeans hideous, with their white skins and prominent noses. The nose is far from being too prominent, according to our ideas, in the natives of Ceylon; yet "the Chinese in the seventh century, accustomed to the flat features of the Mongol races, were surprised at the prominent noses of the Cingalese; and Thsang described them as having 'the beak of a bird, with the body of a man.'"

[56] 'A Journey from Prince of Wales Fort,' 8vo. ed. 1796, p. 89.
[57] Quoted by Prichard, 'Physical History of Mankind,' 3rd ed. vol. iv. 1844, p. 519; Vogt, 'Lectures on Man,' Eng. translat. p. 129. On the opinion of the Chinese on the Cingalese, E. Tennent, 'Ceylon,' 1859, vol. ii. p. 107.

Finlayson, after minutely describing the people of Cochin China, says that their rounded heads and faces are their chief characteristics; and, he adds, "the roundness of the whole countenance is more striking in the women, who are reckoned beautiful in proportion as they display this form of face." The Siamese have small noses with divergent nostrils, a wide mouth, rather thick lips, a remarkably large face, with very high and broad cheek-bones. It is, therefore, not wonderful that "beauty, according to our notion, is a stranger to them. Yet they consider their own females to be much more beautiful than those of Europe." [58]

[58] Prichard, as taken from Crawfurd and Finlayson, 'Phys. Hist. of Mankind,' vol. iv. pp. 534, 535.

It is well known that with many Hottentot women the posterior part of the body projects in a wonderful manner; they are steatopygous; and Sir Andrew Smith is certain that this peculiarity is greatly admired by the men. [59] He once saw a woman who was considered a beauty, and she was so immensely developed behind, that when seated on level ground she could not rise, and had to push herself along until she came to a slope. Some of the women in various negro tribes have the same peculiarity; and, according to Burton, the Somal men are said to choose their wives by ranging them in a line, and by picking her out who projects farthest *a tergo*. Nothing can be more hateful to a negro than the opposite form." [60]

[59] Idem illustrissimus viator dixit mihi præcinctorium vel tabulam fœminæ, quod nobis teterrimum est, quondam permagno æstimari ab hominibus in hâc gente. Nunc res mutata est, et censent talem conformationem minime optandam esse.
[60] The 'Anthropological Review,' November 1864, p. 237. For additional references, see Waitz, 'Introduction to Anthropology,' Eng. translat., 1863, vol. i. p. 105.

With respect to color, the negroes rallied Mungo Park on the whiteness of his skin and the prominence of his nose, both of which they considered as "unsightly and unnatural conformations." He in return praised the glossy jet of their skins and the lovely depression of their noses; this they said was "honeymouth," nevertheless they gave him food. The African Moors, also, "knitted their brows and seemed to shudder" at the whiteness of his skin. On the eastern coast, the negro boys when they saw Burton, cried out, "Look at the white man; does he not look like a white ape?" On the western coast, as Mr. Winwood Reade informs me, the negroes admire a very black skin more than one of a lighter tint. But their horror of whiteness may be attributed, according to this same traveller, partly to the belief held by most negroes that demons and spirits are white, and partly to their thinking it a sign of ill-health.

The Banyai of the more southern part of the continent are negroes, but "a great many of them are of a light coffee-and-milk color, and, indeed, this color is considered handsome throughout the whole country"; so that here we have a different standard of taste. With the Kaffirs, who differ much from negroes, "the skin, except among the tribes near Delagoa Bay, is not usually black, the prevailing color being a mixture of black and red, the most common shade being chocolate. Dark complexions, as being most common, are naturally held in the highest esteem. To be told that he is light-colored, or like a white man, would be deemed a very poor compliment by a Kaffir. I have heard of one unfortunate man who was so very fair that no girl would marry him." One of the titles of the Zulu king is, "You who are black." [61] Mr. Galton, in speaking to me about the natives of S. Africa, remarked that their ideas of beauty seem very different from ours; for in one tribe two slim, slight, and pretty girls were not admired by the natives.

[61] Mungo Park's 'Travels in Africa,' 4to. 1816, pp. 53, 131. Burton's statement is quoted

by Schaaffhausen, 'Archiv. für Anthropolog.' 1866, s. 163. On the Banyai, Livingstone, 'Travels,' p. 64. On the Kaffirs, the Rev. J. Shooter, 'The Kafirs of Natal and the Zulu Country,' 1857, p. 1.

Turning to other quarters of the world; in Java, a yellow, not a white girl, is considered, according to Madame Pfeiffer, a beauty. A man of Cochin China "spoke with contempt of the wife of the English Ambassador, that she had white teeth like a dog, and a rosy color like that of potato-flowers." We have seen that the Chinese dislike our white skin, and that the N. Americans admire "a tawny hide." In S. America, the Yuracaras, who inhabit the wooded, damp slopes of the eastern Cordillera, are remarkably pale-colored, as their name in their own language expresses; nevertheless they consider European women as very inferior to their own. [62]

[62] For the Javans and Cochin-Chinese, see Waitz, 'Introduct. to Anthropology,' Eng. translat. vol. i. p. 305. On the Yuracaras, A. d'Orbigny, as quoted in Prichard, 'Physical History of Mankind,' vol. v. 3rd ed. p. 476.

In several of the tribes of North America the hair on the head grows to a wonderful length; and Catlin gives a curious proof how much this is esteemed, for the chief of the Crows was elected to this office from having the longest hair of any man in the tribe, namely ten feet and seven inches. The Aymaras and Quichuas of S. America, likewise have very long hair; and this, as Mr. D. Forbes informs me, is so much valued as a beauty, that cutting it off was the severest punishment which he could inflict on them. In both the Northern and Southern halves of the continent the natives sometimes increase the apparent length of their hair by weaving into it fibrous substances. Although the hair on the head is thus cherished, that on the face is considered by the North American Indians "as very vulgar," and every hair is carefully eradicated. This practice prevails throughout the American continent from Vancouver's Island in the north to Tierra del Fuego in the south. When York Minster, a Fuegian on board the "Beagle," was taken back to his country, the natives told him be ought to pull out the few short hairs on his face. They also threatened a young missionary, who was left for a time with them, to strip him naked, and pluck the hair from his face and body, yet he was far from being a hairy man. This fashion is carried so far that the Indians of Paraguay eradicate their eyebrows and eyelashes, saying that they do not wish to be like horses. [63]

[63] 'North American Indians,' by G. Catlin, 3rd ed., 1842, vol. i. p. 49; vol. ii, p. 227. On the natives of Vancouver's Island, see Sproat, 'Scenes and Studies of Savage Life,' 1868, p. 25. On the Indians of Paraguay, Azara, 'Voyages,' tom. ii. p. 105.

It is remarkable that throughout the world the races which are almost completely destitute of a beard dislike hairs on the face and body, and take pains to eradicate them. The Kalmucks are beardless, and they are well known, like the Americans, to pluck out all straggling hairs; and so it is with the Polynesians, some of the Malays, and the Siamese. Mr. Veitch states that the Japanese ladies "all objected to our whiskers, considering them very ugly, and told us to cut them off, and be like Japanese men." The New Zealanders have short, curled beards; yet they formerly plucked out the hairs on the face. They had a saying that "there is no woman for a hairy man;" but it would appear that the fashion has changed in New Zealand, perhaps owing to the presence of

Europeans, and I am assured that beards are now admired by the Maories. [64]

[64] On the Siamese, Prichard, ibid. vol. iv. p. 533. On the Japanese, Veitch in 'Gardeners' Chronicle,' 1860, p. 1104. On the New Zealanders, Mantegazza, 'Viaggi e Studi,' 1867, p. 526. For the other nations mentioned, see references in Lawrence, 'Lectures on Physiology,' etc., 1822, p. 272.

On the other hand, bearded races admire and greatly value their beards; among the Anglo-Saxons every part of the body had a recognized value; "the loss of the beard being estimated at twenty shillings, while the breaking of a thigh was fixed at only twelve." [65] In the East men swear solemnly by their beards. We have seen that Chinsurdi, the chief of the Makalolo in Africa, thought that beards were a great ornament. In the Pacific the Fijian's beard is "profuse and bushy, and is his greatest pride"; whilst the inhabitants of the adjacent archipelagoes of Tonga and Samoa are "beardless, and abhor a rough chin." In one island alone of the Ellice group "the men are heavily bearded, and not a little proud thereof." [66]

[65] Lubbock, 'Origin of Civilization,' 1870, p. 321.
[66] Dr. Barnard Davis quotes Mr. Prichard and others for these facts in regard to the Polynesians, in 'Anthropolog. Review,' April 1870, pp. 185, 191.

We thus see how widely the different races of man differ in their taste for the beautiful. In every nation sufficiently advanced to have made effigies of their gods or of their deified rulers, the sculptors no doubt have endeavoured to express their highest ideal of beauty and grandeur. [67] Under this point of view it is well to compare in our mind the Jupiter or Apollo of the Greeks with the Egyptian or Assyrian statues; and these with the hideous bas-reliefs on the ruined buildings of Central America.

[67] Ch. Comte has remarks to this effect in his 'Traité de Législation,' 3rd ed. 1837, p. 136.

I have met with very few statements opposed to this conclusion. Mr. Winwood Reade, however, who has had ample opportunities for observation, not only with the negroes of the West Coast of Africa, but with those of the interior who have never associated with Europeans, is convinced that their ideas of beauty are *on the whole* the same as ours; and Dr. Rohlfs writes to me to the same effect with respect to Bornu and the countries inhabited by the Pullo tribes. Mr. Reade found that he agreed with the negroes in their estimation of the beauty of the native girls; and that their appreciation of the beauty of European women corresponded with ours. They admire long hair, and use artificial means to make it appear abundant; they admire also a beard, though themselves very scantily provided. Mr. Reade feels doubtful what kind of nose is most appreciated; a girl has been heard to say, "I do not want to marry him, he has got no nose"; and this shows that a very flat nose is not admired. We should, however, bear in mind that the depressed, broad noses and projecting jaws of the negroes of the West Coast are exceptional types with the inhabitants of Africa. Notwithstanding the foregoing statements, Mr. Reade admits that negroes "do not like the color of our skin; they look on blue eyes with aversion, and they think our noses too long and our lips too thin." He does not think it probable that negroes would ever prefer the most beautiful European woman, on the mere grounds of physical admiration, to a good-looking negress. [68]

[68] The 'African Sketch Book,' vol. ii. 1873, pp. 253, 394, 521. The Fuegians, as I have been informed by a missionary who long resided with them, consider European women as extremely beautiful; but from what we have seen of the judgment of the other aborigines of America, I cannot but think that this must be a mistake, unless indeed the statement refers to the few Fuegians who have lived for some time with Europeans, and who must consider us as superior beings. I should add that a most experienced observer, Capt. Burton, believes that a woman whom we consider beautiful is admired throughout the world. 'Anthropological Review,' March, 1864, p. 245.

The general truth of the principle, long ago insisted on by Humboldt [69], that man admires and often tries to exaggerate whatever characters nature may have given him, is shown in many ways. The practice of beardless races extirpating every trace of a beard, and often all the hairs on the body affords one illustration. The skull has been greatly modified during ancient and modern times by many nations; and there can be little doubt that this has been practised, especially in N. and S. America, in order to exaggerate some natural and admired peculiarity. Many American Indians are known to admire a head so extremely flattened as to appear to us idiotic. The natives on the north-western coast compress the head into a pointed cone; and it is their constant practice to gather the hair into a knot on the top of the head, for the sake, as Dr. Wilson remarks, "of increasing the apparent elevation of the favourite conoid form." The inhabitants of Arakhan admire a broad, smooth forehead, and in order to produce it, they fasten a plate of lead on the heads of the new-born children. On the other hand, "a broad, well-rounded occiput is considered a great beauty" by the natives of the Fiji Islands. [70]

[69] 'Personal Narrative,' Eng. translat. vol. iv. p. 518, and elsewhere. Mantegazza, in his 'Viaggi e Studi,' strongly insists on this same principle.
[70] On the skulls of the American tribes, see Nott and Gliddon, 'Types of Mankind,' 1854, p. 440; Prichard, 'Physical History of Mankind,' vol. i. 3rd ed. p. 321; on the natives of Arakhan, ibid. vol. iv. p. 537. Wilson, 'Physical Ethnology,' Smithsonian Institution, 1863, p. 288; on the Fijians, p. 290. Sir J. Lubbock ('Prehistoric Times,' 2nd ed. 1869, p. 506) gives an excellent resume on this subject.

As with the skull, so with the nose; the ancient Huns during the age of Attila were accustomed to flatten the noses of their infants with bandages, "for the sake of exaggerating a natural conformation." With the Tahitians, to be called *long-nose* is considered as an insult, and they compress the noses and foreheads of their children for the sake of beauty. The same holds with the Malays of Sumatra, the Hottentots, certain Negroes, and the natives of Brazil. [71] The Chinese have by nature unusually small feet [72]; and it is well known that the women of the upper classes distort their feet to make them still smaller. Lastly, Humboldt thinks that the American Indians prefer coloring their bodies with red paint in order to exaggerate their natural tint; and until recently European women added to their naturally bright colors by rouge and white cosmetics; but it may be doubted whether barbarous nations have generally had any such intention in painting themselves.

[71] On the Huns, Godron, 'De l'Espèce,' tom. ii. 1859, p. 300. On the Tahitians, Waitz, 'Anthropology,' Eng. translat. vol. i. p. 305. Marsden, quoted by Prichard, 'Phys. Hist. of Mankind,' 3rd edit. vol. v. p. 67. Lawrence, 'Lectures on Physiology,' p. 337.
[72] This fact was ascertained in the 'Reise der Novara: Anthropolog. Theil.' Dr. Weisbach, 1867, s. 265.

In the fashions of our own dress we see exactly the same principle and the same desire to carry every point to an extreme; we exhibit, also, the same spirit of emulation. But the fashions of savages are far more permanent than ours; and whenever their bodies are artificially modified, this is necessarily the case. The Arab women of the Upper Nile occupy about three days in dressing their hair; they never imitate other tribes, "but simply vie with each other in the superlativeness of their own style." Dr. Wilson, in speaking of the compressed skulls of various American races, adds, "such usages are among the least eradicable, and long survive the shock of revolutions that change dynasties and efface more important national peculiarities." [73] The same principle comes into play in the art of breeding; and we can thus understand, as I have elsewhere explained [74], the wonderful development of the many races of animals and plants, which have been kept merely for ornament. Fanciers always wish each character to be somewhat increased; they do not admire a medium standard; they certainly do not desire any great and abrupt change in the character of their breeds; they admire solely what they are accustomed to, but they ardently desire to see each characteristic feature a little more developed.

[73] 'Smithsonian Institution,' 1863, p. 289. On the fashions of Arab women, Sir S. Baker, 'The Nile Tributaries,' 1867, p. 121.
[74] The 'Variation of Animals and Plants under Domestication,' vol. i. p. 214; vol. ii. p. 240.

The senses of man and of the lower animals seem to be so constituted that brilliant colors and certain forms, as well as harmonious and rhythmical sounds, give pleasure and are called beautiful; but why this should be so we know not. It is certainly not true that there is in the mind of man any universal standard of beauty with respect to the human body. It is, however, possible that certain tastes may in the course of time become inherited, though there is no evidence in favour of this belief: and if so, each race would possess its own innate ideal standard of beauty. It has been argued [75] that ugliness consists in an approach to the structure of the lower animals, and no doubt this is partly true with the more civilized nations, in which intellect is highly appreciated; but this explanation will hardly apply to all forms of ugliness. The men of each race prefer what they are accustomed to; they cannot endure any great change; but they like variety, and admire each characteristic carried to a moderate extreme. [76] Men accustomed to a nearly oval face, to straight and regular features, and to bright colors, admire, as we Europeans know, these points when strongly developed. On the other hand, men accustomed to a broad face, with high cheek-bones, a depressed nose, and a black skin, admire these peculiarities when strongly marked. No doubt characters of all kinds may be too much developed for beauty. Hence a perfect beauty, which implies many characters modified in a particular manner, will be in every race a prodigy. As the great anatomist Bichat long ago said, if every one were cast in the same mould, there would be no such thing as beauty. If all our women were to become as beautiful as the Venus de' Medici, we should for a time be charmed; but we should soon wish for variety; and as soon as we had obtained variety, we should wish to see certain characters a little exaggerated beyond the then existing common standard.

[75] Schaaffhausen, 'Archiv. für Anthropologie,' 1866, s. 164.
[76] Mr. Bain has collected ('Mental and Moral Science,' 1868, pp. 304-314) about a dozen more or less different theories of the idea of beauty; but none is quite the same as that here given.

CHAPTER XX.

SECONDARY SEXUAL CHARACTERS OF MAN—*continued.*

On the effects of the continued selection of women according to a different standard of beauty in each race—On the causes which interfere with sexual selection in civilized and savage nations—Conditions favourable to sexual selection during primeval times—On the manner of action of sexual selection with mankind—On the women in savage tribes having some power to choose their husbands—Absence of hair on the body, and development of the beard—Color of the skin—Summary.

We have seen in the last chapter that with all barbarous races ornaments, dress, and external appearance are highly valued; and that the men judge of the beauty of their women by widely different standards. We must next inquire whether this preference and the consequent selection during many generations of those women, which appear to the men of each race the most attractive, has altered the character either of the females alone, or of both sexes. With mammals the general rule appears to be that characters of all kinds are inherited equally by the males and females; we might therefore expect that with mankind any characters gained by the females or by the males through sexual selection would commonly be transferred to the offspring of both sexes. If any change has thus been effected, it is almost certain that the different races would be differently modified, as each has its own standard of beauty.

With mankind, especially with savages, many causes interfere with the action of sexual selection as far as the bodily frame is concerned. Civilized men are largely attracted by the mental charms of women, by their wealth, and especially by their social position; for men rarely marry into a much lower rank. The men who succeed in obtaining the more beautiful women will not have a better chance of leaving a long line of descendants than other men with plainer wives, save the few who bequeath their fortunes according to primogeniture. With respect to the opposite form of selection, namely, of the more attractive men by the women, although in civilized nations women have free or almost free choice, which is not the case with barbarous races, yet their choice is largely influenced by the social position and wealth of the men; and the success of the latter in life depends much on their intellectual powers and energy, or on the fruits of these same powers in their forefathers. No excuse is needed for treating this subject in some detail; for, as the German philosopher Schopenhauer remarks, "the final aim of all love intrigues, be they comic or tragic, is really of more importance than all other ends in human life. What it all turns upon is nothing less than the composition of the next generation...It is not the weal or woe of any one individual, but that of the human race to come, which is here at stake." [1]

[1] 'Schopenhauer and Darwinism,' in 'Journal of Anthropology,' Jan. 1871, p. 323.

There is, however, reason to believe that in certain civilized and semi-civilized nations sexual selection has effected something in modifying the bodily frame of some of the members. Many persons are convinced, as it appears to me with justice, that our aristocracy, including under this term all wealthy families in which primogeniture has long prevailed, from having chosen during many generations from all classes the more beautiful women as their wives, have become handsomer, according to the European standard, than the middle classes; yet the middle classes are placed under equally favourable conditions of life for the perfect development of the body. Cook remarks that the superiority in personal appearance "which is observable in the erees or nobles in all the other islands (of the Pacific) is found in the Sandwich Islands"; but this may be chiefly due to their better food and manner of life.

The old traveller Chardin, in describing the Persians, says their "blood is now highly refined by frequent intermixtures with the Georgians and Circassians, two nations which surpass all the world in personal beauty. There is hardly a man of rank in Persia who is not born of a Georgian or Circassian mother." He adds that they inherit their beauty, "not from their ancestors, for without the above mixture, the men of rank in Persia, who are descendants of the Tartars, would be extremely ugly." [2] Here is a more curious case; the priestesses who attended the temple of Venus Erycina at San-Giuliano in Sicily, were selected for their beauty out of the whole of Greece; they were not vestal virgins, and Quatrefages, [3] who states the foregoing fact, says that the women of San-Giuliano are now famous as the most beautiful in the island, and are sought by artists as models. But it is obvious that the evidence in all the above cases is doubtful.

[2] These quotations are taken from Lawrence ('Lectures on Physiology,' etc., 1822, p. 393), who attributes the beauty of the upper classes in England to the men having long selected the more beautiful women.
[3] 'Anthropologie,' 'Revue des Cours Scientifiques,' Oct. 1868, p. 721.

The following case, though relating to savages, is well worth giving for its curiosity. Mr. Winwood Reade informs me that the Jollofs, a tribe of negroes on the west coast of Africa, "are remarkable for their uniformly fine appearance." A friend of his asked one of these men, "How is it that every one whom I meet is so fine looking, not only your men but your women?" The Jollof answered, "It is very easily explained: it has always been our custom to pick out our worst-looking slaves and to sell them." It need hardly be added that with all savages, female slaves serve as concubines. That this negro should have attributed, whether rightly or wrongly, the fine appearance of his tribe to the long-continued elimination of the ugly women is not so surprising as it may at first appear; for I have elsewhere shown [4] that negroes fully appreciate the importance of selection in the breeding of their domestic animals, and I could give from Mr. Reade additional evidence on this head.

[4] 'Variation of Animals and Plants under Domestication,' vol. i. p. 207.

THE CAUSES WHICH PREVENT OR CHECK THE
ACTION OF SEXUAL SELECTION WITH SAVAGES.

The chief causes are, first, so-called communal marriages or promiscuous intercourse; secondly, the consequences of female infanticide; thirdly, early betrothals; and lastly, the low estimation in which women are held, as mere slaves. These four points must be considered in some detail.

It is obvious that as long as the pairing of man, or of any other animal, is left to mere chance, with no choice exerted by either sex, there can be no sexual selection; and no effect will be produced on the offspring by certain individuals having had an advantage over others in their courtship. Now it is asserted that there exist at the present day tribes which practise what Sir J. Lubbock by courtesy calls communal marriages; that is, all the men and women in the tribe are husbands and wives to one another. The licentiousness of many savages is no doubt astonishing, but it seems to me that more evidence is requisite, before we fully admit that their intercourse is in any case promiscuous. Nevertheless all those who have most closely studied the subject, [5] and whose judgment is worth much more than mine, believe that communal marriage (this expression being variously guarded) was the original and universal form throughout the world, including therein the intermarriage of brothers and sisters. The late Sir A. Smith, who had travelled widely in S. Africa, and knew much about the habits of savages there and elsewhere, expressed to me the strongest opinion that no race exists in which woman is considered as the property of the community. I believe that his judgment was largely determined by what is implied by the term marriage. Throughout the following discussion I use the term in the same sense as when naturalists speak of animals as monogamous, meaning thereby that the male is accepted by or chooses a single female, and lives with her either during the breeding-season or for the whole year, keeping possession of her by the law of might; or, as when they speak of a polygamous species, meaning that the male lives with several females. This kind of marriage is all that concerns us here, as it suffices for the work of sexual selection. But I know that some of the writers above referred to imply by the term marriage a recognized right protected by the tribe.

[5] Sir J. Lubbock, 'The Origin of Civilization,' 1870, chap. iii. especially pp. 60-67. Mr. M'Lennan, in his extremely valuable work on 'Primitive Marriage,' 1865, p. 163, speaks of the union of the sexes "in the earliest times as loose, transitory, and in some degree promiscuous." Mr. M'Lennan and Sir J. Lubbock have collected much evidence on the extreme licentiousness of savages at the present time. Mr. L.H. Morgan, in his interesting memoir of the classificatory system of relationship. ('Proceedings of the American Academy of Sciences,' vol. vii. Feb. 1868, p. 475), concludes that polygamy and all forms of marriage during primeval times were essentially unknown. It appears also, from Sir J. Lubbock's work, that Bachofen likewise believes that communal intercourse originally prevailed.

The indirect evidence in favour of the belief of the former prevalence of communal marriages is strong, and rests chiefly on the terms of relationship which are employed between the members of the same tribe, implying a connection with the tribe, and not with either parent. But the subject is too large and complex for even an abstract to be here given, and I will confine myself to a few remarks. It is evident in the case of such marriages, or where the marriage tie is very loose, that the relationship of the child to its father cannot be known. But it seems almost incredible that the relationship of the child to its mother should ever be completely ignored, especially as the women in most savage tribes nurse their infants for a long time. Accordingly, in many cases the lines of descent are traced through the mother alone, to the exclusion of the father. But in other cases the terms employed express a connection with the tribe alone, to the exclusion even of the mother. It seems possible that the connection between the related members of the same barbarous tribe, exposed to all sorts of danger, might be so much more important, owing to the need of mutual protection and aid, than that between the mother and her child, as to lead to the sole use of terms expressive of the former relationships; but Mr. Morgan is convinced that this view is by no means sufficient.

The terms of relationship used in different parts of the world may be divided, according to the author just quoted, into two great classes, the classificatory and descriptive, the latter being employed by us. It is the classificatory system which so strongly leads to the belief that communal and other extremely loose forms of marriage were originally universal. But as far as I can see, there is no necessity on this ground for believing in absolutely promiscuous intercourse; and I am glad to find that this is Sir J. Lubbock's view. Men and women, like many of the lower animals, might formerly have entered into strict though temporary unions for each birth, and in this case nearly as much confusion would have arisen in the terms of relationship as in the case of promiscuous intercourse. As far as sexual selection is concerned, all that is required is that choice should be exerted before the parents unite, and it signifies little whether the unions last for life or only for a season.

Besides the evidence derived from the terms of relationship, other lines of reasoning indicate the former wide prevalence of communal marriage. Sir J. Lubbock accounts for the strange and widely-extended habit of exogamy—that is, the men of one tribe taking wives from a distinct tribe,—by communism having been the original form of intercourse; so that a man never obtained a wife for himself unless he captured her from a neighbouring and hostile tribe, and then she would naturally have become his sole and valuable property. Thus the practice of capturing wives might have arisen; and from the honour so gained it might ultimately have become the universal habit. According to Sir J. Lubbock [6], we can also thus understand "the necessity of expiation for marriage as an infringement of tribal rites, since according to old ideas, a man had no right to appropriate to himself that which belonged to the whole tribe." Sir J. Lubbock further gives a curious body of facts showing that in old times high honour was bestowed on women who were utterly licentious; and this, as he explains, is intelligible, if we admit that promiscuous intercourse was the aboriginal, and therefore long revered custom of the tribe. [7]

[6] 'Address to British Association On the Social and Religious Condition of the Lower Races of Man,' 1870, p. 20.
[7] 'Origin of Civilization,' 1870, p. 86. In the several works above quoted, there will be found copious evidence on relationship through the females alone, or with the tribe alone.

Although the manner of development of the marriage tie is an obscure subject, as we may infer from the divergent opinions on several points between the three authors who have studied it most closely, namely, Mr. Morgan, Mr. M'Lennan, and Sir J. Lubbock, yet from the foregoing and several other lines of evidence it seems probable [8] that the habit of marriage, in any strict sense of the word, has been gradually developed; and that almost promiscuous or very loose intercourse was once extremely common throughout the world. Nevertheless, from the strength of the feeling of jealousy all through the animal kingdom, as well as from the analogy of the lower animals, more particularly of those which come nearest to man, I cannot believe that absolutely promiscuous intercourse prevailed in times past, shortly before man attained to his present rank in the zoological scale. Man, as I have attempted to show, is certainly descended from some ape-like creature. With the existing Quadrumana, as far as their habits are known, the males of some species are monogamous, but live during only a part of the year with the females: of this the orang seems to afford an instance. Several kinds, for example some of the Indian and American monkeys, are strictly monogamous, and associate all the year round with their wives. Others are polygamous, for example the gorilla and several American species, and each family lives separate. Even when this occurs, the families inhabiting the same district are probably somewhat social; the chimpanzee, for instance, is occasionally met with in large bands. Again, other species are polygamous, but several males, each with his own females, live associated in a body, as with several species of baboons. [9] We may indeed conclude from what we know of the jealousy of all male quadrupeds, armed, as many of them are, with special weapons for battling with their rivals, that promiscuous intercourse in a state of nature is extremely improbable. The pairing may not last for life, but only for each birth; yet if the males which are the strongest and best able to defend or otherwise assist their females and young, were to select the more attractive females, this would suffice for sexual selection.

[8] Mr. C. Staniland Wake argues strongly ('Anthropologia,' March, 1874, p. 197) against the views held by these three writers on the former prevalence of almost promiscuous intercourse; and he thinks that the classificatory system of relationship can be otherwise explained.
[9] Brehm ('Thierleben,' B. i. p. 77) says *Cynocephalus hamadryas* lives in great troops containing twice as many adult females as adult males. See Rengger on American polygamous species, and Owen ('Anatomy of Vertebrates,' vol. iii. p. 746) on American monogamous species. Other references might be added.

Therefore, looking far enough back in the stream of time, and judging from the social habits of man as he now exists, the most probable view is that he aboriginally lived in small communities, each with a single wife, or if powerful with several, whom he jealously guarded against all other men. Or he may not have been a social animal, and yet have lived with several wives, like the gorilla; for all the natives "agree that but one adult male is seen in a band; when the young male grows up, a contest takes place for mastery, and the strongest, by killing and driving out the others, establishes himself as the head of the community." [10] The younger males, being thus expelled and wandering about, would, when at last successful in finding a partner, prevent too close interbreeding within the limits of the same family.

[10] Dr. Savage, in 'Boston Journal of Natural History,' vol. v. 1845-47, p. 423.

Although savages are now extremely licentious, and although communal marriages may formerly have largely prevailed, yet many tribes practise some form of marriage, but of a far more lax nature than that of civilized nations. Polygamy, as just stated, is almost universally followed by the leading men in every tribe. Nevertheless there are tribes, standing almost at the bottom of the scale, which are strictly monogamous. This is the case with the Veddahs of Ceylon: they have a saying, according to Sir J. Lubbock, [11] "that death alone can separate husband and wife." An intelligent Kandyan chief, of course a polygamist, "was perfectly scandalized at the utter barbarism of living with only one wife, and never parting until separated by death." It was, he said, "just like the Wanderoo monkeys." Whether savages who now enter into some form of marriage, either polygamous or monogamous, have retained this habit from primeval times, or whether they have returned to some form of marriage, after passing through a stage of promiscuous intercourse, I will not pretend to conjecture.

[11] 'Prehistoric Times,' 1869, p. 424.

INFANTICIDE.

This practice is now very common throughout the world, and there is reason to believe that it prevailed much more extensively during former times. [12] Barbarians find it difficult to support themselves and their children, and it is a simple plan to kill their infants. In South America some tribes, according to Azara, formerly destroyed so many infants of both sexes that they were on the point of extinction. In the Polynesian Islands women have been known to kill from four or five, to even ten of their children; and Ellis could not find a single woman who had not killed at least one. In a village on the eastern frontier of India Colonel MacCulloch found not a single female child. Wherever infanticide [13] prevails the struggle for existence will be in so far less severe, and all the members of the tribe will have an almost equally good chance of rearing their few surviving children. In most cases a larger number of female than of male infants are destroyed, for it is obvious that the latter are of more value to the tribe, as they will, when grown up, aid in defending it, and can support themselves. But the trouble experienced by the women in rearing children, their consequent loss of beauty, the higher estimation set on them when few, and their happier fate, are assigned by the women themselves, and by various observers, as additional motives for infanticide.

[12] Mr. M'Lennan, 'Primitive Marriage,' 1865. See especially on exogamy and infanticide, pp. 130, 138, 165.
[13] Dr. Gerland ('Ueber das Aussterben der Naturvölker,' 1868) has collected much information on infanticide, see especially ss. 27, 51, 54. Azara ('Voyages,' etc., tom. ii. pp. 94, 116) enters in detail on the motives. See also M'Lennan (ibid. p. 139) for cases in India. In the former reprints of the 2nd edition of this book an incorrect quotation from Sir G. Grey was unfortunately given in the above passage and has now been removed from the text.

When, owing to female infanticide, the women of a tribe were few, the habit of capturing wives from neighbouring tribes would naturally arise. Sir J. Lubbock, however, as we have seen, attributes the practice in chief part to the former existence of communal marriage, and to the men having consequently captured women from other tribes to hold as their sole property. Additional causes might be assigned, such as the communities being very small, in which case, marriageable women would often be deficient. That the habit was most extensively practised during former times, even by the ancestors of civilized nations, is clearly shown by the preservation of many curious customs and ceremonies, of which Mr. M'Lennan has given an interesting account. In our own marriages the "best man" seems originally to have been the chief abettor of the bridegroom in the act of capture. Now as long as men habitually procured their wives through violence and craft, they would have been glad to seize on any woman, and would not have selected the more attractive ones. But as soon as the practice of procuring wives from a distinct tribe was effected through barter, as now occurs in many places, the more attractive women would generally have been purchased. The incessant crossing, however, between tribe and tribe, which necessarily follows from any form of this habit, would tend to keep all the people inhabiting the same country nearly uniform in character; and this would interfere with the power of sexual selection in differentiating the tribes.

The scarcity of women, consequent on female infanticide, leads, also, to another practice, that of polyandry, still common in several parts of the world, and which formerly, as Mr. M'Lennan believes, prevailed almost universally: but this latter conclusion is doubted by Mr. Morgan and Sir J. Lubbock. [14] Whenever two or more men are compelled to marry one woman, it is certain that all the women of the tribe will get married, and there will be no selection by the men of the more attractive women. But under these circumstances the women no doubt will have the power of choice, and will prefer the more attractive men. Azara, for instance, describes how carefully a Guana woman bargains for all sorts of privileges, before accepting some one or more husbands; and the men in consequence take unusual care of their personal appearance. So amongst the Todas of India, who practise polyandry, the girls can accept or refuse any man. [15] A very ugly man in these cases would perhaps altogether fail in getting a wife, or get one later in life; but the handsomer men, although more successful in obtaining wives, would not, as far as we can see, leave more offspring to inherit their beauty than the less handsome husbands of the same women.

[14] 'Primitive Marriage,' p. 208; Sir J. Lubbock, 'Origin of Civilization,' p. 100. See also Mr. Morgan, loc. cit., on the former prevalence of polyandry.
[15] Azara, 'Voyages,' etc., tom. ii. pp. 92-95; Colonel Marshall, 'Amongst the Todas,' p. 212.

EARLY BETROTHALS AND SLAVERY OF WOMEN.

With many savages it is the custom to betroth the females whilst mere infants; and this would effectually prevent preference being exerted on either side according to personal appearance. But it would not prevent the more attractive women from being afterwards stolen or taken by force from their husbands by the more powerful men; and this often happens in Australia, America, and elsewhere. The same consequences with reference to sexual selection would to a certain extent follow, when women are valued almost solely as slaves or beasts of burden, as is the case with many savages. The men, however, at all times would prefer the handsomest slaves according to their standard of beauty.

We thus see that several customs prevail with savages which must greatly interfere with, or completely stop, the action of sexual selection. On the other hand, the conditions of life to which savages are exposed, and some of their habits, are favourable to natural selection; and this comes into play at the same time with sexual selection. Savages are known to suffer severely from recurrent famines; they do not increase their food by artificial means; they rarely refrain from marriage [16], and generally marry whilst young. Consequently they must be subjected to occasional hard struggles for existence, and the favoured individuals will alone survive.

[16] Burchell says ('Travels in S. Africa,' vol. ii. 1824, p. 58), that among the wild nations of Southern Africa, neither men nor women ever pass their lives in a state of celibacy. Azara ('Voyages dans l'Amérique Merid.' tom. ii. 1809, p. 21) makes precisely the same remark in regard to the wild Indians of South America.

At a very early period, before man attained to his present rank in the scale, many of his conditions would be different from what now obtains amongst savages. Judging from the analogy of the lower animals, he would then either live with a single female, or be a polygamist. The most powerful and able males would succeed best in obtaining attractive females. They would also succeed best in the general struggle for life, and in defending their females, as well as their offspring, from enemies of all kinds. At this early period the ancestors of man would not be sufficiently advanced in intellect to look forward to distant contingencies; they would not foresee that the rearing of all their children, especially their female children, would make the struggle for life severer for the tribe. They would be governed more by their instincts and less by their reason than are savages at the present day. They would not at that period have partially lost one of the strongest of all instincts, common to all the lower animals, namely the love of their young offspring; and consequently they would not have practised female infanticide. Women would not have been thus rendered scarce, and polyandry would not have been practised; for hardly any other cause, except the scarcity of women seems sufficient to break down the natural and widely prevalent feeling of jealousy, and the desire of each male to possess a female for himself. Polyandry would be a natural stepping-stone to communal marriages or almost promiscuous intercourse; though the best authorities believe that this latter habit preceded polyandry. During primordial times there would be no early betrothals, for this implies foresight. Nor would women be valued merely as useful slaves or beasts of burthen. Both sexes, if the females as well as the males were permitted to exert any choice, would choose their partners not for mental charms, or property, or social position, but almost solely from external appearance. All the adults would marry or pair, and all the offspring, as far as that was possible, would be reared; so that the struggle for existence would be periodically excessively severe. Thus during these times all the conditions for sexual selection would have been more favourable than at a later period, when man had advanced in his intellectual powers but had retrograded in his instincts. Therefore, whatever influence sexual selection may have had in producing the differences between the races of man, and between man and the higher Quadrumana, this influence would have been more powerful at a remote period than at the present day, though probably not yet wholly lost.

THE MANNER OF ACTION OF SEXUAL SELECTION WITH MANKIND.

With primeval man under the favourable conditions just stated, and with those savages who at the present time enter into any marriage tie, sexual selection has probably acted in the following manner, subject to greater or less interference from female infanticide, early betrothals, etc. The strongest and most vigorous men—those who could best defend and hunt for their families, who were provided with the best weapons and possessed the most property, such as a large number of dogs or other animals,—would succeed in rearing a greater average number of offspring than the weaker and poorer members of the same tribes. There can, also, be no doubt that such men would generally be able to select the more attractive women. At present the chiefs of nearly every tribe throughout the world succeed in obtaining more than one wife. I hear from Mr. Mantell that, until recently, almost every girl in New Zealand who was pretty, or promised to be pretty, was *tapu* to some chief. With the Kafirs, as Mr. C. Hamilton states, [17] "the chiefs generally have the pick of the women for many miles round, and are most persevering in establishing or confirming their privilege." We have seen that each race has its own style of beauty, and we know that it is natural to man to admire each characteristic point in his domestic animals, dress, ornaments, and personal appearance, when carried a little beyond the average. If then the several foregoing propositions be admitted, and I cannot see that they are doubtful, it would be an inexplicable circumstance if the selection of the more attractive women by the more powerful men of each tribe, who would rear on an average a greater number of children, did not after the lapse of many generations somewhat modify the character of the tribe.

[17] 'Anthropological Review,' Jan. 1870, p. xvi.

When a foreign breed of our domestic animals is introduced into a new country, or when a native breed is long and carefully attended to, either for use or ornament, it is found after several generations to have undergone a greater or less amount of change whenever the means of comparison exist. This follows from unconscious selection during a long series of generations—that is, the preservation of the most approved individuals— without any wish or expectation of such a result on the part of the breeder. So again, if during many years two careful breeders rear animals of the same family, and do not compare them together or with a common standard, the animals are found to have become, to the surprise of their owners, slightly different. [18] Each breeder has impressed, as von Nathusius well expresses it, the character of his own mind—his own taste and judgment—on his animals. What reason, then, can be assigned why similar results should not follow from the long-continued selection of the most admired women by those men of each tribe who were able to rear the greatest number of children? This would be unconscious selection, for an effect would be produced, independently of any wish or expectation on the part of the men who preferred certain women to others.

[18] The 'Variation of Animals and Plants under Domestication,' vol. ii. pp. 210-217.

Let us suppose the members of a tribe, practising some form of marriage, to spread over an unoccupied continent, they would soon split up into distinct hordes, separated from each other by various barriers, and still more effectually by the incessant wars between all barbarous nations. The hordes would thus be exposed to slightly different conditions and habits of life, and would sooner or later come to differ in some small degree. As soon as this occurred, each isolated tribe would form for itself a slightly different standard of beauty [19]; and then unconscious selection would come into action through the more powerful and leading men preferring certain women to others. Thus the differences between the tribes, at first very slight, would gradually and inevitably be more or less increased.

[19] An ingenious writer argues, from a comparison of the pictures of Raphael, Rubens, and modern French artists, that the idea of beauty is not absolutely the same even throughout Europe: see the 'Lives of Haydn and Mozart,' by Bombet (otherwise M. Beyle), English translation, p. 278.

With animals in a state of nature, many characters proper to the males, such as size, strength, special weapons, courage and pugnacity, have been acquired through the law of battle. The semi-human progenitors of man, like their allies the Quadrumana, will almost certainly have been thus modified; and, as savages still fight for the possession of their women, a similar process of selection has probably gone on in a greater or less degree to the present day. Other characters proper to the males of the lower animals, such as bright colors and various ornaments, have been acquired by the more attractive males having been preferred by the females. There are, however, exceptional cases in which the males are the selectors, instead of having been the selected. We recognize such cases by the females being more highly ornamented than the males,—their ornamental characters having been transmitted exclusively or chiefly to their female offspring. One such case has been described in the order to which man belongs, that of the Rhesus monkey.

Man is more powerful in body and mind than woman, and in the savage state he keeps her in a far more abject state of bondage than does the male of any other animal; therefore it is not surprising that he should have gained the power of selection. Women are everywhere conscious of the value of their own beauty; and when they have the means, they take more delight in decorating themselves with all sorts of ornaments than do men. They borrow the plumes of male birds, with which nature has decked this sex, in order to charm the females. As women have long been selected for beauty, it is not surprising that some of their successive variations should have been transmitted exclusively to the same sex; consequently that they should have transmitted beauty in a somewhat higher degree to their female than to their male offspring, and thus have become more beautiful, according to general opinion, than men. Women, however, certainly transmit most of their characters, including some beauty, to their offspring of both sexes; so that the continued preference by the men of each race for the more attractive women, according to their standard of taste, will have tended to modify in the same manner all the individuals of both sexes belonging to the race.

With respect to the other form of sexual selection (which with the lower animals is much the more common), namely, when the females are the selectors, and accept only those males which excite or charm them most, we have reason to believe that it formerly acted on our progenitors. Man in all probability owes his beard, and perhaps some other characters, to inheritance from an ancient progenitor who thus gained his ornaments. But this form of selection may have occasionally acted during later times; for in utterly barbarous tribes the women have more power in choosing, rejecting, and tempting their lovers, or of afterwards changing their husbands, than might have been expected. As this is a point of some importance, I will give in detail such evidence as I have been able to collect.

Hearne describes how a woman in one of the tribes of Arctic America repeatedly ran away from her husband and joined her lover; and with the Charruas of S. America, according to Azara, divorce is quite optional. Amongst the Abipones, a man on choosing a wife bargains with the parents about the price. But "it frequently happens that the girl rescinds what has been agreed upon between the parents and the bridegroom, obstinately rejecting the very mention of marriage." She often runs away, hides herself, and thus eludes the bridegroom. Captain Musters who lived with the Patagonians, says that their marriages are always settled by inclination; "if the parents make a match contrary to the daughter's will, she refuses and is never compelled to comply." In Tierra del Fuego a young man first obtains the consent of the parents by doing them some service, and then he attempts to carry off the girl; "but if she is unwilling, she hides herself in the woods until her admirer is heartily tired of looking for her, and gives up the pursuit; but this seldom happens." In the Fiji Islands the man seizes on the woman whom he wishes for his wife by actual or pretended force; but "on reaching the home of her abductor, should she not approve of the match, she runs to some one who can protect her; if, however, she is satisfied, the matter is settled forthwith." With the Kalmucks there is a regular race between the bride and bridegroom, the former having a fair start; and Clarke "was assured that no instance occurs of a girl being caught, unless she has a partiality to the pursuer." Amongst the wild tribes of the Malay Archipelago there is also a racing match; and it appears from M. Bourien's account, as Sir J. Lubbock remarks, that "the race, 'is not to the swift, nor the battle to the strong,' but to the young man who has the good fortune to please his intended bride." A similar custom, with the same result, prevails with the Koraks of North-Eastern Asia.

Turning to Africa: the Kafirs buy their wives, and girls are severely beaten by their fathers if they will not accept a chosen husband; but it is manifest from many facts given by the Rev. Mr. Shooter, that they have considerable power of choice. Thus very ugly, though rich men, have been known to fail in getting wives. The girls, before consenting to be betrothed, compel the men to show themselves off first in front and then behind, and "exhibit their paces." They have been known to propose to a man, and they not rarely run away with a favoured lover. So again, Mr. Leslie, who was intimately acquainted with the Kafirs, says, "it is a mistake to imagine that a girl is sold by her father in the same manner, and with the same authority, with which he would dispose of a cow." Amongst the degraded Bushmen of S. Africa, "when a girl has grown up to womanhood without having been betrothed, which, however, does not often happen, her lover must gain her approbation, as well as that of the parents." [20] Mr. Winwood Reade made inquiries for me with respect to the negroes of Western Africa, and he informs me that "the women, at least among the more intelligent Pagan tribes, have no difficulty in getting the husbands whom they may desire, although it is considered unwomanly to ask a man to marry them. They are quite capable of falling in love, and of forming tender, passionate, and faithful attachments." Additional cases could be given.

[20] Azara, 'Voyages,' etc., tom. ii. p. 23. Dobrizhoffer, 'An Account of the Abipones,' vol. ii. 1822, p. 207. Capt. Musters, in 'Proc. R. Geograph. Soc.' vol. xv. p. 47. Williams on the Fiji Islanders, as quoted by Lubbock, 'Origin of Civilization,' 1870, p. 79. On the Fuegians, King and Fitzroy, 'Voyages of the "Adventure" and "Beagle,"' vol. ii. 1839, p. 182. On the Kalmucks, quoted by M'Lennan, 'Primitive Marriage,' 1865, p. 32. On the Malays, Lubbock, ibid. p. 76. The Rev. J. Shooter, 'On the Kafirs of Natal,' 1857, pp. 52-60. Mr. D. Leslie, 'Kafir Character and Customs,' 1871, p. 4. On the Bush-men, Burchell, 'Travels in S. Africa,' ii. 1824, p. 59. On the Koraks by McKennan, as quoted by Mr. Wake, in 'Anthropologia,' Oct. 1873, p. 75.

We thus see that with savages the women are not in quite so abject a state in relation to marriage as has often been supposed. They can tempt the men whom they prefer, and can sometimes reject those whom they dislike, either before or after marriage. Preference on the part of the women, steadily acting in any one direction, would ultimately affect the character of the tribe; for the women would generally choose not merely the handsomest men, according to their standard of taste, but those who were at the same time best able to defend and support them. Such well-endowed pairs would commonly rear a larger number of offspring than the less favoured. The same result would obviously follow in a still more marked manner if there was selection on both sides; that is, if the more attractive, and at the same time more powerful men were to prefer, and were preferred by, the more attractive women. And this double form of selection seems actually to have occurred, especially during the earlier periods of our long history.

We will now examine a little more closely some of the characters which distinguish the several races of man from one another and from the lower animals, namely, the greater or less deficiency of hair on the body, and the color of the skin. We need say nothing about the great diversity in the shape of the features and of the skull between the different races, as we have seen in the last chapter how different is the standard of beauty in these respects. These characters will therefore probably have been acted on through sexual selection; but we have no means of judging whether they have been acted on chiefly from the male or female side. The musical faculties of man have likewise been already discussed.

ABSENCE OF HAIR ON THE BODY, AND ITS DEVELOPMENT ON THE FACE AND HEAD.

From the presence of the woolly hair or lanugo on the human fœtus, and of rudimentary hairs scattered over the body during maturity, we may infer that man is descended from some animal which was born hairy and remained so during life. The loss of hair is an inconvenience and probably an injury to man, even in a hot climate, for he is thus exposed to the scorching of the sun, and to sudden chills, especially during wet weather. As Mr. Wallace remarks, the natives in all countries are glad to protect their naked backs and shoulders with some slight covering. No one supposes that the nakedness of the skin is any direct advantage to man; his body therefore cannot have been divested of hair through natural selection. [21] Nor, as shown in a former chapter, have we any evidence that this can be due to the direct action of climate, or that it is the result of correlated development.

[21] 'Contributions to the Theory of Natural Selection,' 1870, p. 346. Mr. Wallace believes (p. 350) "that some intelligent power has guided or determined the development of man"; and he considers the hairless condition of the skin as coming under this head. The Rev. T.R. Stebbing, in commenting on this view ('Transactions of Devonshire Association for Science,' 1870) remarks, that had Mr. Wallace "employed his usual ingenuity on the question of man's hairless skin, he might have seen the possibility of its selection through its superior beauty or the health attaching to superior cleanliness."

The absence of hair on the body is to a certain extent a secondary sexual character; for in all parts of the world women are less hairy than men. Therefore we may reasonably suspect that this character has been gained through sexual selection. We know that the faces of several species of monkeys, and large surfaces at the posterior end of the body of other species, have been denuded of hair; and this we may safely attribute to sexual selection, for these surfaces are not only vividly colored, but sometimes, as with the male mandrill and female rhesus, much more vividly in the one sex than in the other, especially during the breeding-season. I am informed by Mr. Bartlett that, as these animals gradually reach maturity, the naked surfaces grow larger compared with the size of their bodies. The hair, however, appears to have been removed, not for the sake of nudity, but that the color of the skin may be more fully displayed. So again with many birds, it appears as if the head and neck had been divested of feathers through sexual selection, to exhibit the brightly-colored skin.

As the body in woman is less hairy than in man, and as this character is common to all races, we may conclude that it was our female semi-human ancestors who were first divested of hair, and that this occurred at an extremely remote period before the several races had diverged from a common stock. Whilst our female ancestors were gradually acquiring this new character of nudity, they must have transmitted it almost equally to their offspring of both sexes whilst young; so that its transmission, as with the ornaments of many mammals and birds, has not been limited either by sex or age. There is nothing surprising in a partial loss of hair having been esteemed as an ornament by our ape-like progenitors, for we have seen that innumerable strange characters have been thus esteemed by animals of all kinds, and have consequently been gained through sexual selection. Nor is it surprising that a slightly injurious character should have been thus acquired; for we know that this is the case with the plumes of certain birds, and with the horns of certain stags.

The females of some of the anthropoid apes, as stated in a former chapter, are somewhat less hairy on the under surface than the males; and here we have what might have afforded a commencement for the process of denudation. With respect to the completion of the process through sexual selection, it is well to bear in mind the New Zealand proverb, "There is no woman for a hairy man." All who have seen photographs of the Siamese hairy family will admit how ludicrously hideous is the opposite extreme of excessive hairiness. And the king of Siam had to bribe a man to marry the first hairy woman in the family; and she transmitted this character to her young offspring of both sexes. [22]

[22] The 'Variation of Animals and Plants under Domestication,' vol. ii. 1868, p. 237.

Some races are much more hairy than others, especially the males; but it must not be assumed that the more hairy races, such as the European, have retained their primordial condition more completely than the naked races, such as the Kalmucks or Americans. It is more probable that the hairiness of the former is due to partial reversion; for characters which have been at some former period long inherited are always apt to return. We have seen that idiots are often very hairy, and they are apt to revert in other characters to a lower animal type. It does not appear that a cold climate has been influential in leading to this kind of reversion; excepting perhaps with the negroes, who have been reared during several generations in the United States [23], and possibly with the Ainos, who inhabit the northern islands of the Japan archipelago. But the laws of inheritance are so complex that we can seldom understand their action. If the greater hairiness of certain races be the result of reversion, unchecked by any form of selection, its extreme variability, even within the limits of the same race, ceases to be remarkable. [24]

[23] 'Investigations into Military and Anthropological Statistics of American Soldiers,' by B.A. Gould, 1869, p. 568:—Observations were carefully made on the hairiness of 2129 black and colored soldiers, whilst they were bathing; and by looking to the published table, "it is manifest at a glance that there is but little, if any, difference between the white and the black races in this respect." It is, however, certain that negroes in their native and much hotter land of Africa, have remarkably smooth bodies. It should be particularly observed, that both pure blacks and mulattoes were included in the above enumeration; and this is an unfortunate circumstance, as in accordance with a principle, the truth of which I have elsewhere proved, crossed races of man would be eminently liable to revert to the primordial hairy character of their early ape-like progenitors.

[24] Hardly any view advanced in this work has met with so much disfavour (see for instance, Sprengel, 'Die Fortschritte des Darwinismus,' 1874, p. 80) as the above explanation of the loss of hair in mankind through sexual selection; but none of the opposed arguments seem to me of much weight, in comparison with the facts showing that the nudity of the skin is to a certain extent a secondary sexual character in man and in some of the Quadrumana.

With respect to the beard in man, if we turn to our best guide, the Quadrumana, we find beards equally developed in both sexes of many species, but in some, either confined to the males, or more developed in them than in the females. From this fact and from the curious arrangement, as well as the bright colors of the hair about the heads of many monkeys, it is highly probable, as before explained, that the males first acquired their beards through sexual selection as an ornament, transmitting them in most cases, equally or nearly so, to their offspring of both sexes. We know from Eschricht [25] that with mankind the female as well as the male fœtus is furnished with much hair on the face, especially round the mouth; and this indicates that we are descended from progenitors of whom both sexes were bearded. It appears therefore at first sight probable that man has retained his beard from a very early period, whilst woman lost her beard at the same time that her body became almost completely divested of hair. Even the color of our beards seems to have been inherited from an ape-like progenitor; for when there is any difference in tint between the hair of the head and the beard, the latter is lighter colored in all monkeys and in man. In those Quadrumana in which the male has a larger beard than that of the female, it is fully developed only at maturity, just as with mankind; and it is possible that only the later stages of development have been retained by man. In opposition to this view of the retention of the beard from an early period is the fact of its great variability in different races, and even within the same race; for this indicates reversion,—long lost characters being very apt to vary on re-appearance.

[25] 'Ueber die Richtung der Haare am Menschlichen Korper,' in Müller's 'Archiv. fur Anat. und Phys.' 1837, s. 40.

Nor must we overlook the part which sexual selection may have played in later times; for we know that with savages the men of the beardless races take infinite pains in eradicating every hair from their faces as something odious, whilst the men of the bearded races feel the greatest pride in their beards. The women, no doubt, participate in these feelings, and if so sexual selection can hardly have failed to have effected something in the course of later times. It is also possible that the long-continued habit of eradicating the hair may have produced an inherited effect. Dr. Brown-Séquard has shown that if certain animals are operated on in a particular manner, their offspring are affected. Further evidence could be given of the inheritance of the effects of mutilations; but a fact lately ascertained by Mr. Salvin [26] has a more direct bearing on the present question; for he has shown that the motmots, which are known habitually to bite off the barbs of the two central tail-feathers, have the barbs of these feathers naturally somewhat reduced. [27] Nevertheless, with mankind the habit of eradicating the beard and the hairs on the body would probably not have arisen until these had already become by some means reduced.

[26] On the tail-feathers of Motmots, 'Proceedings of the Zoological Society,' 1873, p. 429.
[27] Mr. Sproat has suggested ('Scenes and Studies of Savage Life,' 1868, p. 25) this same view. Some distinguished ethnologists, amongst others M. Gosse of Geneva, believe that artificial modifications of the skull tend to be inherited.

It is difficult to form any judgment as to how the hair on the head became developed to its present great length in many races. Eschricht [28] states that in the human fœtus the hair on the face during the fifth month is longer than that on the head; and this indicates that our semi-human progenitors were not furnished with long tresses, which must therefore have been a late acquisition. This is likewise indicated by the extraordinary difference in the length of the hair in the different races; in the negro the hair forms a mere curly mat; with us it is of great length, and with the American natives it not rarely reaches to the ground. Some species of Semnopithecus have their heads covered with moderately long hair, and this probably serves as an ornament and was acquired through sexual selection. The same view may perhaps be extended to mankind, for we know that long tresses are now and were formerly much admired, as may be observed in the works of almost every poet; St. Paul says, "if a woman have long hair, it is a glory to her;" and we have seen that in North America a chief was elected solely from the length of his hair.

[28] 'Ueber die Richtung,' ibid. s. 40.

COLOR OF THE SKIN.

The best kind of evidence that in man the color of the skin has been modified through sexual selection is scanty; for in most races the sexes do not differ in this respect, and only slightly, as we have seen, in others. We know, however, from the many facts already given that the color of the skin is regarded by the men of all races as a highly important element in their beauty; so that it is a character which would be likely to have been modified through selection, as has occurred in innumerable instances with the lower animals. It seems at first sight a monstrous supposition that the jet-blackness of the negro should have been gained through sexual selection; but this view is supported by various analogies, and we know that negroes admire their own color. With mammals, when the sexes differ in color, the male is often black or much darker than the female; and it depends merely on the form of inheritance whether this or any other tint is transmitted to both sexes or to one alone. The resemblance to a negro in miniature of *Pithecia satanas* with his jet black skin, white rolling eyeballs, and hair parted on the top of the head, is almost ludicrous.

The color of the face differs much more widely in the various kinds of monkeys than it does in the races of man; and we have some reason to believe that the red, blue, orange, almost white and black tints of their skin, even when common to both sexes, as well as the bright colors of their fur, and the ornamental tufts about the head, have all been acquired through sexual selection. As the order of development during growth, generally indicates the order in which the characters of a species have been developed and modified during previous generations; and as the newly-born infants of the various races of man do not differ nearly as much in color as do the adults, although their bodies are as completely destitute of hair, we have some slight evidence that the tints of the different races were acquired at a period subsequent to the removal of the hair, which must have occurred at a very early period in the history of man.

SUMMARY.

We may conclude that the greater size, strength, courage, pugnacity, and energy of man, in comparison with woman, were acquired during primeval times, and have subsequently been augmented, chiefly through the contests of rival males for the possession of the females. The greater intellectual vigor and power of invention in man is probably due to natural selection, combined with the inherited effects of habit, for the most able men will have succeeded best in defending and providing for themselves and for their wives and offspring. As far as the extreme intricacy of the subject permits us to judge, it appears that our male ape-like progenitors acquired their beards as an ornament to charm or excite the opposite sex, and transmitted them only to their male offspring. The females apparently first had their bodies denuded of hair, also as a sexual ornament; but they transmitted this character almost equally to both sexes. It is not improbable that the females were modified in other respects for the same purpose and by the same means; so that women have acquired sweeter voices and become more beautiful than men.

It deserves attention that with mankind the conditions were in many respects much more favourable for sexual selection, during a very early period, when man had only just attained to the rank of manhood, than during later times. For he would then, as we may safely conclude, have been guided more by his instinctive passions, and less by foresight or reason. He would have jealously guarded his wife or wives. He would not have practised infanticide; nor valued his wives merely as useful slaves; nor have been betrothed to them during infancy. Hence we may infer that the races of men were differentiated, as far as sexual selection is concerned, in chief part at a very remote epoch; and this conclusion throws light on the remarkable fact that at the most ancient period, of which we have not as yet any record, the races of man had already come to differ nearly or quite as much as they do at the present day.

The views here advanced, on the part which sexual selection has played in the history of man, want scientific precision. He who does not admit this agency in the case of the lower animals, will disregard all that I have written in the later chapters on man. We cannot positively say that this character, but not that, has been thus modified; it has, however, been shown that the races of man differ from each other and from their nearest allies, in certain characters which are of no service to them in their daily habits of life, and which it is extremely probable would have been modified through sexual selection. We have seen that with the lowest savages the people of each tribe admire their own characteristic qualities,—the shape of the head and face, the squareness of the cheek-bones, the prominence or depression of the nose, the color of the skin, the length of the hair on the head, the absence of hair on the face and body, or the presence of a great beard, and so forth. Hence these and other such points could hardly fail to be slowly and gradually exaggerated, from the more powerful and able men in each tribe, who would succeed in rearing the largest number of offspring, having selected during many generations for their wives the most strongly characterized and therefore most attractive women. For my own part I conclude that of all the causes which have led to the differences in external appearance between the races of man, and to a certain extent between man and the lower animals, sexual selection has been the most efficient.

CHAPTER XXI.

GENERAL SUMMARY AND CONCLUSION.

Main conclusion that man is descended from some lower form—Manner of development—Genealogy of man—Intellectual and moral faculties—Sexual Selection—Concluding remarks.

A brief summary will be sufficient to recall to the reader's mind the more salient points in this work. Many of the views which have been advanced are highly speculative, and some no doubt will prove erroneous; but I have in every case given the reasons which have led me to one view rather than to another. It seemed worth while to try how far the principle of evolution would throw light on some of the more complex problems in the natural history of man. False facts are highly injurious to the progress of science, for they often endure long; but false views, if supported by some evidence, do little harm, for every one takes a salutary pleasure in proving their falseness: and when this is done, one path towards error is closed and the road to truth is often at the same time opened.

The main conclusion here arrived at, and now held by many naturalists who are well competent to form a sound judgment, is that man is descended from some less highly organized form. The grounds upon which this conclusion rests will never be shaken, for the close similarity between man and the lower animals in embryonic development, as well as in innumerable points of structure and constitution, both of high and of the most trifling importance,—the rudiments which he retains, and the abnormal reversions to which he is occasionally liable,—are facts which cannot be disputed. They have long been known, but until recently they told us nothing with respect to the origin of man. Now when viewed by the light of our knowledge of the whole organic world, their meaning is unmistakable. The great principle of evolution stands up clear and firm, when these groups or facts are considered in connection with others, such as the mutual affinities of the members of the same group, their geographical distribution in past and present times, and their geological succession. It is incredible that all these facts should speak falsely. He who is not content to look, like a savage, at the phenomena of nature as disconnected, cannot any longer believe that man is the work of a separate act of creation. He will be forced to admit that the close resemblance of the embryo of man to that, for instance, of a dog—the construction of his skull, limbs and whole frame on the same plan with that of other mammals, independently of the uses to which the parts may be put—the occasional re-appearance of various structures, for instance of several muscles, which man does not normally possess, but which are common to the Quadrumana—and a crowd of analogous facts—all point in the plainest manner to the conclusion that man is the co-descendant with other mammals of a common progenitor.

We have seen that man incessantly presents individual differences in all parts of his body and in his mental faculties. These differences or variations seem to be induced by the same general causes, and to obey the same laws as with the lower animals. In both cases similar laws of inheritance prevail. Man tends to increase at a greater rate than his means of subsistence; consequently he is occasionally subjected to a severe struggle for existence, and natural selection will have effected whatever lies within its scope. A succession of strongly-marked variations of a similar nature is by no means requisite; slight fluctuating differences in the individual suffice for the work of natural selection; not that we have any reason to suppose that in the same species, all parts of the organization tend to vary to the same degree. We may feel assured that the inherited effects of the long-continued use or disuse of parts will have done much in the same direction with natural selection. Modifications formerly of importance, though no longer of any special use, are long-inherited. When one part is modified, other parts change through the principle of correlation, of which we have instances in many curious cases of correlated monstrosities. Something may be attributed to the direct and definite action of the surrounding conditions of life, such as abundant food, heat or moisture; and lastly, many characters of slight physiological importance, some indeed of considerable importance, have been gained through sexual selection.

No doubt man, as well as every other animal, presents structures, which seem to our limited knowledge, not to be now of any service to him, nor to have been so formerly, either for the general conditions of life, or in the relations of one sex to the other. Such structures cannot be accounted for by any form of selection, or by the inherited effects of the use and disuse of parts. We know, however, that many strange and strongly-marked peculiarities of structure occasionally appear in our domesticated productions, and if their unknown causes were to act more uniformly, they would probably become common to all the individuals of the species. We may hope hereafter to understand something about the causes of such occasional modifications, especially through the study of monstrosities: hence the labors of experimentalists, such as those of M. Camille Dareste, are full of promise for the future. In general we can only say that the cause of each slight variation and of each monstrosity lies much more in the constitution of the organism, than in the nature of the surrounding conditions; though new and changed conditions certainly play an important part in exciting organic changes of many kinds.

Through the means just specified, aided perhaps by others as yet undiscovered, man has been raised to his present state. But since he attained to the rank of manhood, he has diverged into distinct races, or as they may be more fitly called, sub-species. Some of these, such as the Negro and European, are so distinct that, if specimens had been brought to a naturalist without any further information, they would undoubtedly have been considered by him as good and true species. Nevertheless all the races agree in so many unimportant details of structure and in so many mental peculiarities that these can be accounted for only by inheritance from a common progenitor; and a progenitor thus characterized would probably deserve to rank as man.

It must not be supposed that the divergence of each race from the other races, and of all from a common stock, can be traced back to any one pair of progenitors. On the contrary, at every stage in the process of modification, all the individuals which were in any way better fitted for their conditions of life, though in different degrees, would have survived in greater numbers than the less well-fitted. The process would have been like that followed by man, when he does not intentionally select particular individuals, but breeds from all the superior individuals, and neglects the inferior. He thus slowly but surely modifies his stock, and unconsciously forms a new strain. So with respect to modifications acquired independently of selection, and due to variations arising from the nature of the organism and the action of the surrounding conditions, or from changed habits of life, no single pair will have been modified much more than the other pairs inhabiting the same country, for all will have been continually blended through free intercrossing.

By considering the embryological structure of man,—the homologies which he presents with the lower animals,—the rudiments which he retains,—and the reversions to which he is liable, we can partly recall in imagination the former condition of our early progenitors; and can approximately place them in their proper place in the zoological series. We thus learn that man is descended from a hairy, tailed quadruped, probably arboreal in its habits, and an inhabitant of the Old World. This creature, if its whole structure had been examined by a naturalist, would have been classed amongst the Quadrumana, as surely as the still more ancient progenitor of the Old and New World monkeys. The Quadrumana and all the higher mammals are probably derived from an ancient marsupial animal, and this through a long line of diversified forms, from some amphibian-like creature, and this again from some fish-like animal. In the dim obscurity of the past we can see that the early progenitor of all the Vertebrata must have been an aquatic animal, provided with branchiæ, with the two sexes united in the same individual, and with the most important organs of the body (such as the brain and heart) imperfectly or not at all developed. This animal seems to have been more like the larvæ of the existing marine Ascidians than any other known form.

The high standard of our intellectual powers and moral disposition is the greatest difficulty which presents itself, after we have been driven to this conclusion on the origin of man. But every one who admits the principle of evolution, must see that the mental powers of the higher animals, which are the same in kind with those of man, though so different in degree, are capable of advancement. Thus the interval between the mental powers of one of the higher apes and of a fish, or between those of an ant and scale-insect, is immense; yet their development does not offer any special difficulty; for with our domesticated animals, the mental faculties are certainly variable, and the variations are inherited. No one doubts that they are of the utmost importance to animals in a state of nature. Therefore the conditions are favourable for their development through natural selection. The same conclusion may be extended to man; the intellect must have been all-important to him, even at a very remote period, as enabling him to invent and use language, to make weapons, tools, traps, etc., whereby with the aid of his social habits, he long ago became the most dominant of all living creatures.

A great stride in the development of the intellect will have followed, as soon as the half-art and half-instinct of language came into use; for the continued use of language will have reacted on the brain and produced an inherited effect; and this again will have reacted on the improvement of language. As Mr. Chauncey Wright [1] has well remarked, the largeness of the brain in man relatively to his body, compared with the lower animals, may be attributed in chief part to the early use of some simple form of language,—that wonderful engine which affixes signs to all sorts of objects and qualities, and excites trains of thought which would never arise from the mere impression of the senses, or if they did arise could not be followed out. The higher intellectual powers of man, such as those of ratiocination, abstraction, self-consciousness, etc., probably follow from the continued improvement and exercise of the other mental faculties.

[1] 'On the Limits of Natural Selection,' in the 'North American Review,' Oct. 1870, p. 295.

The development of the moral qualities is a more interesting problem. The foundation lies in the social instincts, including under this term the family ties. These instincts are highly complex, and in the case of the lower animals give special tendencies towards certain definite actions; but the more important elements are love, and the distinct emotion of sympathy. Animals endowed with the social instincts take pleasure in one another's company, warn one another of danger, defend and aid one another in many ways. These instincts do not extend to all the individuals of the species, but only to those of the same community. As they are highly beneficial to the species, they have in all probability been acquired through natural selection.

A moral being is one who is capable of reflecting on his past actions and their motives—of approving of some and disapproving of others; and the fact that man is the one being who certainly deserves this designation, is the greatest of all distinctions between him and the lower animals. But in the fourth chapter I have endeavoured to show that the moral sense follows, firstly, from the enduring and ever-present nature of the social instincts; secondly, from man's appreciation of the approbation and disapprobation of his fellows; and thirdly, from the high activity of his mental faculties, with past impressions extremely vivid; and in these latter respects he differs from the lower animals. Owing to this condition of mind, man cannot avoid looking both backwards and forwards, and comparing past impressions. Hence after some temporary desire or passion has mastered his social instincts, he reflects and compares the now weakened impression of such past impulses with the ever-present social instincts; and he then feels that sense of dissatisfaction which all unsatisfied instincts leave behind them, he therefore resolves to act differently for the future,—and this is conscience. Any instinct, permanently stronger or more enduring than another, gives rise to a feeling which we express by saying that it ought to be obeyed. A pointer dog, if able to reflect on his past conduct, would say to himself, I ought (as indeed we say of him) to have pointed at that hare and not have yielded to the passing temptation of hunting it.

Social animals are impelled partly by a wish to aid the members of their community in a general manner, but more commonly to perform certain definite actions. Man is impelled by the same general wish to aid his fellows; but has few or no special instincts. He differs also from the lower animals in the power of expressing his desires by words, which thus become a guide to the aid required and bestowed. The motive to give aid is likewise much modified in man: it no longer consists solely of a blind instinctive impulse, but is much influenced by the praise or blame of his fellows. The appreciation and the bestowal of praise and blame both rest on sympathy; and this emotion, as we have seen, is one of the most important elements of the social instincts. Sympathy, though gained as an instinct, is also much strengthened by exercise or habit. As all men desire their own happiness, praise or blame is bestowed on actions and motives, according as they lead to this end; and as happiness is an essential part of the general good, the greatest-happiness principle indirectly serves as a nearly safe standard of right and wrong. As the reasoning powers advance and experience is gained, the remoter effects of certain lines of conduct on the character of the individual, and on the general good, are perceived; and then the self-regarding virtues come within the scope of public opinion, and receive praise, and their opposites blame. But with the less civilized nations reason often errs, and many bad customs and base superstitions come within the same scope, and are then esteemed as high virtues, and their breach as heavy crimes.

The moral faculties are generally and justly esteemed as of higher value than the intellectual powers. But we should bear in mind that the activity of the mind in vividly recalling past impressions is one of the fundamental though secondary bases of conscience. This affords the strongest argument for educating and stimulating in all possible ways the intellectual faculties of every human being. No doubt a man with a torpid mind, if his social affections and sympathies are well developed, will be led to good actions, and may have a fairly sensitive conscience. But whatever renders the imagination more vivid and strengthens the habit of recalling and comparing past impressions, will make the conscience more sensitive, and may even somewhat compensate for weak social affections and sympathies.

The moral nature of man has reached its present standard, partly through the advancement of his reasoning powers and consequently of a just public opinion, but especially from his sympathies having been rendered more tender and widely diffused through the effects of habit, example, instruction, and reflection. It is not improbable that after long practice virtuous tendencies may be inherited. With the more civilized races, the conviction of the existence of an all-seeing Deity has had a potent influence on the advance of morality. Ultimately man does not accept the praise or blame of his fellows as his sole guide, though few escape this influence, but his habitual convictions, controlled by reason, afford him the safest rule. His conscience then becomes the supreme judge and monitor. Nevertheless the first foundation or origin of the moral sense lies in the social instincts, including sympathy; and these instincts no doubt were primarily gained, as in the case of the lower animals, through natural selection.

The belief in God has often been advanced as not only the greatest, but the most complete of all the distinctions between man and the lower animals. It is however impossible, as we have seen, to maintain that this belief is innate or instinctive in man. On the other hand a belief in all-pervading spiritual agencies seems to be universal; and apparently follows from a considerable advance in man's reason, and from a still greater advance in his faculties of imagination, curiosity and wonder. I am aware that the assumed instinctive belief in God has been used by many persons as an argument for His existence. But this is a rash argument, as we should thus be compelled to believe in the existence of many cruel and malignant spirits, only a little more powerful than man; for the belief in them is far more general than in a beneficent Deity. The idea of a universal and beneficent Creator does not seem to arise in the mind of man, until he has been elevated by long-continued culture.

He who believes in the advancement of man from some low organized form, will naturally ask how does this bear on the belief in the immortality of the soul. The barbarous races of man, as Sir J. Lubbock has shown, possess no clear belief of this kind; but arguments derived from the primeval beliefs of savages are, as we have just seen, of little or no avail. Few persons feel any anxiety from the impossibility of determining at what precise period in the development of the individual, from the first trace of a minute germinal vesicle, man becomes an immortal being; and there is no greater cause for anxiety because the period cannot possibly be determined in the gradually ascending organic scale. [2]

[2] The Rev. J.A. Picton gives a discussion to this effect in his 'New Theories and the Old

Faith,' 1870.

I am aware that the conclusions arrived at in this work will be denounced by some as highly irreligious; but he who denounces them is bound to show why it is more irreligious to explain the origin of man as a distinct species by descent from some lower form, through the laws of variation and natural selection, than to explain the birth of the individual through the laws of ordinary reproduction. The birth both of the species and of the individual are equally parts of that grand sequence of events, which our minds refuse to accept as the result of blind chance. The understanding revolts at such a conclusion, whether or not we are able to believe that every slight variation of structure,—the union of each pair in marriage, the dissemination of each seed,—and other such events, have all been ordained for some special purpose.

Sexual selection has been treated at great length in this work; for, as I have attempted to show, it has played an important part in the history of the organic world. I am aware that much remains doubtful, but I have endeavoured to give a fair view of the whole case. In the lower divisions of the animal kingdom, sexual selection seems to have done nothing: such animals are often affixed for life to the same spot, or have the sexes combined in the same individual, or what is still more important, their perceptive and intellectual faculties are not sufficiently advanced to allow of the feelings of love and jealousy, or of the exertion of choice. When, however, we come to the Arthropoda and Vertebrata, even to the lowest classes in these two great Sub-Kingdoms, sexual selection has effected much.

In the several great classes of the animal kingdom,—in mammals, birds, reptiles, fishes, insects, and even crustaceans,—the differences between the sexes follow nearly the same rules. The males are almost always the wooers; and they alone are armed with special weapons for fighting with their rivals. They are generally stronger and larger than the females, and are endowed with the requisite qualities of courage and pugnacity. They are provided, either exclusively or in a much higher degree than the females, with organs for vocal or instrumental music, and with odoriferous glands. They are ornamented with infinitely diversified appendages, and with the most brilliant or conspicuous colors, often arranged in elegant patterns, whilst the females are unadorned. When the sexes differ in more important structures, it is the male which is provided with special sense-organs for discovering the female, with locomotive organs for reaching her, and often with prehensile organs for holding her. These various structures for charming or securing the female are often developed in the male during only part of the year, namely the breeding-season. They have in many cases been more or less transferred to the females; and in the latter case they often appear in her as mere rudiments. They are lost or never gained by the males after emasculation. Generally they are not developed in the male during early youth, but appear a short time before the age for reproduction. Hence in most cases the young of both sexes resemble each other; and the female somewhat resembles her young offspring throughout life. In almost every great class a few anomalous cases occur, where there has been an almost complete transposition of the characters proper to the two sexes; the females assuming characters which properly belong to the males. This surprising uniformity in the laws regulating the differences between the sexes in so many and such widely separated classes, is intelligible if we admit the action of one common cause, namely sexual selection.

Sexual selection depends on the success of certain individuals over others of the same sex, in relation to the propagation of the species; whilst natural selection depends on the success of both sexes, at all ages, in relation to the general conditions of life. The sexual struggle is of two kinds; in the one it is between individuals of the same sex, generally the males, in order to drive away or kill their rivals, the females remaining passive; whilst in the other, the struggle is likewise between the individuals of the same sex, in order to excite or charm those of the opposite sex, generally the females, which no longer remain passive, but select the more agreeable partners. This latter kind of selection is closely analogous to that which man unintentionally, yet effectually, brings to bear on his domesticated productions, when he preserves during a long period the most pleasing or useful individuals, without any wish to modify the breed.

The laws of inheritance determine whether characters gained through sexual selection by either sex shall be transmitted to the same sex, or to both; as well as the age at which they shall be developed. It appears that variations arising late in life are commonly transmitted to one and the same sex. Variability is the necessary basis for the action of selection, and is wholly independent of it. It follows from this, that variations of the same general nature have often been taken advantage of and accumulated through sexual selection in relation to the propagation of the species, as well as through natural selection in relation to the general purposes of life. Hence secondary sexual characters, when equally transmitted to both sexes can be distinguished from ordinary specific characters only by the light of analogy. The modifications acquired through sexual selection are often so strongly pronounced that the two sexes have frequently been ranked as distinct species, or even as distinct genera. Such strongly-marked differences must be in some manner highly important; and we know that they have been acquired in some instances at the cost not only of inconvenience, but of exposure to actual danger.

The belief in the power of sexual selection rests chiefly on the following considerations. Certain characters are confined to one sex; and this alone renders it probable that in most cases they are connected with the act of reproduction. In innumerable instances these characters are fully developed only at maturity, and often during only a part of the year, which is always the breeding-season. The males (passing over a few exceptional cases) are the more active in courtship; they are the better armed, and are rendered the more attractive in various ways. It is to be especially observed that the males display their attractions with elaborate care in the presence of the females; and that they rarely or never display them excepting during the season of love. It is incredible that all this should be purposeless. Lastly we have distinct evidence with some quadrupeds and birds, that the individuals of one sex are capable of feeling a strong antipathy or preference for certain individuals of the other sex.

Bearing in mind these facts, and the marked results of man's unconscious selection, when applied to domesticated animals and cultivated plants, it seems to me almost certain that if the individuals of one sex were during a long series of generations to prefer pairing with certain individuals of the other sex, characterized in some peculiar manner, the offspring would slowly but surely become modified in this same manner. I have not attempted to conceal that, excepting when the males are more numerous than the females, or when polygamy prevails, it is doubtful how the more attractive males succeed in leaving a large number of offspring to inherit their superiority in ornaments or other charms than the less attractive males; but I have shown that this would probably follow from the females,—especially the more vigorous ones, which would be the first to breed,—preferring not only the more attractive but at the same time the more vigorous and victorious males.

Although we have some positive evidence that birds appreciate bright and beautiful objects, as with the bower-birds of Australia, and although they certainly appreciate the power of song, yet I fully admit that it is astonishing that the females of many birds and some mammals should be endowed with sufficient taste to appreciate ornaments, which we have reason to attribute to sexual selection; and this is even more astonishing in the case of reptiles, fish, and insects. But we really know little about the minds of the lower animals. It cannot be supposed, for instance, that male birds of paradise or peacocks should take such pains in erecting, spreading, and vibrating their beautiful plumes before the females for no purpose. We should remember the fact given on excellent authority in a former chapter, that several peahens, when debarred from an admired male, remained widows during a whole season rather than pair with another bird.

Nevertheless I know of no fact in natural history more wonderful than that the female Argus pheasant should appreciate the exquisite shading of the ball-and-socket ornaments and the elegant patterns on the wing-feather of the male. He who thinks that the male was created as he now exists must admit that the great plumes, which prevent the wings from being used for flight, and which are displayed during courtship and at no other time in a manner quite peculiar to this one species, were given to him as an ornament. If so, he must likewise admit that the female was created and endowed with the capacity of appreciating such ornaments. I differ only in the conviction that the male Argus pheasant acquired his beauty gradually, through the preference of the females during many generations for the more highly ornamented males; the aesthetic capacity of the females having been advanced through exercise or habit, just as our own taste is gradually improved. In the male through the fortunate chance of a few feathers being left unchanged, we can distinctly trace how simple spots with a little fulvous shading on one side may have been developed by small steps into the wonderful ball-and-socket ornaments; and it is probable that they were actually thus developed.

Everyone who admits the principle of evolution, and yet feels great difficulty in admitting that female mammals, birds, reptiles, and fish, could have acquired the high taste implied by the beauty of the males, and which generally coincides with our own standard, should reflect that the nerve-cells of the brain in the highest as well as in the lowest members of the Vertebrate series, are derived from those of the common progenitor of this great Kingdom. For we can thus see how it has come to pass that certain mental faculties, in various and widely distinct groups of animals, have been developed in nearly the same manner and to nearly the same degree.

The reader who has taken the trouble to go through the several chapters devoted to sexual selection, will be able to judge how far the conclusions at which I have arrived are supported by sufficient evidence. If he accepts these conclusions he may, I think, safely extend them to mankind; but it would be superfluous here to repeat what I have so lately said on the manner in which sexual selection apparently has acted on man, both on the male and female side, causing the two sexes to differ in body and mind, and the several races to differ from each other in various characters, as well as from their ancient and lowly-organized progenitors.

He who admits the principle of sexual selection will be led to the remarkable conclusion that the nervous system not only regulates most of the existing functions of the body, but has indirectly influenced the progressive development of various bodily structures and of certain mental qualities. Courage, pugnacity, perseverance, strength and size of body, weapons of all kinds, musical organs, both vocal and instrumental, bright colors and ornamental appendages, have all been indirectly gained by the one sex or the other, through the exertion of choice, the influence of love and jealousy, and the appreciation of the beautiful in sound, color or form; and these powers of the mind manifestly depend on the development of the brain.

Man scans with scrupulous care the character and pedigree of his horses, cattle, and dogs before he matches them; but when he comes to his own marriage he rarely, or never, takes any such care. He is impelled by nearly the same motives as the lower animals, when they are left to their own free choice, though he is in so far superior to them that he highly values mental charms and virtues. On the other hand he is strongly attracted by mere wealth or rank. Yet he might by selection do something not only for the bodily constitution and frame of his offspring, but for their intellectual and moral qualities. Both sexes ought to refrain from marriage if they are in any marked degree inferior in body or mind; but such hopes are Utopian and will never be even partially realised until the laws of inheritance are thoroughly known. Everyone does good service, who aids towards this end. When the principles of breeding and inheritance are better understood, we shall not hear ignorant members of our legislature rejecting with scorn a plan for ascertaining whether or not consanguineous marriages are injurious to man.

The advancement of the welfare of mankind is a most intricate problem: all ought to refrain from marriage who cannot avoid abject poverty for their children; for poverty is not only a great evil, but tends to its own increase by leading to recklessness in marriage. On the other hand, as Mr. Galton has remarked, if the prudent avoid marriage, whilst the reckless marry, the inferior members tend to supplant the better members of society. Man, like every other animal, has no doubt advanced to his present high condition through a struggle for existence consequent on his rapid multiplication; and if he is to advance still higher, it is to be feared that he must remain subject to a severe struggle. Otherwise he would sink into indolence, and the more gifted men would not be more successful in the battle of life than the less gifted. Hence our natural rate of increase, though leading to many and obvious evils, must not be greatly diminished by any means. There should be open competition for all men; and the most able should not be prevented by laws or customs from succeeding best and rearing the largest number of offspring. Important as the struggle for existence has been and even still is, yet as far as the highest part of man's nature is concerned there are other agencies more important. For the moral qualities are advanced, either directly or indirectly, much more through the effects of habit, the reasoning powers, instruction, religion, etc., than through natural selection; though to this latter agency may be safely attributed the social instincts, which afforded the basis for the development of the moral sense.

The main conclusion arrived at in this work, namely, that man is descended from some lowly organized form, will, I regret to think, be highly distasteful to many. But there can hardly be a doubt that we are descended from barbarians. The astonishment which I felt on first seeing a party of Fuegians on a wild and broken shore will never be forgotten by me, for the reflection at once rushed into my mind—such were our ancestors. These men were absolutely naked and bedaubed with paint, their long hair was tangled, their mouths frothed with excitement, and their expression was wild, startled, and distrustful. They possessed hardly any arts, and like wild animals lived on what they could catch; they had no government, and were merciless to every one not of their own small tribe. He who has seen a savage in his native land will not feel much shame, if forced to acknowledge that the blood of some more humble creature flows in his veins. For my own part I would as soon be descended from that heroic little monkey, who braved his dreaded enemy in order to save the life of his keeper, or from that old baboon, who descending from the mountains, carried away in triumph his young comrade from a crowd of astonished dogs—as from a savage who delights to torture his enemies, offers up bloody sacrifices, practices infanticide without remorse, treats his wives like slaves, knows no decency, and is haunted by the grossest superstitions.

Man may be excused for feeling some pride at having risen, though not through his own exertions, to the very summit of the organic scale; and the fact of his having thus risen, instead of having been aboriginally placed there, may give him hope for a still higher destiny in the distant future. But we are not here concerned with hopes or fears, only with the truth as far as our reason permits us to discover it; and I have given the evidence to the best of my ability. We must, however, acknowledge, as it seems to me, that man with all his noble qualities, with sympathy which feels for the most debased, with benevolence which extends not only to other men but to the humblest living creature, with his god-like intellect which has penetrated into the movements and constitution of the solar system—with all these exalted powers—Man still bears in his bodily frame the indelible stamp of his lowly origin.

SUPPLEMENTAL NOTE.

ON SEXUAL SELECTION IN RELATION TO MONKEYS.

(Reprinted from NATURE, *November 2, 1876, p. 18.)*

In the discussion on Sexual Selection in my 'Descent of Man,' no case interested and perplexed me so much as the brightly-colored hinder ends and adjoining parts of certain monkeys. As these parts are more brightly colored in one sex than the other, and as they become more brilliant during the season of love, I concluded that the colors had been gained as a sexual attraction. I was well aware that I thus laid myself open to ridicule; though in fact it is not more surprising that a monkey should display his bright-red hinder end than that a peacock should display his magnificent tail. I had, however, at that time no evidence of monkeys exhibiting this part of their bodies during their courtship; and such display in the case of birds affords the best evidence that the ornaments of the males are of service to them by attracting or exciting the females. I have lately read an article by Joh. von Fischer, of Gotha, published in 'Der Zoologische Garten,' April 1876, on the expression of monkeys under various emotions, which is well worthy of study by any one interested in the subject, and which shows that the author is a careful and acute observer. In this article there is an account of the behaviour of a young male mandrill when he first beheld himself in a looking-glass, and it is added, that after a time he turned round and presented his red hinder end to the glass. Accordingly I wrote to Herr J. von Fischer to ask what he supposed was the meaning of this strange action, and he has sent me two long letters full of new and curious details, which will, I hope, be hereafter published. He says that he was himself at first perplexed by the above action, and was thus led carefully to observe several individuals of various other species of monkeys, which he has long kept in his house. He finds that not only the mandrill (*Cynocephalus mormon*) but the drill (*C. leucophæus*) and three other kinds of baboons (*C. hamadryas, sphinx,* and *babouin*), also *Cynopithecus niger*, and *Macacus rhesus* and *nemestrinus*, turn this part of their bodies, which in all these species is more or less brightly colored, to him when they are pleased, and to other persons as a sort of greeting. He took pains to cure a *Macacus rhesus*, which he had kept for five years, of this indecorous habit, and at last succeeded. These monkeys are particularly apt to act in this manner, grinning at the same time, when first introduced to a new monkey, but often also to their old monkey friends; and after this mutual display they begin to play together. The young mandrill ceased spontaneously after a time to act in this manner towards his master, von Fischer, but continued to do so towards persons who were strangers and to new monkeys. A young *Cynopithecus niger* never acted, excepting on one occasion, in this way towards his master, but frequently towards strangers, and continues to do so up to the present time. From these facts Von Fischer concludes that the monkeys which behaved in this manner before a looking-glass (viz., the mandrill, drill, *Cynopithecus niger*, *Macacus rhesus* and *nemestrinus*) acted as if their reflection were a new acquaintance. The mandrill and drill, which have their hinder ends especially ornamented, display it even whilst quite young, more frequently and more ostentatiously than do the other kinds. Next in order comes *Cynocephalus hamadryas*, whilst the other species act in this manner seldomer. The individuals, however, of the same species vary in this respect, and some which were very shy never displayed their hinder ends. It deserves especial attention that Von Fischer has never seen any species purposely exhibit the hinder part of its body, if not at all colored. This remark applies to many individuals of *Macacus cynomolgus* and *Cercocebus radiatus* (which is closely allied to *M. rhesus*), to three species of Cercopithecus and several American monkeys. The habit of turning the hinder ends as a greeting to an old friend or new acquaintance, which seems to us so odd, is not really more so than the habits of many savages, for instance that of rubbing their bellies with their hands, or rubbing noses together. The habit with the mandrill and drill seems to be instinctive or inherited, as it was followed by very young animals; but it is modified or guided, like so many other instincts, by observation, for Von Fischer says that they take pains to make their display fully; and if made before two observers, they turn to him who seems to pay the most attention.

With respect to the origin of the habit, Von Fischer remarks that his monkeys like to have their naked hinder ends patted or stroked, and that they then grunt with pleasure. They often also turn this part of their bodies to other monkeys to have bits of dirt picked off, and so no doubt it would be with respect to thorns. But the habit with adult animals is connected to a certain extent with sexual feelings, for Von Fischer watched through a glass door a female *Cynopithecus niger*, and she during several days, "umdrehte und dem Männchen mit gurgelnden Tönen die stark geröthete Sitzfläche zeigte, was ich früher nie an diesem Thier bemerkt hatte. Beim Anblick dieses Gegenstandes erregte sich das Männchen sichtlich, denn es polterte heftig an den Stäben, ebenfalls gurgelnde Laute ausstossend." As all the monkeys which have the hinder parts of their bodies more or less brightly colored live, according to Von Fischer, in open rocky places, he thinks that these colors serve to render one sex conspicuous at a distance to the other; but, as monkeys are such gregarious animals, I should have thought that there was no need for the sexes to recognize each other at a distance. It seems to me more probable that the bright colors, whether on the face or hinder end, or, as in the mandrill, on both, serve as a sexual ornament and attraction. Anyhow, as we now know that monkeys have the habit of turning their hinder ends towards other monkeys, it ceases to be at all surprising that it should have been this part of their bodies which has been more or less decorated. The fact that it is only the monkeys thus characterized which, as far as at present known, act in this manner as a greeting towards other monkeys renders it doubtful whether the habit was first acquired from some independent cause, and that afterwards the parts in question were colored as a sexual ornament; or whether the coloring and the habit of turning round were first acquired through variation and sexual selection, and that afterwards the habit was retained as a sign of pleasure or as a greeting, through the principle of inherited association. This principle apparently comes into play on many occasions: thus it is generally admitted that the songs of birds serve mainly as an attraction during the season of love, and that the *leks*, or great congregations of the black-grouse, are connected with their courtship; but the habit of singing has been retained by some birds when they feel happy, for instance by the common robin, and the habit of congregating has been retained by the black-grouse during other seasons of the year.

I beg leave to refer to one other point in relation to sexual selection. It has been objected that this form of selection, as far as the ornaments of the males are concerned, implies that all females within the same district must possess and exercise exactly the same taste. It should, however, be observed, in the first place, that although the range of variation of a species may be very large, it is by no means indefinite. I have elsewhere given a good instance of this fact in the pigeon, of which there are at least a hundred varieties differing widely in their colors, and at least a score of varieties of the fowl differing in the same kind of way; but the range of color in these two species is extremely distinct. Therefore the females of natural species cannot have an unlimited scope for their taste. In the second place, I presume that no supporter of the principle of sexual selection believes that the females select particular points of beauty in the males; they are merely excited or attracted in a greater degree by one male than by another, and this seems often to depend, especially with birds, on brilliant coloring. Even man, excepting perhaps an artist, does not analyze the slight differences in the features of the woman whom he may admire, on which her beauty depends. The male mandrill has not only the hinder end of his body, but his face gorgeously colored and marked with oblique ridges, a yellow beard, and other ornaments. We may infer from what we see of the variation of animals under domestication, that the above several ornaments of the mandrill were gradually acquired by one individual varying a little in one way, and another individual in another way. The males which were the handsomest or the most attractive in any manner to the females would pair oftenest, and would leave rather more offspring than other males. The offspring of the former, although variously intercrossed, would either inherit the peculiarities of their fathers or transmit an increased tendency to vary in the same manner. Consequently the whole body of males inhabiting the same country would tend from the effects of constant intercrossing to become modified almost uniformly, but sometimes a little more in one character and sometimes in another, though at an extremely slow rate; all ultimately being thus rendered more attractive to the females. The process is like that which I have called unconscious selection by man, and of which I have given several instances. In one country the inhabitants value a fleet or light dog or horse, and in another country a heavier and more powerful one; in neither country is there any selection of individual animals with lighter or stronger bodies and limbs; nevertheless after a considerable lapse of time the individuals are found to have been modified in the desired manner almost uniformly, though differently in each country. In two absolutely distinct countries inhabited by the same species, the individuals of which can never during long ages have intermigrated and intercrossed, and where, moreover, the variations will probably not have been identically the same, sexual selection might cause the males to differ. Nor does the belief appear to me altogether fanciful that two sets of females, surrounded by a very different environment, would be apt to acquire somewhat different tastes with respect to form, sound, or color. However this may be, I have given in my 'Descent of Man' instances of closely-allied birds inhabiting distinct countries, of which the young and the females cannot be distinguished, whilst the adult males differ considerably, and this may be attributed with much probability to the action of sexual selection.

THE END

CPSIA information can be obtained at www.ICGtesting.com
Printed in the USA
LVOW11s1808140116

470580LV00001B/44/P